Lecture Notes in Computer Science 9375

Commenced Publication in 1973
Founding and Former Series Editors:
Gerhard Goos, Juris Hartmanis, and Jan van Leeuwen

More information about this series at http://www.springer.com/series/7409

Konrad Jackowski · Robert Burduk
Krzysztof Walkowiak · Michał Woźniak
Hujun Yin (Eds.)

Intelligent Data Engineering and Automated Learning – IDEAL 2015

16th International Conference
Wroclaw, Poland, October 14–16, 2015
Proceedings

Springer

Editors
Konrad Jackowski
Wroclaw University of Technology
Wroclaw
Poland

Robert Burduk
Wroclaw University of Technology
Wroclaw
Poland

Krzysztof Walkowiak
Wroclaw University of Technology
Wroclaw
Poland

Michał Woźniak
Wroclaw University of Technology
Wroclaw
Poland

Hujun Yin
School of Electrical and Electronic
    Engineering
University of Manchester
Manchester
UK

ISSN 0302-9743   ISSN 1611-3349   (electronic)
Lecture Notes in Computer Science
ISBN 978-3-319-24833-2   ISBN 978-3-319-24834-9   (eBook)
DOI 10.1007/978-3-319-24834-9

Library of Congress Control Number: 2015950023

LNCS Sublibrary: SL3 – Information Systems and Applications, incl. Internet/Web, and HCI

Printed on acid-free paper

Springer International Publishing AG Switzerland is part of Springer Science+Business Media
(www.springer.com)

Preface

We are living in a digital world surrounded by various data from numerous sources. Each enterprise collects huge amounts of data; however, manual analysis of these data is virtually impossible. Therefore, one of the timely topics of contemporary computer science is the analytics of big data described by the so-called 4Vs (volume, velocity, variety, and veracity). Nevertheless, we should also think about the fifth and most important V (value), because having access to big data is important, but unless we can turn it into value, it is useless.

The IDEAL conference attracts international experts, researchers, leading academics, practitioners and industrialists from the fields of machine learning, computational intelligence, data mining, knowledge management, biology, neuroscience, bio-inspired systems and agents, and distributed systems. It has enjoyed a vibrant and successful history in the last 17 years and over 12 locations in eight different countries. It continues to evolve to embrace emerging topics and exciting trends, especially in this big data era.

This year IDEAL took place in the vibrant city of Wroclaw, Poland. There were about 127 submissions, which were rigorously peer-reviewed by the Program Committee members. Only the papers judged to be of the highest quality were accepted and included in these proceedings.

This volume contains over 60 papers accepted and presented at the 16th International Conference on Intelligent Data Engineering and Automated Learning (IDEAL 2015), held during October 14–16, 2015, in Wroclaw, Poland. These papers provide a valuable collection of the latest research outcomes in data engineering and automated learning, from methodologies, frameworks and techniques to applications. In addition to various topics such as evolutionary algorithms, neural networks, probabilistic modeling, swarm intelligent, multi-objective optimization, and practical applications in regression, classification, clustering, biological data processing, text processing, and video analysis, IDEAL 2015 also featured a number of special sessions on several emerging topics such as computational intelligence for optimization of communication networks, discovering knowledge from data, simulation-driven DES-like modeling and performance evaluation, and intelligent applications in real-world problems.

IDEAL 2015 enjoyed outstanding keynote speeches by distinguished guest speakers: Prof. Manuel Grana of the University of the Basque Country (Spain), Prof. Leszek Rutkowski of Czestochowa University of Technology (Poland), Prof. Vaclav Snasel of VSB-Technical University of Ostrava (Czech Republic), Prof. Jerzy Stefanowski of Poznań University of Technology (Poland), and Prof. Xin Yao of the University of Birmingham (UK).

We would like to thank all the people who devoted so much time and effort to the successful running of the conference, in particular the members of the Program Committee and reviewers, as well as the authors who contributed to the conference. We are also very grateful to the hard work by the local organizing team at the Department of Systems and Computer Networks, Faculty of Electronics, Wroclaw University of Technology, especially Dr. Konrad Jackowski and Dr. Robert Burduk, for local arrangements, as well as the help of the University of Manchester in various stages. The continued support and collaboration from Springer, in particular from Alfred Hoffman and Anna Kramer, are also greatly appreciated.

July 2015   Konrad Jackowski
                                                               Robert Burduk
                                                          Krzysztof Walkowiak
                                                              Michał Woźniak
                                                                  Hujun Yin

Organization

General Chair

Hujun Yin   University of Manchester, UK
Michal Wozniak   Wroclaw University of Technology, Poland

Program Co-chairs

Robert Burduk   Wroclaw University of Technology, Poland
Krzysztof Walkowiak   Wroclaw University of Technology, Poland

International Advisory Committee

Lei Xu (Chair)   Chinese University of Hong Kong, Hong Kong,
           SAR China
Yaser Abu-Mostafa   CALTECH, USA
Shun-ichi Amari   RIKEN, Japan
Michael Dempster   University of Cambridge, UK
José R. Dorronsoro   Autonomous University of Madrid, Spain
Nick Jennings   University of Southampton, UK
Soo-Young Lee   KAIST, South Korea
Erkki Oja   Helsinki University of Technology, Finland
Latit M. Patnaik   Indian Institute of Science, India
Burkhard Rost   Columbia University, USA
Xin Yao   University of Birmingham, UK

Steering Committee

Hujun Yin (Co-chair)   University of Manchester, UK
Laiwan Chan (Co-chair)   Chinese University of Hong Kong, Hong Kong,
           SAR China
Guilherme Barreto   Federal University of Ceará, Brazil
Yiu-ming Cheung   Hong Kong Baptist University, Hong Kong, SAR
           China
Emilio Corchado   University of Salamanca, Spain
Jose A. Costa   Federal University Natal, Brazil
Colin Fyfe   University of The West of Scotland, UK
Marc van Hulle   K.U. Leuven, Belgium
Samuel Kaski   Helsinki University of Technology, Finland
John Keane   University of Manchester, UK

Jimmy Lee	Chinese University of Hong Kong, Hong Kong, SAR China
Malik Magdon-Ismail	Rensselaer Polytechnic Institute, USA
Vic Rayward-Smith	University of East Anglia, UK
Peter Tino	University of Birmingham, UK
Zheng Rong Yang	University of Exeter, UK
Ning Zhong	Maebashi Institute of Technology, Japan

Publicity Co-chairs

Bin Li	University of Science and Technology of China, China
Jose Alfredo F. Costa	Federal University, UFRN, Brazil
Yang Gao	Nanjing University, China
Minho Lee	Kyungpook National University, Korea
Bartosz Krawczyk	Wrocław University of Technology, Poland

Local Organizing Committee

Konrad Jackowski	Wroclaw University of Technology, Poland — Chair
Bartosz Krawczyk	Wroclaw University of Technology, Poland
Dariusz Jankowski	Wroclaw University of Technology, Poland
Paweł Ksieniewicz	Wroclaw University of Technology, Poland
Paulina Baczyńska	Wroclaw University of Technology, Poland

Program Committee

Aboul Ella Hassanien	Cairo University, Egypt
Adrião Duarte	Federal University, UFRN, Brazil
Agnaldo José da R. Reis	Federal University, UFOP, Brazil
Ajalmar Rêgo da Rocha Neto	Federal University, UFC, Brazil
Ajith Abraham	MirLabs
Alberto Guillen	University of Granada, Spain
Alfredo Cuzzocrea	University of Calabria, Italy
Alfredo Vellido	Universitat Politécnica de Cataluña, Spain
Alicia Troncoso	Universidad Pablo de Olavide, Spain
Álvaro Herrero	University of Burgos, Spain
Ana Belén Gil	University of Salamanca, Spain
Andre Carvalho	University of São Paulo, Brazil
André Coelho	University of Fortaleza, Brazil
Andreas König	University of Kaiserslautern, Germany
Andrzej Cichocki	Brain Science Institute, Japan
Anil Nerode	Cornell University, USA
Anne Canuto	Federal University, UFRN, Brazil
Anne Håkansson	Uppsala University, Sweden
Antônio de P. Braga	Federal University, UFMG, Brazil

Antonio Neme	Universidad Autonoma de la Ciudad de Mexico, Mexico
Ata Kaban	University of Birmingham, UK
Barbara Hammer	University of Bielefeld, Germany
Bernard de Baets	Ghent University, Belgium
Bernardete Ribeiro	University of Coimbra, Portugal
Bin Li	University of Science and Technology of China, China
Bogdan Gabrys	Bournemouth University, UK
Bruno Apolloni	University of Milan, Italy
Bruno Baruque	University of Burgos, Spain
Carla Möller-Levet	University of Manchester, UK
Carlos Pereira	ISEC, Portugal
Carmelo J.A. Bastos Filho	University of Pernambuco, POLI, Brazil
Chung-Ming Ou	Kainan University, Taiwan
Clodoaldo A.M. Lima	University of São Paulo, Brazil
Daniel Glez-Peña	University of Vigo, Spain
Dante I. Tapia	University of Salamanca, Spain
Dariusz Frejlichowski	West Pomeranian University of Technology, Poland
Darryl Charles	University of Ulster, UK
David Camacho	Universidad Autónoma de Madrid, Spain
Davide Anguita	University of Genoa, Italy
Dongqing Wei	Shanghai Jiaotong University, China
Du Zhang	California State University, USA
Eiji Uchino	Yamaguchi University, Japan
Emilio M. Hernandez	University of São Paulo, Brazil
Ernesto Cuadros-Vargas	Universidad Católica San Pablo, Peru
Ernesto Damiani	University of Milan, Italy
Estevam Hruschka Jr.	UFSCar – Federal University of Sao Carlos, Brazil
Eva Lorenzo	University of Vigo, Spain
Fabrice Rossi	National Institute of Research on Computer Science and Automatic, France
Felipe M.G. França	Federal University, UFRJ, Brazil
Fernando Buarque	University of Pernambuco, POLI, Brazil
Fernando Díaz	University of Valladolid, Spain
Fernando Gomide	Unicamp, Brazil
Florentino Fdez-Riverola	University of Vigo, Spain
Francesco Corona	Aalto University, Finland
Francisco Assis	Federal University, UFCG, Brazil
Francisco Ferrer	University of Seville, Spain
Francisco Herrera	University of Granada, Spain
Frank Klawonn	Ostfalia University of Applied Sciences, Germany
Gary Fogel	Natural Selection, USA
Gérard Dreyfus	École Supérieure de Physique et de Chimie Industrielles de Paris, France
Giancarlo Mauri	University of Milano-Bicocca, Italy
Héctor Quintián	University of Salamanca

Heloisa Camargo	Federal University, UFSCar, Brazil
Honghai Liu	University of Portsmouth, UK
Huiyu Zhou	Queen's University Belfast, UK
Hyoseop Shin	Konkuk University Seoul, Korea
Ignacio Rojas	University of Granada, Spain
Igor Farkas	Comenius University in Bratislava, Slovakia
Iñaki Inza	University of the Basque Country, Spain
Ioannis Hatzilygeroudis	University of Patras, Greece
Ivan Silva	Federal University, USP, Brazil
Izabela Rejer	West Pomeranian University of Technology, Poland
J. Michael Herrmann	University of Edinburgh, UK
Jaakko Hollmén	Helsinki University of Technology, Finland
Jaime Cardoso	University of Porto, Portugal
James Hogan	Queensland University of Technology, Australia
Javier Bajo Pérez	Universidad Politécnica de Madrid, Spain
Javier Sedano	Instituto Tecnológico de Castilla y León, Spain
Jerzy Grzymala-Busse	University of Kansas, USA
Jesus Alcala-Fdez	University of Granada, Spain
Jing Liu	Xidian University, China
Joao E. Kogler Jr.	University of São Paulo, Brazil
Jochen Einbeck	Durham University, UK
John Gan	University of Essex, UK
Jongan Park	Chosun University, Korea
Jorge Posada	VICOMTech, Spain
Jose A. Lozano	University of the Basque Country UPV/EHU, Spain
Jose Alfredo F. Costa	Federal University, UFRN, Brazil
José C. Principe	University of Florida, USA
José C. Riquelme	University of Seville, Spain
Jose Dorronsoro	Universidad Autónoma de Madrid, Spain
José Everardo B. Maia	State University of Ceará, Brazil
José F. Martínez	Instituto Nacional de Astrofisica Optica y Electronica, Mexico
José Luis Calvo Rolle	University of A Coruña, Spain
Jose M. Molina	Universidad Carlos III de Madrid, Spain
José Manuel Benítez	University of Granada, Spain
José Ramón Villar	University of Oviedo, Spain
José Riquelme	University of Seville, Spain
Jose Santos	University of A Coruña, Spain
Juan Botía	University of Murcia, Spain
Juan J. Flores	Universidad Michoacana de San Nicolas de Hidalgo, Mexico
Juan Manuel Górriz	University of Granada, Spain
Juán Pavón	Universidad Complutense de Madrid, Spain
Juha Karhunen	Aalto University School of Science, Finland
Ke Tang	University of Science and Technology of China, China
Keshav Dahal	University of Bradford, UK

Kunihiko Fukushima	Kansai University, Japan
Lakhmi Jain	University of South Australia, Australia
Lars Graening	Honda Research Institute Europe, Germany
Leandro Augusto da Silva	Mackenzie University, Brazil
Leandro Coelho	PUCPR/UFPR, Brazil
Lenka Lhotska	Czech Technical University, Czech Republic
Lipo Wang	Nanyang Technological University, Singapore
Lourdes Borrajo	University of Vigo, Spain
Lucía Isabel Passoni	Universidad Nacional de Mar del Plata, Argentina
Luis Alonso	University of Salamanca, Spain
Luiz Pereira Calôba	Federal University, UFRJ, Brazil
Luonan Chen	Shanghai University, China
Maciej Grzenda	Warsaw University of Technology, Poland
Manuel Graña	University of the Basque Country, Spain
Marcelo A. Costa	Universidade Federal de Minas Gerais, Brazil
Marcin Gorawski	Silesian University of Technology, Poland
Márcio Leandro Gonçalves	PUC-MG, Brazil
Marcus Gallagher	The University of Queensland, Australia
Maria Jose Del Jesus	Universidad de Jaén, Spain
Mario Koeppen	Kyushu Institute of Technology, Japan
Marios M. Polycarpou	University of Cyprus, Cyprus
Mark Girolami	University of Glasgow, UK
Marley Vellasco	Pontifical Catholic University of Rio de Janeiro, Brazil
Matjaz Gams	Jozef Stefan Institute Ljubljana, Slovenia
Michael Herrmann	University of Edinburgh, UK
Michael Small	The University of Western Australia, Australia
Michal Wozniak	Wroclaw University of Technology, Poland
Ming Yang	Nanjing Normal University, China
Miroslav Karny	Academy of Sciences of Czech Republic, Czech Republic
Nicoletta Dessì	University of Cagliari, Italy
Olli Simula	Aalto University, Finland
Oscar Castillo	Tijuana Institute of Technology, Mexico
Pablo Estevez	University of Chile, Chile
Paulo Adeodato	Federal University of Pernambuco and NeuroTech Ltd., Brazil
Paulo Cortez	University of Minho, Portugal
Paulo Lisboa	Liverpool John Moores University, UK
Paweł Forczmański	West Pomeranian University of Technology, Poland
Pei Ling Lai	Southern Taiwan University, Taiwan
Perfecto Reguera	University of Leon, Spain
Peter Tino	University of Birmingham, UK
Petro Gopych	Universal Power Systems USA-Ukraine LLC, Ukraine
Rafael Corchuelo	University of Seville, Spain
Ramon Rizo	Universidad de Alicante, Spain
Raúl Cruz-Barbosa	Technological University of the Mixteca, Mexico

Raúl Giráldez	Pablo de Olavide University, Spain
Regivan Santiago	UFRN, Brazil
Renato Tinós	USP, Brazil
Ricardo Del Olmo	Universidad de Burgos, Spain
Ricardo Linden	FSMA, Brazil
Ricardo Tanscheit	PUC-RJ, Brazil
Richard Chbeir	Bourgogne University, France
Roberto Ruiz	Pablo de Olavide University, Spain
Rodolfo Zunino	University of Genoa, Italy
Romis Attux	Unicamp, Brazil
Ron Yang	University of Exeter, UK
Ronald Yager	Machine Intelligence Institute – Iona College, USA
Roque Marín	University of Murcia, Spain
Rudolf Kruse	Otto-von-Guericke-Universität Magdeburg, Germany
Salvador García	University of Jaén, Spain
Saman Halgamuge	The University of Melbourne, Australia
Sarajane M. Peres	University of São Paulo, Brazil
Seungjin Choi	POSTECH, Korea
Songcan Chen	Nanjing University of Aeronautics and Astronautics, China
Stelvio Cimato	University of Milan, Italy
Stephan Pareigis	Hamburg University of Applied Sciences, Germany
Sung-Bae Cho	Yonsei University, Korea
Sung-Ho Kim	KAIST, Korea
Takashi Yoneyama	ITA, Brazil
Tianshi Chen	Chinese Academy of Sciences, China
Tim Nattkemper	University of Bielefeld, Germany
Tomasz Andrysiak	University of Science and Technology in Bydgoszcz, Poland
Tzai-Der Wang	Cheng Shiu University, Taiwan
Urszula Markowska-Kaczmar	Wroclaw University of Technology, Poland
Vahid Jalali	Indiana University, USA
Vasant Honavar	Iowa State University, USA
Vasile Palade	Coventry University, UK
Vicente Botti	Polytechnic University of Valencia, Spain
Vicente Julian	Polytechnic University of Valencia, Spain
Wei-Chiang Samuelson Hong	Oriental Institute of Technology, Taiwan
Weishan Dong	IBM Research, China
Wenjia Wang	University of East Anglia, UK
Wenjian Luo	University of Science and Technology of China, China
Wu Ying	Northwestern University, USA
Yang Gao	Nanjing University, China
Yanira Del Rosario De Paz Santana	Universidad de Salamanca, Spain

| Ying Tan | Peking University, China |
| Yusuke Nojima | Osaka Prefecture University, Japan |

Special Session on Computational Intelligence for Optimization of Communication Networks

Organizers

| Mirosław Klinkowski | National Institute of Telecommunications, Poland |
| Krzysztof Walkowiak | Wroclaw University of Technology, Poland |

Special Session on Discovering Knowledge from Data

Organizers

| Ireneusz Czarnowski | Gdynia Maritime University, Poland |
| Antonio J. Tallón-Ballesteros | University of Seville, Spain |

Simulation-driven DES-like Modeling and Performance Evaluation

Organizers

| Krzysztof Bzdyra | Koszalin University of Technology, Poland |
| Grzegorz Bocewicz | Koszalin University of Technology, Poland |

Special Session on Intelligent Applications in Real-World Problems

Organizers

Marcin Blachnik	Silesian University of Technology, Poland
Anna Burduk	Wroclaw University of Technology, Poland
Krzysztof Halawa	Wroclaw University of Technology, Poland
Antonio J. Tallón-Ballesteros	University of Seville, Spain

Contents

Data Streams Fusion by Frequent Correlations Mining

Radosław Z. Ziembiński[✉]

Institute of Computing Science, Poznan University of Technology,
ul. Piotrowo 2, 60-965 Poznań, Poland
radoslaw.ziembinski@cs.put.poznan.pl

Abstract. Applications acquiring data from multiple sensors have to properly refer data to observables. On-line classification and clustering as basic tools for performing information fusion are computationally viable. However, they poorly exploit temporal relationships in data as patterns mining methods can do. Hence, this paper introduces a new algorithm for the correlations mining in the proposed graph-stream data structure. It can iteratively find relationships in complex data, even if they are partially unsynchronized or disturbed. Retrieved patterns (traces) can be used directly to fuse multi-perspective observations. The algorithm's evaluation was conducted during experiments on artificial data sets while its computational efficiency and results quality were measured.

Keywords: Information fusion · Correlations mining · Complex data

1 Introduction

Once used in big industry and military, multi-sensors installations are basic tools for providing necessary data for automated recognition and control software. Low prices and simplified connectivity caused that they become deployed in many commodities used today by society. Recent explosion of smart devices inevitably changed our landscape to this degree that people begun talking about Internet of Things to describe it. This environment produces huge amount of data that cannot be stored due to their sizes and volatility since observed processes are usually traced by many sensors at once. However, they are providing multi-angle perspective and heterogeneous information required for current software systems effectiveness.

Bad synchronization of information acquired from sensors may hinder the processing quality. Even, if clock of each sensor is synchronized perfectly, various other factors may contribute to distortions. Some sensors are constructed physically in a way that they can detect an event only after some delay e.g., some digital thermometers. In practice, everything that has some inertia inside may contribute to delay. Synchronization can be also violated by various intermediate buffers and transmission technologies rescheduling data e.g., IP network. Lack of synchronization may lead to situation where a new pattern and its support are

© Springer International Publishing Switzerland 2015
K. Jackowski et al. (Eds.): IDEAL 2015, LNCS 9375, pp. 1–8, 2015.
DOI: 10.1007/978-3-319-24834-9_1

similar but not literally the same. Hence, a detection of clear causal relationship in data become difficult.

A method introduced in this paper refers to above concerns. It contributes to the state of art in following areas:

- It proposes a new data structure encapsulating information acquired by many sensors from various angles or a single sensor observing complex process.
- It provides a heuristic algorithm for the mining frequent correlations in data addressing problems with synchronization. Found patterns can be applied to fuse or compress multiple data streams. The proposed data structure and algorithm are new according to author's knowledge.
- This algorithm has "forgetting" capability and can be used in the on-line processing. However, greater performance is usually achievable at cost of lower results quality.

This paper runs as follows. Following section describes the related state of art. Then, the graph-stream data structure and the problem of fusion are introduced. Afterwards, the algorithm for finding bindings between streams is described and evaluated with conclusions at the end.

2   Related Works

A lot of effort is currently put to use on-line classification and clustering methods in the data fusion e.g., [2,6]. Classifiers give a possibility to recognize parts of streams that are related. However, they rely on learned knowledge during the processing. On the other hand, clustering algorithms poorly recognize inter-temporal dependencies and expect well synchronized data.

Fortunately, there are algorithms related to the sequential patterns mining that can mine temporal dependencies e.g., [1,8,9]. However, algorithms from this group are often vulnerable on synchronization errors. Their sensitivity on the data order comes from the fact that they find patters by elicitation of supported combinations of observations. Therefore, even a simple swap of data in the sequence order causes that new subsequence appears and the support for existing patterns drops.

It is particular important in the multi-sensor environment. There, a collapse of multiple data streams into a single one may lead to unpredictable order of elements. A simple way to resolve the issue relies on converting sequences within the processing window to sets [7].

The concept of processing introduced in this paper extends above approach. Instead of using the set perspective within the processing window, this approach introduces a graph-streams data structure as the representation. It allows for causal binding and separation of information acquired from observations of many interplaying subprocesses. The proposed algorithm can mine hidden correlations conveyed by the structure profiting from the better context representation.

Proposed approach also takes some inspirations from research on time series processing [4], approximate string matching [5] and multiple strings alignment [3].

3 Problem Definition

Let's define A as set of heterogeneous attributes. Then, node $n_{p_{type},i}(p_{ts}, A_i) \in N$, where $A_i \subseteq A$ describes an event acquired at time-stamp p_{ts}. Directed edge $e_{i,j}(n_i, n_j) \in E$ connects $n_i, n_j \in N$ and represents causality relation.

The graph-stream uses distinct types of nodes to describe static and dynamic properties of observed objects:

- Interaction node $n_{E,i}(p_{ts}, A_i) \in N$ describes incident where subprocesses mutually interact (exchanging information). It has only the time-stamp.
- State node $n_{S,i}(p_{ts}, A_i) \in N$ represents a boundary state of the episode just after its creation or deletion. It may occur before or after interaction.
- Transformation node $n_{T,i}(p_{ts}, A_i) \in N$ denotes indigenous alteration of the subprocess (in absence of external influences).

Attributes can be freely distributed between state and transformation nodes. However, nodes must be non-empty.

Nodes order is restricted by constraints. An **episode** $s \in S$ is a sequence of transformation nodes beginning and ending with state nodes (creation and deletion nodes). These nodes are connected by single edges to form a list of subsequent events. Episode represents a performance of a subprocess in some period. Further, interaction nodes join input to output episodes by edges according to relation 1..N-1..N. The relation 1-1 may represent a suspension of subprocess.

Considering above, the *graph-stream* $G_{ST}(N, E)$ is directed acyclic graph that can be defined as a set of episodes S connected by interaction nodes according to above constrains imposed on edges (Fig. 1).

Let's now define the problem of the graph-streams fusion for the set of graph-streams. It is assumed that all graph-streams nodes are ordered chronologically according to their time-stamps. Attribute address is a tuple of properties $\langle gidx, sidx, nidx, aidx, p_{ts} \rangle$, where fields identifies appropriately the graph-stream, episode, node, attribute and time-stamp. A **binding pair** connects two attributes $a_1, a_2 \in A$ for which $a_1.gidx \neq a_2.gidx \cup a_1.sidx \neq a_2.sidx$ by undirected edge. Creation of binding pairs for graph-streams undergo some limitations. Primarily, the binding pair is redundant if it connects nodes that are connected by a directed sequence of edges (are causally dependent). Therefore, the pair cannot be created inside the same episode or between causally dependent episodes. Further, two binding pairs are adjacent if at least one attribute address in both pairs is the same. Thereafter, two pairs $a, b \in A$ that are binding the same pair of episodes (the same pairs if $gidx$ and $sidx$) are causally excluding if the following statement is not true $(sgn(a_1.p_{ts} - b_1.p_{ts}) = sgn(a_2.p_{ts} - b_2.p_{ts})) \cup (sgn(a_1.p_{ts} - b_2.p_{ts}) = sgn(a_2.p_{ts} - b_1.p_{ts}))$. It means that two following events observed by two different sensors must be registered in the same order. Then, **trace** $R_i \in R$ is a set of non-adjacent, non-redundant and non-causally excluding binding pairs. Finally, a pair of traces overlaps if their common subset of pairs is non-empty.

A binding pair b **supports** another one a if they bind different episodes' pairs $(\langle a_1.gidx, a_1.sidx, a_2.gidx, a_2.sidx \rangle \neq \langle b_1.gidx, b_1.sidx, b_2.gidx, b_2.sidx \rangle)$,

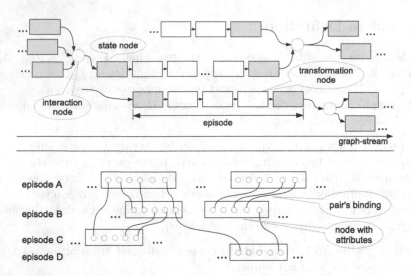

Fig. 1. Graph-stream data structure (top) and example binding pairs joining episodes (bottom)

have the same pairs of attributes ($\langle a_1.aidx, a_2.aidx \rangle \neq b_1. \langle b_1.aidx, b_2.aidx \rangle$) and similarity between values is below thresholds related to attributes ($\sigma(a1.value, b1.value) \leq \Theta_{A1} \cup \sigma(a2.value, b2.value) \leq \Theta_{A2}$). A trace supports (a supporter) another one if all binding pairs of the supporter support in the same order binding pairs of supported trace. The order preservation is crucial because supporter must be causally equivalent to supported trace. Moreover, there is unambiguous (1:1) mapping of supporter episodes to supported trace episodes. A **proper support of the trace** is a maximum number of non-overlapping supporters. Score of the trace $score_R(R_i, t_1, t_2)$ is its size $|R|$ multiplied by its proper support $supp(R)$ and standard deviation from time-stamp differences between nodes of binding pairs within given time frame $[t_1, t_2]$.

Hence, **the problem of the graph-streams fusion** can be defined as finding a set of non-overlapping traces maximizing the $score_R$. The maximization should be considered globally. However, for all practical purposes it is feasible only locally within specified processing window due to costs. In some sense, this problem is similar to the data compression. The difference lies in the fact, that traces with the maximum score are used only once to bind graph-streams while the compression requires repetitions providing the greatest reduction of size.

So defined exploration problem has gigantic search space. Let's consider a simple case with i parallel episodes containing j nodes each and each node has k different attributes. A number of connections between episodes is equal to the number of edges in the complete graph $m = ((i*(i-1))/2)$. Then bindings pairs are $l = (j*k)^2 * m$, what leads to the number of possible traces in order of 2^l. Hence, like for the problem of multiple strings alignment the complexity grows exponentially to the parallel episodes count and lengths. Fortunately, the decomposition, proper data filtering and support constraint can reduce it to more realistic sizes facilitating the mining. Nevertheless, this analysis suggests that any extensive breadth-first search strategies applied without pruning are deemed to failure.

4   Fusion Algorithm

Main loop of the fusion algorithm's is presented on Algorithm 1. The algorithm iteratively mines new traces from acquired episodes moved to S. Found non-overlapping traces are added to winners list. It becomes integrated with the archive, if some remembered traces left beyond the scope of the processing window W (shifted to process new episodes).

Algorithm 1. The graph–streams fusion algorithm cycle

  Data: Set of graph–streams GS, processing window W, set of episodes S,
       processing parameters P
  Result: Updated set of traces TR for fusion
  if $S.newEpisodesArrived()$ **then**
  | $EP = S.getNewEpisodes$
  | $W.shiftWindow(EP)$
  $T = GS.findTraces(T, W, P)$
  $TR.update(T, W)$
  return

The *findTraces* procedure is the most important for the processing because it mines traces. It begins from analysis of attributes values to determine similarity thresholds required for the support counting. Then, it does the enumeration of binding pairs for episodes in the considered processing window and counts support for them. In this phase, it finds the pairs within specified matching window relevant for each GS node alone. This matching window's size depends on difference of nodes time-stamps and sequential distance. Such step filters out binding pairs whose constituents are chronologically or sequentially distant. For each new pair, the algorithm finds supporting binding pairs for episodes available in the history. Pairs without support are removed because represented relationship is not acknowledged in the history. A limited subset of the top supported binding pairs between each pair of episodes is selected as seeds (initial traces' projections) for the mining process. Each projection contains set of binding pairs and their supporting traces.

During the mining, the algorithm extends projections. It maintains population of projections ordered according to their sizes (number of supported binding pairs). A projection is extended always by a single binding pair per iteration. However, it can produce many children with different extending pairs (what implements the breadth-first search tuned by parameter). Produced children are larger thus they are stored in higher layer of child population. In the following processing cycle they will be further extended if it would be feasible. After each cycle, the old population of projections is replaced by the child one.

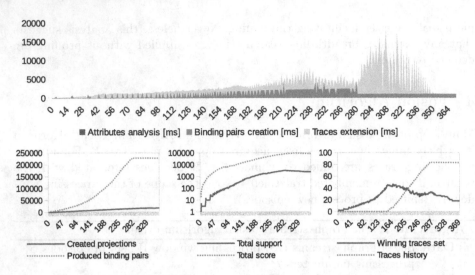

Fig. 2. Duration of mining phases (top) and obtained results (bottom)

If the first layer of the population become empty (it contains projections of the size one), then new projection are created from seeds. Seeds are converted to projections in round robin manner giving an equal chance for all pairs of episodes to develop related traces. If the set of seeds become empty then algorithm again finds new seeds for pairs of recently added episodes (if such have been already added).

Next cycle begin when all projections from population have been processed creating a subsequent one. Cycles end when population and seeds set become empty (no extensions possible and no new episodes available).

If the projection cannot be extended (no available extension or the processing window was moved forward), then it is put to the winner list of non-overlapping traces. Overlapping traces with lower scores are swapped out from the winner list after this operation.

The projection is extended together with supporting traces. Binding pairs that are merged to supporters can significantly enlarge a set of supporters for derived larger projection. This impact can be reduced by pruning of extended supporters. Therefore, the current algorithm implements a strategy of merging a supporting binding pair that is matching and chronologically the closest. After supporters extension, the proper support of child is the same or less than those of the parent (it obeys Apriori Rule).

Proposed algorithm generates projections independently on input availability. If new episodes are not delivered, then it develops older projections making them larger.

5   Experimental Evaluation

Experiments have been conducted on synthetic data obtained from the generator. Generator simulated three separated subprocesses interacting mutually in periodic manner. Thus, they generated episodes containing distinct values and an interaction event for every fixed period. Each episode had 5 nodes, where each node had 4 attributes.

Table 1. Evaluation of clustering results and the processing times (include evaluations)

Result	Experiments					
Generator per event pause [ms]	500	500	500	100	500	1000
Traces extended per cycle	500	500	100	500	500	500
Pop. layer maximum size	5	10	5	5	5	5
Breadth-first fork per iter	2	1	1	1	1	1
Tot. attributes phase cost [ms]	323.27	332.21	361.20	259.66	317.10	319.40
Tot. pairs creation cost [ms]	76895.99	67291.65	66938.89	70245.87	75606.26	68590.09
Tot. traces extension cost [ms]	432913.98	402524.93	267797.83	209174.12	363112.27	531615.41
Traces winner set size	12	36	49	11	18	18
Total support	149	435	850	86	157	159
Total score	4823.02	4795.29	17950.20	9976.01	15834	14524
Alg. Cycles	15	35	95	13	21	30
Projections generated	3266	16500	9096	5148	8538	12636
Pairs generated	74976	74976	74688	62688	71904	70368

Performance experiments were conducted on data stream delivering 2400 events. Results are shown on Fig. 2. Horizontal axis counts processing cycles. During each cycle the algorithm extended 50 traces' projections. Upper plot indicates that the processing costs depend mainly on the traces' extension phase. The mining process receives new episodes until its 280 cycle. It can be observed in a sudden drop in binding pairs creation cost (and also suspension of their production on a plot below). After that, it used following cycles for extending remaining population of projections what raised the traces' extension time. Finally, the seeds set and the population become empty. That has flattened all processing costs, signaling the idle state. On the bottom of Fig. 2, it can be observed that total score grows only to some degree. It indicates the longest and the best supported traces possible to find by the proposed algorithm. A bottom right plot illustrates how found traces were moved from the winning set to the archive (traces history) as the processing window was shifting. The archive contains complete history of non-overlapping and supported traces ready for the fusion.

Experiments measuring quality were conducted on streams containing 800 events (147 complete episodes, 49 hidden traces). Obtained results delivers Table 1 indicating that the breadth-first search during traces' extension phase significantly impaired results. Then, the algorithm wasted iterations on "shallow" searching of the solutions space providing short traces. Results suggest also that the algorithm can be sensitive on the parametrization and the input data rate. Optimal results were obtained at experiment described in the third column (all hidden traces were correctly discovered).

6   Conclusions

The proposed data structure can contain representation of information about the complex process that does not disturb the causal context of events. It can be very helpful when observations from multiple sensors could be distorted leading to synchronization errors. The proposed algorithm is a heuristic that can be used to find frequent correlations (traces) in so represented data. Experiments shown that it is capable to successfully find all hidden traces for example artificial data set. However, the proposed algorithmic solution appears to be sensitive on some mining parameters and the data input rate. Hence, an improvement of its robustness is still a matter of the following research.

This paper is a result of the project financed by National Science Centre in Poland grant no. DEC-2011/03/D/ST6/01621.

References

1. Fricker, D., Zhang, H., Yu, C.: Sequential pattern mining of multimodal data streams in dyadic interactions. In: 2011 IEEE International Conference on Development and Learning (ICDL), vol. 2, pp. 1–6 (2011)
2. Khaleghi, B., Khamis, A., Karray, F.O., Razavi, S.N.: Multisensor data fusion: a review of the state-of-the-art. Inf. Fusion **14**(1), 28–44 (2013)
3. Li, H., Homer, N.: A survey of sequence alignment algorithms for next-generation sequencing. Briefings Bioinform. **11**(5), 473–483 (2010)
4. Matsubara, Y., Sakurai, Y., Faloutsos, C.: Autoplait: automatic mining of co-evolving time sequences. In: Proceedings of the 2014 ACM SIGMOD International Conference on Management of Data, SIGMOD 2014, pp. 193–204. ACM, New York (2014)
5. Navarro, G.: A guided tour to approximate string matching. ACM Comput. Surv. **33**(1), 31–88 (2001)
6. Nguyen, H.L., Woon, Y.K., Ng, W.K.: A survey on data stream clustering and classification. Knowl. Inf. Syst. 1–35 (2014)
7. Oates, T., Cohen, P.R.: Searching for structure in multiple streams of data. In: Proceedings of the Thirteenth International Conference on Machine Learning, pp. 346–354. Morgan Kaufmann (1996)
8. Rassi, C., Plantevit, M.: Mining multidimensional sequential patterns over data streams. In: Song, I.-Y., Eder, J., Nguyen, T.M. (eds.) DaWaK 2008. LNCS, vol. 5182, pp. 263–272. Springer, Heidelberg (2008)
9. Wang, Z., Qian, J., Zhou, M., Dong, Y., Chen, H.: Mining group correlations over data streams. In: 2011 6th International Conference on Computer Science Education (ICCSE), pp. 955–959, August 2011

Web Genre Classification via Hierarchical Multi-label Classification

Gjorgji Madjarov[1], Vedrana Vidulin[2], Ivica Dimitrovski[1],
and Dragi Kocev[3,4(✉)]

[1] Faculty of Computer Science and Engineering,
Ss. Cyril and Methodius University, Skopje, Macedonia
{gjorgji.madjarov,ivica.dimitrovski}@finki.ukim.mk
[2] Ruđer Bošković Institute, Zagreb, Croatia
vedrana.vidulin@irb.hr
[3] Department of Informatics, University of Bari Aldo Moro, Bari, Italy
dragi.kocev@ijs.si
[4] Department of Knowledge Technologies, Jožef Stefan Institute,
Ljubljana, Slovenia

Abstract. The increase of the number of web pages prompts for improvement of the search engines. One such improvement can be by specifying the desired web genre of the result web pages. This opens the need for web genre prediction based on the information on the web page. Typically, this task is addressed as multi-class classification, with some recent studies advocating the use of multi-label classification. In this paper, we propose to exploit the web genres labels by constructing a hierarchy of web genres and then use methods for hierarchical multi-label classification to boost the predictive performance. We use two methods for hierarchy construction: expert-based and data-driven. The evaluation on a benchmark dataset (**20-Genre collection corpus**) reveals that using a hierarchy of web genres significantly improves the predictive performance of the classifiers and that the data-driven hierarchy yields similar performance as the expert-driven with the added value that it was obtained automatically and fast.

Keywords: Web genre classification · Hierarchy construction · Hierarchical multi-label classification

1 Introduction

There is an increasing need for new ways of searching for desired web pages on the Internet (in April 2015 there were $9.4 \cdot 10^8$ websites – http://www. internetlivestats.com). Typically, searching is performed by typing keywords in a search engine that returns web pages of a topic defined by those keywords. The user can, however, obtain more precise results if web page genre is specified

---

The first two authors should be regarded as joint first authors.

© Springer International Publishing Switzerland 2015
K. Jackowski et al. (Eds.): IDEAL 2015, LNCS 9375, pp. 9–17, 2015.
DOI: 10.1007/978-3-319-24834-9_2

in addition to the keywords. *Web genre* represents form and function of a web page thus enabling a user to find a "Scientific" paper about the topic of text mining.

A web page is a complex document that can share conventions of several genres or contain parts from different genres. While this is recognized in the web genre classification community, state-of-the-art genre classifier implementations still attribute a single genre to a web page from a set of predefined genre labels (i.e., address the task as multi-class classification). However, a line of research [1–3] advocates that multi-label classification (MLC) scheme is more suitable for capturing the web page complexity. The rationale is that since several genres are easily combined in a single web page, such hybrid forms thus require attribution of multiple genre labels. For example, a story for children will belong to both "Childrens" and "Prose fiction" genres. Furthermore, web genres naturally form a hierarchy of genres. For example, "Prose fiction" is a type of "Fiction". Aforementioned properties of the web genre classification can be easily mapped to the machine learning task of hierarchical multi-label classification (HMC). HMC is a variant of classification, where a single example may belong to multiple classes at the same time and the classes are organized in the form of a hierarchy. An example that belongs to some class c automatically belongs to all super-classes of c. This is called the hierarchical constraint.

Although it can be easily conceived that the task of web genre classification can be mapped to HMC, the hierarchical and multi-label structure of web genres has not yet been explored. There are two major obstacles for this: lack of a comprehensive genre taxonomy with a controlled vocabulary and meaningful relations between genres and web-page-based corpora labelled with such a taxonomy [4]. In addition to these, from a machine learning point of view, methods that are able to fully exploit the complexity of such data started appearing only recently and have not yet gained much visibility (see [5,6]).

In this work, we aim to address these obstacles. First of all, we propose a hierarchy of web genres that is constructed by an expert and propose to use methods for generating hierarchies using the available data. The use of data-driven methods would bypass the complex process of hierarchy construction by experts: it is difficult (if at all possible) to construct a single hierarchy that would be acceptable for all of the experts. Second, we take a benchmark dataset for genre classification (from [1]) and convert it into a HMC dataset. Finally, we investigate the influence of the hierarchy of web genres on the predictive performance of the predictive models.

For accurately measuring the contribution of the hierarchy and reducing the model bias, we need to consider a predictive modelling method that is able to construct models for both MLC (predicting multiple web genres simultaneously without using a hierarchy of genres) and HMC(predicting multiple web genres simultaneously and expoliting a hierarchy of genres). Such methodology is offered with the predictive clustering trees (PCTs) [6]. PCTs can be seen as a generalization of decision trees towards the task of predicting structured outputs, including the tasks of MLC and HMC.

2   Hierarchical Web Genres Data

State-of-the-art web genre classification approaches mostly deal with feature construction and use benchmark 7-Web and KI-04 multi-class corpora to test the feature sets. The two corpora focus on a set of web genres that are at the same level of hierarchy [3] – experiments in [2] indicated that a mix of genres from different levels may significantly deteriorate multi-class classifier's predictive performance. In a MLC setting, typically used corpus is the 20-Genre Collection benchmark corpus from our previous work [1]. A hierarchical (non multi-label) corpus is presented in [7]: An expert constructed a two-level tree-graph hierarchy composed of 7 top-level and 32 leaf nodes.

In this work, we use the dataset from [1]. It is constructed from 20-Genre Collection corpus and is composed of 2,491 features and 1,539 instances/web pages in English. The features are tailored to cover the different web genre aspects: content (e.g., function words), linguistic form (e.g., part-of-speech trigrams), visual form (e.g., HTML tags) and the context of a web page (e.g., hyperlinks to the same domain). All features, except those pertaining to URL (e.g., appearance of the word blog in a web page URL), are expressed as ratios to eliminate the influence of the page length. The average number of genre labels per page is 1.34. We then converted this dataset to a HMC dataset by expert- and data- driven hierarchy construction methods. We would like to note that the constructed hierarchies are tree-shaped.

Expert-driven hierarchy construction. Expert-based hierarchy was constructed (Fig. 1) by grouping web genres. To this end, we consulted the Web Genre Wiki (http://www.webgenrewiki.org) – it contains results of experts' efforts to construct an unified web genre hierarchy.

Data-driven hierarchy construction. When we build the hierarchy over the label space, there is only one constraint that we should take care of: the original MLC task should be defined by the leaves of the label hierarchy. In particular, the labels from the original MLC problem represent the leaves of the tree hierarchy, while the labels that represent the internal nodes of the tree hierarchy are so-called meta-labels (that model the correlation among the original labels).

In [8], we investigated the use of label hierarchies in multi-label classification, constructed in a data-driven manner. We consider flat label-sets and construct label hierarchies from the label sets that appear in the annotations of the training

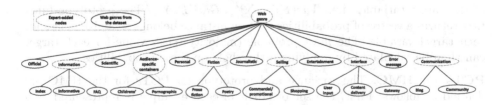

Fig. 1. Web genre hierarchy constructed by an expert.

data by using clustering approaches based on balanced k-means clustering [9], agglomerative clustering with single and complete linkage [10], and clustering performed with PCTs. Multi-branch hierarchy (defined by balanced k-means clustering) appears much more suitable for the global HMC approach (PCTs for HMC) as compared to the binary hierarchies defined by agglomerative clustering with single and complete linkage and PCTs. In this work, for deriving the hierarchy of the (original) MLC problem, we employ balanced k-means.

3   Predictive Modelling for Genre Classification

We present the methodology used to construct predictive models for the task of genre classification using PCTs. We first present general algorithm for constructing PCTs. Next, we otuline the specific PCTs able to predict all of the genres simultaneously but ignore the hierarchical information (i.e., address the task of genre prediction as a multi-label classification task). Furthermore, we give the PCTs able to predict all of the genres simultaneously and exploit the hierarchy information (i.e., address the task of genre prediction as a HMC task).

General algorithm for PCTs. The Predictive Clustering Trees (PCTs) framework views a decision tree as a hierarchy of clusters: the top-node corresponds to one cluster containing all data, which is recursively partitioned into smaller clusters while moving down the tree. The PCT framework is implemented in the CLUS system [6] – available for download at http://clus.sourceforge.net.

PCTs are induced with a standard *top-down induction of decision trees* (TDIDT) algorithm. It takes as input a set of examples and outputs a tree. The heuristic that is used for selecting the tests is the reduction in variance caused by the partitioning of the instances corresponding to the tests. By maximizing the variance reduction, the cluster homogeneity is maximized and the predictive performance is improved. The main difference between the algorithm for learning PCTs and a standard decision tree learner is that the former considers the variance function and the prototype function (that computes a label for each leaf) as *parameters* that can be instantiated for a given learning task. PCTs have been instantiated for both MLC [6,11] and HMC [12]. A detailed computational complexity analysis of PCTs is presented in [6].

PCTs for MLC. These can be considered as PCTs that are able to predict multiple binary (and thus discrete) targets simultaneously. Therefore, the variance function for the PCTs for MLC is computed as the sum of the Gini indices of the target variables, i.e., $Var(E) = \sum_{i=1}^{T} Gini(E, Y_i)$. The prototype function returns a vector of probabilities that an instance belongs to a given class for each target variable. The most probable (majority) class value for each target can then be calculated by applying a threshold on these probabilities.

PCTs for HMC. The variance and prototype for PCTs for the HMC are defined as follows. First, the set of labels of each example is represented as a vector with binary components; the i'th component of the vector is 1 if the example belongs to class c_i and 0 otherwise. The variance of a set of examples E

is defined as the average squared distance between each example's class vector (L_i) and the set's mean class vector $(\overline{L})$: $Var(E) = \frac{1}{|E|} \cdot \sum_{E_i \in E} d(L_i, \overline{L})^2$.

In the HMC context, the similarity at higher levels of the hierarchy is more important than the similarity at lower levels. This is reflected in the distance measure used in the above formula, which is a weighted Euclidean distance: $d(L_1, L_2) = \sqrt{\sum_{l=1}^{|L|} w(c_l) \cdot (L_{1,l} - L_{2,l})^2}$, where $L_{i,l}$ is the lth component of the class vector L_i of an instance E_i, $|L|$ is the size of the class vector, and the class weights $w(c)$ decrease with the depth of the class in the hierarchy. More precisely, $w(c) = w_0 \cdot w(p(c))$, where $p(c)$ denotes the parent of class c and $0 < w_0 < 1$).

In the case of HMC, the mean $\overline{L}$ of the class vectors of the examples in the leaf is stored as a prediction. Note that the value for the ith component of $\overline{L}$ can be interpreted as the probability that an example arriving at the given leaf belongs to class c_i. The prediction for an example that arrives at the leaf can be obtained by applying a user defined threshold τ to the probability. Moreover, when a PCT makes a prediction, it preserves the hierarchy constraint (the predictions comply with the parent-child relationships from the hierarchy).

4   Experimental Design

The comparison of the methods was performed using the CLUS system for predictive clustering implemented in Java. We constructed predictive models corresponding to the two types of modelling tasks, as described in the previous section: multi-label classification (MLC-one model for all of the leaf labels, without using the hierarchy) and hierarchical multi-label classification (HMC-one model for all of the labels by using the hierarchy). For each modeling task, we constructed single tree models.

We used F-test pruning to ensure that the produced models are not overfitted to the training data and have better predictive performance [12]. The exact Fisher test is used to check whether a given split/test in an internal node of the tree results in a statistically significant reduction in variance. If there is no such split/test, the node is converted to a leaf. A significance level is selected from the values 0.125, 0.1, 0.05, 0.01, 0.005 and 0.001 to optimize predictive performance by using internal 3-fold cross validation.

The balanced k-means clustering method that is used for deriving the label hierarchies, requires to be configured the number of clusters k. For this parameter, three different values (2, 3 and 4) were considered [8].

The performance of the predictive models was evaluated using 3-fold cross-validation (as in the study that published the data [1]). We evaluate the predictive performance of the models on the leaf labels in the target hierarchy. In this way, we measure more precisely the influence of the inclusion of the hierarchies in the learning process on the predictive performance of the models.

We used 16 evaluation measures described in detail in [11]. We used six *example-based* evaluation measures (*Hamming loss, accuracy, precision, recall, F_1 score* and *subset accuracy*) and six *label-based* evaluation measures (*micro precision, micro recall, micro F_1, macro precision, macro recall* and *macro*

F_1). These evaluation measures require predictions stating that a given label is present or not (binary 1/0 predictions). However, most predictive models predict a numerical value for each label and the label is predicted as present if that numerical value exceeds some pre-defined threshold τ. To this end, we applied a threshold calibration method by choosing the threshold that minimizes the difference in label cardinality between the training data and the predictions for the test data. In particular, values from 0 to 1 with step 0.05 for τ were considered.

5   Results and Discussion

In this section, we present the results from the experimental evaluation. The evaluation aims to answer three questions: (1) Which data-driven hierarchy construction method yields hierarchy of genres with best performance? (2) Does constructing a hierarchy improves the predictive performance? and (3) Does constructing a data-driven hierarchy yields satisfactory results when compared with expert-constructed hierarchy?

For answering question (1), we compare the performance of three different hierarchies obtained with varying the value of k in the balanced k-means algorithm. For question (2), we compare the performance of the models that exploit the hierarchy information (HMC) with the performance of the flat classification models (MLC). Finally, for addressing question (3), we compare the performance obtained with the expert hierarchy and the data-driven hierarchy.

Table 1 shows the predictive performance of the compared methods. To begin with, the results for the three hierarchies constructed data-driven methods show that the best hierarchy is the one obtained with k set to 4. This reveals that multi-branch hierarchy is more suitable for this domain. Hence, we select this hierarchy for further comparison. Next, we compare the performance of the hierarchical model with the one of the flat classification model. The results clearly show that using a hierarchy of genre labels significantly improve the performance over using the flat genre labels. Moreover, the improvement in performance is across all of the evaluation measures.

Furthermore, we compare the performance of the models obtained with the expert hierarchy and the data-driven hierarchy. We can see that these models

Table 1. The performance of the different approaches in terms of the label, example and ranking based evaluation measures.

	HammingLoss	Accuracy	Precision	Recall	Fmeasure	SubsetAccuracy	MicroPrecision	MicroRecall	MicroF1	MacroPrecision	MacroRecall	MacroF1	OneError	Coverage	RankingLoss	AvgPrecision
HMC - manual hiear.	0.094	**0.276**	**0.327**	**0.341**	**0.334**	0.172	0.31	**0.33**	0.32	**0.424**	**0.296**	**0.297**	0.643	5.561	0.238	0.47
HMC - BkM (k=4)	**0.081**	0.261	0.31	0.3	0.305	**0.177**	**0.368**	0.291	**0.325**	0.368	0.262	0.284	**0.635**	**5.435**	**0.232**	**0.475**
HMC - BkM (k=3)	0.09	0.223	0.273	0.272	0.273	0.131	0.301	0.263	0.281	0.328	0.212	0.211	0.677	5.878	0.254	0.44
HMC - BkM (k=2)	0.084	0.206	0.247	0.247	0.247	0.127	0.328	0.24	0.277	0.361	0.205	0.227	0.682	5.956	0.259	0.433
MLC	0.111	0.136	0.172	0.165	0.168	0.073	0.165	0.163	0.164	0.063	0.1	0.065	0.83	7.955	0.36	0.317

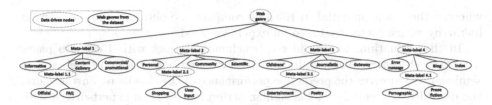

Fig. 2. Web genre hierarchy constructed by balanced k-means algorithm (for k = 4).

have relatively similar predictive performance – each of the models is better than the other according to 8 evaluation measures. It is worth mentioning that the data-driven hierarchy is better on the ranking-based evaluation measures (the last four columns in Table 1). This means that by improving the threshold selection procedure the other evaluation measures will further improve. Nevertheless, even with the results as they are, they convey an important message: The tedious, laborious and expensive method of hierarchy construction by experts can be replaced with a cheap, automatic, fast, data-driven hierarchy construction method without any loss in terms of predictive performance.

The data-driven hierarchy obtained with balanced k-means (and k set to 4) is depicted in Fig. 2. An inspection of the two hierarchies (the first constructed by an expert, Fig. 1, and the second constructed using only data) reveals that these two hierarchies differ to each other completely. Namely, there is no grouping of genres in the expert hierarchy that can be noted in the data-driven hierarchy. This means that there exist a semantic gap between the meaning of the genres and how these meaning are well represented in the data.

Considering that the PCTs are interpretable models, we briefly comment on the attributes selected on the top levels of the trees constructed with the different scenarios: MLC, HMC-manual and HMC-BkM. The MLC and HMC-BkM tree selected first information on the appearance of the word FAQ in the url of the web page and then focus on content related attributes. The HMC-BkM tree also uses the part-of-speech trigrams. Conversely, the HMC-manual tree used mainly content related features on the top levels of the tree-model accompanied with HTML tags information on the lower levels. All in all, the different scenarios exploit different attributes from the dataset.

6 Conclusions

In this paper, we advocated a new approach for resolving the task of web genres classification. Traditionally, this task is treated as a multi-class problem, while there are some recent studies that advise to treat it as a MLC problem. We propose to further exploit the information that is present in the web genres labels by constructing a hierarchy of web genres and then use methods for HMC to boost the predictive performance. Considering that hierarchical benchmark datasets for web genre classification do not exist, we propose to use data-driven methods for hierarchy construction based on balanced k-means. To investigate whether there is a potential in this, we compare the obtained the data-driven hierarchy with a hierarchy based on expert knowledge.

In the evaluation, we consider a benchmark dataset with 1539 web pages with 20 web genres. The results reveal that using a hierarchy of web genres significantly improves the predictive performance of the classifiers. Furthermore, the data-driven hierarchy yields similar performance as the expert-driven with the difference that it was obtained automatically and fast. This means for even larger domains (both in terms of number of examples and number of web genre labels) it would be much simpler and cheaper to use data-driven hierarchies.

We plan to extend this work in two major directions. First, we plan to use more advanced predictive models such as ensembles for predicting structured outputs to see whether the improvement carries over in the ensemble setting. Second, we plan to develop hierarchies of web genres structured as directed acyclic graphs, which seems more natural in modelling relations between genres. It could also be useful to adapt the hierarchy construction algorithm to break down existing genres into sub-genres.

Acknowledgments. We acknowledge the financial support of the European Commission through the grant ICT-2013-612944 MAESTRA.

References

1. Vidulin, V., Luštrek, M., Gams, M.: Multi-label approaches to web genre identification. J. Lang. Tech. Comput. Linguist. **24**(1), 97–114 (2009)
2. Santini, M.: Automatic identification of genre in web pages. Ph.D. thesis, University of Brighton (2007)
3. Santini, M.: Cross-testing a genre classification model for the web. In: Mehler, A., Sharoff, S., Santini, M. (eds.) Genres on the Web, pp. 87–128. Springer, Heidelberg (2011)
4. Crowston, K., Kwaśnik, B., Rubleske, J.: Problems in the use-centered development of a taxonomy of web genres. In: Mehler, A., Sharoff, S., Santini, M. (eds.) Genres on the Web, pp. 69–84. Springer, Heidelberg (2011)
5. Silla Jr., C.N., Freitas, A.A.: A survey of hierarchical classification across different application domains. Data Min. Knowl. Discov. **22**(1–2), 31–72 (2011)
6. Kocev, D., Vens, C., Struyf, J., Džeroski, S.: Tree ensembles for predicting structured outputs. Pattern Recogn. **46**(3), 817–833 (2013)
7. Stubbe, A., Ringlstetter, C., Schulz, K.U.: Genre as noise: noise in genre. Int. J. Doc. Anal. Recogn. **10**(3–4), 199–209 (2007)
8. Madjarov, G., Dimitrovski, I., Gjorgjevikj, D., Džeroski, S.: Evaluation of different data-derived label hierarchies in multi-label classification. In: Appice, A., Ceci, M., Loglisci, C., Manco, G., Masciari, E., Ras, Z.W. (eds.) NFMCP 2014. LNCS, vol. 8983, pp. 19–37. Springer, Heidelberg (2015)
9. Tsoumakas, G., Katakis, I., Vlahavas, I.: Effective and efficient multilabel classification in domains with large number of labels. In: Proceedings of the ECML/PKDD Workshop on Mining Multidimensional Data, pp. 30–44 (2008)
10. Manning, C.D., Raghavan, P., Schütze, H.: An Introduction to Information Retrieval. Cambridge University Press, Cambridge (2009)

11. Madjarov, G., Kocev, D., Gjorgjevikj, D., Džeroski, S.: An extensive experimental comparison of methods for multi-label learning. Pattern Recogn. **45**(9), 3084–3104 (2012)
12. Vens, C., Struyf, J., Schietgat, L., Džeroski, S., Blockeel, H.: Decision trees for hierarchical multi-label classification. Mach. Learn. **73**(2), 185–214 (2008)

Multi-agent Reinforcement Learning for Control Systems: Challenges and Proposals

Manuel Graña[1,2]([⊠]) and Borja Fernandez-Gauna[1]

[1] Grupo de Inteligencia Computacional (GIC), Universidad del País Vasco
(UPV/EHU), San Sebastián, Spain
manuel.granaromay@pwr.edu.pl
[2] ENGINE Centre, Wrocław University of Technology,
Wybrzeże Wyspiańskiego 27, 50-370 Wrocław, Poland

Abstract. Multi-agent Reinforcement Learning (MARL) methods offer
a promising alternative to traditional analytical approaches for the design
of control systems. We review the most important MARL algorithms
from a control perspective focusing on on-line and model-free methods. We review some of sophisticated developments in the state-of-the-
art of single-agent Reinforcement Learning which may be transferred to
MARL, listing the most important remaining challenges. We also propose some ideas for future research aiming to overcome some of these
challenges.

1 Introduction

Reinforcement Learning (RL) [37] methods gaining popularity in the area of control because they allow to build control systems without detailed modeling of
the underlying dynamics, because they learn how to maximize the control objective by means of interacting with the environment. This is quite an advantage
over compared with traditional analytical control techniques requiring a deatiled
formal model, which may be difficult to construct for complex non-linear systems. The quality of these approaches rely on the quality of the model itself
and thus, require a good understanding of the problem at hand. In the RL approach, parameter tuning is substituted by iterative adaptation to an stochastic
environment. Some systems (i.e., Multi-component Robotic Systems [13]) are
best approached from a multi-agent perspective in order to better exploit the
computation capabilities and robustness of distributed control systems. Multi-
Agent Reinforcement Learning (MARL) is the extension of single-agent RL to
multi-agent scenarios. MARL methods have already been successfully applied to
several multi-agent control scenarios [17, 36, 41, 43].

2 Reinforcement Learning

2.1 Single-agent Reinforcement Learning

Markov Decision Process. Single-agent RL methods use Markov Decision Processes (MDPs) to model the interaction between the agent and the environment. An

© Springer International Publishing Switzerland 2015
K. Jackowski et al. (Eds.): IDEAL 2015, LNCS 9375, pp. 18–25, 2015.
DOI: 10.1007/978-3-319-24834-9_3

MDP $< S, A, P, R >$ is defined by the set of *states* (S), the set of *actions* from which the agent can choose (A), a *transition function* (P) that determines state transitions produced by actions, and a *reward function* (R) that gives a numerical value assessing how good a state transition was. S can be a finite set of states (i.e., a cell number in a grid-world) or a vector of real values (i.e., the x and y coordinates read from a GPS receiver). The goal of the agent is to learn a policy π that maximizes the expected *return* R by means of interacting with the environment. The state-action value function $Q^{\pi}(s, a)$ is the value of taking action a in state s.

Policy learning methods. There are basically three classes of RL methods [7]: value iteration, policy iteration and policy search methods. Value iteration methods (such as Q-Learning) generally learn the optimal state-action value-function Q^* and then derive the optimal policy. Policy iteration methods usually follow some policy, evaluate its value by learning V^{π}, and then, aim to improve π. Actor-critic methods belong to this class: an actor implements a parametrized policy, a critic learns its value function (i.e., *Least-Squares Temporal Difference* [6,42]). Value updates are then fed back to the actor, which can us it to improve its policy (i.e., *Natural Policy Gradient* [5]). Finally, policy search methods directly search on the policy space for the optimal policy that maximizes the expected return for any possible initial state.

2.2   Multi-agent Reinforcement Learning

Stochastic Games. The interaction model in Multi-agent Reinforcement Learning (MARL) is the Stochastic Game (SG), that is defined by the number of agents (n) and the tuple $\langle S, A_1, \ldots A_n, P, R_1, \ldots, R_n \rangle$. Each i-th agent chooses actions from its own local action space A_i and receives its own reward R_i. Multi-agent systems can be competitive, cooperative or mixed. In fully cooperative systems, all the agents share the same goals and so, the same reward signals $R_1 = R_2 = \ldots = R_n$. MARL algorithms can also be classified depending on whether they use models of the other agents or not. In this paper, we will focus on model-free methods because we expect them to scale better to multi-dimensional control problems.

   Distributed Q-Learning (D-QL) [26] is an example of independent learning. Each agent assumes that the remaining agents will behave optimally thus projecting the virtual centralized state-action values $Q(s, \mathbf{a})$ $(\mathbf{a} \in \mathbf{A})$ to its own local action space $Q_i(s, a)$, $a \in A_i$. An instance of the MARL algorithms in which agents are aware of other agents' choices, is the *Team Q-Learning* [28], where (Team-QL) agents learn the joint state-action $Q(s, \mathbf{a})$. The algorithm uses the Q-Learning update rule, but using the joint-actions $\mathbf{a}$ and $\mathbf{a}'$ instead of a and a'. This algorithm converges to optimal values under an additional assumption: a unique optimal action exists in each state. This implicit coordination mechanism ensures that agents will exploit the Q-function in a coordinated manner. Some other implicit coordination mechanisms based on heuristics [8,23] or models of the other agents [30,40] can be found in the literature. MARL methods aware of the other agents' actions eliminates the non-stationarity due to other agents' policy changes, but then it becomes a more complex problem.

In order to reduce this complexity, it can be assumed that agents need not coordinate in every state with every other agent, but only with some of them. Under this assumption, the agents first learn when and which agents to coordinate with, and then use an explicit coordination mechanism [20,24] to select a joint-action that maximizes the expected return. *Coordinated Reinforcement Learning* (Coordinated-RL) [20] builds a *Coordination Graph* (CG) for each state that defines with whom agents do need to coordinate. An agent's local state-action values thus only depend on its own local action and those taken by the agents connected to it through the CG. Agents can maximize the global value using a message-based coordination mechanism. An improved version of this algorithm based on an edge-based decomposition of the CG instead of an agent-based decomposition was proposed in [24]. This method scales linearly on the number of agents. The downside of these methods is having to learn and store the CG and the additional processing time introduced by the coordination mechanism. *Hierarchical Reinforcement Learning* (HRL) is another interesting approach to reduce the complexity of a task by decomposing it as a hierarchical structure of subtasks. Single-agent MAXQ [11] allows agents to learn concurrently low-level subtasks and higher-level tasks based on these subtasks. This idea is extended to multi-agent problems in [19]: *Cooperative HRL* assumes costless communication, and *COM-Cooperative HRL* considers communication as part of the problem. A communication level is added to the subtask hierarchy, so that the agent can learn when to communicate. Decomposing the task into a hierarchy of subtasks is not always trivial, and the decomposition of the task itself determines how good the system will approach the optimal solution. This has led research towards automated decomposition of task, both in single-agent [22,29] and multi-agent environments [10].

3   Control Applications of MARL

Last years have seen a number of novel MARL applications: traffic light control [1,4,25,35], robotic hose maneuvering [17], micro-grid control [27], structure prediction of proteins [9], route-choosing [2], supply chains [43], or management of the cognitive radio spectrum [41].

Advantages. MARL-based approaches to control systems offer some inherent advantages over traditional control methods. For example, MARL algorithms can adapt to changes in the environment thanks to their learning nature. Another big advantage over analytical approaches is that model-free algorithms do not require the dynamics of the environment to be fully understood, thus enabling the control of more complex systems. There is still quite a gap between single-agent RL and MARL techniques, but we expect more works currently restricted to the single-agent case to be extended to the multi-agent case in the near future. An example of this is Multi-objective Reinforcement Learning [12], which aims to maximize different possibly conflicting objectives at the same time.

Challenges. We have not found in the literature MARL algorithms able to deal with continuous state-action spaces [21]. Continuous controllers have been shown to outperform algorithms with discretized state and actions in general feedback control tasks [15]. Several methods have been developed in single-agent RL paradigm to deal control tasks involving continuous action spaces: actor-critic learning architectures [5,18,21,32], policy search methods [3] or parametrized controllers [34]. A lot of effort has been devoted in recent years towards obtaining efficient policy gradients [5] and data-efficient value estimation methods [6,31]. For a complete review of the state-of-the-art on single-agent RL using VFA, we refer the reader to [42]. On the other hand, MARL algorithms are mostly based on Q-Learning, hence they estimate the state-action value. General Value Function Approximation (VFA) methods can been used to approximate the state-action value function. This allows continuous states [7], but greedy selection of the action with the highest value will correspond to the center value of one feature. This limits the ability of Q-Learning to output continuous action spaces.

Most of the MARL applications to realistic control problems so far found in the literature are either uncoupled systems of agents operating with no influence on each other [2,41], or loosely coupled tasks, such as traffic light control [1,4]. Some loosely coupled problems may be better approached using systems of unaware agents. Regarding fully-coupled systems in which agents' actions have an effect on other agents' decisions, only a few instances can be found [16,17,25]. This kind of systems require either full observation and learning on the joint state-action space, which does not scale well to real-world environments. Between unaware multi-agent systems and learning on the full joint state-action space, there are alternative approaches, such as exploiting the coordination requirements of the task using CG(i.e., *Coordinated-RL*), or decomposing tasks into a structure of hierarchical subtasks. Both CGs and task hierarchies can be designed by hand in small-scale or clearly structured tasks [25], but manual design is not feasible in more complex or unstructured problems. Some advances have been done towards automatic learning of Coordination Graphs [10] and hierarchies of tasks [27], but none is applicable to continuous state or action spaces. It is not clear either how good these methods will scale to more complex MARL problems. CG-based algorithms require communication each time step. A variable-elimination procedure was proposed in [20] to give an exact solution to the joint-action value maximization process. The number of messages exchanged at each decision step depends on the topology of the specific CG. In order to alleviate this problem, two anytime algorithms were proposed in [39] to approximate the optimal joint-action in a predetermined time: *Coordinated Ascent* and *Max-Plus*. Whether these methods provide an acceptable solution in complex real-time control scenarios within an acceptable time-frame remains an open question.

Another important uncontested challenge is learning an initial policy from scratch in large real-world applications, where it is unaffordable to allow agents thousands of trials before they can start completing the task (i.e., robotic maneuvering tasks [16]). There are several approaches to this problem, all based on some form of Learning Transfer [38]: agents can be first trained in a simulated environment and then allowed to face the real task [14], they can be initialized resembling some initial policy (i.e., a PID controller) available to the system designer [18], or agents may be trained to imitate some expert performing the task [32].

Proposals for Future Research. MARL methods are mostly based on general heterogeneous Stochastic Games and, thus, they work under very broad assumptions. From a control perspective though, one can further assume fully cooperative tasks and homogeneous learning agents. This kind of systems might be better approached from a distributed point of view. Consider a multiple output continuous actor-critic architecture: an approximated value function estimated by the critic and an actor with several VFAs, each representing a different output of the actor. When an improvement in the value function is detected, the actors update its policies towards the last action explored. This same idea can be translated to a multi-agent system in which each agent keeps an instance of the critic learning the value function and an actor with a subset of the system's output. Agents would only require to coordinate exploration and exploitation, which could be achieved by using consensus [33] to share and update the exploration parameters using some preset schedules. This learning structure would allow to use the state-of-the-art in single-agent model-free environments. Full observation of the state could also be alleviated by deferred updates of the critic/actor: agents can follow their policies tracking their local actions and yet incomplete states, and defer the update of the actor and policy until all the state variables have been received.

4   Conclusions

In this paper, we have reviewed the basics of MARL and some recent works in the literature of this field applied to control systems. MARL offers some advantages over traditional analytical control techniques. The most important is that the system designer needs not fully understand or have an accurate model of the system. MARL-based methods also pose some interesting challenges when applied to real-world control problems. Most of the algorithms have been developed with small environments in mind. In this respect, we point out that the main gap between single-agent and MARL algorithms to date is the ability to deal with continuous state and action spaces.

Acknowledgments. This research has been partially funded by grant TIN2011-23823 of the Ministerio de Ciencia e Innovación of the Spanish Government (MINECO), and the Basque Government grant IT874-13 for the research group. Manuel Graña was supported by EC under FP7, Coordination and Support Action, Grant Agreement Number 316097, ENGINE European Research Centre of Network Intelligence for Innovation Enhancement.

References

1. Arel, I., Liu, C., Urbanik, T., Kohls, A.: Reinforcement learning-based multi-agent system for network traffic signal control. Intell. Transport Syst. IET **4**(2), 128–135 (2010)
2. Arokhlo, M., Selamat, A., Hashim, S., Selamat, M.: Route guidance system using multi-agent reinforcement learning. In: 2011 7th International Conference on Information Technology in Asia (CITA 2011), pp. 1–5, July 2011
3. Bagnell, J.A.D., Schneider, J.: Autonomous helicopter control using reinforcement learning policy search methods. In: 2001 Proceedings of the International Conference on Robotics and Automation. IEEE, May 2001
4. Bazzan, A.: Opportunities for multiagent systems and multiagent reinforcement learning in traffic control. Auton. Agents Multi-Agent Syst. **18**(3), 342–375 (2009)
5. Bhatnagar, S., Sutton, R., Ghavamzadeh, M., Lee, M.: Natural actor-critic algorithms. Automatica Int. Fed. Autom. Control **45**(11), 2471–2482 (2009)
6. Boyan, J.A.: Technical update: least-squares temporal difference learning. Mach. Learn. **49**, 233–246 (2002)
7. Bussoniu, L., Babuska, R., Schutter, B.D., Ernst, D.: Reinforcement Learning and Dynamic Programming Using Function Approximators. CRC Press, Boca Raton (2010)
8. Claus, C., Boutilier, C.: The dynamics of reinforcement learning in cooperative multiagent systems. In: Proceedings of the Fifteenth National Conference on Artificial Intelligence, pp. 746–752. AAAI Press (1997)
9. Czibula, G., Bocicor, M.I., Czibula, I.G.: A distributed reinforcement learning approach for solving optimization problems. In: Proceedings of the 5th WSEAS International Conference on Communications and Information Technology, CIT 2011, pp. 25–30. World Scientific and Engineering Academy and Society (WSEAS), Stevens Point (2011)
10. De Hauwere, Y.M., Vrancx, P., Nowé, A.: Learning multi-agent state space representations. In: Proceedings of the 9th International Conference on Autonomous Agents and Multiagent Systems, AAMAS 2010, vol. 1, pp. 715–722. International Foundation for Autonomous Agents and Multiagent Systems, Richland (2010)
11. Dietterich, T.G.: An overview of MAXQ hierarchical reinforcement learning. In: Choueiry, B.Y., Walsh, T. (eds.) SARA 2000. LNCS (LNAI), vol. 1864, p. 26. Springer, Heidelberg (2000)
12. Drugan, M., Nowe, A.: Designing multi-objective multi-armed bandits algorithms: a study. In: The 2013 International Joint Conference on Neural Networks (IJCNN), pp. 1–8, August 2013
13. Duro, R., Graña, M., de Lope, J.: On the potential contributions of hybrid intelligent approaches to multicomponen robotic system development. Inf. Sci. **180**(14), 2635–2648 (2010)
14. Fernandez-Gauna, B., Lopez-Guede, J., Graña, M.: Transfer learning with partially constrained models: application to reinforcement learning of linked multicomponent robot system control. Robot. Auton. Syst. **61**(7), 694–703 (2013)
15. Fernandez-Gauna, B., Ansoategui, I., Etxeberria-Agiriano, I., Graña, M.: Reinforcement learning of ball screw feed drive controllers. Eng. Appl. Artif. Intell. **30**, 107–117 (2014)
16. Fernandez-Gauna, B., Graña, M., Etxeberria-Agiriano, I.: Distributed round-robin q-learning. PLoS ONE **10**(7), e0127129 (2015)

17. Fernandez-Gauna, B., Marques, I., Graña, M.: Undesired state-action prediction in multi-agent reinforcement learning. application to multicomponent robotic system control. Inf. Sci. **232**, 309–324 (2013)
18. Fernandez-Gauna, B., Osa, J.L., Graña, M.: Effect of initial conditioning of reinforcement learning agents on feedback control tasks over continuous state and action spaces. In: de la Puerta, J.G., Ferreira, I.G., Bringas, P.G., Klett, F., Abraham, A., de Carvalho, A.C.P.L.F., Herrero, Á., Baruque, B., Quintián, H., Corchado, E. (eds.) International Joint Conference SOCO'14-CISIS'14-ICEUTE'14. AISC, vol. 299, pp. 125–133. Springer, Heidelberg (2014)
19. Ghavamzadeh, M., Mahadevan, S., Makar, R.: Hierarchical multi-agent reinforcement learning. Auton. Agents Multi-Agent Syst. **13**, 197–229 (2006)
20. Guestrin, C., Lagoudakis, M., Parr, R.: Coordinated reinforcement learning. In: Proceedings of the IXth ICML, pp. 227–234 (2002)
21. van Hasselt, H.: Reinforcement Learning in Continuous State and Action Spaces. In: Wiering, M., van Otterlo, M. (eds.) Reinforcement Learning: State of the Art, pp. 207–246. Springer, Heidelberg (2011)
22. Hengst, B.: Discovering hierarchy in reinforcement learning with HEXQ. In: Maching Learning: Proceedings of the Nineteenth International Conference on Machine Learning, pp. 243–250. Morgan Kaufmann (2002)
23. Kapetanakis, S., Kudenko, D.: Reinforcement learning of coordination in cooperative multi-agent systems. In: AAAI/IAAI 2002, pp. 326–331 (2002)
24. Kok, J.R., Vlassis, N.: Collaborative multiagent reinforcement learning by payoff propagation. J. Mach. Learn. Res. **7**, 1789–1828 (2006)
25. Kuyer, L., Whiteson, S., Bakker, B., Vlassis, N.: Multiagent reinforcement learning for urban traffic control using coordination graphs. In: Daelemans, W., Goethals, B., Morik, K. (eds.) ECML PKDD 2008, Part I. LNCS (LNAI), vol. 5211, pp. 656–671. Springer, Heidelberg (2008)
26. Lauer, M., Riedmiller, M.A.: An algorithm for distributed reinforcement learning in cooperative multi-agent systems. In: Proceedings of the Seventeenth International Conference on Machine Learning, ICML 2000, pp. 535–542. Morgan Kaufmann Publishers Inc., San Francisco (2000)
27. Li, F.D., Wu, M., He, Y., Chen, X.: Optimal control in microgrid using multi-agent reinforcement learning. ISA Trans. **51**(6), 743–751 (2012)
28. Littman, M.L.: Value-function reinforcement learning in Markov games. Cogn. Syst. Res. **2**(1), 55–66 (2001)
29. Mehta, N., Ray, S., Tadepalli, P., Dietterich, T.: Automatic discovery and transfer of MAXQ hierarchies. In: Proceedings of the 25th International Conference on Machine Learning, ICML 2008, pp. 648–655. ACM, New York (2008). http://doi.acm.org/10.1145/1390156.1390238
30. Melo, F., Ribeiro, M.: Coordinated learning in multiagent MDPS with infinite state-space. Auton. Agents Multi-Agent Syst. **21**, 321–367 (2010)
31. Nedic, A., Bertsekas, D.: Least squares policy evaluation algorithms with linear function approximation. Discrete Event Dyn. Syst. **13**(1–2), 79–110 (2003)
32. Peters, J., Schaal, S.: Policy gradient methods for robotics. In: Proceedings of the IEEE/RSJ International Conference on Intelligent Robots and Systems, IROS (2006)
33. Ren, W., Beard, R.W.: Distributed Consensus in Multi-vehicle Cooperative Control: Theory and Applications. Springer, London (2007)
34. Roberts, J.W., Manchester, I.R., Tedrake, R.: Feedback controller parameterizations for reinforcement learning. In: IEEE Symposium on Adaptive Dynamic Programming and Reinforcement Learning (2011)

35. Salkham, A., Cunningham, R., Garg, A., Cahill, V.: A collaborative reinforcement learning approach to urban traffic control optimization. In: Proceedings of the 2008 IEEE/WIC/ACM International Conference on Web Intelligence and Intelligent Agent Technology, WI-IAT 2008, vol. 2, pp. 560–566. IEEE Computer Society, Washington, DC (2008)
36. Servin, A., Kudenko, D.: Multi-agent reinforcement learning for intrusion detection. In: Tuyls, K., Nowe, A., Guessoum, Z., Kudenko, D. (eds.) ALAMAS 2005, ALAMAS 2006, and ALAMAS 2007. LNCS (LNAI), vol. 4865, pp. 211–223. Springer, Heidelberg (2008)
37. Sutton, R.S., Barto, A.G.: Reinforcement Learning I: Introduction. MIT Press, Cambridge (1998)
38. Taylor, M.E., Stone, P.: Transfer learning for reinforcement learning domains: a survey. J. Mach. Learn. Res. **10**(1), 1633–1685 (2009)
39. Vlassis, N., Elhorst, R., Kok, J.R.: Anytime algorithms for multiagent decision making using coordination graphs. In: Proceedings of the International Conference on Systems, Man, and Cybernetics (2004)
40. Wang, X., Sandholm, T.: Reinforcement learning to play an optimal nash equilibrium in team Markov games. In: Advances in Neural Information Processing Systems, pp. 1571–1578. MIT Press (2002)
41. Wu, C., Chowdhury, K., Di Felice, M., Meleis, W.: Spectrum management of cognitive radio using multi-agent reinforcement learning. In: Proceedings of the 9th International Conference on Autonomous Agents and Multiagent Systems: Industry Track, AAMAS 2010, pp. 1705–1712. International Foundation for Autonomous Agents and Multiagent Systems, Richland (2010)
42. Xu, X., Zuo, L., Huang, Z.: Reinforcement learning algorithms with function approximation: recent advances and applications. Inf. Sci. **261**, 1–31 (2014)
43. Zhao, G., Sun, R.: Application of multi-agent reinforcement learning to supply chain ordering management. In: 2010 Sixth International Conference on Natural Computation (ICNC), vol. 7, pp. 3830–3834, August 2010

Optimal Filtering for Time Series Classification

Frank Höppner[✉]

Department of Computer Science, Ostfalia University of Applied Sciences,
38302 Wolfenbüttel, Germany
f.hoeppner@ostfalia.de

Abstract. The application of a (smoothing) filter is common practice in applications where time series are involved. The literature on time series similarity measures, however, seems to completely ignore the possibility of applying a filter first. In this paper, we investigate to what extent the benefit obtained by more complex distance measures may be achieved by simply applying a filter to the original series (while sticking to Euclidean distance). We propose two ways of deriving an optimized filter from classified time series to adopt the similarity measure to a given application. The empirical evaluation shows not only that in many cases a substantial fraction of the performance improvement can also be achieved by filtering, but also that for certain types of time series this simple approach outperforms more complex measures.

1 Motivation

Time series comparison became almost a standard operation just like comparing ordinal or numerical values with tabular data. A broad range of different similarity measures has been proposed in the literature, ranging from simple and straightforward measures such as Euclidean distance (ED) to more complex measures that deal with temporal dilation and translation effects such as dynamic time warping (DTW [1]). Extensive comparative studies have been carried out to compare these measures over a variety of datasets [8,9].

While some scenarios may call for a highly flexible measure, such a measure may perform worse where the flexibility is not needed to solve the task. Just like with classifiers it may misuse its parameters to *overfit*, which is unlikely to happen with simpler measures. Thus, if a comparable performance can be achieved, we should go with the simpler measure (Occam's Razor). The investigation of the literature on time series similarity measures reveals a surprising fact: in this context, *smoothing* of time series is almost completely ignored. The application of a smoothing filter can be considered as a pre-processing step and is applied in many machine learning and data mining applications, but neither plays a role in the comprehensive experimental comparison of time series similarity measures [8,9], nor in a survey on clustering time series data [7] (with distances at the core of any clustering algorithm). This is not to say that the time series in the comparative studies are not pre-processed, in fact they are usually standardized before presented to the similarity measures, or various time series representations

© Springer International Publishing Switzerland 2015
K. Jackowski et al. (Eds.): IDEAL 2015, LNCS 9375, pp. 26–35, 2015.
DOI: 10.1007/978-3-319-24834-9_4

are considered, which implicitly perform smoothing (e.g., piecewise approximations) or simplify noise removal (Fourier or Wavelet transform), but filters are
not explored in general. This is surprising because noise and outliers are very
well perceived as problematic aspects in similarity search. The contribution of
this work is to bring filtering techniques back to the conscience of the time series
similarity measure community, the proposal of two approaches to derive an optimized filter for measuring time series similarity and an experimental evaluation
demonstrating its high potential.

2 Definitions and Discussion

A time series $\mathbf{x}$ of length n is a series $(x_i)_{1 \leq i \leq n}$ of n measurements. The similarity of two series $\mathbf{x}$ and $\mathbf{y}$ is usually measured by means of some distance (or
dissimilarity) measure $d(\mathbf{x}, \mathbf{y}) \geq 0$ where a small distance indicates high similarity. Some distance measures assume that both series $\mathbf{x}$ and $\mathbf{y}$ have the same
length, but in case this is not given one of the series may be stretched to the
length of the other and such measures remain applicable.

Similarity. There are two major groups of similarity measures. *Lock-step measures* directly compare corresponding values of two time series, that is, the ith
value of $\mathbf{x}$ with the ith value of $\mathbf{y}$, as in the Euclidean distance. Simple distortions
such as an offset in the recording time make two observations of the same process
dissimilar under Euclidean distance. Elastic measures identify a *warping path*,
a monotone transformation of time, such that series $\mathbf{x}$ corresponds best to the
warped series $\mathbf{y}$ (e.g. dynamic time warping and variations thereof [1–3]). One
can also find various modifications of these measures, e.g., the authors of [5] try
to "prevent minimum distance distortion by outliers" by giving different weights
to $|x_{p(i)} - y_i|$ depending on the temporal offsets $|p(i) - i|$. But it has also been
pointed out in recent publications that simpler concepts may be hard to beat
[9] or even outperform complex approaches [4]. Some applications seek similar
subsequences of time series only, but this is accomplished by applying the same
range of similarity measures to data from a sliding window, so the fundamental
problem remains the same (only the arguments of the measure change).

Filtering. Filters have a long tradition in signal processing to reduce or enhance
certain aspects of the signal. In data analysis, filters are often applied to smooth
the original series to remove noise or impute missing observations. Here, we
consider (discrete time) linear time-invariant (LTI) filters only. Such a filter may
be described by a vector of coefficients $\alpha = (\alpha_{-m}, \alpha_{-m+1}, \ldots, \alpha_{m-1}, \alpha_m) \in \mathbb{R}^{2m+1}$ and the application of the (discrete) filter α to a (discrete) time series
$\mathbf{x}$ is defined as the convolution $(\mathbf{x} * \alpha)_i = \sum_{j=-m}^{m} \alpha_j \cdot x_{i+j}$. The convolution
$\mathbf{x} * \alpha$ can be considered as a *smoothed version of* $\mathbf{x}$, but for $\mathbf{x} * \alpha$ to have the
same length as $\mathbf{x}$ we need to clarify what x_{i+j} may refer to when $i + j < 1$
or $i + j > n$. Circular discrete convolution is frequently applied (index modulo
time series length), but there is no justification why the last few values of $\mathbf{x}$
should influence the first few values of a smoothed $\mathbf{x}$. So instead of a circular

convolution we define for an arbitrary series $\mathbf{x}$ of length n: $x_i := x_1$ if $i < 1$ and $x_i := x_n$ if $i > n$.

Benefit of Filtering. We argue that the application of filters has not tapped its full potential in the area of time series similarity measures. Such measures are used to compare series against each other with the goal of distinguishing two (or more) types of situations. Any labeled time series dataset may thus be considered as an application domain that defines (by examples) which series should be similar (same class) and which should not (different class). A filter may be tailored to a specific application by focusing on different aspects of time series, which may prove filters to be useful in a broad range of applications.

One important aspect in time series similarity is **temporal locality.** Figure. 1(a) shows a simple, artificial set of time series with data from two classes having a positive peak and a negative peak, resp. The exact position of the peak, however, varies. If the peak positions do not match, we obtain the same distance between examples from the same and different classes with Euclidean distance – it is thus not helpful for discriminating series from different classes. An elastic measure, such as DTW, however, should have no problems with a correct alignment. In yet another applications the peak position may be important. In Fig. 1(b) both classes have a positive peak, but this time the exact position is relevant. This is a simple task for Euclidean distance but nearly unsolvable for DTW, which does not care about the exact temporal location. If we choose our filter *wisely*, the combination of a filter with Euclidean distance (smearing out the singleton peak in the first and leaving the data untouched in the second case) may solve both problems. Secondly, **noise** may easily distort a time series similarity measure, because time series are high dimensional data and suffer from the curse of dimensionality. The right amount of smoothing may help to identify the relevant trends in the series and reduce the impact of incidental differences. Thirdly, a filter is a **versatile preprocessor**, it can be used to approximate the slope or curvature (first or second derivative of the original signal). To distinguish classified time series it may be advantageous to inspect these transformed rather than the original series. If we manage to identify the filter that best discriminates series from different classes we increase the versatility of the measure as it can adopt automatically to a broad range of situations.

3   Optimized Filter for Time Series Similarity

The simple application of the *right* filter may solve a variety of problems for similarity measures. Although it seems like a somewhat obvious idea to apply a filter separately (before applying a distance measure), the potential impact of filtering on the discrimination of time series has not been explored before. In this section we propose ways to automatically find the filter that is best suited to distinguish time series from one another, that is, a filter that emphasizes the *important differences* (between series from different classes) and ignore or attenuate the less important ones (between series from the same class). Apparently we assume that some supervision is available by class labels. We consider two alternative approaches in the following subsections and in both cases we assume that N series $\mathbf{x}_i$, $1 \leq i \leq N$, of length n are given with labels $l_i \in L$, $|L|$ being the number of classes. By $\tilde{\mathbf{x}}_i$ we denote the filtered version of $\mathbf{x}_i$ (after applying a filter $\alpha = (\alpha_{-m}, \ldots, \alpha_0, \ldots, \alpha_m)$ of size $2m+1$, that is, $\tilde{\mathbf{x}} = \mathbf{x} * \alpha$). By $x_{i,j}$ we denote the jth value of series $\mathbf{x}_i$. For the sake of a convenient notation, with $x_{i,j}$ we refer to $x_{i,1}$ for all $j \leq 1$ and $x_{i,n}$ for all $j \geq n$.

3.1   A Filter Derived from Pairwise Comparison

As a first proposal consider the following objective function

$$\text{min. } f(\alpha) = \beta \sum_{l_i = l_j} \|\tilde{\mathbf{x}}_i - \tilde{\mathbf{x}}_j\|^2 - \sum_{l_i \neq l_j} \|\tilde{\mathbf{x}}_i - \tilde{\mathbf{x}}_j\|^2 \quad \text{s.t.} \sum_{k=-m}^{m} \alpha_k = 1 \quad (1)$$

where $\sum_{l_i = l_j}$ is an abbreviation for $\sum_{1 \leq i,j \leq N, i \neq j, l_i = l_j}$ (same for the second sum with $\neq$ rather than $=$). The filter coefficients α are hidden in the smoothed series $\tilde{\mathbf{x}}_i$ on the right hand side of f. Distances between *filtered series* from the same class should be small (first summation), whereas distances between *filtered series* from different classes should be large (second summation). Since the second sum is subtracted, the function has to be minimized overall. The coefficient $\beta \in \mathbb{R}$ is a necessary scaling factor for the first sum, chosen to ensure that f is a convex function ($f \to \infty$ as $\|\alpha\| \to \infty$) that actually has a (global) minimum. (Without the scaling factor the second sum may dominate, turning f into a concave function and the minimum is obtained for $\|\alpha\| \to \infty$.)

Without any constraint on α there is an obvious minimum $\alpha = 0$, but this is apparently an undesired solution because all series would look identical. Here, we require the sum of all filter coefficients to be 1. This constraint ensures that the filtered series stay within the same range as the original series.

Proposition 1. *The optimal filter* $\alpha = (\alpha_{-m}, \ldots, \alpha_0, \ldots, \alpha_m) \in \mathbb{R}^{2m+1}$ *minimizing (1) is obtained from a linear equation system* $A\alpha' = b$ *with* $A \in \mathbb{R}^{2m+2 \times 2m+2}$, $b = (0, \ldots, 0, 1) \in \mathbb{R}^{2m+2}$, $\alpha' = (\alpha_{-m}, \ldots, \alpha_m, \lambda)$ *where*

$$A = \begin{pmatrix} M & 1 \\ 1 & 0 \end{pmatrix} \quad (2)$$

$$M = 2 \sum_{(i,j)} s_{i,j} \sum_{l=1}^{n} \pi_l(\mathbf{x}_i - \mathbf{x}_j) \pi_l(\mathbf{x}_i - \mathbf{x}_j)^\top \in \mathbb{R}^{2m+1 \times 2m+1} \quad (3)$$

$$\pi_l(\mathbf{z}) = (z_{l-m}, \ldots, z_{l+m}) \in \mathbb{R}^{2m+1} \quad (4)$$

with $s_{i,j} = \beta$ *for* $l_i = l_j$ *and* $s_{i,j} = -1$ *for* $l_i \neq l_j$. *(The notation* $\sum_{(i,j)}$ *is an abbreviation for* $\sum_{1 \leq i,j \leq N, i \neq j}$.)

Finally, we have to choose β such that f is guaranteed to be convex. This is accomplished by setting

$$\beta = 1.5 \cdot \max_{-m \leq k \leq m} \left\{ \frac{M_{k,k}^{\neq}}{M_{k,k}^{=}} \right\} \quad \text{where } M^= = \sum_{l_i = l_j} M_{\mathbf{x}_i, \mathbf{x}_j}, M^{\neq} = \sum_{l_i \neq l_j} M_{\mathbf{x}_i, \mathbf{x}_j}$$

where $M_{\mathbf{x},\mathbf{y}} := \sum_{l=1}^{n} \pi_l(\mathbf{x} - \mathbf{y})\pi_l(\mathbf{x} - \mathbf{y})^\top$. [Proofs are omitted due to lack of space.] The factor of 1.5 ensures that the coefficients of the quadratic terms are strictly positive, other choices are possible as long as the factor is larger than one (but the larger the factor, the more the first sum of (1) dominates). We obtained satisfactory results with a factor of 1.5 and stick to it throughout the paper.

If we scale the resulting filter α by some scalar, the (squared) Euclidean distances scale by the same factor. While the minimization of (1) yields a unique filter, to discriminate series from different classes we will subsequently order time series by distance (to find the closest match) and this order is not affected by a factor. Therefore we divide α by its norm and arrive at a filter with $\|\alpha\| = 1$.

3.2   A Filter Derived from Groupwise Comparisons

If we assume that all time series from the same class label are similar, we may consider the groupwise (or classwise) means as prototypical time series for their respective class. We define the *mean smoothed series* $\bar{\tilde{\mathbf{x}}}_l$ for a class label l as:

$$\bar{\tilde{\mathbf{x}}}_l = \frac{\sum_{1 \le i \le n, l_i = l} \tilde{\mathbf{x}}_i}{\sum_{1 \le i \le n, l_i = l} 1}$$

The rationale for finding a filter is then that the distance of any series $\mathbf{x}$ with label l to the mean series $\bar{\tilde{\mathbf{x}}}_l$ of its own class should be small, but distances to the mean series $\bar{\tilde{\mathbf{x}}}_k$ of other classes, $k \ne l$, should be large. This time we enforce a unit length constraint on the filter to allow for filter types whose coefficients sum up to zero.

$$\text{max. } f(\alpha) = \frac{\sum_{l,k \in L} \|\bar{\tilde{\mathbf{x}}}_l - \bar{\tilde{\mathbf{x}}}_k\|^2}{\sum_{1 \le i \le N} \|\tilde{\mathbf{x}}_i - \bar{\tilde{\mathbf{x}}}_{l_i}\|^2} \qquad \text{s.t. } \|\alpha\| = 1 \qquad (5)$$

This objective function has to be maximized (subject to the unit length constraint): The nominator has to be maximized (distances between class means), the denominator has to be minimized (distance of individual series to its own class mean).

Proposition 2. *The optimal filter* $\alpha = (\alpha_{-m}, \dots, \alpha_0, \dots, \alpha_m) \in \mathbb{R}^{2m+1}$ *maximizing (5) is the eigenvector with the largest eigenvalue of the matrix* $Q^{-1}P \in \mathbb{R}^{2m+1 \times 2m+1}$*, where*

$$P = \sum_{l,k \in L} M_{\bar{\mathbf{x}}_l, \bar{\mathbf{y}}_k}, \quad Q = \sum_{1 \le i \le n} M_{\mathbf{x}_i, \bar{\mathbf{x}}_{l_i}}, \quad M_{\mathbf{x},\mathbf{y}} = \sum_{l=1}^{n} \pi_l(\mathbf{x} - \mathbf{y})\pi_l(\mathbf{x} - \mathbf{y})^\top$$

and $\pi_l(\mathbf{z}) = (z_{l-m}, \dots, z_{l+m}) \in \mathbb{R}^{2m+1}$.

[Again, the proof is omitted due to lack of space.]

3.3   Computational Complexity

Lock-step measures such as Euclidean distance are computationally inexpensive ($O(n)$ for the comparison of two series of length n). The nature of most elastic measures, such as dynamic time warping, calls for a quadratic complexity $O(n^2)$. The advantage of the filter approaches is that we can spend some computational effort beforehand to determine the optimal filter and then stick to a lock-step measure, taking advantage of the low linear complexity $O(n)$ when comparing time series. The computation of the optimal filter involves the pairwise combination of time series in both corollaries. While Proposition 1 requires to combine all series with the same or different label ($O(N^2)$), with Proposition 2 we combine only mean series for each class label which drastically reduces the computational effort ($O(|L|^2)$ with $|L|$ being much smaller than N).

4   Experimental Evaluation

Time series similarity measures are typically evaluated against each other by examining their performance in a one-nearest-neighbor (1-NN) classification task (cf. [8,9]). A dataset is subdivided into train and test data and for each series from the test dataset, the closest series from the training set is sought. The class label from the closest match is then used for class prediction. The accuracy (or error rate) reports the number of correct (or incorrect) predictions. We report cross-validated results in Table 1 and (in contrast to the typical cross-validation for classifiers) use only one fold for training and $k-1$ for testing. As ED and DTW are the most prominent representatives of lock-step and elastic measures we compare them to three types of filters: (a) filter constraint "sum = 1" (ED-FS) as defined by Proposition 1, (b) filter constraint "norm = 1" (ED-FN) as defined by Proposition 2, and (c) a standard Gaussian filter (ED-FG). The filter is *always determined using the training data only*, then the filter is applied to the test and training data and the 1-NN classifier is carried out with Euclidean distance.

Filter Width. Both proposals for determining an optimized smoothing filter require the filter size m to be given a priori. We ran experiments with varying the filter size to observe the effect on the classification rate. There are differences in the performance, but the variation is quite limited. For all experiments in this paper, however, we have consistently chosen $2m$ to be 10 % of the time series length but not larger than $m = 7$.

4.1   Artificial Data

As a sanity check, we revisit the two datasets from Sect. 2 that are particularly difficult for either DTW or ED: In Fig. 1(a) the series from two groups differ in the orientation of a single peak (up vs down), but the exact location of the peak varies. A small amount of warping compensates for the offset in the peak position, but for ED two series from the same group are (most of the time) almost as similar as two series from different groups. Smoothing the series blurs

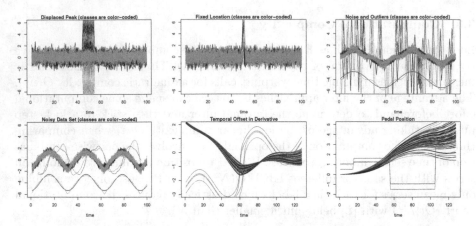

Fig. 1. Some data sets used. Top row from left to right (a)–(c), bottom row (d)–(f). All datasets consist of 100 time series. Class information is color-coded (green/black) (Color figure online).

the local peak, which makes it easier to detect the similarity between peaks of the same direction even if no warping is possible. As we can see from Table 1 the filter approaches perform (equally) well (about 94 % accuracy).

The second example is shown in Fig. 1(b): the peak orientation is always identical (up), but the exact location of the peak makes the difference between both classes. As expected DTW is not able to perceive a difference between both classes, but for ED this is a very simple task (100 % accuracy). The proposed approaches manage to adopt automatically to this situation, FS and FN perform very well (close to 100 %, cf. Table 1), but apparently the standard Gaussian filter cannot help here, its performance is close to the poor DTW performance.

Response to Noise. A second set of examples is shown in Fig. 1(c and d), where the dataset (c) consists of sinusoidal curves plus Gaussian noise and a few outliers. One example from each class (without noise and outliers) is shown near the bottom (being not part of the dataset). As we can see from Table 1, DTW performance drops below 60 %; ED performs much better, but all the smoothing approaches outperform ED. The second example (Fig. 1(d)) consists of noisy series, which do not contain any outliers. Again, one example from each class without any noise is shown at the bottom. No warping is necessary, but the noise seems to prevent DTW from performing as well as ED. The most prominent differences of the two classes are the rounded versus angular minima. When applying a smoothing filter we risk to smear out the angular edge and to loose an important feature for class discrimination. But actually the FN and FS filter manage to keep an accuracy close to 100 %. The chosen filter does not only denoise but delivers series where the peak positions are displaced for series from different classes (examples of smoothed series shown in blue and pink).

4.2   Results for Data from the UCR Time Series Repository

We also report results on datasets from the UCR time series repository [6] in Table 1. All datasets were used as they were provided, no changes were made

Table 1. Mean accuracy and standard deviation of cross-validated 1-NN-classifier (no. of folds in column #, for UCR data the same number of folds was used as in [9]). Euclidean distance (ED) with: Gaussian filter (ED-FG), filter obtained from sum constraint (ED-FS), filter obtained from norm constraint (ED-FN).

	#	ED	DTW	ED-FG	ED-FN	ED-FS
Figure 1(a)	11	0.68|0.06	1.00|0.00	0.93|0.04	0.94|0.03	0.94|0.03
Figure 1(b)	11	1.00|0.00	0.50|0.05	0.56|0.04	0.99|0.01	0.99|0.01
Figure 1(c)	11	0.72|0.15	0.58|0.07	0.84|0.08	0.84|0.04	0.81|0.08
Figure 1(d)	11	0.90|0.06	0.84|0.08	0.96|0.02	0.98|0.02	0.98|0.02
Figure 1(e)	11	0.78|0.09	0.62|0.09	0.78|0.08	0.98|0.05	0.95|0.08
Figure 1(f)	11	0.78|0.08	0.61|0.07	0.78|0.08	0.97|0.04	0.97|0.04
ECG200	5	0.85|0.03	0.77|0.03	0.83|0.03	0.83|0.03	0.82|0.04
ECGFiveDays	32	0.86|0.04	0.80|0.04	0.92|0.03	0.90|0.06	0.90|0.04
FISH	5	0.73|0.03	0.69|0.03	0.72|0.03	0.72|0.04	0.68|0.04
GunPoint	5	0.87|0.03	0.85|0.03	0.86|0.03	0.87|0.04	0.88|0.04
OliveOil	2	0.86|0.05	0.86|0.04	0.85|0.05	0.86|0.05	0.84|0.05
Beef	2	0.46|0.07	0.46|0.07	0.45|0.07	0.44|0.07	0.45|0.07
Adiac	5	0.54|0.02	0.54|0.02	0.60|0.02	0.42|0.16	0.52|0.02
Coffee	2	0.82|0.07	0.84|0.08	0.80|0.07	0.97|0.03	0.86|0.09
50words	5	0.59|0.02	0.62|0.02	0.60|0.02	0.42|0.16	0.53|0.02
SwedishLeaf	5	0.70|0.02	0.75|0.01	0.72|0.02	0.70|0.01	0.69|0.02
CBF	12	0.94|0.02	0.99|0.00	0.98|0.01	0.99|0.01	0.99|0.01
OSULeaf	5	0.52|0.03	0.58|0.04	0.53|0.02	0.48|0.05	0.51|0.03
FaceFour	5	0.80|0.05	0.87|0.04	0.81|0.05	0.82|0.05	0.82|0.05
Lighting7	2	0.61|0.05	0.71|0.04	0.65|0.04	0.68|0.04	0.68|0.05
Lighting2	5	0.67|0.05	0.78|0.06	0.72|0.05	0.68|0.08	0.72|0.05
Synth. control	5	0.86|0.02	0.99|0.00	0.99|0.00	0.96|0.01	0.96|0.00
Trace	5	0.64|0.03	0.98|0.02	0.59|0.04	0.63|0.04	0.63|0.04

to them. We are interested in how much of the performance increase of elastic measures can be achieved by filtering (when sticking to a lock-step measure). By scanning through the Table we can see that quite often a substantial fraction of the performance increase obtained by switching from ED to DTW is also obtained when switching to a filter approach – in various cases the achieved accuracy is identical. To our surprise, we can also identify cases where filtering actually outperforms DTW (e.g. Coffee and Adiac), which is a remarkable result.

4.3 Accumulated Signals

Finally, we consider a situation where the FS/FN filter approaches perform particularly well. Many real series record a physical property reacting to some external input, for example, the temperature during a heating period, the rise and fall of the water level as inlets or outlets are opened, the distance covered when driving a car, etc. What is actually changing is some input variable (power of heating element, valve position, throttle control), but instead of capturing this parameter directly, some other accumulated physical property is measured (temperature, water level, covered distance).

The following examples are artificially generated but were created to reproduce a real case.[1] In the dataset of Fig. 1(e) all the series appear very similar and a discrimination between the classes seems close to impossible. The series correspond to an accumulation of some physical property, whose derivative is very similar for all examples, but differs in the temporal location of a steep increase near $t = 60$ (two examples from each class in red/blue). As the integrated values are actually measured, this difference is hardly recognized in the original series. But both filters, FS and FN, manage to transform the raw data such that the accuracy increases dramatically (cf. Table 1).

For the second example in Fig. 1(f) we have a similar situation, the time series look very similar across the two classes. One may think of a gas pedal position as the actually controlled input (two examples from each class in red/blue), which influences speed, but only the mileage is measured. Again, both filters manage to identify filters that separate both classes very well. In both cases, a Gaussian filter does not help (data is not noisy) and the elastic DTW performs worst.

5   Conclusions

When seeking for the best similarity measure for a specific application, among equally well performing solutions we should prefer the simplest approach (Occam's Razor). Filtering time series and measuring their Euclidean distance is one of the most simple things one can possibly think of, however, this option has not received much attention in the literature. Two approaches have been proposed to derive such a filter from training data and the experimental results have shown that they turn Euclidean distance into a much more versatile tool, as it can adapt to specific properties of the time series. For various datasets a substantial increase in the performance has been observed and for a specific class of problems (discrimination of series that represent accumulating physical properties) this approach outperforms elastic measures. Together with the simplicity of the Euclidean measure, which has a computational advantage over complex elastic measures, this approach is a worthwhile alternative to existing measures. How to identify an optimal filter in combination with elastic measures remains an open question for future work.

---

[1] Only the artificial data can be shared: public.ostfalia.de/~hoeppnef/tsfilter.html.

References

1. Berndt, D.J., Clifford, J.: Finding patterns in time series: a dynamic programming approach. In: Advances in Knowledge Discovery and Data Mining, pp. 229–248. MITP (1996)
2. Chen, L., Ng, R.: On the marriage of Lp-norms and edit distance. In: International Conference on Very Large Databases, pp. 792–803 (2004)
3. Chen, L., Özsu, M., Oria, V.: Robust and fast similarity search for moving object trajectories. In: Proceedings of ACM SIGMOD, pp. 491–502 (2005)
4. Höppner, F.: Less is more: similarity of time series under linear transformations. In: Proceedings of SIAM International Conference on Data Mining (SDM), pp. 560–568 (2014)
5. Jeong, Y.-S., Jeong, M.K., Omitaomu, O.A.: Weighted dynamic time warping for time series classification. Pattern Recogn. **44**(9), 2231–2240 (2011)
6. Keogh, E., Zhu, Q., Hu, B., Hao, Y., Xi, X., Wei, L., Ratanamahatana, C.A.: The UCR Time Series Classification/Clustering Homepage (2011)
7. Liao, W.T.: Clustering of time series data - a survey. Pattern Recogn. **38**(11), 1857–1874 (2005)
8. Serrà, J., Arcos, J.L.: An empirical evaluation of similarity measures for time series classification. Knowl. Based Syst. **67**, 305–314 (2014)
9. Wang, X., Mueen, A., Ding, H., Trajcevski, G., Scheuermann, P., Keogh, E.: Experimental comparison of representation methods and distance measures for time series data. Data Min. Knowl. Discov. **26**(2), 275–309 (2012)

Addressing Overlapping in Classification with Imbalanced Datasets: A First Multi-objective Approach for Feature and Instance Selection

Alberto Fernández[1]([✉]), Maria Jose del Jesus[1], and Francisco Herrera[2]

[1] Department of Computer Science, University of Jaén, Jaén, Spain
{alberto.fernandez,mjjesus}@ujaen.es
[2] Department of Computer Science and Artificial Intelligence,
University of Granada, Granada, Spain
herrera@decsai.ugr.es

Abstract. In classification tasks with imbalanced datasets the distribution of examples between the classes is uneven. However, it is not the imbalance itself which hinders the performance, but there are other related intrinsic data characteristics which have a significance in the final accuracy. Among all, the overlapping between the classes is possibly the most significant one for a correct discrimination between the classes.

In this contribution we develop a novel proposal to deal with the former problem developing a multi-objective evolutionary algorithm that optimizes both the number of variables and instances of the problem. Feature selection will allow to simplify the overlapping areas easing the generation of rules to distinguish between the classes, whereas instance selection of samples from both classes will address the imbalance itself by finding the most appropriate class distribution for the learning task, as well as removing noise and difficult borderline examples.

Our experimental results, carried out using C4.5 decision tree as baseline classifier, show that this approach is very promising. Our proposal outperforms, with statistical differences, the results obtained with the SMOTE + ENN oversampling technique, which was shown to be a baseline methodology for classification with imbalanced datasets.

Keywords: Imbalanced classification · Overlapping · Feature selection · Instance selection · Multiobjective evolutionary algorithms

1 Introduction

The imbalanced class problem is one of the new challenges that arose when Machine Learning reached its maturity [6], being widely present in the fields of businesses, industry and scientific research. This issue grew up in importance at the same time that researchers realize that the datasets they analyzed hold more instances or examples from one class than that of the remaining ones, and they

© Springer International Publishing Switzerland 2015
K. Jackowski et al. (Eds.): IDEAL 2015, LNCS 9375, pp. 36–44, 2015.
DOI: 10.1007/978-3-319-24834-9_5

standard classification algorithms achieved a model below a desired accuracy threshold for the underrepresented class.

One of the main drawbacks for the correct identification of the minority or positive class of the problem, is related to overlapping between classes [8]. Rules with a low confidence and/or coverage, because they are associated with an overlapped boundary area, will be discarded.

The former fact is related with the attributes that represent the problem. It is well known that a large number of features can degrade the discovery of the borderline areas of the problem, either because some of these variables might be redundant or because they do not show a good synergy among them. Therefore, the use of feature selection can ease to diminish the effect of overlapping [4].

However, the imbalance class problem cannot be addressed by itself just by carrying out a feature selection. For this reason, it is also mandatory to perform a preprocessing of instances by resampling the training data distribution, avoiding a bias of the learning algorithm does towards the majority class.

In accordance with the above, in this work contribution we aim at improving current classification models in the framework of imbalanced datasets by developing both a feature and instance selection. This process will be carried out means of a multi-objective evolutionary algorithm (MOEA) optimization procedure. The multi-objective methodology will allow us to perform an exhaustive search by means of the optimization of several measures which, on a whole, are expected to be capable of giving a quality answer to the learnt system. In this sense, this wrapper approach will be designed to take advantage of the exploration of the full search space, as well as providing a set of different solutions for selecting the best suited for the final user/task.

Specifically, we will make use of the well known NSGA2 approach [3] as the optimization procedure, and the C4.5 decision tree [10] as baseline classifier. We must stress that, although the C4.5 algorithm carries out itself an inner feature selection process, our aim is to 'ease' the classifier by carrying out a pre-selection of the variables with respect to the intrinsic characteristics of the problem, mainly referring the overlapping between the classes.

This contribution is arranged as follows. Section 2 introduces the problem of classification with imbalanced datasets and overlapping. Section 3 describes our MOEA approach for addressing this problem. Next, Sect. 4 contains the experimental results and the analysis. Finally, Sect. 5 will conclude the paper.

2 Imbalanced Datasets in Classification

In this section, we will first introduce the problem of imbalanced datasets. Then, we will focus on the presence of overlapping between the classes.

2.1 Basic Concepts

Most of the standard learning algorithms consider a balanced training set for the learning stage. Therefore, addressing problems with imbalanced data may cause obtaining of suboptimal classification models, i.e. a good coverage of the majority examples whereas the minority ones are misclassified frequently [8]. There are several reasons behind this behaviour which are enumerated below:

- The use of global performance measures for guiding the search process, such as standard accuracy rate, may benefit the covering of the majority examples.
- Classification rules that predict the positive class are often highly specialized, and they are discarded in favour of more general rules.
- Very small clusters of minority class examples can be identified as noise, and therefore they could be wrongly discarded by the classifier.

In order to overcome the class imbalance problem, we may find a large number of proposed approaches, which can be categorized in three groups [8]:

1. Data level solutions: the objective consists of rebalancing the class distribution via preprocessing of instances [2].
2. Algorithmic level solutions: these solutions try to adapt several classification algorithms to reinforce the learning towards the positive class [1].
3. Cost-sensitive solutions: they consider higher costs for the misclassification of examples of the positive class with respect to the negative class [5].

2.2 Overlapping or Class Separability

The problem of overlapping between classes appears when a region of the data space contains a similar quantity of training data from each class, imposing a hard restriction to finding discrimination functions.

In previous studies on the topic [9], authors depicted the performance of the different datasets ordered according to different data complexity measures (including IR) in order to search for some regions of interesting good or bad behaviour. They could not characterize any interesting behaviour according IR, but they do for example according the so called metric $F1$ or *maximum Fisher's discriminant ratio* [7], which measures the overlap of individual feature values.

This metric for one feature dimension is defined as: $f = \frac{(\mu_1 - \mu_2)^2}{\sigma_1^2 + \sigma_2^2}$ where μ_1, μ_2, σ_1^2, σ_2^2 are the means and variances of the two classes respectively, in that feature dimension. We compute f for each feature and take the maximum as measure F1. For a multidimensional problem, not all features have to contribute to class discrimination. Therefore, we can just take the maximum f over all feature dimensions when discussing class separability. Datasets with a small value for the F1 metric will have a high degree of overlapping.

Finally, a closely related issue is the impact of noisy and borderline examples on the classifier performance in imbalanced classification [11]. Regarding this fact, a preprocessing cleaning procedure can help the learning algorithm to better discriminate the classes, especially in the overlapped areas.

3   Addressing Overlapping in Imbalanced Domains by a Multi-objective Feature and Instance Selection

In this work, our contribution is to introduce a new methodology that makes use of a MOEA to determine the best subset of attributes and instances in imbalanced classification. Instance selection aims at both balancing the data distribution between the positive and negative classes, and removing noisy and borderline examples that hinder the classification ability of the learning algorithm. Feature selection will simplify the boundaries of the problem by limiting the influence of those features create difficulties for the discrimination process.

However, the estimation of the best suited subset of instances and features is not trivial. In accordance with the former, an optimization search procedure must be carried out in order to determine the former values. Among the different techniques that can be used for this task, genetic algorithms excel due to their ability to perform a good exploration and exploitation of the solution space. Our ultimate goal is to build the simplest classifier with the highest accuracy in the context of imbalanced classification. Regarding this issue, the first objective can be overcome by *maximizing the reduction of instances*, whereas the second one is achieved by *maximizing the recognition of both the positive and negative classes*. In accordance with the former, we propose the use of the "Area Under the ROC Curve" (AUC), as it provides a good trade-off between the individual performance for each individual class (Eq. 1).

$$AUC = \frac{1 + TP_{rate} - FP_{rate}}{2} \tag{1}$$

Taking into account the objectives previously outlined, we propose the design of a work methodology using as basis a MOEA. This way, we can take advantage of both the exploration capabilities of this type of technique, as well as allowing the selection among a set of different solutions, depending on the user's requirements. We will name this approach as *IS + FS-MOEA*.

Specifically, we will make use of the NSGA-II algorithm [3] for implementing our model, as it is widely known for being a high-performance MOEA. Its two main features are first the fitness evaluation of each solution based on both the Pareto ranking and a crowding measure, and the other is an elitist generation update procedure.

In order to codify the solutions, we will make use of a chromosome with two well differentiate parts: one (FS) for the feature selection and another one (IS) for the instance selection. Both parts will have a binary codification, in such a way that a 0 means that the variable (or instance) will not take part for generating the classification model, whereas a 1 value stands for the opposite case. Chromosomes will be evaluated jointly with aims at obtaining the best synergy between both characteristics, instead of optimizing them separately. This issue is based on the fact that it is not clearly defined which the best order for carrying our both processes is. An initial chromosome will be built with all genes equal to '1' in order to implement the standard case study, i.e. the full training set, whereas the remaining individuals will be generated at random.

As baseline classifier, we will make use of the C4.5 decision tree [10] for several reasons. The first one is its wide use in classification with imbalanced data, so that we may carry out a fair comparative versus the state-of-the-art. The second one is its efficiency; since we need to perform a large number of evaluations throughout the search process, it is important the base model to be particularly quick for not biasing the global complexity of the methodology.

We must stress that, the C4.5 algorithm carries out itself an inner feature selection process. However, our aim to "ease" the classifier by carrying out a pre-selection of the variables with respect to the intrinsic characteristics of the problem, mainly referring the overlapping between the classes, so that we can improve the classification of both classes together.

For the evaluation of the chromosomes, we carry out the preprocessing of the training set codified in the phenotype, and then the C4.5 classifier is executed with the modified dataset. Then, the objective functions to be maximized are computed as stated in Eq. 2, being N the number of initial training instances, and IS_i the value of the chromosome for the instance selection part.

$$OBJ_1 : AUC$$
$$OBJ_2 : RED = N - \sum_{i=0}^{N-1} IS_i; \tag{2}$$

4 Experimental Study

This section includes the experimental analysis of the proposed approach. With this aim, we first present the experimental framework including the datasets selected for the study, as well as the parameters of the algorithms, and the use of statistical test. Then, we show the complete results and the comparison with the state-of-the-art to determine the goodness of our proposal.

4.1 Experimental Framework

Table 1 shows the benchmark problems selected for our study, in which the name, number of examples, number of attributes, and IR (ratio between the majority and minority class instances) are shown. Datasets are ordered with respect to their degree of overlapping. A wider description for these problems can be found at http://www.keel.es/datasets.php. The estimates of AUC measure are obtained by means of a 5 fold Cross-Validation, aiming to include enough positive class instances in the different folds.

The parameters of the NSGA-II MOEA have been set up as follows: 60 individuals as population size, with 100 generations. The crossover and the mutation (per gen) probabilities are 0.8 and 0.025 respectively. For the C4.5 decision tree we use a confidence level at 0.25, with 2 as the minimum number of item-sets per leaf, and the application of pruning will be used to obtain the final tree. As state-of-the-art approach for the sake of a fair comparison we have selected the SMOTE + ENN preprocessing technique [2], which has shown a good synergy with the C4.5 algorithm [8]. This approach creates synthetic examples of the

Table 1. Summary of imbalanced datasets used

Name	#Ex.	#Atts.	IR	F1	Name	#Ex.	#Atts.	IR	F1
glass4	214	9	15.47	1.4690	pimaImb	768	8	1.90	0.5760
ecoli01vs5	240	6	11.00	1.3900	abalone19	4174	8	128.87	0.5295
cleveland0vs4	177	113	12.62	1.3500	ecoli0147vs2356	336	7	10.59	0.5275
ecoli0146vs5	280	6	13.00	1.3400	pageblocks0	5472	10	8.77	0.5087
yeast2vs8	482	8	23.10	1.1420	glass2	214	9	10.39	0.3952
ecoli0347vs56	257	7	9.28	1.1300	vehicle2	846	18	2.52	0.3805
vehicle0	846	18	3.23	1.1240	yeast1289vs7	947	8	30.56	0.3660
ecoli01vs235	244	7	9.17	1.1030	yeast1vs7	459	8	13.87	0.3534
yeast05679vs4	528	8	9.35	1.0510	glass0146vs2	205	9	11.06	0.3487
glass06vs5	108	9	11.00	1.0490	yeast0359vs78	506	8	9.12	0.3113
glass5	214	9	22.81	1.0190	glass016vs2	192	9	10.29	0.2692
ecoli067vs35	222	7	9.09	0.9205	yeast1	1484	8	2.46	0.2422
ecoli0267vs35	244	7	9.18	0.9129	glass1	214	9	1.82	0.1897
ecoli0147vs56	332	6	12.28	0.9124	vehicle3	846	18	2.52	0.1855
yeast4	1484	8	28.41	0.7412	habermanImb	306	3	2.68	0.1850
yeast0256vs3789	1004	8	9.14	0.6939	yeast1458vs7	693	8	22.10	0.1757
glass0	214	9	2.06	0.6492	vehicle1	846	18	2.52	0.1691
abalone918	731	8	16.68	0.6320	glass015vs2	172	9	9.12	0.1375

minority class by means of interpolation to balance the data distribution, and then it removes noise by means of the ENN cleaning procedure. Its configuration will be the standard with a 50 % class distribution, 5 neighbors for generating the synthetic samples and 3 for the ENN cleaning procedure, and Euclidean Metric for computing the distance among the examples.

Finally, we will make use of Wilcoxon signed-rank test [12] to find out whether significant differences exist between a pair of algorithms, thus providing statistical support for the analysis of the results.

4.2   Analysis of the Results

In this case study, the final aim is to obtain the highest precision for both classes of the problem in the test set. In this way, we will always select the one solution of the Pareto with the best performance with respect to the AUC metric. In this case, a comparison with the optimal Pareto front is not possible since for classification functions this is often unavailable.

Average values for the experimental results are shown in Table 2, where datasets are ordered from low to high overlapping. From these results we may highlight the goodness of our approach, as it achieves the highest average value among all problems. Additionally, we must stress that in the case study of the higher overlapped problems, i.e. from "ecoli0147vs2356", that our proposed approach outperforms the baseline SMOTE + ENN technique in 12 out of 16 datasets. Finally, it is worth to point out that our IS + FS-MOEA does not show

Table 2. Experimental results for C4.5 with SMOTE + ENN (C4.5 + S_ENN) and our C4.5 with IS + FS-MOEA approach (C4.5 + MOEA) in training and test with AUC metric.

Dataset	IR	F1	C4.5+S_ENN		C4.5+MOEA		Dataset	IR	F1	C4.5+S_ENN		C4.5+MOEA	
glass4	15.47	1.4690	.9813	**.8292**	.9838	.8225	pima	1.90	0.5760	.7976	**.7311**	.8293	.7084
ecoli01vs5	11.00	1.3900	.9676	**.8477**	.9795	.8455	abalone19	128.87	0.5295	.9009	**.5185**	.5000	.5000
cleveland0vs4	12.62	1.3500	.9922	.7179	.9828	**.8582**	ecoli0147vs2356	10.59	0.5275	.9561	.8529	.9571	**.8755**
ecoli0146vs5	13.00	1.3400	.9861	.8923	.9856	**.8981**	page-blocks0	8.77	0.5087	.9792	.9437	.9798	**.9442**
yeast2vs8	23.10	1.1420	.9115	**.8012**	.8359	.7664	glass2	10.39	0.3952	.9402	.6819	.9364	**.7797**
ecoli0347vs56	9.28	1.1300	.9540	.8502	.9718	**.8541**	vehicle2	2.52	0.3805	.9846	.9396	.9842	**.9512**
vehicle0	3.23	1.1240	.9716	.9160	.9761	**.9448**	yeast1289vs7	30.56	0.3660	.9359	.6397	.7931	**.6733**
ecoli01vs235	9.17	1.1030	.9720	.8218	.9527	**.8873**	yeast1vs7	13.87	0.3534	.9107	.6968	.8890	**.7759**
yeast05679vs4	9.35	1.0510	.9276	**.7725**	.9207	.7674	glass0146vs2	11.06	0.3487	.9157	**.7344**	.9553	.7274
glass06vs5	11.00	1.0490	.9912	**.9647**	.9975	.9350	yeast0359vs78	9.12	0.3113	.9214	**.7078**	.8628	.6978
glass5	22.81	1.0190	.9480	.8232	.9988	**.9951**	glass016vs2	10.29	0.2692	.9237	.6667	.9947	**.9572**
ecoli067vs35	9.09	0.9205	.9700	.7875	.9632	**.8450**	yeast1	2.46	0.2422	.7781	**.6957**	.7857	.6677
ecoli0267vs35	9.18	0.9129	.9851	**.7854**	.9642	.7827	glass1	1.82	0.1897	.8601	.6668	.8912	**.7420**
ecoli0147vs56	12.28	0.9124	.9598	.8457	.9738	**.8538**	vehicle3	2.52	0.1855	.8892	**.7675**	.8894	.7206
yeast4	28.41	0.7412	.9113	**.7157**	.8648	.7089	haberman	2.68	0.1850	.7428	.6076	.7326	**.6178**
yeast0256vs3789	9.14	0.6939	.9121	**.7649**	.8140	.7581	yeast1458vs7	22.10	0.1757	.8719	.5192	.7996	**.5745**
glass0	2.06	0.6492	.8862	.7748	.8917	**.8103**	vehicle1	2.52	0.1691	.8881	.7170	.8960	**.7340**
abalone9-18	16.68	0.6320	.9302	**.7332**	.8425	.7122	glass015vs2	9.12	0.1375	.9342	.7226	.9429	**.7433**
			C4.5-SMOTE+ENN							C4.5-MOEA			
Average			.9247	.7626			Average			.9033	**.7899**		

Table 3. Wilcoxon test for the comparison between C4.5 + MOEA $[R^+]$ and C4.5 + S_ENN $[R^-]$.

Comparison	R^+	R^-	p-value	W/T/L
C4.5 + MOEA vs C4.5 + S_ENN	460.0	206.0	0.044745	21/0/15

the curse of over-fitting, as the training performance is even lower than that of the standard preprocessing approach.

In order to determine statistically the best suited metric, we carry out a Wilcoxon pairwise test in Table 3. Results of this test agree with our previous remarks, since significant differences are found in favour of our IS + FS-MOEA approach with a confidence degree above the 95 %.

Finally, we must remark that the IS-FS-MOEA approach has a greater computational cost in terms of both memory and CPU time than the C4.5 + S_ENN algorithm, as it carries out an evolutionary process. However, its advantage over the former is twofold: (1) it has been shown to clearly outperform the former in terms of precision; and (2) it allows the final user to apply several solutions in order to select the one that better suites to the problem that is being addressed.

5   Concluding Remarks

In this work we have proposed a novel MOEA in the framework of classification with imbalanced datasets. This approach has been designed under a double perspective: (1) to carry out an instance selection for compensating the example distribution between the classes, as well as removing those examples which include noise, or which difficult the discrimination of the classes; and (2) to perform a feature selection to remove those attributes that may imply a high degree of overlapping in the borderline areas.

The goodness in the use of the MOEA is related to its high exploration abilities, the capability of using several metrics to guide the search, and the availability of several solutions so that they any of them can be selected depending on the problem requirements.

Our experimental results have shown the robustness of our novel proposal in contrast with the state-of-the-art, and confirms the significance of this topic for future research. Among others, we plan to study the use of different objectives to guide the search, the use of the solutions of the MOEA as an ensemble approach, or even to develop a heuristic rule to select the best suited solution overall. Finally, we will test the behaviour of our model with problems with a higher complexity, including both a wider number of instances and/or features.

Acknowledgments. This work was supported by the Spanish Ministry of Science and Technology under projects TIN-2011-28488, TIN-2012-33856; the Andalusian Research Plans P11-TIC-7765 and P10-TIC-6858; and both the University of Jaén and Caja Rural Provincial de Jaén under project UJA2014/06/15.

References

1. Barandela, R., Sánchez, J.S., García, V., Rangel, E.: Strategies for learning in class imbalance problems. Pattern Recogn. **36**(3), 849–851 (2003)
2. Batista, G., Prati, R.C., Monard, M.C.: A study of the behaviour of several methods for balancing machine learning training data. SIGKDD Explor. **6**(1), 20–29 (2004)
3. Deb, K., Pratap, A., Agarwal, S., Meyarivan, T.: A fast and elitist multiobjective genetic algorithm: Nsga-II. IEEE Trans. Evol. Comput. **6**(2), 182–197 (2002)
4. Denil, M., Trappenberg, T.: Overlap versus imbalance. In: Farzindar, A., Kešelj, V. (eds.) Canadian AI 2010. LNCS, vol. 6085, pp. 220–231. Springer, Heidelberg (2010)
5. Domingos, P.: Metacost: A general method for making classifiers cost-sensitive. In: Proceedings of the 5th International Conference on Knowledge Discovery and Data Mining (KDD 1999), pp. 155–164 (1999)
6. He, H., Garcia, E.A.: Learning from imbalanced data. IEEE Trans. Knowl. Data Eng. **21**(9), 1263–1284 (2009)
7. Ho, T., Basu, M.: Complexity measures of supervised classification problems. IEEE Trans. Pattern Anal. Mach. Intell. **24**(3), 289–300 (2002)
8. López, V., Fernández, A., García, S., Palade, V., Herrera, F.: An insight into classification with imbalanced data: empirical results and current trends on using data intrinsic characteristics. Inf. Sci. **250**(20), 113–141 (2013)
9. Luengo, J., Fernández, A., García, S., Herrera, F.: Addressing data complexity for imbalanced data sets: analysis of SMOTE-based oversampling and evolutionary undersampling. Soft Comput. **15**(10), 1909–1936 (2011)
10. Quinlan, J.: C4.5: Programs for Machine Learning. Morgan Kauffmann, San Francisco (1993)

11. Sáez, J., Luengo, J., Stefanowski, J., Herrera, F.: Smote-IPF: addressing the noisy and borderline examples problem in imbalanced classification by a re-sampling method with filtering. Inf. Sci. **291**, 184–203 (2015)
12. Sheskin, D.: Handbook of Parametric and Nonparametric Statistical Procedures. Chapman & Hall/CRC, Boca Raton (2006)

Cost-Sensitive Neural Network with ROC-Based Moving Threshold for Imbalanced Classification

Bartosz Krawczyk$^{(\boxtimes)}$ and Michał Woźniak

Department of Systems and Computer Networks,
Wroclaw University of Technology, Wrocław, Poland
{bartosz.krawczyk,michal.wozniak}@pwr.edu.pl

Abstract. Pattern classification algorithms usually assume, that the distribution of examples in classes is roughly balanced. However, in many cases one of the classes is dominant in comparison with others. Here, the classifier will become biased towards the majority class. This scenario is known as imbalanced classification. As the minority class is usually the one more valuable, we need to counter the imbalance effect by using one of several dedicated techniques. Cost-sensitive methods assume a penalty factor for misclassifying the minority objects. This way, by assuming a higher cost to minority objects we boost their importance for the classification process. In this paper, we propose a model of cost-sensitive neural network with moving threshold. It relies on scaling the output of the classifier with a given cost function. This way, we adjust our support functions towards the minority class. We propose a novel method for automatically determining the cost, based on the *Receiver Operating Characteristic* (ROC) curve analysis. It allows us to select the most efficient cost factor for a given dataset. Experimental comparison with state-of-the-art methods for imbalanced classification and backed-up by a statistical analysis prove the effectiveness of our proposal.

Keywords: Machine learning · Neural networks · Imbalanced classification · Cost-sensitive · Moving threshold

1 Introduction

Machine learning algorithms have been effectively used for classification purposes in last decades. However, new problems continuously emerge that pose challenge to pattern classifiers. Often these difficulties are embedded in the nature of the data. They may be connected with the volume, non-stationary nature, some specific characteristics, or differing quality of examples. One of such problems is the imbalance between the class representatives.

Standard classifiers assume, that the distribution of objects between classes is roughly equal. When this assumption is violated, they tend to get biased towards the class with higher quantity of examples. This deteriorates the performance over the minority class. Therefore, in order to get a well-balanced classifier that is competent over all of classes, we need to counter the imbalanced distribution.

© Springer International Publishing Switzerland 2015
K. Jackowski et al. (Eds.): IDEAL 2015, LNCS 9375, pp. 45–52, 2015.
DOI: 10.1007/978-3-319-24834-9_6

There is a plethora of methods proposed, usually focusing on some data pre-processing methods or guiding the training process towards the minority examples. Former group concentrates on operations on the dataset in order to re-balance it, and then uses standard classifiers. Latter group works on the original dataset, but changes the training procedure of the classifier. Both of these approaches are considered as an effective aid in the imbalanced classification domain.

Between these two groups of methods lies the cost-sensitive solution. Usually during the design of the classifier one assumes that all classes are identically important and calculate the training error on the basis of quantity of misclassified examples (e.g., as in 0–1 loss function). However, in imbalanced domain usually the minority class is the more interesting one. Therefore, one may associate a higher misclassification cost with the minority class examples in order to boost their recognition rate. This will penalize the training process for errors on the minority objects and counter the problem of uneven class representations. Cost-sensitive paradigm has been successfully introduced to some types of classifiers, like decision trees or neural networks.

In this paper, we concentrate on the design of cost-sensitive neural networks. We propose to work with the neural classifiers that use the moving threshold approach for incorporating the classification cost. Here instead of re-balancing the training set or modifying the learning procedure, we scale the continuous output of a neural network. Therefore, we modify the classification phase instead of the training phase. Such a scaling forces the classification boundary to be moved towards the objects with higher cost, thus alleviating the bias towards the majority class.

However, the cost factor has a crucial influence on the performance of such a neural classifier and the problem lies in establishing its value. Some problems have the cost supplied by an expert (like in medical domains), but mainly we have no prior information on how to set it. We propose a fully automatic method, based on *Receiver Operating Characteristic* (ROC) curve [4] analysis. We use it to select the best cost factor for a given dataset, that returns balanced performance on both classes.

Our ROC-based cost-sensitive neural network is compared with a set of reference approaches for handling the class imbalance problem. Experimental analysis carried over a set of datasets with varying imbalance ratio proves the usefulness of our approach.

2   Imbalanced Classification

The performance and quality of machine learning algorithms is conventionally evaluated using predictive accuracy. However, this is not appropriate when the data under consideration is strongly imbalanced, since the decision boundary may be strongly biased towards the majority class, leading to poor recognition of the minority class. Disproportion in the number of class examples makes the learning task more complex [12], but is not the sole source of difficulties for machine learning algorithms. It is usually accompanied by difficulties embedded in the structure of data such as small sample size (very limited availability of minority examples), small disjuncts (minority class can consist of several sub-concepts) or class overlapping.

Over the last decade there was developed a number of dedicated techniques for handling such difficulties [1]. They can be divided into two major categories. First one consist of data-level approaches that in the pre-processing stage aim at re-balancing the original distribution [3]. Classifier-level approaches try to adapt existing algorithms to the problem of imbalanced datasets and alleviate bias towards the majority class. Third group relies on cost-sensitive classification and assign higher misclassification cost for minority class, while classification is performed so as to reduce the overall learning cost.

Ensemble systems have also been successfully applied to this domain, and mainly combine a committee learning algorithm (such as Bagging or Boosting) with one of the above mentioned methods [5]. One may also propose a hybrid training procedure for such a combined classifier that will apply cost-sensitive learning locally (for each base classifier) and globally (in the ensemble pruning step) [7].

3 Proposed Cost-Sensitive Neural Network

In this paper, we propose to investigate the cost-sensitive neural network model, based on moving threshold [13].This algorithm concentrates on modifying the output of a classifier, instead of changing the structure of the training data or the training algorithm. The cost-sensitive modification is introduced during the classification step - this the model is being trained in a traditional manner.

Let us assume, that we have a binary imbalanced problem with *majority* and *minority* classes. Then the continuous output of two neurons in the final layer of a neural network for object x can be denoted as $O_{maj}(x)$ and $O_{min}(x)$, where $O_{maj}(x) + O_{min}(x) = 1$ and both outputs are bounded within [0, 1]. In canonical neural network models, the final class is selected according to *winner-takes-all* (WTA) procedure: $\Psi(x) = \arg\max_{m \in \{maj, min\}} O_m(x)$.

However, in moving-threshold model we modify the outputs of the neural network, thus denoting them as $O^*_{maj}(x)$ and $O^*_{min}(x)$. In cost-sensitive threshold-moving model, we compute the output as follows:

$$O^*_{maj}(x) = \eta O_{maj} \text{Cost}[maj, min], \tag{1}$$

and

$$O^*_{min}(x) = \eta O_{min} \text{Cost}[min, maj], \tag{2}$$

where $\text{Cost}[m, n]$ is the misclassification cost between m-th and n-th class, and η is a normalization parameter such that $O^*_{maj}(x) + O^*_{min}(x) = 1$ and both outputs are bounded within [0,1].

One should note, that threshold-moving approaches for neural networks have been overlooked for a long time and is not even close in popularity to sampling-based methods for class imbalance [5]. However, some studies report its high usefulness for dealing with datasets with skewed distributions [9]. Other works

---

Algorithm 1. Cost-sensitive neural network with moving threshold.

Require: training set $\mathcal{TS}$,
    validation set $\mathcal{VS}$,
    neural network training procedure ()
    cost matrix

1: **Training phase:**
2: Train a neural network on $\mathcal{TS}$ using supplied classifier training procedure
3: Optimize cost parameters using ROC curve analysis on the basis of $\mathcal{VS}$

4: **Classification phase:**
5: obtain continuous outputs $O_{maj}(x)$ and $O_{min}(x)$
6: $O^*_{maj}(x) \leftarrow$ modify $O_{maj}(x)$ according to Eq. (1)
7: $O^*_{min}(x) \leftarrow$ modify $O_{min}(x)$ according to Eq. (2)
8: Apply the WTA procedure on cost-sensitive outputs

---

report, that simply changing the data distribution without considering the imbalance effect on the classification threshold (and thus adjusting it properly) may be misleading [10].

As we can see, the proper settings of cost values has a crucial effect on the performance of this algorithm. Too low cost would lead to insignificant improvements over the standard methods, while too high cost would degrade the performance over the majority class. One must remember, that we cannot sacrifice the majority class, as our ultimate goal is to obtain a classifier with good performance on both classes. In optimal scenario the cost would be supplied by a domain expert according to his/her knowledge. However, in most of real-life imbalanced applications we do not have an access to a pre-defined cost and thus must set it manually. This procedure can be time-consuming, difficult and may lead to an increased classification error when conducted erroneously. So far only simple heuristics were used to calculate the cost, like setting it equal to class imbalance ratio [8].

In this paper, we propose a new method for cost-sensitive neural network training based on ROC curve analysis [7]. Here, we use different values of cost parameter as cut-off points for plotting a ROC curve. Then, we select such a setting of neural classifier that offers the best ratio between the *True Positive rate* and *False Positive rate* (in practice - point located closest to the top left corner of the ROC plot). This allows us for an automatic cost selection that offers a balanced performance on both classes. User only needs to supply a search range and the tuning procedure is conducted in a fully automatic manner.

The detailed steps of the cost-sensitive moving-threshold classifier are presented in a form of pseudo-code in Algorithm 1.

4   Experimental Study

The aims of the experiments were to establish the usefulness of the cost-sensitive moving-threshold neural network, compare it with-state-of-the-art methods for imbalanced classification, and asses the quality of ROC-based cost parameter selection.

In the experiments we have used 10 binary imbalanced datasets from the KEEL repository[1]. Their details are given in Table 1.

Table 1. Details of datasets used in the experimental investigations, with the respect to no. of objects, no. of features, no. of classes, and imbalance ratio (IR).

No	Name	Objects	Features	Classes	IR
1	Haberman	306	3	2	2.78
2	Segment0	2308	19	2	6.02
3	Page-blocks0	5472	10	2	8.79
4	Vowel0	988	13	2	9.98
5	Glass4	214	9	2	15.47
6	Abalone9-18	731	8	2	16.40
7	Yeast6	1484	8	2	41.40
8	Ecoli-0-1-3-7_vs_2-6	281	7	2	39.14
9	Poker-8_vs_6	1477	10	2	85.88
10	Kddcup-rootkit-imap_vs_back	2225	41	2	100.14

As a base classifier we use a single-layer neural network trained with resilient backpropagation algorithm [11]. The number of input neurons is equal to the number of features, output neurons to the number of classes, and the number of neurons in the hidden layer is equal to $\frac{neurons_{input} + neurons_{output}}{2}$. Each neural network is trained for 1000 iterations.

As reference methods for dealing with class imbalance, we combine neural networks with Random Oversampling (NN + OS), Random Undersampling (NN + US), Synthetic minority over-sampling technique (NN + SMOTE) [3], and cost-sensitive moving threshold method with $Cost[minority, majority] = IR$ [8] (NN + MV(IR)).

The proposed method with ROC-based cost optimization (NN + MV(ROC)) uses [5, 200] as a possible range of possible cost parameter values.

We use 5×2 CV F-test for training / testing and pairwise statistical analysis, while Friedman ranking test and Shaffer post-hoc tests are applied for statistical comparison over multiple datasets [6].

The results are given in Table 2. The output of Shaffer post-hoc test is reported in Table 3.

From the obtained results one may see, that the proposed ROC-based cost-sensitive neural network outperforms all other methods in 6 out of 10 cases. What is highly interesting it always delivers superior performance in comparison with the cost selected on the basis of the imbalance ratio. This shows, how important is the proper selection of the cost parameter and that imbalance ratio is not

---

[1] http://sci2s.ugr.es/keel/imbalanced.php.

Table 2. Results according to G-mean [%] for each examined methods over 10 datasets. Small numbers under proposed method stands for the indexes of reference classifiers that were statistically inferior in the pairwise 5×2 CV F-test. Last row represents the ranks after Friedman test.

Data	NN + OS[1]	NN + US[2]	NN + SMOTE[3]	NN + MT(IR)[4]	NN + MT(ROC)[5]
1	61.18	61.76	**62.27**	60.46	62.02 1,2,4
2	98.87	98.36	**99.18**	97.00	97.97 −
3	92.73	94.11	93.88	93.64	**94.82** 1,3,4
4	93.18	94.36	94.84	93.88	**95.80** ALL
5	81.18	82.74	**83.98**	81.06	82.39 1,4
6	57.44	58.92	57.87	58.14	**60.31** ALL
7	82.44	81.89	83.59	83.20	**85.07** ALL
8	74.18	76.39	77.99	76.82	**78.38** 1,2,4
9	83.17	**85.66**	85.13	82.78	84.19 1,4
10	71.87	73.22	74.04	73.86	**76.08** ALL
Rank	4.25	3.50	2.15	3.80	1.40

the best indicator of the proper misclassification cost. This can be explained by the fact, that IR is not the sole reason behind the imbalanced difficulty. There are other factors, embedded in the nature of data [2]. Therefore, one can easily imagine two hypothetical datasets with identical IR, but completely different classification difficulty. In such situations misclassification cost based purely on IR will definitely fail. This is further confirmed by Shaffer post-hoc test.

When comparing the proposed ROC-based neural network to other solutions, we can clearly see that it easily outperforms Random Oversampling. This is because our cost parameter was optimized to balance the performance on both classes, while RO multiplies the minority class without considering the importance of objects.

On some datasets, the proposed method was inferior to Random Undersampling and SMOTE. This can be explained by the lack of selectiveness of our procedure - it modifies the output for all of examples. Thus it may happen that a correctly recognized examples is weighted towards the incorrect class, which

Table 3. Shaffer test for comparison between the proposed ROC-based cost-sensitive neural network and reference methods over multiple datasets. Symbol '+' stands for a situation in which the proposed method is superior, '−' for vice versa, and '=' represents a lack of statistically significant differences.

Hypothesis	p-value
NN + MT(ROC) vs NN + OS	+ (0.0196)
NN + MT(ROC) vs NN + US	+ (0.0348)
NN + MT(ROC) vs NN + SMOTE	+ (0.0402)
NN + MT(ROC) vs NN + MT(IR)	+ (0.0097)

may result in an increased rate of false positives (majority examples misclassified as minority ones). To counter this problem, one would need to introduce a sample selection mechanism, that would apply the cost modification only on examples that are potentially uncertain.

5   Conclusions and Future Works

In this paper we have presented a modification of cost-sensitive neural network classifier based on moving threshold for imbalanced learning domain. We proposed to augment this model with automatic procedure for parameter selection. We applied a ROC-based parameter selection to chose an optimal cut-off point that determined the selected value of misclassification penalty. This allowed for selecting such a parameter, that would offer a balanced performance on both classes.

Experimental evaluation carried out on a number of datasets with varying imbalance ratio confirmed the usefulness of the proposed approach. Our method was always better than the normally used approach for cost parameter estimation. Additionally, in 6 out of 10 cases it was able to outperform the reference methods based on data sampling. This was further backed-up with a thorough statistical analysis. However, the experiments revealed the weak side of our method, that is lack of selectiveness when modifying the output of the classifier.

In future works we plan to develop an active learning solution for selecting important samples to modify the threshold, and propose a dynamic ensemble system based on this classifier.

Acknowledgments. This work was supported by the Polish National Science Center under the grant no. DEC-2013/09/B/ST6/02264.

References

1. Antonelli, M., Ducange, P., Marcelloni, F.: An experimental study on evolutionary fuzzy classifiers designed for managing imbalanced datasets. Neurocomputing **146**, 125–136 (2014)

2. Błaszczyński, J., Stefanowski, J.: Neighbourhood sampling in bagging for imbalanced data. Neurocomputing **150**, 529–542 (2015)
3. Chawla, N.V., Bowyer, K.W., Hall, L.O., Kegelmeyer, W.P.: Smote: synthetic minority over-sampling technique. J. Artif. Intell. Res. **16**, 321–357 (2002)
4. Flach, P.A.: The geometry of ROC space: understanding machine learning metrics through ROC isometrics. In: Proceedings of the Twentieth International Conference on Machine Learning, ICML 2003, 21–24 August 2003, Washington, DC, USA, pp. 194–201 (2003)
5. Galar, M., Fernandez, A., Barrenechea, E., Bustince, H., Herrera, F.: A review on ensembles for the class imbalance problem: bagging-, boosting- and hybrid-based approaches. IEEE Trans. Syst. Man Cybern. C Appl. Rev. **42**(4), 463–484 (2012)
6. García, S., Fernández, A., Luengo, J., Herrera, F.: Advanced nonparametric tests for multiple comparisons in the design of experiments in computational intelligence and data mining: Experimental analysis of power. Inf. Sci. **180**(10), 2044–2064 (2010)
7. Krawczyk, B., Woźniak, M., Schaefer, G.: Cost-sensitive decision tree ensembles for effective imbalanced classification. Appl. Soft Comput. **14**, 554–562 (2014)
8. Lopez, V., Fernandez, A., Moreno-Torres, J.G., Herrera, F.: Analysis of preprocessing vs. cost-sensitive learning for imbalanced classification open problems on intrinsic data characteristics. Expert Syst. Appl. **39**(7), 6585–6608 (2012)
9. Maloof, M.A.: Learning when data sets are imbalanced and when costs are unequal and unknown. In: ICML-2003 Workshop on Learning from Imbalanced Data Sets II (2003)
10. Provost, F.: Machine learning from imbalanced data sets 101. In: Proceedings of the AAAI 2000 Workshop on Imbalanced Data Sets, pp. 1–3 (2000)
11. Riedmiller, M., Braun, H.: A direct adaptive method for faster backpropagation learning: the RPROP algorithm. In: 1993 IEEE International Conference on Neural Networks, pp. 586–591 (1993)
12. Sun, Y., Wong, A.K.C., Kamel, M.S.: Classification of imbalanced data: a review. Int. J. Pattern Recogn. Artif. Intell. **23**(4), 687–719 (2009)
13. Zhou, Z.-H., Liu, X.-Y.: Training cost-sensitive neural networks with methods addressing the class imbalance problem. IEEE Trans. Knowl. Data Eng. **18**(1), 63–77 (2006)

Managing Monotonicity in Classification by a Pruned Random Forest

Sergio González, Francisco Herrera, and Salvador García[✉]

Department of Computer Science and Artificial Intelligence,
University of Granada, 18071 Granada, Spain
sergio.gvz@gmail.com, {herrera,salvagl}@decsai.ugr.es

Abstract. In ordinal monotonic classification problems, the class variable should increase according to a subset of explanatory variables. Standard classifiers do not guarantee to produce model that satisfy the monotonicity constraints. Some algorithms have been developed to manage this issue, such as decision trees which have modified the growing and pruning mechanisms. In this contribution we study the suitability of using these mechanisms in the generation of Random Forests. We introduce a simple ensemble pruning mechanism based on the degree of monotonicity. After an exhaustive experimental analysis, we deduce that a Random Forest applied over these problems is able to achieve a slightly better predictive performance than standard algorithms.

Keywords: Monotonic classification · Decision tree induction · Random forest · Ensemble pruning

1 Introduction

The classification of examples in ordered categories is a popular problem which has drawn attention in data mining practitioners over the last years. This problem has been given with different names, such as ordinal classification, ordinal regression or ranking labelling, but all they share a common property in the data: the output attribute or class is ordinal. Classification with monotonicity constraints, also known as monotonic classification [1], is an ordinal classification problem where monotonic restriction is clear: a higher value of an attribute in an example, fixing other values, should not decrease its class assignment [2].

Decision trees [3] and rule induction [4] constitute two of the most promising techniques to tackle monotonic classification. Any approach for classification can be integrated into an ensemble-type classifier, thus empowering the achieved performance [5]. However, a classifier selection is needed to enhance its performance, and it is known as ensemble pruning [6]. Random Forests (RFs) is a well-known form of ensembles of decision trees based on bagging.

The general goals pursued in this contribution are to promote the application of the RF approach in monotonic classification tasks and to introduce a monotonicity ordering-based pruning mechanism for RFs based on the non-monotonicity index.

© Springer International Publishing Switzerland 2015
K. Jackowski et al. (Eds.): IDEAL 2015, LNCS 9375, pp. 53–60, 2015.
DOI: 10.1007/978-3-319-24834-9_7

This contribution is organized as follows. In Sect. 2 we present the ordinal classification with monotonic constraints. Section 3 is devoted to describing our proposal of RF and its adaptation to satisfy the monotonicity constraints. Section 4 describes the experimental framework and examines the results obtained in the empirical study, presenting a discussion and analysis. Finally, Sect. 5 concludes the contribution.

2   Monotonic Classification

Ordinal classification problems are those in which the class is neither numeric nor nominal. Instead, the class values are ordered. For instance, a worker can be described as "excellent", "good" or "bad", and a bond can be evaluated as "AAA", "AA", "A", "A-", etc. Similar to a numeric scale, an ordinal scale has an order, but it does not posses a precise notion of distance. Ordinal classification problems are important, since they are fairly common in our daily life.

A monotonic classifier is one that will not violate monotonicity constraints. Informally, the monotonic classification implies that the assigned class values are monotonically nondecreasing (in ordinal order) with the attribute values. More formally, let $\{\mathbf{x}_i, \text{class}(\mathbf{x}_i)\}$ denote a set of examples with attribute vector $\mathbf{x}_i = (\mathbf{x}_{i,1}, \ldots, \mathbf{x}_{i,m})$ and a class, $\text{class}(\mathbf{x}_i)$, being n the number of instances and m the number of attributes. Let $\mathbf{x}_i \succeq \mathbf{x}_h$ if $\forall_{j=1,\ldots,m}, \mathbf{x}_{i,j} \geq \mathbf{x}_{h,j}$. A data set $\{\mathbf{x}_i, \text{class}(\mathbf{x}_i)\}$ is monotonic if and only if all the pairs of examples i, h are monotonic with respect to each other [7].

3   Monotonic Random Forest

In this section, we explain our proposal to tackle monotonic classification. The modifications introduced to the standard RF are mainly focused on the way the splitting is made for every tree, the promotion of the diversity by a new random factor and the aggregation of the results with the pruning mechanism proposed, maintaining the bootstrap sample method untouched.

First of all, we define the Non Monotonic Index (NMI) as the rate of number of violations of monotonicity divided by the total number of examples in a data set. Previously, we have introduced the MID based algorithms in the process of building the trees. With this change, we accomplish the initial objective of adapting the well-known ensemble to monotonic classification. We choose MID-C4.5 to build every random tree of the forest. This method selects the best attribute to perform the split using the *total-ambiguity*-score as a criterion. This measurement was defined by Ben-David in [7] as the sum of the E-score of the ID3 algorithm and the *order-ambiguity*-score weighted by the parameter R. The $order - ambiguity - score$ is computed, as shown in Eq. 1, using the concept of the non-monotonicity index, which is the ratio between the actual number of non-monotonic branch pairs and the maximum number of pairs that could have been non-monotonic. In the MID-C4.5, the entropy of the ID3 is substituted by the gain information of the C4.5 decision tree.

Algorithm 1. Monotonic RF algorithm.

---

function MONRF(D - dataset, $nTrees$ - number of random trees built, R_{limit} - importance factor for monotonic constrains, T - Threshold used in the pruning procedure, S - the predicted version of D)

initialize: $S = \{\}$, $Trees[1..nTrees]$, $D_{bootstraps}[1..nTrees]$, $NMIs[1..nTrees]$

for i in $[1,nTrees]$ **do**

    $D_{bootstraps}[i] = Bootstrap\_Sampler(nTrees, D)$

    $rand = Random(1, R_{limit})$

    $Trees[i] = Build\_Tree(D_{bootstraps}[i], rand)$

    $NMIs[i] = Compute\_NMI(Trees[i])$

end for

$Trees = Sort(Trees, NMIs)$

for i in $[1, \lceil nTrees * T \rceil]$ **do**

    $\widehat{Trees} \leftarrow Trees[i]$

end for

for d in D **do**

    $S \leftarrow Predict\_Majority\_Voting(\widehat{Trees}, d)$

end for

return S

end function

$$A = \begin{cases} 0 & \text{if } NMI = 0 \\ -(\log_2 NMI)^{-1} & \text{otherwise} \end{cases} \qquad (1)$$

The factor R was first introduced by Ben-David [7] as an importance factor of the *order-ambiguity-score* in the decision of the splitting with the calculation of the *total-ambiguity-score*. As higher as R was set, more relevant were the monotonicity constraints considered. We use this parameter as a way to further randomise and diversify the different trees built in the RF and at the same time, we force the tree building procedure to be dominated by the monotonicity considerations. In order to fulfill this, each tree is built from the beginning with a different factor R, picked as a random number from 1 to R_{limit}, set as a parameter shown in Algorithm 1.

Furthermore, we did not consider for our proposal the maximum depth imposed to all the random trees of the standard RF. We have decided this, due to the fact that monotonic decision tree classifiers already highly reduce the complexity of the built tree compared with the traditional ones.

Finally, we design a pruning threshold mechanism in the final combination of the different results to predict the class of each example. Instead of using all the decision trees built, to form the class through the majority vote of the predictions, we choose the best trees in term of monotonicity constraints within a certain threshold, latest lines of the Algorithm 1. With this objective, our Monotonic RF sorts the different trees built by the Non-Monotonic-Index in increasing order and the pruning method selects the first n trees, where n is the number of trees computed by product of the total number of trees built and the threshold T within the range (0,1]. We recommend to set it at 0.50, due to the results obtained in the next section.

4   Experimental Framework, Results and Analysis

In this section, we present the experimental framework followed to compare and analyze the application of RFs to monotonic classification.

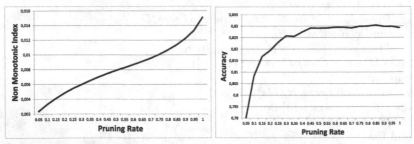

(a) Average NMI of the Random Tree depending on the pruning rate.

(b) Average accuracy of the Random Tree depending on the pruning rate.

Fig. 1. Effect of the pruning rate in the random forest

4.1   Experimental Methodology

The experimental methodology is described next by specifying some basic elements:

- Data Sets: 50 monotonic data sets are used in this study. Most of the monotonic data sets are standard data sets used in the classification scope and extracted from KEEL repository [8] which have been relabeled following the procedure used in the experimental design of [9].
- Algorithms to compare: We will compare RFs with three decision trees: MID-C4.5 [7], MID-CART [3] and MID-RankTree [10]; and two classical algorithms in this field: OLM [11] and OSDL [12].
- Evaluation metrics: Several measures will be used in order to estimate the performance of the algorithms compared: Accuracy (Acc); Mean Absolute Error (MAE) [13]; Non Monotonic Index (NMI); Number of Leaves (NL).
- Parameters configuration: The parameters of the baseline algorithms are the standard in KEEL software [8]. The maximum depth for CART and Rank-Tree is 90. The value of R for MID is 1. RF also uses the standard parameters, but for monotonic classifications it requires to set the $R_{limit} = 100$ and Threshold = 0.5. The number of trees built is 100 in all cases.

4.2   Results

This section is dedicated to present the results gathered from the runs of the algorithms using the configuration described in the previous subsection.

Table 1. Accuracy and MAE results reported.

	Accuracy						Mean Absolute Error					
	MID RF	MID C4.5	MID CART	MID RankTree	OLM	OSDL	MID RF	MID C4.5	MID CART	MID RankTree	OLM	OSDL
appendicitis	0.8667	0.8964	0.8864	**0.9064**	0.8109	0.6418	0.1333	0.1036	0.1136	**0.0936**	0.1891	0.3582
australian	0.8261	0.8029	0.8319	**0.8362**	0.7232	0.8319	0.1739	0.1971	0.1681	**0.1638**	0.2768	0.1681
auto-mpg	0.6529	0.6247	0.4489	0.6223	**0.6812**	0.3854	0.4630	0.4851	1.0335	0.5001	**0.3879**	0.8291
automobile	0.8039	**0.8250**	0.7304	0.7429	0.2333	0.3758	0.3069	**0.2688**	0.4696	0.4263	2.1046	0.8958
balance	0.9830	0.9777	0.9777	**0.9856**	0.9776	0.9777	0.0186	0.0239	0.0271	**0.0176**	0.0272	0.0271
bostonhousing	**0.6483**	0.5674	0.4982	0.5237	0.3003	0.2569	**0.4958**	0.6102	0.7504	0.6856	1.3045	1.0099
breast	0.7597	0.7337	0.6933	0.7440	**0.8409**	0.8015	0.2403	0.2663	0.3067	0.2560	**0.1591**	0.1985
bupa	0.7981	0.7508	0.7534	0.7879	**0.8375**	0.7625	0.2019	0.2492	0.2466	0.2121	**0.1625**	0.2375
car	0.8731	0.9433	0.8183	0.9386	**0.9705**	**0.9705**	0.1609	0.0666	0.2396	0.0735	**0.0324**	**0.0324**
cleveland	0.5644	0.4909	0.5284	0.5253	**0.5793**	0.5421	**0.6893**	0.8332	0.8014	0.8586	0.8311	0.7848
contraceptive	0.8185	0.7991	0.5601	0.7719	**0.8799**	0.8398	0.2351	0.2552	0.6449	0.2844	**0.1534**	0.1602
crx	**0.8290**	0.7903	0.7933	0.7839	0.6110	0.7058	**0.1710**	0.2097	0.2067	0.2161	0.3890	0.2942
dermatology	**0.8633**	0.8437	0.8408	0.7512	0.4499	0.1593	**0.2810**	0.3325	0.3465	0.5339	1.3821	1.6421
ecoli	**0.6441**	0.6074	0.5750	0.5779	0.6368	0.0652	1.0802	1.0549	1.5201	1.1250	**0.9467**	2.1998
ERA	**1.0000**	**1.0000**	**1.0000**	**1.0000**	**1.0000**	**1.0000**	**0.0000**	**0.0000**	**0.0000**	**0.0000**	**0.0000**	**0.0000**
ESL	0.9043	0.9159	0.6162	0.9344	0.9179	**0.9364**	0.1107	0.1026	0.5788	0.0738	0.0923	**0.0656**
flare	0.9025	0.9165	0.6380	0.9456	**0.9738**	0.9606	0.1256	0.1191	0.6479	0.0826	**0.0318**	0.0572
glass	**0.7464**	0.6773	0.6258	0.6883	0.3175	0.3223	**0.4934**	0.6929	0.7747	0.6747	1.7994	1.8000
haberman	0.9291	0.9177	0.9312	0.9537	0.9310	**0.9606**	0.0709	0.0823	0.0688	0.0463	0.0690	**0.0394**
hayes-roth	0.9042	0.9438	0.7688	0.8500	**0.9500**	0.9438	0.1104	**0.0563**	0.2500	0.1688	0.0750	**0.0563**
heart	**0.8235**	0.7926	0.7593	0.8111	0.6704	0.6259	**0.1765**	0.2074	0.2407	0.1889	0.3296	0.3741
hepatitis	0.8917	0.7750	0.8375	**0.9000**	0.2375	0.8000	0.1083	0.2250	0.1625	**0.1000**	0.7625	0.2000
housevotes	0.9528	**0.9741**	0.8750	0.9266	0.9047	0.9096	0.0472	**0.0259**	0.1250	0.0734	0.0953	0.0904
ionosphere	**0.8832**	0.8348	0.7863	0.7810	0.6580	0.7237	**0.1168**	0.1652	0.2137	0.2190	0.3420	0.2763
iris	0.9711	0.9667	0.9733	**0.9867**	0.9000	0.3733	0.0289	0.0333	0.0267	**0.0133**	0.1000	0.9067
led7digit	0.8600	0.9520	0.7880	0.9660	**0.9820**	0.9740	0.3800	0.1140	0.6780	0.0700	**0.0340**	0.0340
LEV	0.9993	**1.0000**	0.6990	**1.0000**	**1.0000**	**1.0000**	0.0007	**0.0000**	0.4450	**0.0000**	**0.0000**	**0.0000**
lymphography	**0.7819**	0.7705	0.6767	0.6900	0.7100	0.7029	**0.2314**	0.2567	0.3633	0.3714	0.3571	0.2971
machinecpu	**0.6520**	0.5638	0.4398	0.6369	0.6267	0.3662	**0.4741**	0.6086	0.7564	0.4726	0.5031	0.4067
mammographic	0.9763	0.9831	0.9735	**0.9904**	0.9892	0.9831	0.0237	0.0169	0.0265	**0.0096**	0.0108	0.0169
monk-2	0.9807	0.9746	0.9792	0.9769	0.9721	**0.9908**	0.0193	0.0254	0.0208	0.0232	0.0279	**0.0092**
movement_libras	**0.6796**	0.5194	0.5583	0.5333	0.3139	0.1194	**1.1602**	2.0306	1.7861	1.8528	4.3472	5.7000
newthyroid	**0.8621**	0.8279	0.8511	0.8329	0.6223	0.1818	**0.1905**	0.2186	0.1909	0.2139	0.5504	0.8413
pima	**0.8702**	0.8242	0.7837	0.8007	0.8151	0.6224	**0.1298**	0.1758	0.2163	0.1993	0.1849	0.3776
post-operative	0.6968	0.6333	0.4403	0.7403	**0.8292**	0.7569	0.3773	0.4806	0.7292	0.3153	**0.2042**	0.2431
saheart	**0.7302**	0.6627	0.6645	0.6624	0.6862	0.6839	**0.2698**	0.3373	0.3355	0.3376	0.3138	0.3161
segment	**0.9759**	0.9649	0.9632	0.9602	0.3061	0.1684	**0.0447**	0.0610	0.0671	0.0723	2.4022	2.8597
sonar	**0.8042**	0.7681	0.7250	0.7648	0.4662	0.5724	**0.1958**	0.2319	0.2750	0.2352	0.5338	0.4276
spectfheart	**0.8028**	0.7379	0.7412	0.7339	0.2095	0.8016	**0.1972**	0.2621	0.2588	0.2661	0.7905	0.1984
SWD	0.9993	**1.0000**	0.3820	**1.0000**	**1.0000**	**1.0000**	0.0007	**0.0000**	1.0240	**0.0000**	**0.0000**	**0.0000**
tae	0.8278	0.8483	0.5904	0.8417	0.8546	**0.8946**	0.1921	0.1650	0.5283	0.1913	0.1850	**0.1054**
titanic	**1.0000**	**1.0000**	**1.0000**	**1.0000**	**1.0000**	**1.0000**	**0.0000**	**0.0000**	**0.0000**	**0.0000**	**0.0000**	**0.0000**
vehicle	**0.7409**	0.6904	0.6384	0.6644	0.2530	0.2588	**0.4929**	0.5661	0.6475	0.6334	1.4917	0.9717
vowel	**0.9525**	0.7758	0.2182	0.7778	0.0909	0.0859	**0.1010**	0.6242	2.9313	0.5232	5.0000	4.8273
wdbc	**0.7183**	0.6768	0.6749	0.6713	0.6451	0.5607	**0.2817**	0.3232	0.3251	0.3287	0.3549	0.4393
windsorhousing	0.8932	0.8939	0.8738	0.8664	**0.9174**	0.8847	0.1068	0.1061	0.1262	0.1336	**0.0826**	0.1153
wine	**0.7926**	0.6794	0.7297	0.7578	0.3484	0.3314	**0.2882**	0.4219	0.3719	0.3154	0.9660	0.9667
wisconsin	**0.9747**	0.9591	0.9693	0.9591	0.8815	0.9547	**0.0253**	0.0409	0.0307	0.0409	0.1185	0.0453
yeast	0.4095	0.3659	0.2811	0.3639	**0.4596**	0.0836	1.7143	1.8605	3.1107	1.8111	**1.6300**	3.0862
zoo	0.7427	0.8127	0.4564	0.8127	**0.8409**	0.7727	0.7000	0.4727	1.8073	0.4336	**0.3682**	0.3745

First of all, we present the study that allows us to determine the best choice of the monotonicity pruning parameter value for the RF proposal. The trees built from the ensemble are sorted by their NMI in increasing order. The pruning mechanism selects the trees using a threshold coming from 0.05 to 1. This represents the rate of trees that will belong to the ensemble. In this way, if the rate is 1 all the trees will belong to the ensemble and if the rate is 0.3, only 30 % of the most monotonic trees will form the ensemble.

Table 2. Summary table for statistical inference outcome: ranks and APVs

	Acc		MAE		NL		NMI	
	Ranks	APVs	Ranks	APVs	Ranks	APVs	Ranks	APVs
MID-RF	**2.480**	–	**2.540**	–	**1.469**	–	2.300	0.423
MID-C4.5	3.360	0.037	3.310	0.079	2.041	0.028	**2.000**	–
MID-CART	4.310	0.000	4.320	0.000	2.345	0.000	4.130	0.000
MID-RankTree	3.200	0.054	3.230	0.079	3.255	0.000	5.600	0.000
OLM	3.650	0.005	3.800	0.003	–	–	3.780	0.000
OSDL	4.000	0.000	3.800	0.003	–	–	3.190	0.001

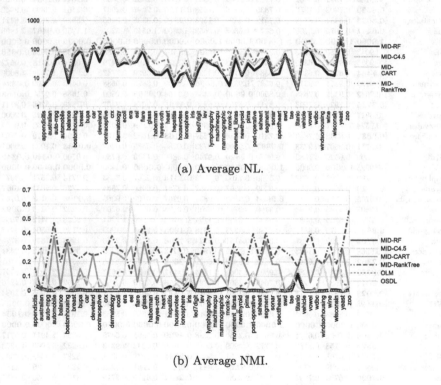

(a) Average NL.

(b) Average NMI.

Fig. 2. Number of leaves and non monotonic index results reported.

Figure 1 shows the effect of the pruning rate explained above in RF. The values represented for both graphics are associated with the average values of Acc and NMI of the 50 data sets. Observing Fig. 1a, we can see that there is a turning point in the growth curve surrounding the rate value of 0.5. Simultaneously, in Fig. 1b there is a limit in which the improvement registered in accuracy stops decreasing and this limit matches with the same turning point indicated previously: the rate of 0.5. Hence, it seems logical that this rate could be an interesting value to be adopted as the monotonicity pruning rate used for RFs.

Henceforth, we will consider 0.5 as the pruning rate used in RFs. Under our recommended configuration of pruning and random choice of R parameter, we compare RF with the other contestant methods. Table 1 and Fig. 2 exhibit the results obtained for the algorithms over monotonic data sets, in terms of average values of the three runs of 10-fcv.

In order to support the results, we include a statistical analysis based on non parametric tests. The results obtained by the application of the Friedman test and the Holm post-hoc procedure are depicted in Table 2. Furthermore, the Adjusted P-Value (APV) [14] computed by the Holm procedure is reported in the algorithms whose ranking is not the best in each group.

4.3 Analysis

From this study, we may stress the following conclusions:

- In terms of accuracy, the goodness of the Monotonic RF the with pruning threshold mechanism is clear. In all cases, the RF outperforms the other 5 algorithms by a significant difference, a fact that can be noticed in Tables 1 and 2, where the p-value is smaller than 0.10.
- With the same results, the superiority of RF in relation to the MAE over the other algorithms is overwhelming. This outcome was expected, when such a difference in terms of accuracy was obtained.

  Furthermore, RF succeeds to obtain less complex trees, as can be seen with a smaller number of leaves, in Fig. 2. A remarkable fact keeping in mind that the maximum depth of the standard RF was not used. This is all due to the variability caused by the pruning procedure, which allows the most monotonic and simple trees to be selected.
- Finally, referring to the NMI, Fig. 2 and Table 2 reflect better results for MID-C4.5. However, the difference between MID-RF and MID-C4.5 is not pointed out as significant.

5 Concluding Remarks

The purpose of this contribution is to present and to analyse a Random forest proposal for classification with monotonicity constraints. In order to be adapted to this problem, it includes the rate of monotonicity as a parameter to be randomised during the growth of the trees. After building of all the decision trees, an ensemble pruning mechanism based on the monotonicity index of each tree is used to select the subset of the most monotonic decision trees to constitute the forest. The results show that Random Forests are promising models to address this problem obtaining very accurate results involving trees with a low non monotonic index.

Acknowledgments. This work is supported by the research project TIN2014-57251-P and by a research scholarship, given to the author Sergio Gonzalez by the University of Granada.

References

1. Ben-David, A., Sterling, L., Pao, Y.H.: Learning, classification of monotonic ordinal concepts. Comput. Intell. **5**, 45–49 (1989)
2. Kotłowski, W., Słowiński, R.: On nonparametric ordinal classification with monotonicity constraints. IEEE Trans. Knowl. Data Eng. **25**, 2576–2589 (2013)
3. Rokach, L., Maimon, O.: Data Mining with Decision Trees: Theory and Applications, 2nd edn. World Scientific, Singapore (2014)
4. Furnkranz, J., Gamberger, D., Lavrac, N.: Foundations of Rule Learning. Springer, Heidelberg (2012)
5. Wozniak, M., Graña, M., Corchado, E.: A survey of multiple classifier systems as hybrid systems. Inf. Fusion **16**, 3–17 (2014)
6. Martínez-Muñoz, G., Hernández-Lobato, D., Suárez, A.: An analysis of ensemble pruning techniques based on ordered aggregation. IEEE Trans. Pattern Anal. Mach. Intell. **31**, 245–259 (2009)
7. Ben-David, A.: Monotonicity maintenance in information-theoretic machine learning algorithms. Mach. Learn. **19**, 29–43 (1995)
8. Alcala-Fdez, J., Fernández, A., Luengo, J., Derrac, J., García, S., Sánchez, L., Herrera, F.: KEEL data-mining software tool: data set repository, integration of algorithms and experimental analysis framework. J. Multiple Valued Logic Soft Comput. **17**, 255–287 (2011)
9. Duivesteijn, W., Feelders, A.: Nearest neighbour classification with monotonicity constraints. In: Daelemans, W., Goethals, B., Morik, K. (eds.) ECML PKDD 2008, Part I. LNCS (LNAI), vol. 5211, pp. 301–316. Springer, Heidelberg (2008)
10. Xia, F., Zhang, W., Li, F., Yang, Y.: Ranking with decision tree. Knowl. Inf. Syst. **17**, 381–395 (2008)
11. Ben-David, A.: Automatic generation of symbolic multiattribute ordinal knowledge-based DSSs: methodology and applications. Decis. Sci. **23**, 1357–1372 (1992)
12. Lievens, S., Baets, B.D., Cao-Van, K.: A probabilistic framework for the design of instance-based supervised ranking algorithms in an ordinal setting. Ann. Operational Res. **163**, 115–142 (2008)
13. Japkowicz, N., Shah, M. (eds.): Evaluating Learning Algorithms: A Classification Perspective. Cambridge University Press, Cambridge (2011)
14. García, S., Fernández, A., Luengo, J., Herrera, F.: Advanced nonparametric tests for multiple comparisons in the design of experiments in computational intelligence and data mining: experimental analysis of power. Inf. Sci. **180**, 2044–2064 (2010)

Ensemble Selection Based on Discriminant Functions in Binary Classification Task

Paulina Baczyńska and Robert Burduk[✉]

Department of Systems and Computer Networks, Wroclaw University of Technology,
Wybrzeze Wyspianskiego 27, 50-370 Wroclaw, Poland
robert.burduk@pwr.edu.pl

Abstract. The paper describes the dynamic ensemble selection. The proposed algorithm uses values of the discriminant functions and it is dedicated to the binary classification task. The proposed algorithm of the ensemble selection uses decision profiles and the normalization of the discrimination functions is carried out. Additionally, the difference of the discriminant functions is used as one condition of selection. The reported results based on the ten data sets from the UCI repository show that the proposed dynamic ensemble selection is a promising method for the development of multiple classifiers systems.

Keywords: Ensemble selection · Multiple classifier system · Binary classification task

1 Introduction

Supervised learning is one of the types of machine learning [1]. Generally, the recognition algorithm maps the feature space to the set of class labels. The output of an individual (base) classifier can be divided into three types [19].

- The abstract level – classifier ψ assigns the unique label j to a given input x [20,26].
- The rank level – in this case for each input x, each classifier produces an integer rank array. Each element within this array corresponds to one of the defined class labels [14]. The array is usually sorted and the label at the top is the first choice.
- The measurement level – the output of a classifier is represented by a discriminant function value that addresses the degree of assigning the class label to the given output x [17,18]. An example of such a representation of the output is a posteriori probability returned by Bayes classifier.

For several years, in the field of supervised learning a number of base classifiers have been used in order to solve one classification task. The use of multiple base classifier for a decision problem is known as an ensemble of classifiers (EoC) or as multiple classifiers systems (MCSs) [5,11,28]. The building of MCSs consists of three phases: generation, selection and integration [3]. For example, in the

© Springer International Publishing Switzerland 2015
K. Jackowski et al. (Eds.): IDEAL 2015, LNCS 9375, pp. 61–68, 2015.
DOI: 10.1007/978-3-319-24834-9_8

third phase and for abstract the level of classifier outputs the simple majority voting scheme [24] is most popular. Generally, the final decision which is made in the third phase uses the prediction of the base classifiers and it is the popular for their ability to fuse together multiple classification outputs for the better accuracy of classification. If the outputs of all base classifiers are used in the third phase then this method is called classifier fusion. Formally, then there is no second phase in building MCSs.

The second phase of building MCSs is one of the important problems in the creation of these recognition systems [15,25]. This phase is related to the choice of a set of classifiers from the whole available pool of classifiers. Formally, if we choose one classifier then it is called the classifier selection. But if we choose a subset of classifiers from the pool then is called the ensemble selection or ensemble pruning. Here you can distinguish between the static or dynamic selection [13,22,27]. In the static classifier selection one set of classifiers is selected to create EoC. This EoC is used in the classification of all the objects from the testing set. The main problem in this case is to find a pertinent objective function for selecting the classifiers. In the dynamic classifier selection, also called instance-based, for each unknown sample a specific subset of classifiers is selected [2]. It means that we are selecting different EoCs for different objects from the testing set. In this type of the classifier selection, the classifier is chosen and assigned to the sample based on different features [29] or different decision regions [7,16]. Therefore, the rationale of using the dynamic classifier selection approach is that different base classifiers have different areas of expertise in the instance space.

In this work we will consider the dynamic ensemble selection. In detail we propose the new selection method based on the analysis of the discriminant functions in the contents of the binary classification task. The proposed algorithm of the ensemble selection uses the decision profiles and the difference of the discriminant functions is used as a condition of selection. In the proposed algorithm the normalization of the discrimination functions is carried out.

The text is organized as follows: after the Introduction, in Sect. 2 the concept of the ensemble of classifiers is presented. Section 3 contains the new method for the dynamic ensemble selection. Section 4 includes the description of research experiments comparing the proposed algorithm with base classifiers. Finally, the discussion and conclusions from the experiments are presented.

2   Ensemble of Classifiers

Let us consider the binary classification task. It means that we have two class labels $M = \{1, 2\}$. Each pattern is characterized by a feature vector X. The recognition algorithm maps the feature space X to the set of class labels M according to the general formula:

$$\Psi : X \to M. \tag{1}$$

Let us assume that K different classifiers $\Psi_1, \Psi_2, \ldots, \Psi_K$ are available to solve the classification task. In MCSs these classifiers are called base classifiers. In the binary classification task K is assumed to be an odd number. As a result, of all the classifiers' actions, their K responses are obtained. The output information from all K component classifiers is applied to make the ultimate decision of MCSs. This decision is made based on the predictions of all the base classifiers.

One of the possible methods for integrating the output of the base classifier is the sum rule. In this method the score of MCSs is based on the application of the following sums:

$$s_i(x) = \sum_{k=1}^{K} \hat{p}_k(i|x), \qquad i \in M, \tag{2}$$

where $\hat{p}_k(i|x)$ is an estimate of the discrimination functions for class label i returned by classifier k.

The final decision of MCSs is made following the maximum rule:

$$\Psi_S(x) = \arg\max_i s_i(x). \tag{3}$$

In the presented method (3) the discrimination functions obtained from the individual classifiers take an equal part in building MCSs. This is the simplest situation in which we do not need additional information on the testing process of the base classifiers except for the models of these classifiers. One of the possible methods in which weights of the base classifier are used is presented in [4].

2.1 Ensemble Selection Algorithm

The proposed algorithm of ensemble selection uses the decision profiles [19]. The decision profile is a matrix containing DFs for each base classifier. In the binary classification task it is as follows:

$$DP(x) = \begin{bmatrix} \hat{p}_1(1|x) & \hat{p}_1(2|x) \\ \vdots & \vdots \\ \hat{p}_K(1|x) & \hat{p}_K(2|x) \end{bmatrix}, \tag{4}$$

In the first step of the algorithm the normalization of DFs is carried out. The normalization is performed for each label class i according to the rule:

$$\widehat{p}'_k(i|x) = \frac{\hat{p}_k(i|x) - \min(\hat{p}_1(i|x), ..., \hat{p}_k(i|x))}{\max(\hat{p}_1(i|x), ..., \hat{p}_k(i|x)) - \min(\hat{p}_1(i|x), ..., \hat{p}_k(i|x))}, \qquad k \in K. \tag{5}$$

Then, the decision scheme DS is calculated according to the formula:

$$DS = \begin{bmatrix} \widehat{ds}_{11} & \widehat{ds}_{12} \\ \vdots & \vdots \\ \widehat{ds}_{K1} & \widehat{ds}_{K2} \end{bmatrix}, \tag{6}$$

where

$$\widehat{ds}_{k\omega} = \frac{\sum_{n=1}^{m} I(\Psi_k(x_n) = \omega_n) \, \widehat{p}'_k(\omega_n|x_n)}{\sum_{n=1}^{m} I(\Psi_k(x_n) = \omega_n)}. \tag{7}$$

The DS is calculated with the use of a test set. For the new object being recognized, the outputs of the base classifiers create the decision profile. In receipt of the decision profile from the outputs of the base classifiers the normalization is carried out similarly to the formula (5). After the normalization the selection DFs process is performed, which consists of two stages. During the first phase the following formula is used:

$$\text{if } |\widehat{p}'_k(1|x) - \widehat{p}'_k(2|x)| < \alpha \text{ then } \widehat{p}'_k(\omega|x) = null, k = 1, ..., K, \ \omega = 1, 2. \quad (8)$$

The parameter α determines the size of the difference of DFs. The values are derived from the interval $\alpha \in [0, 1)$. In the second phase the decision scheme is used according to the formula:

$$\text{if } \widehat{p}'_k(\omega|x) < \widehat{ds}_{k\omega} \text{ then } \widehat{p}_k(\omega|x) = null, k = 1, ..., K, \ \omega = 1, 2. \quad (9)$$

The obtained decision profile using the formulas (8) and (9) is applied to make the final decision of the classifiers ensemble. The algorithm using this method is denoted as Ψ^α_{SDF}. Since the conditions from the Eqs. (8) and (9) can be satisfied we can talk about the selection of DFs process. In experimental studies we use the sum method (3) to make the final decision of the selected classifiers ensemble.

3   Experimental Studies

In the experiment 9 base classifiers were used. Three of them work according to $k - NN$ rule where k parameter is equal to 3, 5 or 7. Three base classifiers use Support Vector Machines models. One of them uses Least Squares SVM, the second Decomposed Quadratic Programming and the third Quadratic Programming modelling method. The other three base classifiers use the decision trees algorithms, with the various number of branches and splitting rule.

In the experiential research 10 benchmark data sets were used. Eight of them come from the UCI repository [9] and the other two were generated randomly - they are the so called Banana and Higleyman sets. The description of data sets used in the experiments is included in Table 1. The studies did not include the impact of the feature selection process on the quality of classifications. Therefore, the feature selection process [12, 23] was not performed. The results are obtained via 10-fold-cross-validation method.

Table 2 shows the results of the classification for the proposed ensemble selection with normalization of the posteriori probability functions. Additionally, the mean ranks obtained by the Friedman test were presented. The values of the mean ranks show that the best value of the parameter α is 0.4 for that ensemble selection method.

Classifier $\Psi^{0.4}_{ES}$ with the selected value of parameter α was compared with the base classifiers and the ensemble methods were based on the sum methods. The results of the classification with the mean ranks obtained by the Friedman test are presented in Table 3. To compare the results the post-hoc Nemenyi test was used [25]. The critical difference for this test at $p = 0.05$ is equal to $CD = 4.75$.

Table 1. Description of data sets selected for the experiments

Data set	Example	Attribute	Ration (0/1)
Banana	400	2	1.0
Blood	748	5	3.2
Breast cancer wisconsin	699	10	1.9
Haberman's survival	306	3	0.4
Highleyman	400	2	1.0
Ionosphere	351	34	1.8
Indian liver patient	583	10	0.4
Mammographic mass	961	6	1.2
Parkinson	197	23	0.3
Pima indians diabetes	768	8	1.9

Table 2. Classification error and mean rank positions for the proposed selection algorithm produced by the Friedman test

Data set	Ψ_{ES}^{α} with $\alpha =$						
	0	0.1	0.2	0.3	0.4	0.5	0.6
Banana	0.030	0.032	0.032	0.032	0.032	0.032	0.032
Blood	0.175	0.184	0.189	0.191	0.196	0.200	0.219
Cancer	0.085	0.082	0.082	0.082	0.082	0.082	0.082
Haber.	0.260	0.264	0.260	0.260	0.273	0.290	0.316
Hig.	0.072	0.070	0.072	0.067	0.057	0.050	0.047
Ion.	0.085	0.082	0.082	0.073	0.073	0.073	0.082
Liver	0.085	0.082	0.082	0.073	0.073	0.073	0.082
Mam.	0.139	0.138	0.143	0.149	0.152	0.160	0.172
Park.	0.104	0.097	0.092	0.082	0.082	0.071	0.066
Pima	0.199	0.200	0.203	0.212	0.199	0.200	0.200
Mean rank	3.6	3.2	2.8	3.5	3.7	3.4	2.7

Since the difference between the best algorithm $\Psi_{ES}^{0.4}$ and the worst algorithm Ψ_5 is greater than CD. We can conclude that the post-hoc Nemenyi test detects significant differences between mean ranks. Additionally, the same situation is for $\Psi_{ES}^{0.4}$ and Ψ_5 classifiers. The post-hoc Nemenyi test is not powerful enough to detect any significant differences between the ensemble methods based on the sum methods Ψ_S and other tested classifiers. For comparison only Ψ_S and $\Psi_{ES}^{0.4}$ classifiers a Wilcoxon signed-rank test was used. Since the signed-rank test has p-value of 0.0391 we reject the null hypothesis. It means, that the distributions of results of the investigated two classifiers differ only with respect to the median.

Table 3. Classification error and mean rank positions for the base classifiers ($\Psi_1, ..., \Psi_9$), sum method algorithm Ψ_S and proposed algorithm $\Psi_{ES}^{0.4}$ produced by the Friedman test

Data set	Ψ_1	Ψ_2	Ψ_3	Ψ_4	Ψ_5	Ψ_6	Ψ_7	Ψ_8	Ψ_9	Ψ_S	$\Psi_{ES}^{0.4}$
Banana	0.015	0.012	0.020	0.132	0.172	0.092	0.042	0.037	0.027	0.027	0.032
Blood	0.271	0.240	0.221	0.240	0.272	0.237	0.220	0.217	0.247	0.215	0.196
Cancer	0.038	0.040	0.038	0.042	0.037	0.041	0.071	0.078	0.062	0.070	0.082
Haber.	0.274	0.264	0.251	0.267	0.271	0.264	0.284	0.277	0.323	0.261	0.273
Hig.	0.075	0.080	0.072	0.172	0.182	0.185	0.087	0.092	0.087	0.067	0.057
Ion.	0.128	0.150	0.147	0.113	0.212	0.261	0.116	0.119	0.133	0.135	0.073
Liver	0.128	0.150	0.147	0.113	0.212	0.261	0.116	0.119	0.133	0.135	0.073
Mam.	0.217	0.196	0.193	0.200	0.211	0.196	0.182	0.178	0.197	0.184	0.152
Park.	0.123	0.133	0.122	0.112	0.114	0.218	0.108	0.138	0.138	0.132	0.082
Pima	0.282	0.272	0.270	0.232	0.286	0.248	0.261	0.256	0.275	0.235	0.199
Mean rank	5.6	5.7	7.2	6.4	3.6	4.1	6.5	5.8	4.2	7.3	8.9

4 Conclusion

This paper discusses the dynamic classifier selection based on the discriminant functions. The dynamic ensemble selection algorithm is dedicated to the binary classification task. The proposed algorithm of the ensemble selection uses the decision profiles and the normalization of the discrimination functions is carried out. The presented algorithm uses a single parameter, which quantifies the magnitude in the difference of the discriminant functions. This difference is used as one of the conditions of the selection.

In the paper several experiments on data sets from UCI repository were carried out. The aim of the experiments was to compare the proposed selection algorithm with the nine base classifiers and ensemble classifiers based on the sum methods. For the proposed selection method with experimentally selected value of the parameter α, we obtained improvement of the classification quality measured by average values from the Friedman test. Additionally, the proposed selection algorithm with the $\alpha = 0.4$ obtains better results than classifier Ψ_S which uses the sum methods. The basis for this statement is carried out using the Wilcoxon signed-rank test.

The paper presents the dynamic classifier selection which can be applied in various practical tasks involving multiple elementary classification tasks [6,8,10]. Additionally, the advantage of the proposed algorithm is to work in the parallel and distributed environment. The classification systems with multiple classifiers are used in this type of environment [21]. The parallel processing provides a possibility to speed up the selection of the posteriori probability functions which results are needed to make the decision by the classifier ensemble.

Acknowledgments. This work was supported by the Polish National Science Center under the grant no. DEC-2013/09/B/ST6/02264 and by the statutory funds of the Department of Systems and Computer Networks, Wroclaw University of Technology.

References

1. Bishop, C.M.: Pattern Recognition and Machine Learning (Information Science and Statistics). Springer, Heidelberg (2006)
2. Cavalin, P.R., Sabourin, R., Suen, C.Y.: Dynamic selection approaches for multiple classifier systems. Neural Comput. Appl. **22**(3–4), 673–688 (2013)
3. Britto, A.S., Sabourin, R., Oliveira, L.E.S.: Dynamic selection of classifiers a comprehensive review. Pattern Recognit. **47**(11), 3665–3680 (2014)
4. Burduk, R.: Classifier fusion with interval-valued weights. Pattern Recognit. Lett. **34**(14), 1623–1629 (2013)
5. Cyganek, B.: One-class support vector ensembles for image segmentation and classification. J. Math. Imaging Vis. **42**(2–3), 103–117 (2012)
6. Cyganek, B., Woźniak, M.: Vehicle logo recognition with an ensemble of classifiers. In: Nguyen, N.T., Attachoo, B., Trawiński, B., Somboonviwat, K. (eds.) ACIIDS 2014, Part II. LNCS, vol. 8398, pp. 117–126. Springer, Heidelberg (2014)
7. Didaci, L., Giacinto, G., Roli, F., Marcialis, G.L.: A study on the performances of dynamic classifier selection based on local accuracy estimation. Pattern Recognit. **38**, 2188–2191 (2005)
8. Forczmański, P., Łabędź, P.: Recognition of occluded faces based on multi-subspace classification. In: Saeed, K., Chaki, R., Cortesi, A., Wierzchoń, S. (eds.) CISIM 2013. LNCS, vol. 8104, pp. 148–157. Springer, Heidelberg (2013)
9. Frank, A., Asuncion, A.: UCI machine learning repository (2010)
10. Frejlichowski, D.: An algorithm for the automatic analysis of characters located on car license plates. In: Kamel, M., Campilho, A. (eds.) ICIAR 2013. LNCS, vol. 7950, pp. 774–781. Springer, Heidelberg (2013)
11. Giacinto, G., Roli, F.: An approach to the automatic design of multiple classifier systems. Pattern Recognit. Lett. **22**, 25–33 (2001)
12. Guyon, I., Elisseeff, A.: An introduction to variable and feature selection. J. Mach. Learn. Res. **3**, 1157–1182 (2003)
13. Markatopoulou, F., Tsoumakas, G., Vlahavas, I.: Dynamic ensemble pruning based on multi-label classification. Neurocomputing **150**, 501–512 (2015)
14. Ho, T.K., Hull, J.J., Srihari, S.N.: Decision combination in multiple classifier systems. IEEE Trans. Pattern Anal. Mach. Intell. **16**(1), 66–75 (1994)
15. Jackowski, K., Krawczyk, B., Woźniak, M.: Improved adaptive splitting and selection: the hybrid training method of a classifier based on a feature space partitioning. Int. J. Neural Syst. **24**(3), 1430007 (2014)
16. Jackowski, K., Woźniak, M.: Method of classifier selection using the genetic approach. Expert Syst. **27**(2), 114–128 (2010)
17. Kittler, J., Alkoot, F.M.: Sum versus vote fusion in multiple classifier systems. IEEE Trans. Pattern Anal. Mach. Intell. **25**(1), 110–115 (2003)
18. Kuncheva, L.I.: A theoretical study on six classifier fusion strategies. IEEE Trans. Pattern Anal. Mach. Intell. **24**(2), 281–286 (2002)
19. Kuncheva, L.I.: Combining Pattern Classifiers: Methods and Algorithms. Wiley, Hoboken (2004)

20. Lam, L., Suen, C.Y.: Application of majority voting to pattern recognition: an analysis of its behavior and performance. IEEE Trans. Syst. Man Cybern. Part A **27**(5), 553–568 (1997)
21. Przewoźniczek, M., Walkowiak, K., Woźniak, M.: Optimizing distributed computing systems for k-nearest neighbours classifiers evolutionary approach. Log. J. IGPL **19**(2), 357–372 (2010)
22. Ranawana, R., Palade, V.: Multi-classifier systems: review and a roadmap for developers. Int. J. Hybrid Intell. Syst. **3**(1), 35–61 (2006)
23. Rejer, I.: Genetic algorithms in EEG feature selection for the classification of movements of the left and right hand. In: Burduk, R., Jackowski, K., Kurzynski, M., Wozniak, M., Zolnierek, A. (eds.) CORES 2013. AISC, vol. 226, pp. 581–590. Springer, Heidelberg (2013)
24. Ruta, D., Gabrys, B.: Classifier selection for majority voting. Inf. Fus. **6**(1), 63–81 (2005)
25. Smętek, M., Trawiński, B.: Selection of heterogeneous fuzzy model ensembles using self-adaptive genetic algorithms. New Gener. Comput. **29**(3), 309–327 (2011)
26. Suen, C.Y., Legault, R., Nadal, C.P., Cheriet, M., Lam, L.: Building a new generation of handwriting recognition systems. Pattern Recognit. Lett. **14**(4), 303–315 (1993)
27. Trawiński, K., Cordon, O., Quirin, A.: A study on the use of multiobjective genetic algorithms for classifier selection in furia-based fuzzy multiclassifiers. Int. J. Comput. Intell. Syst. **5**(2), 231–253 (2012)
28. Ulas, A., Semerci, M., Yildiz, O.T., Alpaydin, E.: Incremental construction of classifier and discriminant ensembles. Inf. Sci. **179**(9), 1298–1318 (2009)
29. Woloszyński, T., Kurzyński, M.: A probabilistic model of classifier competence for dynamic ensemble selection. Pattern Recognit. **44**(10–11), 2656–2668 (2011)

An Extension of Multi-label Binary Relevance Models Based on Randomized Reference Classifier and Local Fuzzy Confusion Matrix

Pawel Trajdos$^{(\boxtimes)}$ and Marek Kurzynski

Department of Systems and Computer Networks, Wroclaw University of Technology,
Wybrzeze Wyspianskiego 27, 50-370 Wroclaw, Poland
pawel.trajdos@pwr.wroc.pl

Abstract. In this paper we addressed the issue of applying a stochastic classifier and a local, fuzzy confusion matrix under the framework of multi-label classification. We proposed a novel solution to the problem of correcting Binary Relevance ensembles. The main step of the correction procedure is to compute label-wise competence and cross-competence measures, which model error pattern of the underlying classifier. The method was evaluated using 20 benchmark datasets. In order to assess the efficiency of the introduced model, it was compared against 3 state-of-the-art approaches. The comparison was performed using 4 different evaluation measures. Although the introduced algorithm, as its base algorithm – Binary Relevance, is insensitive to dependencies between labels, the conducted experimental study reveals that the proposed algorithm outperform other methods in terms of Hamming-loss and False Discovery Rate.

Keywords: Multi-label classification · Binary relevance · Confusion matrix

1 Introduction

In many real-world recognition task, there emerges a situation when an object is simultaneously assigned to multiple categories. For example an image may be described using such tags as sea, beach and sunset. This is an example of so called multi-label data [1]. Unfortunately, traditional single-label classification methods cannot directly be employed to solve this problem. A solution to this issue is a generalization of classical classification task called multi-label classification which assumes that object is described by a set of tags. Multi-label learning was employed in a wide range of practical applications including text classification [2], multimedia classification [3] and bioinformatics [4].

Our study explores the application of Random Reference Classifier and local fuzzy confusion matrix to improve the classification quality of the Binary Relevance ensembles. The procedure computes label specific competence and cross-competence measures which are used to correct predictions of the classifiers

© Springer International Publishing Switzerland 2015
K. Jackowski et al. (Eds.): IDEAL 2015, LNCS 9375, pp. 69–76, 2015.
DOI: 10.1007/978-3-319-24834-9_9

constituting the BR ensemble. The outcome of each member of the BR is individually modified according to the confusion pattern obtained during the validation stage.

This paper is organized as follows. The next section (Sect. 2) shows the work related to the issue which is considered throughout this paper. The subsequent section (Sect. 3) provide a formal notation used throughout this article, and introduces the proposed algorithm. Section 4 contains a description of experimental setup. In Sect. 5 the experimental results are presented and discussed. Finally, Sect. 6 concludes the paper.

2   Related Work

Multi-label classification algorithms can be broadly divided into two main groups: set transformation and algorithm adaptation approaches [1]. Algorithm adaptation methods are based upon existing multi-class methods which are tailored to solve multi-label classification problem directly. A great example of such methods is Multi-label back propagation method for artificial neuron networks [2]. On the other hand, methods from the former group transform original multi-label problem into a set of single-label classification problems and then combine their output into multi-label prediction [1]. The simplest method from this group is the *Binary Relevance* approach that decomposes multi-label classification into a set of binary classification problems. The method assigns an one-vs-rest classifier to each label. The second technique of decomposition of multi-label classification task into a set of binary classifiers is the *label-pairwise* scheme. The approach assigns a classifier for each pair of labels. Another approach is *label power set* method that encodes each combination of labels into a separate meta-class.

In this paper we concentrated the Binary Relevance (BR) approach. The conditional independence of labels, which lies at the heart of this method, results in a decomposition of multi-label classification problem into a set of L (where L is the number of labels) independent single-label binary classification tasks. This approach can be proven to be optimal in terms of the Hamming loss optimization [5]. Notwithstanding the underlying assumption, which do not hold in most of real-life recognition problems, the BR framework and its modifications are one of the most widespread multi-label classification methods [1]. This is due to their excellent scalability and acceptable classification quality. What is more this approach still offers a room for improvement. Nowadays, most efforts are focused on incorporating model of inter-label dependence into the BR framework without affecting its scalability. The focal point of another branch of research is to fit the method to achieve an optimal solution in terms of loss functions other than the Hamming loss [6].

Under the BR framework, inter-label dependencies can be modelled in various ways. Let us begin with a stacking-based two-level architecture proposed in [7]. The first-level classifier is an BR ensemble learned with the original input space, whereas the input space of the second one is extended with using the label-set. The second-level BR classifier implicitly extracts between-label relations using predictions of first-level classifiers. However,it is also possible to model these relations explicitly.

Read et al. [8] provided us with an another solution to this issue. They developed the Classifier Chain model (CC) which establish a linked chain of binary classifiers in which each classifier is responsible for learning and predicting one label. For a given label sequence, the feature space of each classifier along the chain is extended with a set of binary variables corresponding to the labels that precedes the given one. This model implies that during the training phase, input space of given classifier is extended using the ground-truth labels, although during the inference step, due to lack of the ground-truth labels, we have to employ binary labels predicted by preceding classifiers. The described method passes label information along the chain, allowing CC to take into account label correlations. Nevertheless, it also allow the errors to propagate along the chain [8]. Moreover, the experimental study revealed that the performance of a chain classifier strongly depends on chain configuration. To overcome these effects, the authors suggested to generate an ensemble of chain classifiers. The ensemble consists of classifiers trained using different label sequences. An alternative solution to this issue is to find a chain structure that allows the model to improve the classification quality [9].

3 Proposed Methods

Under the ML formalism a $d-$ dimenstional object $x = [x_1, x_2, \ldots, x_d] \in \mathcal{X}$ is assigned to a set of labels indicated by a binary vector of length L: $y = [y_1, y_2, \ldots, y_L] \in \mathcal{Y} = \{0, 1\}^L$. Each element of the vector is related to a single label and $y_i = 1$ denotes that $i-$ th label is relevant for the object x. Further, it is assumed that there exist a relationship $f : \mathcal{X} \mapsto \mathcal{Y}$ which maps each element of the input space $\mathcal{X}$ to a corresponding element in the output space $\mathcal{Y}$ and a classifier ψ is an approximation of the mapping. Throughout this paper we follow the statistical classification framework, so vectors x and y are treated as realisations of random vectors $X = [X_1, X_2, \ldots, X_d]$ and $Y = [Y_1, Y_2, \ldots, Y_L]$ respectively.

Now, we introduce a correction scheme for a single member ψ_i ($i \in 1, 2, \ldots, L$) of the BR system, which is a common binary classifier. The proposed method consists of two main steps. The first is a binary classification, and the other is a correcting procedure that improve the outcome of the classifier. During the learning phase, the transformed learning set is split into two equal parts. One is used to build the base classifier, the other is saved for the further use during the inference phase. At the recognition phase the outcome of the classifier is corrected using the statistical model of competence and cross-competence.

The proposed algorithm performs the inference process using the probabilistic framework which is described in previous subsection. In addition to formerly made assumptions, the procedure requires the outcome of classifier $\psi_i(x)$ to be a realization of a random variable $\Psi_i(x)$. As a consequence, the classifier is considered to be a stochastic one which randomly assigns given instance x

for class '0' or '1' according to the probability distribution $P(\psi_i(\boldsymbol{x}) = h) = P_i(h|\boldsymbol{x}), h \in \{0,1\}$. The presence of the additional random variable allows us to define the posteriori probability $P(Y_i = y|\boldsymbol{x}) = P_i(y|\boldsymbol{x})$ of the label i as:

$$P_i(y|\boldsymbol{x}) = \sum_{h=0}^{1} P_i(y, h|\boldsymbol{x}) = \sum_{h=0}^{1} P_i(h|\boldsymbol{x})P_i(y|h, \boldsymbol{x}) \qquad (1)$$

where $P(h_i|\boldsymbol{x})$ is interpreted as a support that the classifier gives to the hypothesis that $\boldsymbol{x}$ belongs to class h. Additionally, $P_i(y|h, \boldsymbol{x})$ denotes the probability that the true assignment of object $\boldsymbol{x}$ to $i-$th label is y given that the stochastic classifier ψ_i has predicted it as h. Aforesaid probability can be interpreted as pointwise indicator of competence and cross-competence of the stochastic classifier. The competence index is proportional to the probability of correct classification, whereas the cross-competence follows the probability of miss-classification.

Unfortunately, at the core of the proposed method, we put rather an impractical assumption that the classifier assigns a label in a stochastic manner. We dealt with this issue by harnessing deterministic binary classifiers whose statistical properties were modelled using the RRC procedure [10]. The RRC model calculates the probability that the underlying classifier assigns an instance to class h_i $P_i(h|\boldsymbol{x}) \approx P_i^{(RRC)}(h|\boldsymbol{x})$.

The probability $P_i(y|h, \boldsymbol{x})$ was estimated using a lazy-learning procedure based on local, fuzzy confusion matrix. The rows of the matrix corresponds to the ground-truth classes, whereas the columns match the outcome of a classifier. Each entry of the matrix is an estimation of the probability $\varepsilon_{y,h}^i \approx P(y_i = y, h_i = h|x)$. The fuzzy nature of the confusion matrix arises directly from the fact that a stochastic model has been employed. In other words, decision regions of the random classifier must be described using Fuzzy set formalism [11]. In order to provide an accurate estimation, we have also defined our confusion matrix as local what means that the matrix is build using neighbouring points of the instance $\boldsymbol{x}$. The neighbourhood is defined using the Gaussian potential function to assign a membership coefficients:.

$$\mathcal{N}(\boldsymbol{x}) = \left\{ (\exp(-\beta\delta(\boldsymbol{x}^{(k)}, \boldsymbol{x})^2), \boldsymbol{x}^{(k)}) : \boldsymbol{x}^{(k)} \in \mathcal{V} \right\}. \qquad (2)$$

where $\beta \in \mathbb{R}^+$ and $\delta(\boldsymbol{z}, \boldsymbol{x})$ is the Euclidean distance between $\boldsymbol{z}$ and $\boldsymbol{x}$. During the experimental study β was set to 1.

The matrix is estimated using a validation set:

$$\mathcal{V} = \left\{ (\boldsymbol{x}^{(1)}, \boldsymbol{y}^{(1)}), (\boldsymbol{x}^{(2)}, \boldsymbol{y}^{(2)}), \dots, (\boldsymbol{x}^{(k)}, \boldsymbol{y}^{(N)}) \right\}; \quad \boldsymbol{x}^{(k)} \in \mathcal{X}, \ \boldsymbol{y}^{(k)} \in \mathcal{Y}. \qquad (3)$$

On the basis of this set we define the BR subset of validation set, fuzzy decision regions of ψ_i and set of neighbours of $\boldsymbol{x}$:

$$\mathcal{V}_y^i = \left\{ (\boldsymbol{y}_i^{(k)} = y, \boldsymbol{x}^{(k)}) : \boldsymbol{x}^{(k)} \in \mathcal{V} \right\}, \qquad (4)$$

$$\overline{\mathcal{D}}_h^i = \left\{ (\mu_{\mathcal{D}_h^i}(\boldsymbol{x}^{(k)}) = P_i^{(RRC)}(h|\boldsymbol{x}^{(k)}), \boldsymbol{x}) : \boldsymbol{x}^{(k)} \in \mathcal{V} \right\}, \qquad (5)$$

These sets are employed to approximate entries of the local confusion matrix:

$$\varepsilon_{y,h}^i(\boldsymbol{x}) = \frac{\left| \mathcal{V}_y^i \cap \overline{\mathcal{D}}_h^i \cap \mathcal{N}(\boldsymbol{x}) \right|}{|\mathcal{N}(\boldsymbol{x})|}, \tag{6}$$

where $|.|$ is the cardinality of a fuzzy set.

4   Experimental Setup

During the experimental study, we compared the proposed method (FC) against three state-of-the-art, Binary-Relevance-based procedures, namely Binary Relevance classifier (BR), Classifier Chain [8] (CC) and Stacking [7] (ST). The algorithms were implemented using the Naïve Bayes classifier [12] as a base classifier (the experimental code was implemented using mulan [13] and meka [14]). With a view to improving the robustness of the considered classifiers, we build a multi-classifier (MC) systems based on the aforementioned algorithms. Each homogeneous MC system consists of 30 base multi-label classifiers and their training sets are constructed using bagging approach [15]. The label specific results are combined on the support level using simple mean combiner [15].

The experiments were conducted using 20 multi-label benchmark sets. The main characteristics of the datasets are summarized in Table 1. The first two columns of the table contain the number of the set, set name and its source. Next three columns are filled with the number of instances, dimensionality of the input space and the number of labels respectively. Another column provide us with measure of multi-label-specific characteristics of given set i.e. label density (LD). [1]. The extraction of training and test datasets was performed using $10 - CV$. Some of the employed sets needed preprocessing transformation. That is, multi label regression sets (solar_flare1/2 and water-quality) were binarized and multi-label multi-instance [3] sets were transformed using the procedure described by Wu et al. [4].

The algorithms were compared in terms of 4 different quality criteria coming from two groups: instance-based (Hamming loss) and micro-averaged (F-measure, False Negative Rate (FNR) and False Discovery Rate (FDR)) [16]. Statistical evaluation of the results was performed using the Friedman test and the Nemenyi post-hoc test [17]. Additionally, we applied the Wilcoxon signed-rank test [17] and the family-wise error rates were controlled using the Holm's procedure [17]. For all statistical tests, the significance level was set to $\alpha = 0.05$.

5   Results and Discussion

The summarised results of the experimental procedure are presented in Table 2. Additionally, the table also show the outcome of the conducted statistical evaluation. Excluding the table header, which contains names of the evaluation criteria and algorithm names, the table is basically divided into three main sections.

Table 1. Dataset summary.

No	Set name	N	d	L	LD	No	Set name	N	d	L	LD
1	birds [13]	645	279	19	.053	11	Solar_flare2 [13]	1066	30	3	.070
2	Caenorhab. [4]	3509	47	27	.084	12	Genbase [13]	662	1213	27	.046
3	CAL500 [13]	502	242	174	.150	13	Llog [14]	1460	1079	75	.016
4	Corel16k1 [13]	13766	653	153	.019	14	Mediamill [13]	43907	221	101	.043
5	Corel16k2 [13]	13761	664	164	.018	15	Medical [13]	978	1494	45	.028
6	Corel16k3 [13]	13760	654	154	.018	16	MimlImg [3]	2000	140	5	.247
7	Emotions [13]	593	78	6	.311	17	Scene [13]	2407	300	6	.179
8	Enron [13]	1702	1054	53	.064	18	Slashdot [14]	3782	1101	22	.054
9	Flags [13]	194	50	7	.485	19	Water-quality [13]	1060	30	14	.362
10	Solar_flare1 [13]	323	28	3	.077	20	Yeast [13]	2417	117	14	.303

The first one consists of 20 rows and each of them shows set specific result averaged over the CV folds. The best results for each set was highlighted in bold type. Additionally, the section also contains outcome of the Wilcoxon test that compares, for the given set, the proposed procedure against remaining methods. Statistical improvement, decrease was marked by ↑ and ↓ respectively. The second part delivers average ranks obtained by the algorithms throughout the benchmark sets (Rnk), and the outcome of the Nemenyi test. The outcome is reduced to only two groups, namely a group of classifiers whose results do not significantly differ from the results obtained by the best performing (BG) and worst performing (WG) classifiers respectively. The third section of the results table displays p-values (Wp) achieved during the pairwise application of the Wilcoxon test which was used to provide a global comparison of the proposed approach against the reference methods.

The presented results reveals that the proposed method turned out to be the best performing algorithm under the Hamming-loss. This is a promising result since the approach was designed to minimize that loss. What is more important, FC outperformed not only BR algorithm but also CC and Stacking. On the other hand, the results also showed that there is a major discrepancy between results obtained by different criteria. Contrary to high performance under the Hamming-loss, the efficiency described by the micro-averaged F-measure is rather poor. This phenomenon can easily be explained by analysis of classification assessment in terms of micro-averaged FNR and FDR. Namely, low sensitivity and high precision are characteristic of the FC classifier, consequently the system achieved low performance under the F-loss. In other words, the coefficients suggest that base classifiers of the FC system are biased toward the majority class '0' (label is irrelevant). The presence of the tendency to prefer majority class combined with low label density of benchmark datasets may have led to an overoptimistic performance assessment under the Hamming loss. To be more precise, when the average cardinality of a set of relevant labels is low and a BR system, whose base classifiers are biased to '0', may achieve high Hamming score by setting a majority of labels as irrelevant. This fact confirms the foregoing observations that FC manifest the bias towards the majority class. The cause of such behaviour is a tendency to underestimate the conditional probability $P_i(y|x)$ for rare labels which results in increased performance on those labels and decreased performance for frequent labels.

Table 2. Summarized results for micro-averaged FNR, FDR and F-measure.

	FNR				FDR				F-loss				Hamming			
	BR	CC	ST	FC	BR	CC	ST	FC	BR	CC	ST	FC	BR	CC	ST	FC
1	.44↑	.43↑	.54↑	.89	.91↓	.91↓	.90↓	.28	.85	.85	.83	.81	.34↓	.35↓	.25↓	.05
2	.73↑	.76↑	.75↑	.92	.47↓	.40↓	.46↓	.07	.64↑	.66↑	.66↑	.86	.11	.11	.11	.11
3	.49↑	.50↑	.52↑	.83	.76↓	.76↓	.76↓	.37	.67↑	.68↑	.68↑	.73	.31↓	.32↓	.30↓	.14
4	.77	.76	.75	.89	.77	.78	.79	.67	.77	.77	.77	.84	.03	.03	.03	.02
5	.76	.76	.75	.90	.77	.78	.79	.66	.76	.76	.77	.84	.03	.03	.03	.02
6	.75	.75	.74	.89	.77	.77	.79	.66	.76	.76	.77	.84	.03	.03	.03	.02
7	.24↑	.25↑	.46↑	.94	.44	.43	.43	.32	.36↑	.35↑	.45↑	.90	.26	.25↑	.27	.30
8	.34↑	.35↑	.35↑	.70	.81↓	.81↓	.80↓	.41	.71↓	.70↓	.70↓	.60	.21↓	.20↓	.19↓	.06
9	.33	.31	.31	.39	.34↓	.33	.34↓	.26	.34	.32	.33	.35	.33	.31	.32	.30
10	.66↑	.66↑	.66↑	.95	.78	.75	.75	.88	.75	.74	.74	.93	.14↓	.13↓	.13↓	.08
11	.49↑	.49↑	.56↑	.93	.73	.73	.74	.51	.66↑	.66↑	.68↑	.88	.13↓	.13↓	.13↓	.07
12	.71↓	.71↓	.71↓	.49	.00↑	.00↑	.00↑	.39	.56↓	.55↓	.55↓	.45	.03↑	.03↑	.03↑	.04
13	.27↑	.26↑	.26↑	1.00	.94	.95	.95	.90	.90↑	.90↑	.91↑	1.0	.20↓	.22↓	.23↓	.02
14	.29	.32	.33	.26	.91	.93	.93	.92	.85	.88	.88	.86	.34	.43	.41	.38
15	.56	.55	.54	.56	.45↓	.45↓	.45↓	.19	.51↓	.51↓	.50↓	.43	.03↓	.03↓	.03↓	.02
16	.36↑	.37↑	.40↑	.94	.55↓	.55↓	.52↓	.18	.47↑	.47↑	.50↑	.88	.28↓	.28↓	.26↓	.23
17	.16↑	.16↑	.16↑	.78	.58↓	.58↓	.57↓	.18	.44↑	.44↑	.44↑	.65	.24↓	.24↓	.23↓	.15
18	.47↑	.44↑	.44↑	.87	.62↑	.58↑	.57↑	.82	.56↑	.52↑	.51↑	.85	.07↑	.06↑	.06↑	.08
19	.28↑	.27↑	.25↑	.77	.56↓	.56↓	.57↓	.43	.46↑	.46↑	.45↑	.68	.44↓	.44↓	.45↓	.34
20	.40↑	.41↑	.41↑	.61	.49↓	.49↓	.48↓	.24	.45↑	.45↑	.45↑	.49	.30↓	.30↓	.29↓	.22
Rnk	2.10	1.98	2.23	3.70	2.90	2.85	2.75	1.50	2.25	2.25	2.20	3.30	3.00	2.80	2.70	1.55
BG	+	+	+		+	+	+	+	+	+	+	+	+	+	+	+
WG		+	+		+	+	+	+	+	+	+	+	+	+	+	+
Wp	.010	.001	.001	–	.025	.026	.026	–	.026	.026	.025	–	.010	.007	.006	–

6   Conclusion

The problem addressed in the paper was solved with modest success. We obtained promising experimental results which suggest that the proposed correction scheme for the BR system (FC) achieved the best results in terms of the Hamming-loss and FDR. The algorithm is a very conservative procedure that predicts as relevant only labels which are strongly related to the recognised object. That property may be desirable under certain practical circumstances. Additionally, it is safe to say that the performance of FC does not decrease in terms of zero-one loss. However, the proposed system suffer from a major drawback which is a bias towards majority class. This bias effects in poor performance under F-loss and False Negative Rate.

Although the results do not lead to excessive optimism, we are willing to continue our research because the proposed methodology is at an early stage of development and it still offers a room for improvement. The further research is going to be focused on a procedure of bias reduction for the FC method. In near future we are going to develop an alternative neighbourhood definition that prevent the confusion matrix to be skewed towards the majority class.

Acknowledgements. Computational resources were provided by PL-Grid Infrastructure.

References

1. Tsoumakas, G., Katakis, I., Vlahavas, I.: Mining multi-label data. In: Maimon, O., Rokach, L. (eds.) Data Mining and Knowledge Discovery Handbook, pp. 667–685. Springer, Heidelberg (2009)
2. Zhang, M.L., Zhou, Z.H.: Multilabel neural networks with applications to functional genomics and text categorization. IEEE Trans. Knowl. Data Eng. **18**, 1338–1351 (2006)
3. Zhou, Z.-h., Zhang, M.-l.: Multi-instance multilabel learning with application to scene classification. In: Advances in Neural Information Processing Systems 19 (2007)
4. Wu, J.S., Huang, S.J., Zhou, Z.H.: Genome-wide protein function prediction through multi-instance multi-label learning. IEEE/ACM Trans. Comput. Biol. and Bioinf. **11**, 891–902 (2014)
5. Dembczyński, K., Waegeman, W., Cheng, W., Hüllermeier, E.: On label dependence and loss minimization in multi-label classification. Mach. Learn. **88**, 5–45 (2012)
6. Pillai, I., Fumera, G., Roli, F.: Threshold optimisation for multi-label classifiers. Pattern Recognit. **46**, 2055–2065 (2013)
7. Cherman, A.E., Metz, J., Monard, M.C.: A simple approach to incorporate label dependency in multi-label classification. In: Sidorov, G., Hernández Aguirre, A., García, C.A.R. (eds.) MICAI 2010, Part II. LNCS, vol. 6438, pp. 33–43. Springer, Heidelberg (2010)
8. Read, J., Pfahringer, B., Holmes, G., Frank, E.: Classifier chains for multi-label classification. Mach. Learn. **85**, 333–359 (2011)
9. Read, J., Martino, L., Luengo, D.: Efficient monte carlo methods for multidimensional learning with classifier chains. Pattern Recognit. **47**, 1535–1546 (2014)
10. Woloszynski, T., Kurzynski, M.: A probabilistic model of classifier competence for dynamic ensemble selection. Pattern Recognit. **44**, 2656–2668 (2011)
11. Zadeh, L.: Fuzzy sets. Inf. Control **8**, 338–353 (1965)
12. Hand, D.J., Yu, K.: Idiot's bayes: not so stupid after all? Int. Stat. Rev. Revue Internationale de Statistique **69**, 385 (2001)
13. Tsoumakas, G., Spyromitros-Xioufis, E., Vilcek, J., Vlahavas, I.: Mulan: a java library for multi-label learning. J. Mach. Learn. Res. **12**, 2411–2414 (2011)
14. Read, J., Peter, R.: (Meka. http://meka.sourceforge.net/). Accessed 29-03-2015
15. Kuncheva, L.I.: Combining Pattern Classifiers: Methods and Algorithms, 1st edn. Wiley Interscience, Hoboken (2004)
16. Luaces, O., Díez, J., Barranquero, J., del Coz, J.J., Bahamonde, A.: Binary relevance efficacy for multilabel classification. Prog. Artif. Intell. **1**, 303–313 (2012)
17. Demšar, J.: Statistical comparisons of classifiers over multiple data sets. J. Mach. Learn. Res. **7**, 1–30 (2006)

Fusion of Self-Organizing Maps
with Different Sizes

Leandro Antonio Pasa[1,2]($\boxtimes$), José Alfredo F. Costa[2],
and Marcial Guerra de Medeiros[2]

[1] Federal University of Technology – Paraná, UTFPR, Medianeira, Brazil
pasa@utfpr.edu.br
[2] Federal University of Rio Grande do Norte, UFRN, Natal, Brazil
{jafcosta,marcial.guerra}@gmail.com

Abstract. An ensemble consists of several neural networks whose outputs are fused to produce a single output, which usually will be better than the individual results of each network. This work presents a methodology to aggregate the results of several Kohonen Self-Organizing Maps in an ensemble. Computational simulations demonstrate an increase in the accuracy classification and the proposed method effectiveness was evidenced by the Wilcoxon Signed Rank Test.

Keywords: Ensemble · Self-Organizing Maps · Classification

1 Introduction

An ensemble is defined as a collection of individual classifiers which has differences between each other which is able to achieve higher generalization than when working separately, as well as a decrease in variance model and higher noise tolerance when compared to a single component. Each classifier operates independently and generates a solution that is combined by the ensemble producing a single output [1]. It is essential the errors introduced by each component are uncorrelated for a successful outcome of the ensemble [2]. For Kohonen Self-Organizing Maps, this task is simple because different networks can be trained from the same set of feature vectors - varying some training parameters - or different training sets, for example. The most difficult point is to find a way to combine the maps and generate a single output. This can be done by merging the neurons of the maps to be fused. These neurons must represent the same region of the data input space, in other words, the weight vectors to be fused should be quite similar. This work presents a methodology to fuse different size Kohonen Self-Organizing Maps using two different equations in order to increase the classification accuracy.

The paper is organized as follows: Sect. 2 presents basic concepts. Section 3 presents the proposed fusion algorithm. Section 4 shows and discuss the results. Section 5 presents the conclusions and some proposals for future works.

© Springer International Publishing Switzerland 2015
K. Jackowski et al. (Eds.): IDEAL 2015, LNCS 9375, pp. 77–86, 2015.
DOI: 10.1007/978-3-319-24834-9_10

2   Background

2.1   Self-Organizing Maps

Kohonen Self-Organizing Maps is known as a method for dimensionality reduction, data visualization and also for data classification. The Self-Organizing Maps (SOM), developed by Kohonen [3] has became a popular neural network model. It has competitive and unsupervised learning, performing a non-linear projection of the input space $\Re^p$, with $p \gg 2$, in a grid of neurons arranged in two-dimensional array, having only two layers: an input and an output layer. The network inputs x, correspond to the p-dimensional vector space. The neuron i of the output layer is connected to all the inputs of the network, being represented by a vector of synaptic weights, also in p-dimensional space, $w_i = [wi_1, wi_2, ..., wi_p]^T$. These neurons are connected to adjacent neurons by a neighbourhood relation that describes the topological structure of the map. During the training phase, following a random sequence, input patterns x are compared to the neurons of the output layer. Through the Euclidean distance criterion a winning neuron, called BMU (Best Match Unit), is chosen and will represent the weight vector with the smallest distance to the input pattern, i.e. the BMU will be the most similar to the input x.

Assigning the winner neuron index by c, The BMU can be formally defined as the neuron according to the 1.

$$\|x - w_c\| = argmin_i \|x - w_i\| \tag{1}$$

The Eq. 2 adjust the BMU weights and the neighbouring neurons.

$$w_i(t + 1) = w_i(t) + h_{ci}(t)[x(t) - w_i(t)] \tag{2}$$

where t indicates the iteration of the training process, $x(t)$ is the input pattern and $h_{ci}(t)$ is the nucleus of neighbourhood around the winner neuron c.

2.2   Cluster Validity Indexes

To make sure that an algorithm found the groups that best fit to the data, a very important task is to evaluate the clustering algorithm results using Cluster Validation Indexes (CVI). The literature presents several CVI to evaluate clusters results and most of them have high computational complexity, which can be a complicating factor in applications involving large data volumes. A modification in the CVI calculations was proposed in [4] using a vector quantization produced by Kohonen Map. The synaptic weight vectors (prototypes) are used instead of the original data. Thus, it causes the decrease of the amount of data and therefore the computational complexity for calculating the CVI decreases too. Also to avoid possible differences between the values calculated with all data and only the prototypes, the author proposed that hits should be used in conjunction with the prototypes. The following example illustrates the proposed change in the calculation of the distance between two clusters, C_i and C_j:

$$\delta_{i,j} = \frac{1}{|C_i||C_j|} \sum_{x \in C_i, y \in C_j} d(x,y) \tag{3}$$

In Eq. 3, $d(x,y)$ is a distance measure, and $|C_i| e |C_j|$ refers to the clusters' amount of points C_i and C_j, respectively. When the amount of those points is high, the computational complexity is also high. Equation 4 shows the modification proposed in [4]:

$$\delta_{i,j}^{SOM} = \frac{1}{|C_i||C_j|} \sum_{w_i \in W_i, w_j \in W_j} h(w_i) \cdot h(w_j) \cdot d(w_i, w_j) \tag{4}$$

Where W_i and W_j are the SOM prototype sets that represent the clusters C_i e C_j, respectively; $d(x,y)$ is the same distance measure type (Euclidian, for example) of Eq. 3, $h(w_i)$ is the prototype's hits w_i belonging to W_i and $h(w_j)$ is the prototype's hits w_j belonging to W_j.

The Eq. 4 presents a lower computational cost, since the quantities involved, w_i and w_j are lower than C_i and C_j. The inclusion of the prototypes' hits $h(.)$ leads to error minimization caused by the vector quantization that Kohonen Map produces, since it introduces in the calculation the approach for the points density in the input space, here represented by prototypes.

2.3   Related Work

Ensemble methods became popular and there are several approaches and applications in diverse knowledge areas. In [5] they compared merged maps with the traditional SOM for document organization and retrieval. As a criterion for combining maps, the Euclidean distance between neurons was used in order to select the neurons were aligned (allowing the fusion), working with two maps each time, until all maps are fused into one. The ensemble of SOM obtained a better result than the traditional application. In *Fusion*-SOM [6] was based on Voronoi polygons. The proposed method outperforms the performance of the SOM in MSQE (Mean Squared Quantization Error) and topology preservation, by effectively locating the prototypes and relating the neighbour nodes. In a weighted voting process, called WeVoS-ViSOM [7], the purpose was the preservation of the map topology, in order to obtain the most truthful visualization of datasets. This algorithm was used in a hybrid system to predict business failure [8]. This methodology succeeds in reducing the distortion error of single models. SOM ensemble methods applications are found in diverse areas such as image segmentation [9], robotic [10], identification and characterization of computer attacks [11], unsupervised analysis of outliers on astronomical data [12] and financial distress model [13], among others.

3 Proposed Fusion Method

This work introduces a novel method to fuse Kohonen Self-organizing Maps. The purpose is to improve the classification accuracy by fusing at most forty nine different sized maps. The fusion process has seven steps: data split, initialization, training, segmentation, performance, ranking and fusion. Each step is explained in the following.

3.1 Fusion Algorithm

The fusion algorithm can be described as follows:

Step 1: Split the data in two: training set (80 %) and test set (20 %);

Step 2: Calculate the size of the Base Map (from now on abbreviated as BM), i.e. the SOM which will be improved upon by consecutive fusions to other maps of different sizes. The BM dimensions x and y are proportional to the two largest covariance matrix eigenvalues of the data: λ_1 and λ_2:

$$\frac{x}{y} \approx \sqrt{\frac{\lambda_1}{\lambda_2}} \tag{5}$$

Step 3: From the BM there will be created another 48 maps, which differ in a combination of three values up and down of x and y, as show in Fig. 1. As some components (maps) are removed from the ensemble because it did not contribute to the performance increasing [14], it was chosen the variation of three units up and down to generate maps in sufficient number for the ensemble. The variation of one or two units up and down could lead to a few maps, that do not contribute to the increase of the ensemble performance;

Step 4: The SOM algorithm is then applied on the training set to create all different size maps mentioned on Step 3;

Step 5: Segment all maps with k-means clustering, which is repeated 10 times and the result with smallest sum of squared errors is selected due to the random initialization nature of k-means;

Step 6: Calculate map performances for each map. The performances include MSQE and CVI, namely: Generalized Dunn, Davies-Bouldin, CDbw, PBM and Calinski and Harabasz;

Step 7: All maps are sorted according to their calculated map performances, from the best to worst, e.g. from lowest error (best) to highest error (worst);

Step 8: Each map is sequentially fused to the base map according to a fusion formula and based on a criteria this fusion will or will not be discarded. This step is explained thoroughly in the next subsection.

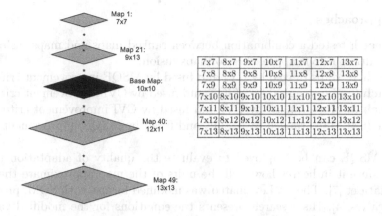

7x7	8x7	9x7	10x7	11x7	12x7	13x7
7x8	8x8	9x8	10x8	11x8	12x8	13x8
7x9	8x9	9x9	10x9	11x9	12x9	13x9
7x10	8x10	9x10	10x10	11x10	12x10	13x10
7x11	8x11	9x11	10x11	11x11	12x11	13x11
7x12	8x12	9x12	10x12	11x12	12x12	13x12
7x13	8x13	9x13	10x13	11x13	12x13	13x13

Fig. 1. The maps created from the Base Map.

3.2 Maps Fusion

The inclusion of all candidates in the ensemble may degrade its performance [14], because not all components will contribute to the overall ensemble performance. So, it is necessary to identify and discard these components. The base map is the first component of ensemble. The 48 remaining maps are candidates to be ensemble components. These maps shall be tested according to their ranked performances. The maps are fused in pairs and the fusion occurs between two neurons that have the minimum Euclidean distance between them, indicating that they represent the same region of the input space.

In this study two different equations for fusion were tested to find out each neuron weight vector: fusion considering the hits of each neuron (Eq. 6) and fusion considering the hits of each neuron and the CVI for each considered map (Eq. 7).

$$w_c = \frac{w_i \cdot h_i + w_j \cdot h_j}{h_i + h_j} \tag{6}$$

$$w_c = \frac{w_i \cdot h_i \cdot VI_i + w_j \cdot h_j \cdot VI_j}{h_i \cdot VI_i + h_j \cdot VI_j} \tag{7}$$

where: w_c represents each neuron of fused map; w_i and w_j are the neurons to be fused; h_i and h_j are the prototype's hits and VI_i and VI_j are the CVI for each SOM map to be fused.

The process begins with the base map being fused with the best performance map, according with performance value (CVI or MSQE) considered in this work, among the other 48 maps. If the performance criteria of the resulting fused map has improved, this fusion is maintained. Otherwise, it is discarded and the next map, with the second best performance, is fused to the base map. Each time the fusion improves the considered performance, the next map (in descending order of performance) is fused to the resulting fused map and so on, until all maps are tested. Four different approaches were tested in this work to rank the maps and to establish the fusion criterion. It is explained in Sect. 3.3.

3.3 Approaches

This research tested a combination between ranked maps and maps fusion criteria, resulting in four approaches to maps fusion:

Approach 1: Maps ranked by CVI and fused by MSQE improvement criterion.

Approach 2: Maps ranked by MSQE and fused by CVI improvement criterion.

Approach 3: Maps ranked by CVI and fused by CVI improvement criterion.

Approach 4: Maps ranked by MSQE and fused by MSQE improvement criterion.

The MSQE can be employed to evaluate the quality of adaptation to the data [6] since it indicates how well the units of the map approximate the data on the dataset [7]. The CVI evaluated was modified for use with SOM, proposed by Gonçalves [4]. His research presents the equations for the modified validity indexes used in this work.

In the first approach the maps were ranked by five CVI, but the fusion process maps was controlled by MSQE improvement of the fused map. In the second approach the maps were ranked by MSQE and the fusion process was validated by the CVI improvement. In third approach, maps were ranked by each CVI and the fusion was controlled by these CVI increase. At last, in the fourth approach the maps were ranked by MSQE and fused by MSQE improvement.

3.4 Maps Fusion Requirement

In this proposed method the hits and BMUs percentage was varied from 10 % to 100 %, with a step of 10 % in order to better evaluate the influence of these measures in fusion process. When a neuron is fused to another neuron (from another map), it will take into the account the minimum hits percentage (how many times that neuron was the winner - BMU) this neuron must have when compared to the base map neuron. This percentage was varied from 10 % to 100 %. The limitation of BMUs percentage is very important in this method. When the base map has smaller dimensions than the map to be merged with it, there will be more neurons in the second map than in the base map. Thus, this procedure limit the neurons number to be fused at 100 % of the smallest map.

4 Experimental Results

4.1 Datasets

The proposed algorithm was tested with datasets from the UCI Repository [15] and from Fundamental Clustering Problems Suite (FCPS) [16]. Their characteristics are shown in Table 1. All missing values were removed from the datasets.

4.2 Results

Tables 2 and 3 show the best experimental results for each dataset, i.e. which approach and which CVI resulted in the best accuracy value for each dataset.

Table 1. Dataset collection

Dataset	Repository	Instances	Attributes	Classes
BC Wisconsin	UCI	699	10	2
Chainlink	FCPS	1000	3	2
Column	UCI	310	6	2
Engytime	FCPS	4096	2	2
Heart	UCI	303	75	2
Hepatitis	UCI	155	19	2
Ionosphere	UCI	351	34	2
Iris	UCI	150	4	3
Lsun	FCPS	400	2	3
Pima Indians	UCI	768	8	2
Seeds	UCI	210	7	3
Tetra	FCPS	400	3	4
Two Diamonds	FCPS	800	2	2
Wine	UCI	178	13	3
Wingnut	FCPS	1070	2	2

The column Approach refers to the way the maps were ranked and fused, as specified in Sect. 3.3. As can be seen, approaches 1 and 2 produced the bests results in this experiment.

The column Index shows which of the CVI was the best for each dataset (DB means Davies-Bouldin index and CH means Calinski and Harabasz index). In approach 1, the indexes were used for ranking and in approach 2, the indexes were used as fusion criteria. The column Map Size refers to the map size defined by the Eq. 5. The Hits and BMUs percentage columns (explained in Sect. 3.4), shows the value of these variables to achieve the maximum accuracy value.

The fusion accuracy results for each dataset was compared with the accuracy results for a single Self-Organizing Map in the last two columns. As can be observed, for all tested datasets the classification accuracy of the proposed model was higher than or equal to the accuracy obtained by a single map. The nonparametric Wilcoxon signed rank test [17] was employed in order to evaluate the statistical significance of the fusion results, i.e. if the accuracies values was statistically different from a single Kohonen map. The H_0 hypothesis was that there is no difference between the accuracies of a Single SOM and the Fused Maps. The Wilcoxon test rejected the null hypothesis and the *p-value* were 0.00097656 for both fusion equations used in this work.

The Figs. 2 and 3 summarize the accuracy results for this proposed method, for each Fusion Equation. The red line represents the single SOM accuracy, while green line shows the accuracy results for fused maps.

Table 2. Experimental results - Hits Fusion Equation (Eq. 6)

Dataset	Approach	Index	Map Size	Hits (%)	BMUs (%)	Fusion accuracy	Single SOM accuracy
BC Wisconsin	1	CH	22 × 6	10	60	**0.9852**	0.9778
Chainlink	2	DB	14 × 11	10	90	**0.7800**	0.7200
Column	1	CH	13 × 7	30	10	**0.7097**	0.6774
Engytime	1	CDbw	20 × 16	60	30	0.9731	0.9731
Heart	1	CDbw	11 × 8	10	10	**0.8305**	0.8136
Hepatitis	1	CDbw	8 × 5	10	10	0.9333	0.9333
Ionosphere	1	Dunn	13 × 7	10	10	**0.7429**	0.7286
Iris	1	CDbw	13 × 5	10	20	0.9667	0.9667
Lsun	2	Dunn	11 × 9	20	90	**0.9125**	0.7750
Pima Indians	1	CH	14 × 10	30	70	**0.6209**	0.5948
Seeds	1	CDbw	14 × 5	10	10	**0.9524**	0.9286
Tetra	1	CDbw	11 × 9	10	10	1	1
Two Diamonds	1	Dunn	14 × 10	10	100	**1**	0.9938
Wine	2	CH	11 × 6	30	20	**0.9412**	0.8529
Wingnut	2	Dunn	16 × 10	20	90	**0.8119**	0.7970

Table 3. Experimental results - Hits and Indexes Fusion Equation (Eq. 7)

Dataset	Approach	Index	Map Size	Hits (%)	BMUs (%)	Fusion accuracy	Single SOM accuracy
BC Wisconsin	1	CH	22 × 6	10	60	**0.9852**	0.9778
Chainlink	2	DB	14 × 11	70	90	**0.7450**	0.7200
Column	2	DB	13 × 7	10	100	**0.7419**	0.6774
Engytime	1	PBM	20 × 16	20	20	**0.9743**	0.9731
Heart	1	Dunn	11 × 8	10	40	**0.8305**	0.8136
Hepatitis	1	CDbw	8 × 5	10	10	0.9333	0.9333
Ionosphere	1	Dunn	13 × 7	10	10	**0.7429**	0.7286
Iris	1	CDbw	13 × 5	10	10	0.9667	0.9667
Lsun	2	Dunn	11 × 9	100	60	**0.8625**	0.7750
Pima Indians	2	DB	14 × 10	70	100	**0.6209**	0.5948
Seeds	1	CDbw	14 × 5	10	10	**0.9524**	0.9286
Tetra	1	CDbw	11 × 9	10	10	1	1
Two Diamonds	1	CDbw	14 × 10	90	20	0.9938	0.9938
Wine	2	PBM	11 × 6	10	90	**0.9412**	0.8529
Wingnut	2	CH	16 × 10	40	30	**0.8020**	0.7970

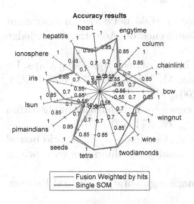

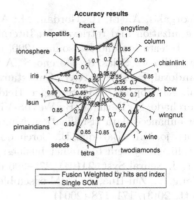

Fig. 2. Fusion by Eq. 6 (Color figure online)   **Fig. 3.** Fusion by Eq. 7 (Color figure online)

5   Conclusion and Future Work

This work presented a methodology to aggregate the results of several Kohonen different sized maps in an ensemble to increase the classification accuracy. Two different equations were tested for the neurons fusion and both achieved significant results. Four different approaches were used and we found that approaches 1 and 2 produce the best results. Although for some datasets the accuracy gain has not been great, the Wilcoxon test showed that the accuracies obtained by the proposed model are significantly different and better than the accuracies obtained from a single Kohonen network. The achieved results demonstrate the effectiveness of the proposed method to fuse different maps size.

Future work includes checking the influence of parameters such as the number of maps, the adjustments in SOM training, the segmentation method, among others.

Acknowledgments. Authors would like to thank the support of CAPES Foundation, Ministry of Education of Brazil, Brasilia - DF, Zip Code 70.040-020.

References

1. Dietterich, T.G.: Ensemble methods in machine learning. In: Kittler, J., Roli, F. (eds.) MCS 2000. LNCS, vol. 1857, pp. 1–15. Springer, Heidelberg (2000)
2. Perrone, M.P., Cooper, L.N.: When networks disagree: ensemble methods for hybrid neural networks. In: Mammone, R.J. (ed.) Neural Networks for Speech and Image Processing, pp. 126–142. Chapman and Hall, New York (1993)
3. Kohonen, T.: Self-Organized Maps, 2nd edn. Springer, Berlin (1997)
4. Gonalves, M.L., De Andrade Netto, M.L., Costa, J.A.F., Zullo, J.: Data clustering using self-organizing maps segmented by mathematic morphology and simplified cluster validity indexes: an application in remotely sensed images. In: IJCNN 2006, International Joint Conference on Neural Networks, pp. 4421–4428 (2006)

5. Georgakis, A., Li, H., Gordan, M.: An ensemble of SOM networks for document organization and retrieval. In: International Conference on Adaptive Knowledge Representation and Reasoning (2005)
6. Saavedra, C., Salas, R., Moreno, S., Allende, H.: Fusion of Self Organizing Maps. In: Sandoval, F., Prieto, A.G., Cabestany, J., Graña, M. (eds.) IWANN 2007. LNCS, vol. 4507, pp. 227–234. Springer, Heidelberg (2007)
7. Corchado, E., Baruque, B.: WeVoS-ViSOM: an ensemble summarization algorithm for enhanced data visualization. Neurocomputing **75**, 171–184 (2012)
8. Borrajo, M.L., Baruque, B., Corchado, E., Bajo, J., Corchado, J.M.: Hybrid neural intelligent system to predict business failure in small-to-medium-size enterprises. Int. J. Neural Syst. **21**(04), 277–296 (2011)
9. Jiang, Y., Zhi-Hua, Z.: SOM ensemble-based image segmentation. Neural Process. Lett. **20**(3), 171–178 (2004)
10. Low, K.H., Wee, K.L., Marcelo, H.A.: An ensemble of cooperative extended Kohonen maps for complex robot motion tasks. Neural Comput. **17**, 1411–1445 (2005)
11. DeLooze, L.L.: Attack characterization and intrusion detection using an ensemble of self-organizing maps. In: 2006 IEEE Information Assurance Workshop, pp. 108–115 (2006)
12. Fustes, D., Dafonte, C., Arcay, B., Manteiga, M., Smith, K., Vallenari, A., Luri, X.: SOM ensemble for unsupervised outlier analysis. Application to outlier identification in the Gaia astronomical survey. Expert Syst. Appl. **40**(5), 1530–1541 (2013)
13. Tsai, C.-F.: Combining cluster analysis with classifier ensembles to predict financial distress. Inf. Fusion **16**, 46–58 (2014)
14. Zhou, Z.-H., Wu, J., Tang, W.: Ensembling neural networks: many could be better than all. Artif. Intell. **137**(1–2), 239–263 (2002)
15. Bache, K., Lichman, M.: Machine Learning Repository. University of California, School of Information and Computer Sciences, Irvine (2013)
16. Ultsch, A.: Clustering with SOM: U*C. In: Proceedings Workshop on Self-Organizing Maps (WSOM 2005), pp. 75–82 (2005)
17. Demšar, J.: Statistical comparisons of classifiers over multiple data sets. J. Mach. Learn. Res. **7**, 1–30 (2006)

A Particle Swarm Clustering Algorithm with Fuzzy Weighted Step Sizes

Alexandre Szabo[1,3(✉)], Myriam Regattieri Delgado[2],
and Leandro Nunes de Castro[3]

[1] Federal University of ABC, Santo André, Brazil
alexandreszabo@gmail.com
[2] Federal University of Technology of Paraná, Curitiba, Brazil
myriamdelg@utfpr.edu.br
[3] Natural Computing Laboratory, Mackenzie University, São Paulo, Brazil
lnunes@mackenzie.br

Abstract. This paper proposes a modification in the Fuzzy Particle Swarm Clustering (FPSC) algorithm such that membership degrees are used to weight the step size in the direction of the local and global best particles, and in its movement in the direction of the input data at every iteration. This results in the so-called Membership Weighted Fuzzy Particle Swarm Clustering (MWFPSC). The modified algorithm was applied to six benchmark datasets from the literature and its results compared to that of the standard FPSC and FCM algorithms. By introducing these modifications it could be observed a gain in accuracy, representativeness of the clusters found and the final Xie-Beni index, at the expense of a slight increase in the practical computational time of the algorithm.

Keywords: Fuzzy clustering · Particle swarm · Data mining · FPSC

1 Introduction

Clustering is the organization of objects in groups (clusters) such that similar objects are assigned to the same cluster [1]. One of the major challenges in clustering is to find the natural clusters in the data, that is, those that satisfy two particular conditions [2]: (1) presence of continuous and relatively densely populated regions of the space; and (2) presence of empty regions around these more dense areas.

As real datasets are often characterized by overlapping clusters [3], which intrinsically break the natural cluster conditions, these are not easily identified by traditional clustering methods. In such cases, fuzzy clustering approaches are more suitable, for they allow each object to belong to more than one cluster simultaneously with varying membership degrees, bringing a more realistic perspective of the clusters distribution [4]: the closer an object to a cluster, the higher its membership to that cluster.

During the learning phase of most clustering algorithms, those objects belonging to the cluster (and sometimes a small neighborhood around them) are the ones used for updating the model. In fuzzy clustering, as all objects belong to all groups, but with varying degrees, all objects contribute to the model update, promoting a more robust

© Springer International Publishing Switzerland 2015
K. Jackowski et al. (Eds.): IDEAL 2015, LNCS 9375, pp. 87–95, 2015.
DOI: 10.1007/978-3-319-24834-9_11

learning scheme. The most well-known fuzzy clustering algorithm is the *Fuzzy c-Means* (FCM) [5], which is a fuzzy version of *k*-Means [6].

According to [7, 8], clustering is a type of combinatorial optimization problem in which each object can be allocated to one of the clusters. Thus, any heuristic method, such as evolutionary, immune and swarm algorithms [9, 10] can be used to guide the clustering process without exhaustively exploring the search space.

Among the many bio-inspired techniques applied to clustering tasks, those based on the *Particle Swarm Optimization* (PSO) algorithm [11] have been given a great deal of attention over the past decade [12–15]. PSO uses the analogy of the interaction of social agents among themselves and with the environment so as to guide the search for a satisfactory solution. In the original PSO, individuals are represented by real-valued vectors, called *particles*, which represent candidate solutions to the problem. Particles are positioned in a vector space, usually an Euclidean space within the domain [0,1], with random initial positions and velocities. The basic PSO moves a set of particles over the space, such that a memory of the best of each particle and of a set of neighbors are used to guide the next movement of the particles.

There are several clustering algorithms based on PSO, including both traditional and fuzzy approaches. This paper proposes a modification in the *Fuzzy Particle Swarm Clustering* (FPSC) algorithm [15], such that membership degrees are directly added to the particle velocity equation. The resultant algorithm is named *Membership Weighted Fuzzy Particle Swarm Clustering* (MWFPSC). The modified algorithm was applied to six benchmark datasets from the literature and its results were compared to that of the original FPSC and FCM. The evaluation measures include the Xie-Beni index, accuracy, cluster representativeness and computational cost.

The paper is organized as follows. Section 2 presents the related works, Sect. 3 introduces the FPSC algorithm, Sect. 4 describes the proposed modification that led to MWFPSC. Section 5 presents the experimental results and the paper is concluded in Sect. 6 with a general discussion about the proposal and future research.

2   Related Works

Pang et al. [16] proposed the *Fuzzy Particle Swarm Optimization* (FPSO) algorithm to solve optimization problems. Over the past years FPSO has been adapted to solve data clustering problems. Some of these adaptations are briefly discussed below.

Runkler and Katz [17] proposed two methods for minimizing FCM objective-functions, called PSO-V and PSO-U. They differ by the form of representing each particle in the swarm: in PSO-V each particle corresponds to a prototype; whilst in PSO-U each particle corresponds to a membership degree. Experiments showed that PSO-V produces better results than PSO-U when the number of objects is greater than the number of attributes, whilst PSO-U is better when the number of attributes is greater than the number of objects.

In [8] a hybrid algorithm was proposed in which each particle in the swarm represents a set of prototypes. The algorithm, named HFPSO, combines the FCM and FPSO algorithms, where FCM is first run until a stopping criterion is met, and then the resulting prototypes will compose a part of the swarm in FPSO. Results concluded that the hybrid algorithm outperformed its original counterparts (FCM and FPSO).

A similar proposal was presented in [18], where a hybrid algorithm was proposed, named Hybrid FCM-PSO. In FCM-PSO the FPSO algorithm is run until convergence, as a first stage, and then the solution is refined by the FCM algorithm as a second and final stage. These stages are repeated until a given stopping criterion is met.

In the present paper we propose a modification in FPSC [15] such that membership degrees are directly used to weight the step size of particles at every iteration. Generally, none information about membership degrees is directly applied in equations of fuzzy PSO for clustering problems, which suggest an hybridization with some fuzzy algorithm. Thus, the MWFPSC makes use of fuzzy theory to weight the representativeness of prototypes to respective clusters found, weighting the importance of each term from velocity equation.

3 FPSC: The Fuzzy Particle Swarm Clustering Algorithm

The PSC [14] and FPSC [15] algorithms move a set of particles over the space such that they become prototypes of the clusters that exist in the data. For every input object, the most similar particle is selected to be evaluated and its local and global memories are updated. Then, its velocity and position are updated following Eqs. (1) and (2), respectively. This process is repeated until a stopping criterion is met:

$$
\mathbf{V}_j(t+1) = \omega(t)\mathbf{V}_j(t) + \boldsymbol{\varphi}_1 \otimes (\mathbf{P}_j^i(t) - \mathbf{C}_j(t)) + \boldsymbol{\varphi}_2 \otimes (\mathbf{G}_i(t) - \mathbf{C}_j(t)) + \boldsymbol{\varphi}_3 \otimes (\mathbf{Y}^i(t) - \mathbf{C}_j(t))
$$

$$(1)$$

Equation (1) is responsible for updating the velocity (step size) of the winner particle, that is, the one closer to the input pattern. Parameter ω is called *inertia moment* and contributes to the convergence of the algorithm. The cognitive term, $(\mathbf{P}_j^i(t) - \mathbf{C}_j(t))$, is associated with the particles experience, where $\mathbf{P}_j^i(t)$ represents the best particle position achieved so far in relation to the input object $(\mathbf{Y}^i(t))$ and $\mathbf{C}_j(t)$ is the position of the particle in the space at time t. The term $(\mathbf{G}_i(t) - \mathbf{C}_j(t))$ is associated with the social component, where $(\mathbf{G}_i(t))$ is the nearest particle to the object $\mathbf{Y}^i(t)$ at time t. The self-organizing term $(\mathbf{Y}^i(t) - \mathbf{C}_j(t))$ moves the particle towards the input object. The values φ_1, φ_2 and φ_3 are random vectors that influence the cognitive, social and self-organizing terms, respectively. The symbol $\otimes$ represents an element-wise vector multiplication. Equation (2) updates the position of the *winner* particle:

$$
\mathbf{C}_j(t+1) = \mathbf{C}_j(t) + \mathbf{V}_j(t+1) \tag{2}
$$

After each iteration the algorithm checks for particle stagnations and moves stagnated particles towards the one that most won at the current iteration ($\mathbf{C}_{\text{most\_win}}$).

To adapt the original PSC so as to accommodate fuzzy clusters, leading to the FPSC algorithm, the following structural modifications were made:

- **Selection of particles:** In PSC the *winner* particle ($\mathbf{C}_j$) for an object ($\mathbf{Y}^i$) is obtained by calculating the smallest Euclidean distance between the input object and the particles in the swarm. In FPSC the *winner* particle corresponds to the centroid to which the input pattern i has the maximal membership degree.
- **Evaluation of particles:** In PSC the position of the *winner* particle ($\mathbf{C}_j$) is compared with that of its memories ($\mathbf{P}_j^i$ and $\mathbf{G}^i$) in relation to an input object ($\mathbf{Y}^i$) using an Euclidean distance. In FPSC the comparison between the *winner* and its memories considers the membership degree of the object to such particles.

The membership degrees in FPSC are subject to some constraints [5]:

$$\mu_{ij} \in [0, 1] \tag{3}$$

$$\sum_{j=1}^{k} \mu_{ij} = 1, \quad i = 1, \ldots, n \tag{4}$$

where μ_{ij} is the membership degree of object i to group j. Equation (4) guarantees that the sum of the membership degrees of an object to all groups is one. Equation (5) guarantees that each group has at least one object with membership degree greater than zero:

$$0 < \sum_{i=1}^{n} \mu_{ij} < n \tag{5}$$

The membership matrix is initialized and iteratively updated by:

$$\mu_{ij} = \frac{1}{\sum_{p=1}^{k} \left(\frac{d_{ij}}{d_{ip}}\right)^{\frac{2}{m-1}}} \tag{6}$$

where d is the distance between object i and prototypes, and m is the weighting exponent $(1, \infty)$.

4   MWFPSC: The Membership Weighted Fuzzy Particle Swarm Clustering Algorithm

The *Membership Weighted Function Fuzzy Particle Swarm Clustering* algorithm (MWFPSC) is a modification of FPSC in which membership degrees are directly introduced into the particle velocity equation. The idea behind MWFPSC is to weight the representativeness of particles to the clusters centroid by weighting the step size of particles. Thus, particles move toward its memories and input object with an uncertainty degree that takes into account their membership degrees, as follows:

$$\begin{aligned} \mathbf{V}_j(t+1) = {}& \omega(t) * \mathbf{V}_j(t) + \mu_1 * \boldsymbol{\varphi}_1 \otimes \left(\mathbf{P}_j^i(t) - \mathbf{C}_j(t)\right) \\ & + \mu_2 * \boldsymbol{\varphi}_2 \otimes \left(\mathbf{G}_i(t) - \mathbf{C}_j(t)\right) + \mu_3 * \boldsymbol{\varphi}_3 \otimes \left(\mathbf{Y}^i(t) - \mathbf{C}_j(t)\right) \end{aligned} \tag{7}$$

where μ_1 is the membership degree of object $\mathbf{Y}^i(t)$ to the cluster represented by particle $\mathbf{P}_j^i(t)$; μ_2 is the membership degree of object $\mathbf{Y}^i(t)$ to the cluster represented by particle $\mathbf{G}_i(t)$ and μ_3 is the membership degree of object $\mathbf{Y}^i(t)$ to the cluster represented by particle $\mathbf{C}_j(t)$. In both FPSC and MWFPSC particles interact with one another by comparing their current states with their memories, and interact with the dataset by comparing them directly with the input data. By adding the membership degrees into the updating equation it is expected that a more refined step is given by the particles at each iteration, leading to a more precise location of the clusters centers.

The difference between MWFPSC and its predecessor in principle does not change its time computational complexity. In FPSC (and PSC), the main loop is that corresponding to the dissimilarity evaluation between the input object and all particles, which has a cost of k operations (k is the number of particles) to the n input objects. Thus, the complexity of these algorithms can be reduced to $O(nk)$.

5 Performance Assessment

To assess the performance of the proposed algorithm it was applied to six datasets from the UCI Machine Learning Repository[1]: Wine; Ionosphere; Soybean; S. Heart; Glass Id.; and Iris. Its performance was compared to that of the original FPSC and FCM, considering different parametric settings to evaluate the best solution. Each experiment was run 30 times for each algorithm with various parametric values:

1. Number of prototypes equals to the number of classes (labelled clusters) in each dataset.
2. Weighting exponent $m = \{1.50, 2.00, 2.50, 3.00, 3.50, 4.00\}$.
3. Inertia moment $\omega = \{0.80, 0.90, 1.00\}$, with a decay rate of 0.95 at each iteration.
4. Maximal velocity $v_{max} = \{0.01, 0.05, 0.10, 0.20\}$.

The number of solutions is determined by combining the distinct values of m, ω and v_{max}, totaling 72 combinations. Each value of m was applied 12 times over 72 solutions (3 different values of ω and 4 different values of v_{max}). Thus, for 30 runs, each value of m was applied 360 times.

The stopping criterion is either a minimal variation in the cost function ($|xb(t) - xb(t+1)| < 10^{-3}$), a consecutive increase of the objective-function in 10 iterations from iteration 500, or a maximum of 1,000 iterations.

The following measures were applied to assess the quality of the solutions found: the representativeness of prototypes to the centroids (*dist*) obtained by averaging the Euclidean distance between particles and the respective centroids; number of iterations for convergence (*iter*); processing time (*t(s)*), in seconds; the Xie-Beni index [19]; and the accuracy (*pcc %*). The Xie-Beni index evaluates how compact and separated are the clusters. Note that the clusters labels are used to determine the accuracy of the final partitions, but are not used for learning.

---

[1] http://archive.ics.uci.edu/ml/datasets.html.

The results obtained by the clustering algorithms are summarized in Table 1. We performed two different statistic tests over results for evaluating the significantly in among results of experiments for every dataset. The Kolmogorov-Smirnov normality test [20] was applied, with 95 % of confidence, to test data normality and showed that the null hypothesis should be rejected for all measures, that is, the results do not come from a normal distribution. Therefore, a nonparametric method was applied to evaluate whether the difference among results from all runs is significant. For that, the Friedman test [21] was used with 95 % confidence, showing a significant difference among the results for all measures. When looking at the results presented in Table 1, it is interesting to observe that whilst FCM shows a slightly superior performance in terms of accuracy, cluster representativeness, and computational time, it generates clusters that are much less compact and separated than the PSO-based methods, as measured by the Xie-Beni index. Also, the datasets with high xb values (Soybean, S. Heart and Glass Id.) are consequence of both the sensitivity of the algorithm to the prototype initialization and the high values for m. It is known that high values of m tend to produce homogeneous membership matrices, that is, with elements close to $1/k$, which culminates in overlapping clusters and, consequently, high values of xb. FPSC produces more compact and separable clusters (lower xb values) than the MWFPSC and FCM algorithms, on average. The sensitivity to the prototypes initialization and a search local are the main drawbacks of FCM, which are clarified by results of xb showed in Table 1, and can affect the correct classification.

Table 1. Mean/best value produced by FPSC, MWFPSC and FCM for different evaluation measures.

Id	Algorithm	xb	pcc (%)	$dist$	Iter	$t(s)$
Wine	FPSC	0,41/0,19	87,18/97,75	0,30/0,08	188,63/2,00	13,85/0,15
	MWFPSC	0,44/0,19	88,66/97,19	0,26/0,09	191,64/2,00	14,28/0,15
	FCM	8,27/0,36	89,92/95,51	0,20/0,05	110,40/4,00	3,84/0,14
Ionos.	FPSC	0,55/0,22	68,53/73,22	0,78/0,40	224,78/2,00	22,73/0,20
	MWFPSC	0,56/0,24	68,57/73,22	0,76/0,40	229,92/2,00	23,98/0,21
	FCM	1,19/0,62	70,56/70,94	0,43/0,33	25,63/8,00	1,12/0,34
Soyb.	FPSC	0,42/0,19	87,26/100,00	0,85/0,28	88,20/2,00	2,55/0,06
	MWFPSC	0,48/0,19	87,72/100,00	0,80/0,27	105,12/2,00	2,44/0,05
	FCM	$1,82*10^{30}$/0,47	76,65/100,00	0,99/0,13	234,06/7,00	3,50/0,11
Heart	FPSC	0,44/0,27	77,20/84,44	0,53/0,32	143,23/2,00	10,82/0,15
	MWFPSC	0,46/0,27	77,37/84,44	0,51/0,25	156,74/2,00	12,16/0,16
	FCM	$1,68*10^{27}$/0,83	75,04/80,00	0,34/0,12	521,17/3,00	13,10/0,07
Glass	FPSC	0,37/0,09	51,39/65,42	0,61/0,22	222,83/2,00	54,51/0,48
	MWFPSC	0,44/0,09	52,11/64,95	0,53/0,23	268,05/2,00	66,66/0,49
	FCM	$1,27*10^{27}$/0,33	53,67/57,01	0,24/0,21	328,35/9,00	39,67/1,09
Iris	FPSC	0,25/0,08	86,48/96,67	0,15/0,02	196,79/2,00	11,75/0,12
	MWFPSC	0,25/0,08	86,52/96,67	0,14/0,02	185,63/2,00	12,52/0,13
	FCM	0,29/0,17	89,37/92,67	0,03/0,02	12,27/3,00	0,38/0,09

When comparing FPSC and MWFPSC, the Kolmogorov-Smirnov test showed that the null hypothesis should be rejected for all measures. Thus, the Friedman test was applied and showed that the difference among the results is significant for all measures for the Wine, Soybean, S. Heart and Glass Id. For the Ionosphere data, the difference among measures is not significant for the number of iterations (*iter*) and the processing time (*t(s)*). For the Iris data, the difference is not significant for *xb*. MWFPSC presented the best classification accuracy *pcc* (%) and representativeness of prototypes (*dist*), on average, for most datasets when compared with FPSC. On the other hand, FPSC was better in terms of the iterations (*iter*) and processing time (*t (s)*).

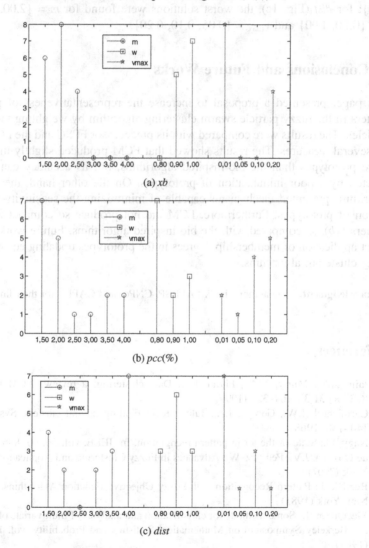

Fig. 1. Frequency of parametric values for the best *xb* (a), *pcc* (%) (b) and *dist* (c) values, among all 30 runs for 3 algorithms and 6 datasets. Weighting exponent (*m*), inertia moment (ω) and maximal velocity (v_{max}).

Figure 1 presents a brief sensitivity analysis of the proposed algorithm in relation to m, ω and v_{max}. It shows for the Xie-Beni index, the accuracy and the representativeness of clusters varied in the experiments. There are 3 algorithms (FPSC, MWFPSC and FCM) and 6 datasets, so 18 occurrences for m, and 12 for ω and v_{max} (6 datasets and 2 algorithms). As can be observed in Fig. 1, the best solutions were found for $m = \{1.50; 2.00\}$, $\omega = \{0.90, 1.00\}$ and $v_{max} = 0.01$, which is consistent with the range for m suggested in [22]. For the values of xb (Fig. 1a), the worst solutions were found for $m = \{3.00, 3.50, 4.00\}$, $\omega = 0.80$ and $V = \{0.05, 0.10\}$; for pcc (%) (Fig. 1b), the worst values were found for $m = \{2.50, 3.00, 3.50, 4.00\}$, $\omega = \{0.90, 1.00\}$ and $v_{max} = \{0.01, 0.05\}$; for $dist$ (Fig. 1c), the worst solutions were found for $m = \{2.00, 2.50, 3.00\}$, $\omega = \{0.80, 1.00\}$ and $v_{max} = \{0.05, 0.10, 0.20\}$.

6 Conclusions and Future Works

This paper presented a proposal to increase the representativeness of prototypes to clusters in the fuzzy particle swarm clustering algorithm by weighting the step size of particles. The results were compared with its predecessor FPSC and the FCM algorithm for several measures. The results showed that FCM produced slightly more representative prototypes than the bio-inspired algorithms, but its accuracy can be seriously affected by a poor initialization of prototypes. On the other hand, the swarm-based algorithms presented mechanisms capable of minimizing the sensitivity to the initialization of prototypes. Furthermore, FCM did not produce so compact and separable clusters (xb), as compared with the bio-inspired algorithms. Future works include the direct application of membership degrees in the prototypes updating equations of other fuzzy clustering algorithms.

Acknowledgment. The authors thank FAPESP, CNPq, and CAPES for their financial support.

References

1. Jain, A.K., Murty, M.N., Flynn, P.J.: Data clustering: a review. ACM Comput. Surv. (CSUR) **31**(3), 264–323 (1999)
2. Carmichael, J.W., George, J.A., Julius, R.S.: Finding natural clusters. Syst. Zool. **17**(2), 144–150 (1968)
3. Nagy, G.: State of the art in pattern recognition. In: IEEE, vol. 56, pp. 836–863 (1968)
4. de Oliveira, J.V., Pedrycz, W.: Advances in Fuzzy Clustering and Applications. Wiley, New York (2007)
5. Bezdek, J.: Pattern Recognition with Fuzzy Objective Function Algorithms. Plenum Press, New York (1981)
6. MacQueen, J.: Some methods for classification and analysis of multivariate observations. In: 5th Berkeley Symposium on Mathematical Statistics and Probability, vol. 1, pp. 281–297 (1967)
7. Cui, X., Potok, T., Palathingal, P.: Document clustering using particle swarm optimization. In: Swarm Intelligence Symposium, SIS 2005, pp. 185–191. IEEE (2005)

8. Mehdizadeh, E., Tavakkoli-Moghaddam, R.: A hybrid fuzzy clustering PSO algorithm for a clustering supplier problem. In: International Conference on Industrial Engineering and Engineering Management, pp. 1466–1470 (2007)
9. de Castro, L.N., Von Zuben, F.J.: aiNet: an artificial immune network for data analysis, chapter XII. In: Abbass, H.A., Saker, R.A., Newton, C.S. (eds.) Data Mining: A Heuristic Approach, pp. 231–259. Idea Group Publishing, Hershey (2001)
10. da Cruz, D.P.F., Maia, R.D., Szabo, A., de Castro, L.N.: A bee-inspired algorithm for optimal data clustering. In: IEEE Congress on Evolutionary Computation (CEC), pp. 3140–3147 (2013)
11. Kennedy, J., Eberhart, R.C.: Particle swarm optimization. In: IEEE International Conference on Neural Networks, pp. 1942–1948, Perth, Australia (1995)
12. van der Merwe, D.W., Engelbrecht, A.P.: Data clustering using particle swarm optimization. In: IEEE Congress on Evolutionary Computation, pp. 215–220 (2003)
13. Omran, M.G.H., Salman, A., Engelbrecht, A.P.: Dynamic clustering using particle swarm optimization with application in image segmentation. Pat. Anal. Appl. **8**, 332–344 (2006)
14. Cohen, S.C.M., de Castro, L.N.: Data clustering with particle swarms. In: IEEE World Congress on Computational Intelligence, pp. 6256–6262 (2006)
15. Szabo, A., de Castro, L.N., Delgado, M.R.: The proposal of a fuzzy clustering algorithm based on particle swarm. In: IEEE 3rd NABIC, pp. 459–465 (2011)
16. Pang, W., Wang, K-P, Zhou, C-G, Dong, L-J: Fuzzy discrete particle swarm optimization for solving traveling salesman problem. In: 4th International Conference on Computer and Information Technology, pp. 796–800 (2004)
17. Runkler, T.A., Katz, C.: Fuzzy Clustering by Particle Swarm Optimization. In: IEEE International Conference on Fuzzy Systems, pp. 601–608 (2006)
18. Izakian, H., Abraham, A.: Fuzzy C-means and fuzzy swarm for fuzzy clustering problem. Expert Syst. Appl. **38**, 1835–1838 (2011)
19. Xie, X.L., Beni, G.: A validity measure for fuzzy clustering. IEEE Trans. Pattern Anal. Mach. Intell. **8**, 841–847 (1991)
20. Massey, F.J.: The Kolmogorov-Smirnov test for goodness of fit. J. Am. Stat. Assoc. **46**(253), 68–78 (1951)
21. Derrac, J., Garcia, S., Molina, D., Herrera, F.: A practical tutorial on the use of nonparametric statistical tests as a methodology for comparing evolutionary and swarm intelligence algorithms. Swarm Evol. Comput. **1**, 3–18 (2011)
22. Cox, E.: Fuzzy Modeling and Genetic Algorithms for Data Mining and Exploration. Morgan Kaufmann, San Francisco (2005)

A Bacterial Colony Algorithm for Association Rule Mining

Danilo Souza da Cunha[✉], Rafael Silveira Xavier, and Leandro Nunes de Castro

Natural Computing Laboratory - LCoN, Mackenzie Presbyterian University, São Paulo, Brazil
{danilocunha85,rsixavier}@gmail.com, lnunes@mackenzie.br

Abstract. Bacterial colonies perform a cooperative distributed exploration of the environment. This paper describes bacterial colony networks and their skills to explore resources as a tool for mining association rules in databases. The proposed algorithm is designed to maintain diverse solutions to the problem at hand, and its performance is compared to other well-known bio-inspired algorithms, including a genetic and an immune algorithm (CLONALG) and, also, to Apriori over some benchmarks from the literature.

Keywords: Bio-inspired algorithm · Bacterial colony · Association rules · Data mining

1 Introduction

Bacterial colonies can be seen as complex adaptive systems that perform distributed information processing to solve complex problems. They use a collaborative system of chemical signals to exploit the resources of a given environment [1].

The collective and collaborative activities carried out by a bacterial colony can be understood as a type of social intelligence [2], where each bacterium is able to sense and communicate with other bacteria in the colony to perform its task in a coordinated manner. This enables the colony to learn about the environment. Thus, a colony can be seen as an adaptive computational system that processes information on different levels of abstraction [3]. Parallel computational characteristics and social behavior of bacteria colonies are presented in [3].

This paper introduces an algorithm inspired by the exploratory behavior of environmental resources by a colony of bacteria for mining association rules of items in transactional databases. Furthermore, the proposed bacteria algorithm is able to avoid the genic conversion problem presented in [4].

The bacterial colony algorithm is compared to other two bio-inspired heuristics, an elitist genetic algorithm [4] and a clonal selection algorithm [4]. Also, the proposed algorithm is compared to the well-known Apriori algorithm. The following performance measures are accounted for: support, confidence, interestingness, number of rules and processing time.

The paper is organized as follows. Section 2 has a theoretical background about association rule mining problem and bacteria colony communication networks, and Sect. 3 proposes an algorithm for association rule mining based on colonies of

© Springer International Publishing Switzerland 2015
K. Jackowski et al. (Eds.): IDEAL 2015, LNCS 9375, pp. 96–103, 2015.
DOI: 10.1007/978-3-319-24834-9_12

bacteria. The experimental results are shown in Sect. 4 and final considerations are presented in Sect. 5.

2   Theoretical Background

This section provides a brief review of the two main topics covered by this paper: association rule mining; and bacteria colonies.

2.1   Association Rule Mining

Association rule mining, originally known as market-basket analysis, is one of the main data mining tasks. It is a descriptive task that uses unsupervised learning and focuses on the identification of associations of items that occur together, known as a transaction, in a given data set [5–8]. In the scenario described in the original market-basket analysis, items in a transaction are those that are purchased together by a consumer [8].

The first set A is called the body (or antecedent part) and the other one C is called the head (or consequent part) of the association rule. The intersection between these two sets is empty ($A \cap C = \emptyset$), because it would be redundant for an item to imply itself. The rule means that the presence of (all items in) A in a transaction implies the presence of (all items in) C in the same transaction with some associated probability [6, 8]. The problem with the previous definition is that the number N of possible association rules, given a number d of items, grows exponentially and the problem is placed within the NP-complete set [5, 6, 8]:

$$N = 3^d - 2^{d+1} + 1. \tag{1}$$

Therefore, it is compulsory to somehow filter or prune the association rules built before trying to analyze their usefulness.

1. Measures of Interest

The *Confidence* and *Support*, proposed in [5, 6], are the most studied measures of interest in the association rule mining literature. The support of an association rule is a measure of its relative frequency in the set of transactions:

$$Support(A \rightarrow C) = Supp(A \rightarrow C) = P(A \cup C). \tag{2}$$

On the other hand, the confidence of a rule is a measure of its satisfiability when its antecedent part is found in T; that is, from all the occurrences of A, how often C also occurs in the rule:

$$Confidence(A \rightarrow C) = Conf(A \rightarrow C) = P(A|C). \tag{3}$$

Whilst confidence is a measure of the strength of a rule, the support corresponds to its statistical significance over the database. The interestingness of a rule, $I(A \rightarrow C)$, is calculated as follows [7]:

$$I(A \rightarrow C) = (|A \cup C|/|A|) * (|A \cup C|/|C|) * (1 - (|A \cup C|/|T|)), \tag{4}$$

where A and C are defined as previously, and T is the number of transactions in the database. This measure of interest looks for low frequency rules inside the database. It is interesting to note that the position of an item in a rule, either in the antecedent or consequent part of it, influences the confidence of the rule, but does not influence its support and interestingness.

2.2   Bacterial Colonies

Typical behaviors of *Escherichia Coli* bacteria during their whole lifecycle, including chemotaxis, communication, elimination, reproduction, and migration are used by the Bacterial Colony Optimization (BCO) algorithm to solve problems in different areas [9]. Another well-known algorithm inspired by *E. Coli* bacteria is the Bacterial Foraging Optimization Algorithm (BFOA) that simulates the social behavior of the *E. Coli* colony [9]. Despite these initiatives, there are other phenomena in a bacterial colony to be observed as well.

Bacterial communication occurs via chemical signals. The main entities around this communication are the *signaling cell*, the *target cell*, the *signal molecule* and the *receiver protein*. The signaling cell sends the chemical signal, presented by the signal molecule, to one or more target cells. The target cells read the message contained in the signaling molecule via protein receptors and then send the message to the intracellular gel. Note that the signaling molecule does not enter the bacteria, the responsible one for decoding and sending each message to the intercellular plasma is the receiver protein [10]. This communication of a bacterium with others via chemical signals corresponds to the *extracellular communication* of the proposed algorithm.

The most studied bacterial communication process in the literature is the *quorum sensing*, which depends on the concentration of a diffusible molecule called *autoinducer* [11, 12]. The concentration of autoinducers increases in the environment with the growth of the number of cells that produce them, thus promoting the activation or suppression of gene expression that are responsible for generating certain behaviors in bacteria. Collectively, the bacteria might extract latent information from the environment and other organisms in order to interpret the contextual information, develop a common knowledge and learn from past experiences [11]. The *latent information* refers to the information contained in the environment around the colony, which, once processed, trigger changes in the function or behavior of the organism [12].

Quorum sensing is the mechanism used by the proposed algorithm to perform exploitation of the search space. It works as an intracellular communication network that analyzes and interprets the data read from the environment. Each bacterium has a probability to activate or suppress its genes according with its position in the environment. In others words, this bacterial phenomenon looks for better solutions by exchanging parts of the candidate solution. In association rule mining, an antecedent part is stochastically exchanged with the consequent part. The bacterial colony networks are represented in Fig. 1.

The following section introduces the algorithm proposed by using the two bacterial phenomena summarized previously: the communication via chemical signals, which represents the *extracellular communication*; and quorum sensing, which represents the *intracellular communication*.

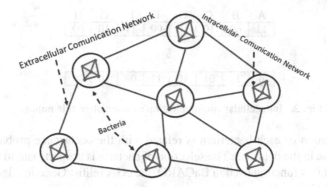

Fig. 1. Intracellular and extracellular communication networks [3].

3 Bacterial Colony in Association Rule Mining

The proposed algorithm is inspired by the biological process of extra- and intracellular communication networks of bacterial colonies. It is called Bacterial Colony Association Rule Optimization (BaCARO) and its steps and variables are detailed in the following pseudocode.

```
procedure [P] = BaCARO(ec,ic,DATA)
    initialize P
    t ← 1
    while not_stopping_criterion do,
        f ← evaluate(P,DATA)
        rf ← intCellular(P,ic,DATA)
        f ← update(f,rf)
        f ← extCellular(ec)
        P ← select(P,f)
        t ← t+1
    end while
end procedure
```

Each item in a transaction is represented by a pair of bits, where items present in the association rule are represented by a bit pair of 00 or 11. Pair 00 represents the antecedent part and pair 11 represents the consequent part. Elements out of a rule are composed of the other combinations: 01 or 10.

The probability of extracellular communication is accounted for in parameter ec and the probability of intracellular communication is represented by parameter ic. In the experiments to be reported here $ec = 0.6$ and $ic = 0.9$. DATA is the database to be mined, and can be seen as the environment to be explored by the bacterial colony. Figure 2 illustrates two bacteria, one in the top of the picture representing the rule B ∧ E ∧ H → A ∧ F and the other in the bottom representing rule A ∧ F ∧ H → B ∧ E. The bacteria perform a local search in the solutions space with its intracellular communication networks, as illustrated in Fig. 2.

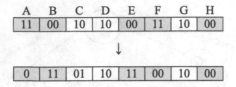

Fig. 2. Intracellular communication in association rule mining.

The evaluation of each bactetium is related with the occurrence probability of an association rule in the database. The selection of bacteria is proportional to their fitness values. The fitness function used in BaCARO, the eGA (elitist Genetic Algorithm) and also CLONALG is:

$$Fitness(A \to C) = w_1 * Supp(A \to C) + w_2 * Conf(A \to C), \tag{5}$$

where, $w_1 = w_2 = 0.5$ and $w_1 + w_2 = 1$. The algorithms use support and confidence to calculate the fitness value, and the interestingness measure is used to compare them from a different perspective. Finally, the other performance measures are the number of rules and average running time. The number of rules is calculated based on all valid and unique rules in the final bacteria repertoire.

4 Experimental Results

Three different binary datasets were chosen for this investigation. The first one was the *SPECT Heart* database from the machine learning repository of the University of California at Irvine. The sparsity of this dataset is around 70%, which means that the majority of the elements of the matrix have no associated values. The second database used was *Supermarket* from Weka, we used 10% of the first transactions and products from it. The sparsity of this dataset is around 90%. The *E-Commerce Database* used has 430 products and 372 shopping transactions in which a total of 1.612 products were sold or accessed, and with a measure of sparsity equals to 98.99%.

All the values taken over ten simulations of eGA, CLONALG and BaCARO are summarized in Tables 1, 2 and 3, and still compared with the Apriori, the single deterministic algorithm used for comparison. The eGA is a genetic algorithm with elitism rate equals to 0.5, crossover probability equals to 0.6 and mutation rate equals to 0.01 for both databases. The results from the competitors are based on [13–16]. The values presented are the average ± standard deviation for the set of rules found in the final population of each algorithm over ten simulations. However, the Apriori was performed only once due to its determinism. In Tables 1, 2 and 3, S means support, C confidence, I interestingness, U unique rules found over the last set of candidate solutions, and P is the processing time.

Another bioinspired algorithm used in this paper was the CLONALG [17], developed with inspiration in the Clonal Selection Theory, and part of the Artificial Immune Systems field of research.

Table 1. Average results for the SPECT data.

	Apriori	eGA	CLONALG	BaCARO
S	0.35 ± 0.04	0.37 ± 0.03	**0.46 ± 0.02**	0.45 ± 0.14
C	0.65 ± 0.16	0.86 ± 0.05	0.94 ± 0.01	**0.99 ± 0.00**
I	0.35 ± 0.08	0.35 ± 0.08	0.30 ± 0.00	**0.41 ± 0.03**
U	17 ± 0.00	1.60 ± 0.60	1.50 ± 1.50	**53.40 ± 5.69**
P	6.5 s ± 0.00	4.5 s ± 1.01	9.3 s ± 1.13	**3.9 s ± 0.05**

Table 2. Average results for the Supermarket data.

	Apriori	eGA	CLONALG	BaCARO
S	0.01 ± 0.34	**0.57 ± 0.02**	0.47 ± 0.04	0.52 ± 0.02
C	0.68 ± 0.30	**0.88 ± 0.02**	0.79 ± 0.02	0.80 ± 0.02
I	0.01 ± 0.03	**0.28 ± 0.01**	0.26 ± 0.01	**0.28 ± 0.01**
U	**3.545 ± 0.00**	6.40 ± 1.89	3.70 ± 1.49	2.00 ± 0.00
P	12.8 s ± 0.0	4.6 s ± 0.05	8.61 s ± 0.73	**3.53 s ± 0.06**

Table 3. Average results for the E-commerce data.

	Apriori	eGA	CLONALG	BaCARO
S	0.004 ± 0.01	**0.017 ± 0.01**	0.005 ± 0.01	0.004 ± 0.01
C	0.631 ± 0.11	0.605 ± 0.18	**1.000 ± 0.00**	0.623 ± 0.19
I	0.241 ± 0.12	0.370 ± 0.15	**0.688 ± 0.32**	0.321 ± 0.13
U	**1.776 ± 0.00**	1.600 ± 1.08	4.60 ± 7.67	2.00 ± 0.00
P	653.1 s ± 0.0	**80.1 s ± 1.61**	110.4 s ± 3.54	82.1 s ± 1.21

The BaCARO presented competitive results for all datasets. The best performance of the bacterial algorithm was for the *SPECT* database. The average values of support, confidence and interestingness are too similar. However, the number of rules generated by BaCARO is larger than that of eGA. It is also important to stress that BaCARO presented a low processing time, smaller than most of its competitors for almost all datasets. Nevertheless, BaCARO performed worse than the other algorithms for two databases, *Supermarket* and *E-Commerce*. The eGA presented a better Support, Confidence and Interestingness for the *Supermarket* data, whilst CLONALG presented a better Confidence and Interestingness for the *E-Commerce* data.

Figure 3 shows one random execution of BaCARO and CLONALG to illustrate the behavior of the best and average solutions of each algorithm. Note that both were able to achieve the maximum fitness value for this dataset, also found by the Apriori algorithm. Interestingly, the average BaCARO fitness is always lower than that of its competitors, suggesting a greater diversity.

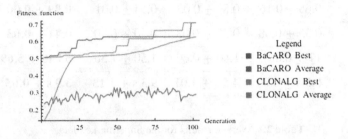

Fig. 3. BaCARO and CLONALG's fitness for the SPECT database.

5   Final Considerations

There are many phenomena happening in a bacterial colony. We proposed a new bacteria-inspired algorithm by taking a look at two features observed in a colony: intracellular communication networks; and extracellular communication networks.

The preliminary experiments performed here suggest a competitive performance, but require more tests with different databases to assess the proposed technique in association rule mining. The BaCARO performance over a consolidated database from the literature, *SPECT Heart* database suggest that bacterial colony deserve more attention as a useful data mining metaphor. The BaCARO should be applied to other optimization tasks to prove its robustness as a search-heuristic. Future works will involve understanding other behaviors present in bacteria colonies to build novel bio-inspired algorithms for solving real world data mining and optimization problems.

Acknowledgement.   The authors thank CAPES, Fapesp, CNPq and MackPesquisa for the financial support.

References

1. Matsushita, M., Fujikawa, H.: Diffusion-limited growth in bacterial colony formation. Physica A **168**, 498–506 (1990)
2. Ben-Jacob, E.: "Learning from bacteria about natural information processing", natural genetic engineering and natural genome editing. Ann. N.Y. Acad. Sci. **1178**, 78–90 (2009)
3. Xavier, R.S., Omar, N., de Castro, L.N.: Bacterial colony: information processing and computation behavior. In: NaBIC, pp. 439–443 (2011)

4. da Cunha, D.S., de Castro, L.N.: Evolutionary and immune algorithms applied to association rule mining. In: Panigrahi, B.K., Das, S., Suganthan, P.N., Nanda, P.K. (eds.) SEMCCO 2012. LNCS, vol. 7677, pp. 628–635. Springer, Heidelberg (2012)
5. Agrawal, R., Imielinski, T., Swami, A.: Mining association rules between sets of items in large databases. In: Proceedings of the 1993 ACM SIGMOD International Conference on Management of Data Ser. SIGMOD 1993, pp. 207–216. ACM, New York, NY, USA (1993). http://doi.acm.org/10.1145/170035.170072
6. Agrawal, R., Srikant, R.: Fast algorithms for mining association rules. In: Proceedings of 20th International Conference on Very Large Data Bases VLDB, pp. 487–499 (1994)
7. Dehuri, S., Jagadev, A.K., Ghosh, A., Mall, R.: Multi-objective genetic algorithm for association rule mining using a homogeneous dedicated cluster of workstations. Am. J. Appl. Sci. 3(11), 2086–2095 (2006)
8. Cios, K.J., Pedrycz, W., Swiniarski, R.W.: Data Mining: A Knowledge Discovery Approach, vol. XV. Springer, Berlin (2007)
9. Ming, L.: A novel swarm intelligence optimization inspired by evolution process of a bacterial colony. In: 2012 10th World Congress on Intelligence Control and Automation (WCICA), pp. 450–453, July 2012
10. Das, S., Biswas, A., Dasgupta, S., Abraham, A.: Bacterial foraging optimization algorithm: theoretical foundations, analysis, and applications. In: Abraham, A., Hassanien, A.-E., Siarry, P., Engelbrecht, A. (eds.) Foundations of Computational Intelligence Volume 3. SCI, vol. 203, pp. 23–55. Springer, Heidelberg (2009)
11. Alberts, B., Johnson, A., Lewis, J., Raff, M., Roberts, K., Walter, P.: Molecular Biology of the Cell, 4th edn. Garland Science, New York (2002)
12. Ben-Jacob, E., Shapira, Y., Tauber, A.I.: Seeking the foundations of cognition in bacteria: from schrdinger's negative entropy to latent information. Physica A 359(1–4), 495–524 (2006)
13. del Jesus, M.J., Gamez, J.A., Gonzalez, P., Puerta, J.M.: On the discovery of association rules by means of evolutionary algorithms. Wiley Interdisc. Rev. Data Min. Knowl. Discov. 1(5), 397–415 (2011)
14. Dehuri, S., Jagadev, A.K., Ghosh, A., Mall, R.: Multiobjective genetic algorithm for association rule mining using a homogeneous dedicated cluster of workstations. Am. J. Appl. Sci. 3, 2086–2095 (2006)
15. Shenoy, P.D., Srinivasa, K.G., Venugopal, K.R., Patnaik, L.M.: Evolutionary approach for mining association rules on dynamic databases. In: Whang, K.-Y., Jeon, J., Shim, K., Srivastava, J. (eds.) PAKDD 2003. LNCS, vol. 2637, pp. 325–336. Springer, Heidelberg (2003)
16. Venugopal, K.R., Srinivasa, K.G., Patnaik, L.M.: Dynamic association rule mining using genetic algorithms. In: Venugopal, K.R., Srinivasa, K.G., Patnaik, L.M. (eds.) Soft Computing for Data Mining Applications. SCI, vol. 190, pp. 63–80. Springer, Heidelberg (2009)
17. de Castro, L.N., Von Zuben, F.V.: Learning and optimization using the clonal selection principle. IEEE Trans. Evol. Comput. 6(3), 239–251 (2002)

Information Retrieval and Data Forecasting via Probabilistic Nodes Combination

Dariusz Jacek Jakóbczak[(✉)]

Department of Electronics and Computer Science,
Technical University of Koszalin, Koszalin, Poland
dariusz.jakobczak@tu.koszalin.pl

Abstract. Proposed method, called Probabilistic Nodes Combination (PNC), is the method of 2D data interpolation and extrapolation. Nodes are treated as characteristic points of information retrieval and data forecasting. PNC modeling via nodes combination and parameter γ as probability distribution function enables 2D point extrapolation and interpolation. Two-dimensional information is modeled via nodes combination and some functions as continuous probability distribution functions: polynomial, sine, cosine, tangent, cotangent, logarithm, exponent, arc sin, arc cos, arc tan, arc cot or power function. Extrapolated values are used as the support in data forecasting.

Keywords: Information retrieval · Data extrapolation · Curve interpolation · PNC method · Probabilistic modeling · Forecasting

1 Introduction

Information retrieval and data forecasting are still the opened questions not only in mathematics and computer science. For example the process of planning can meet such a problem: what is the next value that is out of our knowledge, for example any wanted value by tomorrow. This planning may deal with buying or selling, with anticipating costs, expenses or with foreseeing any important value. The key questions in planning and scheduling, also in decision making and knowledge representation [1] are dealing with appropriate information modeling and forecasting. Two-dimensional data can be regarded as points on the curve. Classical polynomial interpolations and extrapolations (Lagrange, Newton, Hermite) are useless for data forecasting, because values that are extrapolated (for example the stock quotations or the market prices) represent continuous or discrete data and they do not preserve a shape of the polynomial. This paper is dealing with data forecasting by using the method of Probabilistic Nodes Combination (PNC) and extrapolation as the extension of interpolation. The values which are retrieved, represented by curve points, consist of information which allows us to extrapolate and to forecast some data for example before making a decision [2].

If the probabilities of possible actions are known, then some criteria are to be applied: Laplace, Bayes, Wald, Hurwicz, Savage, Hodge-Lehmann [3] and others [4]. But this paper considers information retrieval and data forecasting based only on 2D nodes. Proposed method of Probabilistic Nodes Combination (PNC) is used in data reconstruction and forecasting. PNC method uses two-dimensional data for knowledge

K. Jackowski et al. (Eds.): IDEAL 2015, LNCS 9375, pp. 104–112, 2015.
DOI: 10.1007/978-3-319-24834-9_13

representation [5] and computational foundations [6]. Also medicine [7], industry and manufacturing are looking for the methods connected with geometry of the curves [8]. So suitable data representation and precise reconstruction or extrapolation [9] of the curve is a key factor in many applications of artificial intelligence: forecasting, planning, scheduling and decision making.

The author wants to approach a problem of curve interpolation [10–12] and data forecasting by characteristic points. Proposed method relies on nodes combination and functional modeling of curve points situated between the basic set of key points and outside of this set. The functions that are used in computations represent whole family of elementary functions together with inverse functions: polynomials, trigonometric, cyclometric, logarithmic, exponential and power function. These functions are treated as probability distribution functions in the range [0;1]. Nowadays methods apply mainly polynomial functions, for example Bernstein polynomials in Bezier curves, splines and NURBS [13]. But Bezier curves do not represent the interpolation method and cannot be used for extrapolation. Numerical methods for data reconstruction are based on polynomial or trigonometric functions, for example Lagrange, Newton, Aitken and Hermite methods. These methods have some weak sides [14] and are not sufficient for curve interpolation and extrapolation in the situations when the curve cannot be build by polynomials or trigonometric functions. Proposed 2D point retrieval and forecasting method is the functional modeling via any elementary functions and it helps us to fit the curve.

Author presents novel Probabilistic Nodes Combination (PNC) method of curve interpolation-extrapolation and takes up PNC method of two-dimensional curve modeling via the examples using the family of Hurwitz-Radon matrices (MHR method) [15], but not only (other nodes combinations). The method of PNC requires minimal assumptions: the only information about a curve is the set of two or more nodes $p_i = (x_i, y_i) \in R^2$, $i = 1, 2, \ldots n$, $n \geq 2$. Proposed PNC method is applied in data forecasting and information retrieval via different coefficients: polynomial, sinusoidal, cosinusoidal, tangent, cotangent, logarithmic, exponential, arc sin, arc cos, arc tan, arc cot or power. Function for PNC calculations is chosen individually at each modeling and it represents probability distribution function of parameter $\alpha \in [0;1]$ for every point situated between two successive interpolation knots. For more precise modeling knots ought to be settled at key points of the curve, for example local minimum or maximum, highest point of the curve in a particular orientation, convexity changing or curvature extrema.

The goal of this paper is to answer the question: how to build the data model by a set of knots and how to extrapolate the points?

2   Data Simulation and Extrapolation

The method of PNC is computing points between two successive nodes of the curve: calculated points are interpolated and parameterized for real number $\alpha \in [0;1]$ in the range of two successive nodes. PNC method uses the combinations of nodes $p_1 = (x_1, y_1)$, $p_2 = (x_2, y_2), \ldots, p_n = (x_n, y_n)$ as $h(p_1, p_2, \ldots, p_m)$ and $m = 1, 2, \ldots n$ to interpolate second coordinate y as (2) for first coordinate c in (1):

$$c = \alpha \cdot x_i + (1 - \alpha) \cdot x_{i+1}, \quad i = 1, 2, \ldots n - 1, \tag{1}$$

$$y(c) = \gamma \cdot y_i + (1 - \gamma)y_{i+1} + \gamma(1 - \gamma) \cdot h(p_1, p_2, \ldots, p_m), \tag{2}$$

$\alpha \in [0;1]$, $\gamma = F(\alpha) \in [0;1]$, $F:[0;1] \rightarrow [0;1]$, $F(0) = 0$, $F(1) = 1$ and F is strictly monotonic.

PNC extrapolation requires α outside of $[0;1]$: $\alpha < 0$ (anticipating points right of last node for $c > x_n$) or $\alpha > 1$ (extrapolating values left of first node for $c < x_1$), $\gamma = F(\alpha)$, F: $P \rightarrow \mathbf{R}$, $P \supset [0;1]$, $F(0) = 0$, $F(1) = 1$. Here are the examples of h computed for MHR method [15]:

$$h(p_1, p_2) = \frac{y_1}{x_1}x_2 + \frac{y_2}{x_2}x_1 \tag{3}$$

or

$$h(p_1, p_2, p_3, p_4) = \frac{1}{x_1^2 + x_3^2}(x_1x_2y_1 + x_2x_3y_3 + x_3x_4y_1 - x_1x_4y_3) + \frac{1}{x_2^2 + x_4^2}(x_1x_2y_2 + x_1x_4y_4 + x_3x_4y_2 - x_2x_3y_4).$$

Three other examples of nodes combinations:
$h(p_1, p_2) = \frac{y_1x_2}{x_1y_2} + \frac{y_2x_1}{x_2y_1}$ or $h(p_1, p_2) = x_1x_2 + y_1y_2$ or   the   simplest
$h(p_1, p_2, \ldots, p_m) = 0$.

Nodes combination is chosen individually for each data, based on specific knowledge and experience, and it depends on the type of information modeling. Formulas (1)–(2) represent curve parameterization as $\alpha \in P$:

$$x(\alpha) = \alpha \cdot x_i + (1 - \alpha) \cdot x_{i+1}$$

and

$$y(\alpha) = F(\alpha) \cdot y_i + (1 - F(\alpha))y_{i+1} + F(\alpha)(1 - F(\alpha)) \cdot h(p_1, p_2, \ldots, p_m),$$

$$y(\alpha) = F(\alpha) \cdot (y_i - y_{i+1} + (1 - F(\alpha)) \cdot h(p_1, p_2, \ldots, p_m)) + y_{i+1}.$$

Proposed parameterization gives us the infinite number of possibilities for calculations (determined by choice of F and h) as there is the infinite number of data for reconstruction and forecasting. Nodes combination is the individual feature of each modeled data. Coefficient $\gamma = F(\alpha)$ and nodes combination h are key factors in PNC interpolation and forecasting.

2.1   Extended Distribution Functions in PNC Forecasting

Points settled between the nodes are computed using PNC method. Each real number $c \in [a;b]$ is calculated by a convex combination $c = \alpha \cdot a + (1 - \alpha) \cdot b$ for

$\alpha = \frac{b-c}{b-a} \in [0;1]$. Key question is dealing with coefficient γ in (2). The simplest way of PNC calculation means $h = 0$ and $\gamma = \alpha$ (basic probability distribution). Then PNC represents a linear interpolation and extrapolation. MHR method is not a linear interpolation. MHR is the example of PNC modeling [15]. Each interpolation requires specific distribution of parameter α and γ (1)–(2) depends on parameter $\alpha \in [0;1]$. Coefficient γ is calculated using different functions (polynomials, power functions, sine, cosine, tangent, cotangent, logarithm, exponent, arc sin, arc cos, arc tan or arc cot, also inverse functions) and choice of function is connected with initial requirements and data specifications. Different values of coefficient γ are connected with applied functions $F(\alpha)$. These functions $\gamma = F(\alpha)$ represent the examples of probability distribution functions for random variable $\alpha \in [0;1]$ and real number $s > 0$: $\gamma = \alpha^s$, $\gamma = sin$ $(\alpha^s \cdot \pi/2)$, $\gamma = sin^s(\alpha \cdot \pi/2)$, $\gamma = 1-cos(\alpha^s \cdot \pi/2)$, $\gamma = 1-cos^s(\alpha \cdot \pi/2)$, $\gamma = tan(\alpha^s \cdot \pi/4)$, $\gamma = tan^s(\alpha \cdot \pi/4)$, $\gamma = log_2(\alpha^s+1)$, $\gamma = log_2^s(\alpha+1)$, $\gamma = (2^\alpha-1)^s$, $\gamma = 2/\pi \cdot arcsin(\alpha^s)$, $\gamma = (2/\pi \cdot arcsin\alpha)^s$, $\gamma = 1-2/\pi \cdot arccos(\alpha^s)$, $\gamma = 1-(2/\pi \cdot arccos\alpha)^s$, $\gamma = 4/\pi \cdot arctan(\alpha^s)$, $\gamma = (4/\pi \cdot arctan\alpha)^s$, $\gamma = ctg(\pi/2-\alpha^s \cdot \pi/4)$, $\gamma = ctg^s(\pi/2-\alpha \cdot \pi/4)$, $\gamma = 2-4/\pi \cdot arcctg(\alpha^s)$, $\gamma = (2-4/\pi \cdot arcctg\alpha)^s$, $\gamma = \beta \cdot \alpha^2+(1-\beta) \cdot \alpha$, $\gamma = \beta \cdot \alpha^4+(1-\beta) \cdot \alpha$,…, $\gamma = \beta \cdot \alpha^{2k}+(1-\beta) \cdot \alpha$ for $\beta \in [0;1]$ and $k \in N$ or $\gamma = 1 - (1 - \alpha) \cdot s^\alpha$.

Functions above, used in γ calculations, are strictly monotonic for random variable $\alpha \in [0;1]$ as $\gamma = F(\alpha)$ is probability distribution function. There is one important probability distribution in mathematics: beta distribution where for example $\gamma = 3\alpha^2$ $-2\alpha^3$, $\gamma = 4\alpha^3-3\alpha^4$ or $\gamma = 2\alpha-\alpha^2$. Also inverse functions F^{-1} are appropriate for γ calculations. Choice of function and value s depends on data specifications and individual requirements during data interpolation. The optimal selection of the value of these parameters can be done via comparing with learning set.

Extrapolation demands that α is out of range [0;1], for example $\alpha \in (1;2]$ or $\alpha \in [-1;0)$, with $\gamma = F(\alpha)$ as probability distribution function and then F is called extended distribution function in the case of extrapolation. Some of these functions γ are useless for data forecasting because they do not exist ($\gamma = \alpha^{\frac{1}{2}}$, $\gamma = \alpha^{\frac{1}{4}}$) if $\alpha < 0$ in (1). Then it is possible to change parameter $\alpha < 0$ into corresponding $\alpha > 1$ and formulas (1)–(2) turn to equivalent equations:

$$c = \alpha \cdot x_{i+1} + (1 - \alpha) \cdot x_i, \quad i = 1, 2, \ldots n - 1, \qquad (4)$$

$$y(c) = \gamma \cdot y_{i+1} + (1 - \gamma)y_i + \gamma(1 - \gamma) \cdot h(p_1, p_2, \ldots, p_m). \qquad (5)$$

PNC forecasting for $\alpha < 0$ or $\alpha > 1$ uses function F as extended distribution function for the arguments from $P \supset [0;1]$, $\gamma = F(\alpha)$, $F:P \rightarrow \mathbf{R}$, F(0) = 0, F(1) = 1 and F has to be strictly monotonic only for $\alpha \in [0;1]$. Data simulation and modeling for $\alpha < 0$ or $\alpha > 1$ is done using the same function $\gamma = F(\alpha)$ that is earlier defined for $\alpha \in [0;1]$.

3 PNC Extrapolation and Data Trends

Probabilistic Nodes Combination is a novel approach of interpolation and extrapolation via data probability distribution and appropriate nodes combinations. Unknown data are modeled (interpolated or extrapolated) by the choice of nodes, determining specific nodes combination and probabilistic distribution function to show trend of values: increasing, decreasing or stable. Less complicated models take $h(p_1,p_2,...,p_m) = 0$ and then the formula of interpolation (2) looks as follows:

$$y(c) = \gamma \cdot y_i + (1 - \gamma)y_{i+1}.$$

It is linear interpolation for basic probability distribution ($\gamma = \alpha$).

Example 1

Nodes are (1;3), (3;1), (5;3) and (7;3), $h = 0$, extended distribution $\gamma = \alpha^2$, extrapolation is computed with (4)–(5) for $\alpha > 1$. And for all graphs in this paper: first coordinate of points is dealing with horizontal axis and second coordinate with vertical axis.

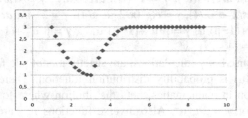

Fig. 1. PNC for 9 interpolated points between nodes and 9 extrapolated points.

Anticipated points (stable trend): (7.2;3), (7.4;3), (7.6;3), (7.8;3), (8;3), (8.2;3), (8.4;3), (8.6;3), (8.8;3) for $\alpha = 1.1, 1.2, ..., 1.9$.

Example 2

Nodes (1;3), (3;1), (5;3) and (7;2), $h = 0$, extended distribution $\gamma = F(\alpha) = \alpha^2$. Forecasting is computed as (4)–(5) with $\alpha > 1$.

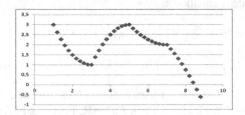

Fig. 2. PNC with 9 interpolated points between nodes and 9 extrapolated points.

Extrapolated points (decreasing trend): (7.2;1.79), (7.4;1.56), (7.6;1.31), (7.8;1.04), (8;0.75), (8.2;0.44), (8.4;0.11), (8.6;−0.24), (8.8;−0.61) for $\alpha = 1.1, 1.2, ..., 1.9$.

Example 3

Nodes (1;3), (3;1), (5;3) and (7;4), $h = 0$, extended distribution $\gamma = F(\alpha) = \alpha^3$.

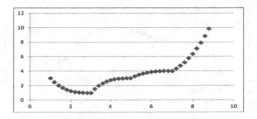

Fig. 3. PNC for 9 interpolated points between nodes and 9 extrapolated points.

Forecast (increasing trend): (7.2;4.331), (7.4;4.728), (7.6;5.197), (7.8;5.744), (8;6.375), (8.2;7.096), (8.4;7.913), (8.6;8.832), (8.8;9.859) for $\alpha = 1.1, 1.2, \ldots, 1.9$.

These three examples 1–3 (Figs. 1, 2, 3) with nodes combination $h = 0$ differ at fourth node and extended probability distribution functions $\gamma = F(\alpha)$. Much more possibilities of modeling are connected with a choice of nodes combination $h(p_1, p_2, \ldots, p_m)$. MHR method uses the combination (3) with good features connected with orthogonal rows and columns at Hurwitz-Radon family of matrices [15]:

$$h(p_i, p_{i+1}) = \frac{y_i}{x_i} x_{i+1} + \frac{y_{i+1}}{x_{i+1}} x_i$$

and then (2): $y(c) = \gamma \cdot y_i + (1 - \gamma) y_{i+1} + \gamma(1 - \gamma) \cdot h(p_i, p_{i+1})$.

Here are two examples 4 and 5 of PNC method with MHR combination (3).

Example 4

Nodes are (1;3), (3;1) and (5;3), extended distribution $\gamma = F(\alpha) = \alpha^2$. Forecasting is computed with (4)–(5) for $\alpha > 1$ (Fig. 4).

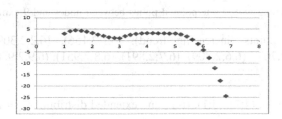

Fig. 4. PNC modeling with 9 interpolated points between nodes and 9 extrapolated points.

Extrapolation (decreasing trend): (5.2;2.539), (5.4;1.684), (5.6;0.338), (5.8; −1.603), (6;−4.25), (6.2;−7.724), (6.4;−12.155), (6.6;−17.68), (6.8;−24.443) for $\alpha = 1.1, 1.2, \ldots, 1.9$.

Example 5

Nodes (1;3), (3;1) and (5;3), extended distribution $\gamma = F(\alpha) = \alpha^{1.5}$. This forecasting is computed with (4)–(5) for $\alpha > 1$ (Fig. 5).

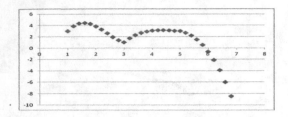

Fig. 5. PNC modeling with 9 interpolated points between nodes and 9 extrapolated points.

Value forecasting (decreasing trend): (5.2;2.693), (5.4;2.196), (5.6;1.487), (5.8;0.543), (6;−0.657), (6.2;−2.136), (6.4;−3.915), (6.6;−6.016), (6.8;−8.461) for α = 1.1, 1.2, …, 1.9.

Now let us consider PNC method with other functions F than power functions, α < 0 for extrapolation (1)–(2) and nodes combination $h = 0$.

Example 6

Nodes (2;2), (3;1), (4;2), (5;1), (6;2) and extended distribution $F(\alpha) = sin(\alpha \cdot \pi/2)$, $h = 0$:

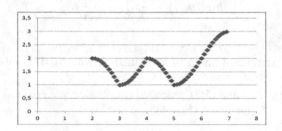

Fig. 6. PNC modeling with 9 interpolated points between nodes and 9 extrapolated points.

Extrapolation points (increasing trend): (6.1;2.156), (6.2;2.309), (6.3;2.454), (6.4;2.588), (6.5;2.707), (6.6;2.809), (6.7;2.891), (6.8;2.951), (6.9;2.988) for α = −0.1, −0.2, …, −0.9.

Example 7

Nodes (2;2), (3;1), (4;2), (5;1), (6;2) and extended distribution $\gamma = F(\alpha) = sin^3(\alpha \cdot \pi/2)$, $h = 0$:

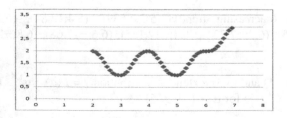

Fig. 7. PNC modeling with nine interpolated points between successive nodes and nine extrapolated points right of the last node.

Forecast points (increasing trend): (6.1;2.004), (6.2;2.03), (6.3;2.094), (6.4;2.203), (6.5;2.354), (6.6;2.53), (6.7;2.707), (6.8;2.86), (6.9;2.964) for $\alpha = -0.1, -0.2, \ldots, -0.9$.

These two examples 6 and 7 (Figs. 6, 7) with nodes combination $h = 0$ and the same set of nodes differ only at extended probability distribution functions $\gamma = F(\alpha)$. Figure 8 is the example of nodes combination h as (3) in MHR method.

Example 8

Nodes (2;2), (3;1), (4;1), (5;1), (6;2) and extended distribution function $\gamma = F(\alpha) = 2^{\alpha} - 1$:

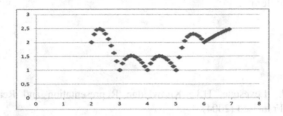

Fig. 8. PNC modeling with nine interpolated points between successive nodes and nine extrapolated points right of the last node.

Forecast points (increasing trend): (6.1;2.067), (6.2;2.129), (6.3;2.188), (6.4;2.242), (6.5;2.293), (6.6;2.34), (6.7;2.384), (6.8;2.426), (6.9;2.464) for $\alpha = -0.1, -0.2, \ldots, -0.9$.

Examples that are calculated above have one function $\gamma = F(\alpha)$ and one combination h for all ranges between nodes. But it is possible to create a model with functions $\gamma_i = F_i(\alpha)$ and combinations h_i individually for every range of nodes $(p_i;p_{i+1})$. Then it enables very precise modeling of data between each successive pair of nodes. Each data point is interpolated or extrapolated by PNC via three factors: the set of nodes, probability distribution function $\gamma = F(\alpha)$ and nodes combination h. These three factors are chosen individually for each data, therefore this information about modeled points seems to be enough for specific PNC data retrieval and forecasting. Function γ is selected via the analysis of known points before extrapolation, we may assume $h = 0$ at the beginning and after some time exchange h by more adequate.

These eight examples illustrate the forecasting of some values in planning process, for example anticipation of some costs or expenses and foreseeing the prices or other significant data in the process of planning.

4   Conclusions

The paper is dealing with information retrieval and data forecasting. The method of Probabilistic Nodes Combination (PNC) enables interpolation and extrapolation of two-dimensional curves using nodes combinations and different coefficients γ: polynomial, sinusoidal, cosinusoidal, tangent, cotangent, logarithmic, exponential, arc sin, arc cos, arc tan, arc cot or power function, also inverse functions. Function for γ calculations is chosen individually at each case and it is treated as probability distribution function: γ depends on initial requirements and data specifications. PNC method leads to point extrapolation and interpolation via discrete set of fixed knots. Main features of PNC method are: PNC method develops a linear interpolation and extrapolation into other functions as probability distribution functions; PNC is a generalization of MHR method via different nodes combinations; nodes combination and coefficient γ are crucial in the process of data probabilistic retrieval and forecasting. Future works are going to precise the choice and features of nodes combinations and coefficient γ, also to implementation of PNC in handwriting and signature recognition. PNC method for higher knowledge dimensionality in real-world data mining problems is an important challenge.

References

1. Brachman, R.J., Levesque, H.J.: Knowledge Representation and Reasoning. Morgan Kaufman, San Francisco (2004)
2. Fagin, R., Halpern, J.Y., Moses, Y., Vardi, M.Y.: Reasoning About Knowledge. MIT Press, Cambridge (1995)
3. Straffin, P.D.: Game Theory and Strategy. Mathematical Association of America, Washington, D.C. (1993)
4. Watson, J.: Strategy – An Introduction to Game Theory. University of California, San Diego (2002)
5. Markman, A.B.: Knowledge Representation. Lawrence Erlbaum Associates, New Jersey (1998)
6. Sowa, J.F.: Knowledge Representation: Logical, Philosophical and Computational Foundations. Brooks/Cole, New York (2000)
7. Soussen, C., Mohammad-Djafari, A.: Polygonal and polyhedral contour reconstruction in computed tomography. IEEE Trans. Image Process. 11(13), 1507–1523 (2004)
8. Tang, K.: Geometric optimization algorithms in manufacturing. Comput. Aided Des. Appl. 2 (6), 747–757 (2005)
9. Kozera, R.: Curve Modeling via Interpolation Based on Multidimensional Reduced Data. Silesian University of Technology Press, Gliwice (2004)
10. Collins II, G.W.: Fundamental Numerical Methods and Data Analysis. Case Western Reserve University, Cleveland (2003)
11. Chapra, S.C.: Applied Numerical Methods. McGraw-Hill, Columbus (2012)
12. Ralston, A., Rabinowitz, P.: A First Course in Numerical Analysis, 2nd edn. Dover Publications, New York (2001)
13. Schumaker, L.L.: Spline Functions: Basic Theory. Cambridge Mathematical Library, Cambridge (2007)
14. Dahlquist, G., Bjoerck, A.: Numerical Methods. Prentice Hall, New York (1974)
15. Jakóbczak, D.J.: 2D Curve Modeling via the Method of Probabilistic Nodes Combination - Shape Representation, Object Modeling and Curve Interpolation-Extrapolation with the Applications. LAP Lambert Academic Publishing, Saarbrucken (2014)

EVIDIST: A Similarity Measure
for Uncertain Data Streams

Abdelwaheb Ferchichi[(⊠)], Mohamed Salah Gouider,
and Lamjed Ben Said

SOIE, University of Tunis, Avenue de la liberté, 2000 Le Bardo, Tunisia
{ferchichiabdelwaheb, bensaid_lamjed}@yahoo.fr,
ms.gouider@isg.rnu.tn

Abstract. Large amount of data generated by sensors, and increased use of
privacy-preserving techniques have led to an increasing interest in mining
uncertain data streams. Traditional distance measures such as the Euclidean
distance do not always work well for uncertain data streams. In this paper, we
present EVIDIST, a new distance measure for uncertain data streams, where
uncertainty is modeled as sample observations at each time slot. We conduct an
extensive experimental evaluation of EVIDIST (Evidential Distance) on the
1-NN classification task with 15 real datasets. The results show that, compared
with Euclidean distance, EVIDIST increases the classification accuracy by about
13 % and is also far more resilient to error.

Keywords: Data mining · Distance measure · Similarity · Uncertain data
streams

1 Introduction

The similarity search in uncertain data streams has attracted the attention of the
research community in data mining during the last decade [1, 6, 11]. The uncertainty of
the data has two main reasons. First, the measurement tools, such as sensors, are
imperfect and their accuracy is always associated with a certain error distribution. The
readings could be the temperature measurements, or the location of moving objects.
Second, to preserve privacy, some degree of uncertainty is sometimes intentionally
introduced. In the literature, there are two models for representing uncertainty in data
streams: (1) the pdf-based model where each element at each timestamp is represented
as a random variable [6, 11], and (2) the multiset-based model where, at each time-
stamp, there is a multiset of observed values [1]. In this work we assume that uncertainty is modeled using sample observations rather than probability density functions.

Traditional distance measures such as the Euclidean distance do not always work
well for uncertain data streams. Section 5 will validate this for a large variety of data
sets. In this paper, we present EVIDIST, a new distance measure for uncertain data
streams, where uncertainty is modeled as sample observations at each time slot. In
Sect. 5 we extensively evaluate the EVIDIST distance on the UCR datasets [4]. We
perform experiments on 1-NN classification task. The EVIDIST distance outperforms

© Springer International Publishing Switzerland 2015
K. Jackowski et al. (Eds.): IDEAL 2015, LNCS 9375, pp. 113–120, 2015.
DOI: 10.1007/978-3-319-24834-9_14

Euclidean distance. It increases the classification accuracy by about 13 %. It is also far more resilient to error as compared to Euclidean distance.

The remainder of this paper is organized as follows. A literature review is presented in Sect. 2. The problem definition is given in Sect. 3. In Sect. 4, we prove that EVIDIST is a metric and describe how to compute it. The experimental studies are presented in Sect. 5. Finally, the work is concluded in Sect. 6.

2 Literature Review

2.1 Similarity Measures for Uncertain Data Streams

For similarity search on standard data streams, efficient algorithms and similarity measures have been developed [8]. The standard measures however are inadequate for uncertain data streams, and new suitable ones are required to be developed [5]. In the current literature, we identify two models for uncertain Data Streams:

1. **PDF-based model:** In this model, each element at each timestamp is represented as a random variable [6, 11]. In [11], an approach for processing queries on proba-bilistic uncertain Data Streams (PROUD) is presented. Inspired by the Euclidean distance, PROUD distance is modeled as the sum of the differences of the random variables of the data stream, where each random variable represents the uncertainty about the value of the corresponding timestamp. In [6], the authors propose a new distance measure, DUST. As in [11], DUST is inspired by the Euclidean distance, but works under the assumption that all the values of uncertain Data Streams follow a specific distribution. Given two uncertain Data Streams X and Y, the distance between two uncertain values x_i and y_i is defined as the distance between their true (unknown) values $r(x_i)$ and $r(y_i)$:

$$dist(x_i, y_i) = L^1(r(x_i), r(y_i))\tag{1}$$

   This distance may then be used to define a function ϕ which measures the similarity between two uncertain values:

$$\phi(|x_i - y_i|) = Pr(dist(0, |r(x_i) - r(y_i)|) = 0)\tag{2}$$

   This basic similarity function is then used in the dissimilarity function DUST:

$$dust(x, y) = \sqrt{-\log(\phi(|x - y|)) - k}\tag{3}$$

   where $k = -\log(\varphi(0))$.
   The constant k was introduced to support reflexivity.
2. **Multiset-based model:** In this model, at each timestamp, there is a multiset of observed values [5]. Given two uncertain data streams, X and Y, the authors in [1] proceeds as follows. First, the two streams X, Y are represented by all certain possible data streams:

$$S_X = \{\langle s_{11}, s_{21}, \ldots, s_{n1}\rangle, \ldots, \langle s_{1s}, s_{2s}, \ldots, s_{ns}\rangle\} \tag{4}$$

where s_{ij} is the jth observation at timestamp i. Then we proceed in the same manner for S_Y. The set of all possible distances between X and Y is then defined as follows:

$$dists(X, Y) = \{ L^P(x, y) \, | x \in S_X, y \in S_Y\}. \tag{5}$$

The uncertain L^P distance is then formulated by counting the number of distances less than or equal to the data threshold ε given by the user:

$$Pr(distance(X, Y) \le \mathcal{E}) = \frac{|\{d \in dists(X, Y) | d \le \varepsilon\}|}{|dists(X, Y)|} \tag{6}$$

2.2 Theory of Belief Functions

In this section, we present the works of Dempster [2], resumed by Shafer [7] and by Smets [9, 10] on the theory of belief functions.

Let $\Theta = \{C_1, C_2, \ldots, C_n\}$ corresponding to the set of possible classes for a classification problem. The mass functions are defined on 2^Θ (set of all the disjunctions of Θ) and with values in [0,1]. By construction, the mass function m verifies $\sum_{A \in 2^\Theta} m(A) = 1$. When $m(A) > 0$, we say that A is a focal element. From these mass functions, it is possible to build different functions:

- The credibility function, denoted *bel*, brings together the masses of all proposals $B \subseteq A$. It indicates the total confidence in the subset A. It is determined by:

$$bel(A) = \sum_{B \subseteq A, B \neq \emptyset} m(B) \qquad \forall A \in 2^\Theta \tag{7}$$

- The plausibility function, denoted *pl*, indicates the maximum confidence in A. It is determined by:

$$pl(A) = \sum_{A \cap B \neq \emptyset} m(B) \qquad \forall A \in 2^\Theta \tag{8}$$

3 Problem Statement

In this section, we first describe the problem of similarity search over multiple uncertain data streams. The basic notations and definitions are introduced as follows.

The uncertainty is represented by a set of sample observations at each time slot.

Definition 1 (Uncertain Data Stream). *An uncertain data stream S of length n consists of a sequence* $(S_1, \ldots, S_n)$ *of n elements, where each element* S_t *contains a set of sample observations, i.e.* $S_t = \{x_{t1}, x_{t2}, \ldots, x_{ts}\}$, *with s the sample size of S.*

We assume that the data at each time slot t follow a normal distribution with a mean μ_t and a standard deviation σ_t. Based on this assumption, we can use the standard deviation to construct confidence intervals for which we allocate most of the mass of belief. Thus a sigma deviation from either side of the mean covers 68.2 % of the values at time slot t, two sigma deviation covers 95.4 %, three sigma deviation covers 99.6 % and so on. The confidence interval will take the form $[\mu_t - p*\sigma_t, \mu_t + p*\sigma_t]$ where p is a parameter chosen by the user.

Let x_{tmin} be the minimum value at the time slot t and x_{tmax} be the maximum value at the same time slot. As proposed in [8], the uncertainty at each time slot will be modeled based on the theory of belief functions where the entire mass will be allocated to the interval $[\mu_t - p*\sigma_t, \mu_t + p*\sigma_t]$ as follows:

$$\begin{cases} m([\mu_t - p * \sigma_t, \mu_t + p * \sigma_t]) = h \\ m([x_{tmin}, x_{tmax}]) = 1 - h \end{cases} \tag{9}$$

where $h = \frac{2*p*\sigma_t}{x_{tmax} - x_{tmin}}$.

For the similarity measure between two samples of observations at each time slot, we adopt the Hausdorff distance between two sets defined as follows.

Definition 2 (Hausdorff Distance). *Given two finite sets $A = \{a_1, a_2, ..., a_p\}$ and $B = \{b_1, b_2, ..., b_q\}$:*

$$H(A, B) = max(h(A, B), h(B, A)) \tag{10}$$

where $h(A, B) = max_{a \in A} \min_{b \in B} |a - b|$

To take in evidence the effect of uncertainty at each time slot, the Hausdorff distance is weighted by the belief mass as follows:

$$H_C(A, B) = max(m_A h(A, B), m_B h(B, A)) \tag{11}$$

where m_A and m_B are the belief masses attributed to A and B and H_C is the evidential Hausdorff distance.

For the similarity measure between two uncertain data streams, we define EVIDIST (Evidential Distance), inspired by the Euclidean distance and based on the evidential Hausdorff distance, as follows.

Definition 3 (EVIDIST Distance). *Given two uncertain data streams S_u and S_v and a specified time range $T = [t_s, t_e]$, we define the EVIDIST distance as:*

$$EVIDIST(S_u, S_v)\big|_{ts}^{te}$$

$$= \sqrt{\sum_{i=t_s}^{t_e} H_c(S_u[i], S_v[i])} \tag{12}$$

where $S_u[i]$ is the set of sample observations of the data stream S_u at time slot i covered by the interval $[\mu_t - p*\sigma_t, \mu_t + p*\sigma_t]$ and $S_v[i]$ is the set of sample observations of the data stream S_v at time slot i covered by the interval $[\mu_t - p*\sigma_t, \mu_t + p*\sigma_t]$.

4 Distance Between Uncertain Data Streams

We first present some desirable properties for a distance measure, then prove that the EVIDIST distance is a metric and finally describe how to compute the EVIDIST distance.

4.1 Desired Properties of a Distance Measure

A distance measure should be a metric and fulfill the following conditions:

1. Non-negativity: $d(A,B) \geq 0$
2. Identity of indiscernibles: $d(A,B) = 0$ iff $A = B$
3. Symmetry: $d(A,B) = d(B,A)$
4. Triangle inequality: $d(A,B) + d(A,C) \geq d(B,C)$

In [3], the authors proved that the Hausdorff distance is a metric.

Proposition 1. *Let S_u and S_v two uncertain data streams: $EVIDIST(S_u, S_v) \geq 0$.*

Proof:

$$EVIDIST(S_u, S_v)|_{ts}^{te} = \sqrt{\sum_{i=t_s}^{t_e} H_c(S_u[i], S_v[i])}$$

so by construction, we can conclude that $EVIDIST(S_u, S_v) \geq 0$

Proposition 2. *$EVIDIST(S_u, S_v) = 0$ iff $S_u = S_v$*

Proof:
If $EVIDIST(S_u, S_v) = 0$ and we have $H_c(S_u[i], S_v[i]) \geq 0$ (because the Hausdorff distance is a metric)
so $\forall i \in [t_e, t_s]$, $H_c(S_u[i], S_v[i]) = 0$, *thus* $S_u[i] = S_v[i]$ *and finally* $S_u = S_v$

Proposition 3. *$EVIDIST(S_u, S_v) = EVIDIST(S_v, S_u)$*

Proof: Since the Hausdorff distance is a metric, we have

$$EVIDIST(S_u, S_v)|_{ts}^{te} = \sqrt{\sum_{i=t_s}^{t_e} H_c(S_u[i], S_v[i])} = \sqrt{\sum_{i=t_s}^{t_e} H_c(S_v[i], S_u[i])}$$
$$= EVIDIST(S_v, S_u)|_{ts}^{te}$$

Proposition 4. *Let S_u, S_v and S_w three uncertain data streams:*

$$EVIDIST(S_u, S_v) + EVIDIST(S_u, S_w) \geq EVIDIST(S_v, S_w)$$

Proof: $\forall i \in [t_e, t_s]$

$$H_c(S_u[i], S_v[i]) + H_c(S_u[i], S_w[i]) \geq H_c(S_v[i], S_w[i])$$

Then, $\sum_{i=t_s}^{t_e} H_c(S_u[i], S_v[i]) + \sum_{i=t_s}^{t_e} H_c(S_u[i], S_w[i]) \geq \sum_{i=t_s}^{t_e} H_c(S_v[i], S_w[i])$

Then, $\sqrt{\sum_{i=t_s}^{t_e} H_c(S_u[i], S_v[i]) + \sum_{i=t_s}^{t_e} H_c(S_u[i], S_w[i])} \geq \sqrt{\sum_{i=t_s}^{t_e} H_c(S_v[i], S_w[i])}$

Or $\forall$ a and b two reals: $\sqrt{a+b} \leq \sqrt{a} + \sqrt{b}$

So $\sqrt{\sum_{i=t_s}^{t_e} H_c(S_u[i], S_v[i])} + \sqrt{\sum_{i=t_s}^{t_e} H_c(S_u[i], S_w[i])} \geq \sqrt{\sum_{i=t_s}^{t_e} H_c(S_v[i], S_w[i])}$

And

$$EVIDIST(S_u, S_v) + EVIDIST(S_u, S_w) \geq EVIDIST(S_v, S_w)$$

4.2   Computing the EVIDIST Distance

We now describe how to compute the EVIDIST distance given two uncertain data streams S_u and S_v and a specified time range $T = [t_s, t_e]$. We first compute the mean μ_t and the standard deviation σ_t of the sample observations at each time slot t in T for S_u and S_v. We then extract the two sets $S_u[t]$ and $S_v[t]$ of values covered by the interval $[\mu_t - p * \sigma_t, \mu_t + p * \sigma_t]$, where p is a value given by the user and compute the mass of belief $m = \frac{2 * p * \sigma_t}{x_{tmax} - x_{tmin}}$ attributed to this interval. We need then to compute the evidential Hausdorff distance at each time slot t:

$$H_C(S_u[t], S_v[t]) = max(m_u h(S_u[t], S_v[t]), m_v h(S_v[t], S_u[t])).$$

Finally we compute the EVIDIST distance using these H_C distances as follows:

$$EVIDIST(S_u, S_v)|_{ts}^{te} = \sqrt{\sum_{i=t_s}^{t_e} H_c(S_u[i], S_v[i])}$$

5   Experimental Validation

5.1   Overview

We evaluate the effectiveness of the EVIDIST distance on the 1-NN classification task. We use the UCR classification datasets [4]. Because all of the datasets contain exact

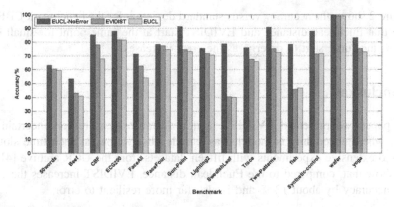

Fig. 1. Classification accuracy for EVIDIST vs Euclidean Distance

measurements, we generated probabilistic data streams by generating samples normally distributed around the given exact values and plot the accuracy of the classification for EVIDIST and the Euclidean distance.

5.2 Evaluation

In this section, we consider fifteen UCR datasets [4]. These datasets represent data streams, where the data streams have been classified into a few classes. For each dataset there is a training set and a test set. As in [6], we evaluate the accuracy for three configurations: on original data using Euclidean distance (EUCL-NoError), on perturbed data using Euclidean (EUCL) and EVIDIST distances. For all experiments, we computed the average over 10 different random runs. The results are shown in Fig. 1.

We observe that EVIDIST performs 2–13 % better than Euclidean distance. Wafer benchmark is error resilient in the sense that the classification accuracy does not decrease significantly. For all the other benchmarks there is close to a 5–50 % loss in accuracy between No Error and EUCL. We observe also that for CBF and FaceAll, EVIDIST performs so well that it almost completely makes up for the introduced error.

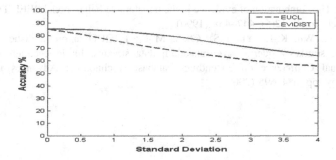

Fig. 2. Accuracy vs error for CBF benchmark

Figure 2 shows the accuracy versus standard deviation for the benchmark CBF. We observe that Euclidean distance and EVIDIST start at the same point for small error. However, the curve start to diverge as the error increases.

6 Conclusion

In this paper, we presented EVIDIST, a new distance measure for uncertain data streams, where uncertainty is modeled as sample observations at each time slot. We conducted extensive experiments with fifteen datasets from the UCR archive [4]. The results show that, compared to the Euclidean distance, EVIDIST increases the classification accuracy by about 13 % and is also far more resilient to error.

References

1. Aßfalg, J., Kriegel, H.-P., Kröger, P., Renz, M.: Probabilistic similarity search for uncertain time series. In: Winslett, M. (ed.) SSDBM 2009. LNCS, vol. 5566, pp. 435–443. Springer, Heidelberg (2009)
2. Dempster, A.P.: Upper and lower probabilities induced by a multivalued mapping. Ann. Math. Stat **219**, 325–339 (1967)
3. Henrikson, J.: Completeness and total boundedness of the Hausdorff metric. MIT Undergraduate J. Math. **1**, 69–80 (1999)
4. Keogh, E., Xi, X., Wei, L., Ratanamahatana, C.A.: The UCR time series classification/ clustering homepage. www.cs.ucr.edu/~eamonn/time_series_data. Accessed 5 March 2015
5. Orang, M., Shiri, N.: An experimental evaluation of similarity measures for uncertain time series. In: Proceedings of the 18th International Database Engineering and Applications Symposium, pp. 261–264 (2014)
6. Sarangi, S.R., Murthy, K.: DUST: a generalized notion of similarity between uncertain time series. In: Proceedings of the 16th ACM SIGKDD International Conference on Knowledge Discovery and Data Mining, pp. 383–392 (2010)
7. Shafer, G., et al.: A Mathematical Theory of Evidence, vol. 1. Princeton University Press, Princeton (1976)
8. Shasha, D.E., Zhu, Y.: High Performance Discovery in Time Series: Techniques and Case Studies. Springer Science & Business Media, Berlin (2004)
9. Smets, P., Kennes, R.: The transferable belief model. Artif. Intell. **66**(2), 191–234 (1994)
10. Smets, P.: The combination of evidence in the transferable belief model. IEEE Trans. Pattern Anal. Mach. Intell. **12**(5), 447–458 (1990)
11. Yeh, M.-Y., Wu, K.-L., Yu, P.S., Chen, M.-S.: PROUD: a probabilistic approach to processing similarity queries over uncertain data streams. In: Proceedings of the 12th International Conference on Extending Database Technology: Advances in Database Technology, pp. 684–695 (2009)

Knowledge Discovery in Enterprise Databases for Forecasting New Product Success

Marcin Relich[1(✉)] and Krzysztof Bzdyra[2]

[1] Faculty of Economics and Management, University of Zielona Gora, Zielona Gora, Poland
m.relich@wez.uz.zgora.pl
[2] Faculty of Electronic and Computer Engineering, Koszalin University of Technology,
Koszalin, Poland
krzysztof.bzdyra@tu.koszalin.pl

Abstract. This paper presents the knowledge discovery process that aims to improve the forecast quality of the success of new product development projects. The critical success factors for new product development are identified on the basis of information acquired from an enterprise system, including the fields of sales and marketing, research and development, production, and project management. The proposed knowledge discovery process consists of stages such as data selection from enterprise databases, data preprocessing, data mining, and the use of the discovered patterns for forecasting new product success. The illustrative example presents the use of fuzzy neural networks for forecasting net profit from new products.

Keywords: Knowledge retrieval · Data mining · Fuzzy neural systems · New product development · Rule-based systems

1 Introduction

Popularity of data mining and knowledge discovery is caused by an increasing demand for tools that help reveal and comprehend information hidden in huge amount of data [1]. Such data is routinely generated in scientific domains (e.g. astronomical, biological data) as well as through business transactions in banks, credit card companies, supermarkets, and other organizations. Data reflecting business transactions is stored increasingly in an integrated enterprise information system such as enterprise resource planning or customer relationship management system. As the amount of available data in companies becomes greater and greater, companies have become aware of an opportunity to derive valuable information from their databases, which can be further used to improve their business [2]. One of the most important aspects of continuing business is the successful new product development (NPD) and launch.

The NPD process allows the company to maintain its competitive position and continue business success. This process consists of the stages such as idea generation, idea screening, concept testing, development of the selected concepts, testing and commercialisation of new products [3]. As the success of the NPD projects is closely connected with the success of the entire company, the NPD projects play an important

© Springer International Publishing Switzerland 2015
K. Jackowski et al. (Eds.): IDEAL 2015, LNCS 9375, pp. 121–129, 2015.
DOI: 10.1007/978-3-319-24834-9_15

role in an organisation's growth and development [4]. In turn, the unsuccessful products may decrease the volume sales, profit, and finally, lead the company to bankruptcy. To survive and succeed in the dynamic business environment, companies usually focus on several areas to improve their new product development, such as identifying customer needs for continuous new product development, improving product quality, and accelerating the process of commercialization [5].

The new product development literature focuses on identifying success factors of new products with the use of multi-sectors surveys that often base on evaluating NPD process, organization, culture, and company strategy [6–8]. These surveys are useful to identify the key success factors, which bring positive influence on NPD (e.g. clear definition of product before development begins, high-quality preparatory work), and trends in the context of the various industries. However, the usefulness of multi-sectors surveys is limited for forecasting the success of a new product in the specific company. Therefore, the proposed approach bases on the success factors of NPD that are identified with the use of an enterprise system. The advancement of information technology helps today's organisations in business management processes and collecting data that is potential source of information [9]. A main task faced by NPD projects is how to acquire knowledge and sustain success rate among the products.

The previous research has focused mainly on the use of the econometric models for forecasting the success of a new product [10, 11]. This paper aims to present the use of fuzzy neural network as data mining techniques to forecast the success of new products and compare the forecast quality with the econometric models. The proposed approach takes into account data of the previous projects that is stored into an enterprise system. The remaining sections of this paper are organised as follows. Section 2 presents the knowledge discovery and data mining process. The proposed method of forecasting new product success is presented in Sect. 3. An example of the proposed approach is illustrated in Sect. 4. Finally, some concluding remarks are contained in Sect. 5.

2   Knowledge Discovery and Data Mining Process

Knowledge discovery in databases is a nontrivial process of identifying valid, novel, potentially useful, and ultimately understandable patterns in data [12]. Knowledge discovery (KD) is concerned with the development of methods and techniques for making sense of data. The basic problem addressed by KD is mapping low-level data (which is typically too voluminous to understand and digest easily) into other forms that might be more compact (e.g. a short report), more abstract (e.g. a descriptive approximation or model of the process that generated the data), or more useful (e.g. a predictive model for estimating the value of future cases) [12].

A knowledge discovery process can be described as a series of steps that differ in their number and range. Fayyad et al. presented the nine-step model that includes developing and understanding of the application domain, creating a target data set, data cleaning and preprocessing, data reduction and projection, choosing the data mining task, choosing the data mining algorithm, data mining, interpreting mined patterns, and finally consolidating discovered knowledge [12]. Cabena et al. described five steps of the KD process: business objectives determination, data preparation, data mining, domain knowledge elicitation, and assimilation of knowledge [13]. Cross-Industry Standard Process for Data Mining (CRISP-DM) states which tasks have to be carried out to complete a data mining project. These tasks contain business understanding, data understanding, data preparation, modeling, evaluation, and deployment [14]. Cios et al. adapted the CRISP-DM towards providing a more general, research-oriented description of the KD model through the following steps: understanding the domain, understanding the data, preparation of the data, data mining, evaluation of the discovered knowledge, and using the discovered knowledge [1, 14]. In general, the presented KD models include the stages such as data understanding and selection, data preprocessing, data mining, and implementation of the discovered patterns [15]. This framework of the KD process is further considered in this study (see Fig. 1).

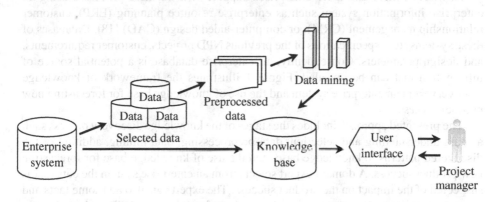

Fig. 1. Framework of decision support system for forecasting new product success

At the core of the KD process is the application of specific data mining methods for pattern discovery and extraction [12]. Data mining tasks can be classified as descriptive and predictive [15]. While the descriptive techniques provide a summary of the data, the predictive techniques learn from the current data in order to make predictions about the behaviour of new data sets. The most commonly used tasks in data mining include classification, clustering, associations, visualization, and forecasting that is further considered. Among data mining techniques, a fuzzy neural system has been chosen for identifying patterns and forecasting new product success.

The fuzzy neural system has the advantages of both neural networks (e.g. learning abilities, optimization abilities and connectionist structures) and fuzzy systems (e.g. if-then reasoning, simplicity of incorporating expert knowledge). The outcome of a fuzzy neural system is a set of if-then rules that can be used to perform nonlinear predictive modelling, classification, simulation, and forecasting [16, 17]. One well-known structure of fuzzy neural networks is the adaptive neuro-fuzzy inference system (ANFIS) that has been used in this study.

The methodology of the proposed approach is based on CRISP-DM model in which understanding the business and data is provided by a domain expert, the variables to analysis are preprocessed according to principal component analysis, data mining is implemented in the ANFIS, and discovered knowledge is used for forecasting the success of a new product in order to support the project manager.

3   The Proposed Method of Forecasting New Product Success

The success of a new product depends on several external (e.g. customer demand, changes in technology, environmental regulations) and internal factors (resources and processes used during new product development). The challenge is to increase the forecast quality of product success on the basis of an accessible enterprise database. The forecasts of the product success can help the project manager in taking decision of continuing or deleting NPD projects.

Nowadays, more and more enterprises support their business processes using an enterprise information system such as enterprise resource planning (ERP), customer relationship management (CRM), or computer-aided design (CAD) [18]. Databases of these systems store specifications of the previous NPD projects, customer requirements, and design parameters. Consequently, an enterprise database is a potential source of information that can be revealed. Figure 1 illustrates the framework of knowledge discovery from an enterprise system and the use of knowledge base for forecasting new product success.

The presented approach includes the stages of the knowledge discovery process such as data understanding and selection, data preprocessing, data mining, addition of the discovered patterns to knowledge base, and the use of knowledge base for forecasting new product success. A domain expert selects from an enterprise system the data that is suspected of the impact on the product success. The expert can also add some facts and constraints (e.g. amounts of resources, the number of activities in a NPD project) directly to knowledge base. In the next step, the chosen variables are reduced and preprocessed according to principal component analysis. Taking into account good forecasting properties and the possibility of obtaining if-then rules that can be stored in knowledge base, among the data mining techniques, the ANFIS has been chosen in this study. The identified patterns are stored in the form of rules in knowledge base that helps the project manager forecast the success of a new product. The forecasts can be further used for selecting a set of the most promising new products [19].

Taking into account the large number of potential new products, there is a need to use some techniques that reduce a search space. The processing time of calculations can be significantly reduced with the use of constraints programming techniques [20]. Constraints programming has embedded ways to solve constraints satisfaction problems with greatly reduction of the amount of search needed [21]. This is sufficient to solve many practical problems such as supply chain problem [22, 23] or scheduling problem [24].

Selection of new products for further development usually bases on metrics of the product success, and it should also take into account the company's resources. A successful NPD project can be considered as the achievement of an acceptable level of performance such as sales, market share, return on investment, profit, or customer satisfaction [8]. Taking into account the product lifetime and return on product development expense, the average net profit from a product per month is considered as a metric of the product success in this study. This metric is estimated on the basis of the previous developed products which specification can be retrieved from an enterprise system. The next section presents an example of using the proposed decision support system for forecasting the success of a new product.

4   Illustrative Example

The example refers to the above-described KD process and includes the steps such as data selection, data preprocessing, data mining, and the use of the retrieved patterns for forecasting the success of a new product. Table 1 presents the input variables of four fields that are suspected of the impact on the success of the NPD project.

Table 1. Input variables

Marketing and sales	R&D	Production	Project management
Number of customer requirements for the NPD project	Percentage of existing parts used in the NPD project	Number of resource overloads in the production phase of NPD	Number of project team members
Number of customer complaints for the NPD project	Percentage of customer requirements translated into speci-fication	Number of work orders	Average time spent on communication in project team members
Duration of marketing campaign of the product	Number of standard activities in the NPD project	Number of subcontractors for the NPD project	Duration of the NPD project
Cost of marketing campaign of the product	Number of activities in the NPD project	Unit cost of production for the product	Cost of the NPD project

Data selection can be divided into two phases. First, the expert selects the variables from the enterprise database according to his/her experience (Table 1). In the second phase, the optimal number of variables is sought with the use of a variable selection method. In this study, the stepwise procedure has been used, in which the size of the input vectors may be reduced by retaining only those components that contribute more than a specified fraction of the total variation in the data set. After using the stepwise procedure, the data set has been reduced from 16 input variables to 5 following variables: percentage of customer requirements translated into technical specification (CR), duration of the NPD project (D), cost of the NPD project (C), unit cost of production for the product (UC), and cost of marketing campaign for the product (MC).

The considered example includes data from 33 previous product development projects (P1-P33) that derive from 3 separate product lines. The collected data has a numerical format but different units, for example, cost of the NPD project is measured in monetary unit, duration of the NPD project in days, whereas the share of customer requirements translated into technical specification is measured in percent. As a result, the data requires transformation in order to use data mining techniques more effective. Data preprocessing facilitates the improvement of modeling accuracy of a fuzzy neural system (ANFIS). In this study, the principal component analysis has been used to transform data before the learning phase of ANFIS.

Data mining stage is connected with revealing the hidden relationships with the use of ANFIS. In order to reduce the overtraining of ANFIS and to increase the forecast quality, the data set has been divided into learning (P1-P27) and testing sets (P28-P33). Fuzzy inference system has been generated with the use of two methods: grid partition and subtractive clustering. The grid partition (GP) method has used three triangular membership functions for each input variables, and constant and linear type of output membership function (MF). In turn, the subtractive clustering (SC) method has used Gaussian membership functions and different values of the parameter concerning the range of influence (RI). After learning phase, the testing data has been led to input of the fuzzy neural system to compare the RMSE for ANFIS and other models. The results have been obtained in the Matlab® software and presented in Table 2 as the root mean square errors (RMSE) for the learning set (LS) and testing set (TS), as well as the number of rules generated. The comparison also includes the average and linear regression model.

Table 2. Comparison of forecasting models

Model	RMSE for LS	RMSE for TS	Number of rules
ANFIS (SC, RI = 0.3)	0.001	68.78	15
ANFIS (SC, RI = 0.5)	0.134	21.35	6
ANFIS (SC, RI = 0.7)	1.58	6.97	3
ANFIS (GP, MF = constant)	0.005	29.18	243
ANFIS (GP, MF = linear)	0.0001	29.25	243
Linear regression	20.57	20.88	1
Average	69.85	74.32	1

The results presented in Table 2 indicate the importance of parameter adjustment of specific data mining techniques. All the RMSE generated by ANFIS are less than the average, but only one case (ANFIS trained with the use of subtractive clustering method and range of influence equals 0.7) has obtained the RMSE less than linear regression model. The presented forecasting models also differ in the number of rules that influence the performance of decision support system. The identified relationships can be described as if-then rules and used for forecasting net profit from a new product. Let us assume that for the actual NPD project the following values are considered: customer requirements translated into technical specification – 70 %, duration of the NPD project – 30 weeks, cost of the NPD project – 65 monetary units, unit cost of production for the product – 0.35 monetary unit, and cost of marketing campaign for the product – 90 monetary units. Figure 2 presents three rules generated by ANFIS (case: SC, RI = 0.7) that have been used for forecasting net profit which reaches 227 monetary units for the considered product. The forecasts of net profits from the NDP projects can be further used for seeking the NPD project portfolio that ensures the maximal total net profit from all products.

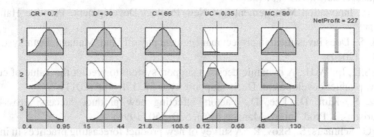

Fig. 2. Forecast of net profit with the use of ANFIS

5 Conclusions

The characteristics of the presented approach includes the use of an enterprise system database to knowledge retrieval and fuzzy neural networks to seek the patterns in the context of forecasting new product success, as well as the description of the identified patterns in the form of if-then rules that are suitable for human-like reasoning. The knowledge base includes the rules identified by fuzzy neural network or/and an expert, facts (including company's resources), and it allows the project managers to obtain the forecast of net profit from the NPD projects.

This research includes the comparison of the different methods in the context of the forecast quality and the number of rules generated. The results indicate that the forecast quality with the use of fuzzy neural networks can outperform the forecasting models that base on the average and linear regression. However, there is needed the adjustment of learning parameters of a fuzzy neural network what can be treated as a disadvantage. Nevertheless, fuzzy neural networks are able to identify the nonlinear and complex patterns (if there are any), develop the knowledge base, and finally, use these patterns to perform nonlinear predictive modelling, simulation, and forecasting.

The identified patterns are used in the decision support system to help the managers in forecasting the success of the NPD projects, selecting the most promising products, and reducing the risk of unsuccessful product development. On the other hand, the application of the proposed approach encounters some difficulties by collecting enough amount of data of the past similar NPD projects and ambiguous principles to build the structure of a fuzzy neural network. Nevertheless, the presented approach seems to have the promising properties for acquiring additional information from an enterprise system and using them to forecasting new product success. The further research aims to develop an interactive decision support system for identifying the most promising product portfolio according to the user's preferences.

References

1. Cios, K.J., Kurgan, L.A.: Trends in data mining and knowledge discovery. In: Pal, N., Jain, L. (eds.) Advanced Techniques in Knowledge Discovery, pp. 1–26. Springer, London (2005)
2. Li, T., Ruan, D.: An extended process model of knowledge discovery in database. J. Enterp. Inf. Manage. **20**(2), 169–177 (2007)
3. Trott, P.: Innovation Management and New Product Development. Prentice Hall, Essex (2005)
4. Spalek, S.: Does investment in project management pay off? Ind. Manage. Data Syst. **114**(5), 832–856 (2014)
5. Chan, S.L., Ip, W.H.: A dynamic decision support system to predict the value of customer for new product development. Decis. Support Syst. **52**, 178–188 (2011)
6. Mishra, S., Kim, D., Lee, D.: Factors affecting new product success: cross-country comparisons. J. Prod. Innov. Manage **13**(6), 530–550 (1996)
7. Lynn, G., Schnaars, S., Skov, R.: A survey of new product forecasting practices in industrial high technology and low technology businesses. Ind. Mark. Manage. **28**(6), 565–571 (1999)
8. Ernst, H.: Success factors of new product development: a review of the empirical literature. Int. J. Manage. Rev. **4**(1), 1–40 (2002)
9. Relich, M.: Knowledge acquisition for new product development with the use of an ERP database. In: Federated Conference on Computer Science and Information Systems (FedCSIS), pp. 1285–1290 (2013)
10. Hardie, B., Fader, P., Wisniewski, M.: An empirical comparison of new product trial forecasting models. J. Forecast. **17**, 209–229 (1998)
11. Kahn, K.: An exploratory investigation of new product forecasting practices. J. Prod. Innov. Manage **19**, 133–143 (2002)
12. Fayyad, U., Piatetsky-Shapiro, G., Smith, P.: From data mining to knowledge discovery in databases. Am. Assoc. Artif. Intell. 37–54 (1996). Fall
13. Cabena, P., Hadjinian, P., Stadler, R., Verhees, J., Zanasi, A.: Discovering Data Mining: From Concepts to Implementation. Prentice Hall, Saddle River (1998)
14. Marban, O., Mariscal, G., Segovia, J.: A data mining & knowledge discovery process model. In: Data Mining and Knowledge Discovery in Real Life Applications. I-Tech (2009)
15. Han, J., Kamber, M.: Data Mining. Concepts and Techniques. Morgan Kaufmann Publishers, San Francisco (2006)
16. Hudec, M., Vujosevic, M.: Integration of data selection and classification by fuzzy logic. Expert Syst. Appl. **39**, 8817–8823 (2012)
17. Relich, M., Muszynski, W.: The use of intelligent systems for planning and scheduling of product development projects. Procedia Comput. Sci. **35**, 1586–1595 (2014)
18. Gola, A., Świć, A.: Computer-aided machine tool selection for focused flexibility manufacturing systems using economical criteria. Actual Probl. Econ. **10**(124), 383–389 (2011)
19. Relich, M., Pawlewski, P.: A multi-agent system for selecting portfolio of new product development projects. In: Bajo, J., Hallenborg, K., Pawlewski, P., Botti, V., Sánchez-Pi, N., Duque Méndez, N.D., Lopes, F., Vicente, J. (eds.) PAAMS 2015 Workshops. CCIS, vol. 524, pp. 102–114. Springer, Heidelberg (2015)
20. Sitek, P.: A hybrid CP/MP approach to supply chain modelling, optimization and analysis. In: Federated Conference on Computer Science and Information Systems, pp. 1345–1352 (2014)

21. Van Roy, P., Haridi, S.: Concepts, Techniques and Models of Computer Programming. Massachusetts Institute of Technology, Cambridge (2004)
22. Grzybowska, K., Kovács, G.: Logistics process modelling in supply chain – algorithm of coordination in the supply chain – contracting. In: de la Puerta, J.G., Ferreira, I.G., Bringas, P.G., Klett, F., Abraham, A., de Carvalho, A.C.P.L.F., Herrero, Á., Baruque, B., Quintián, H., Corchado, E. (eds.) International Joint Conference SOCO'14-CISIS'14-ICEUTE'14. AISC, vol. 299, pp. 311–320. Springer, Heidelberg (2014)
23. Grzybowska, K.: Selected activity coordination mechanisms in complex systems. In: Bajo, J., Hallenborg, K., Pawlewski, P., Botti, V., Sánchez-Pi, N., Duque Méndez, N.D., Lopes, F., Vicente, J. (eds.) PAAMS 2015 Workshops. CCIS, vol. 524, pp. 69–79. Springer, Heidelberg (2015)
24. Bzdyra, K., Banaszak, Z., Bocewicz, G.: Multiple project portfolio scheduling subject to mass customized service. In: Szewczyk, R., Zieliński, C., Kaliczyńska, M. (eds.) Progress in Automation, Robotics and Measuring Techniques. AISC, vol. 350, pp. 11–22. Springer, Heidelberg (2015)

Effective Realizations of Biorthogonal Wavelet Transforms of Lengths $2K + 1/2K - 1$ with Lattice Structures on GPU and CPU

Dariusz Puchala[1(✉)], Bartomiej Szczepaniak[2], and Mykhaylo Yatsymirskyy[1]

[1] Institute of Information Technology, Lodz University of Technology,
Wólczańska Str. 215, 90-924 Lodz, Poland
{dariusz.puchala,mykhaylo.yatsymirskyy}@p.lodz.pl
[2] Institute of Applied Computer Science, Lodz University of Technology,
Stefanowskiego Str. 18/22, 90-924 Lodz, Poland
bartlomiej.szczepaniak@p.lodz.pl

Abstract. The paper presents comparative results in times of calculation of biorthogonal wavelet transform of lengths $2K + 1/2K - 1$ implemented with aid of two variants of lattice structures on parallel graphics processors (GPU) and on CPU. The aim of the research is to indicate lattice structure which allows to obtain higher efficiency of computations in case of both GPU and CPU architectures.

1 Introduction

Discrete wavelet transform (DWT) finds wide practical applications in many engineering problems. Exemplary fields of application of DWT include, e.g.: lossy data compression, multiband filtration of signals, data recognition and data analysis, identification of systems, segmentation and classification of data, problems of big data clusterization or watermarking of images [1–3]. Due to the popularity of wavelet transform we can still observe a non-decreasing interest of scientific community in synthesis of novel and more computationally effective algorithms for its calculation. The proposed in literature variants of effective algorithms are based on various optimization concepts, e.g.: polyphase decomposition [4], lattice structures of different types [5–7], lifting scheme [8], etc.

In the latter years a dynamic growth of interest in parallel computations on graphics processing units (GPU) could be observed. The popularization of high-performance GPUs together with publishing the dedicated libraries for general purpose parallel computing (GPGPU) by the leading manufacturers make it now possible to carry out massively parallel computations on typical personal computers. It is not surprising that an increasing research effort is devoted to adaptation of well-known algorithms to the architectures of consumer segment GPUs, including the algorithms for calculation of DWT (see e.g. [9,10]).

In the view of above, we present in this paper a novel variant of known lattice structure dedicated for calculation of biorthogonal wavelet transform with filters of lengths $2K + 1/2K - 1$ (see [7]). Then, the proposed variant is compared with

© Springer International Publishing Switzerland 2015
K. Jackowski et al. (Eds.): IDEAL 2015, LNCS 9375, pp. 130–137, 2015.
DOI: 10.1007/978-3-319-24834-9_16

the known structure with regard to the computational efficiency. The parallel implementations of lattice structures on GPU as well as sequential ones on CPU are considered. The aims of this comparison are: to indicate structures more computationally effective for both CPU and GPU architectures, to verify the advisability of DWT implementations on GPUs. The choice of $2K+1/2K-1$ filters is dictated by popularity and in particular by their application to JPEG2000 image compression stream (CDF filters 7/5 and 9/7 [11]).

2   Wavelet Transform as a Two-Channel Filter Banks

Typically, DWT is calculated by the successive repetition of operations within an analysis (or synthesis for inverse DWT) stage of two-channel banks of filters in the so-called pyramid scheme [1]. Both stages of analysis and synthesis for two-channel filter banks (see Fig. 1) consist of four finite response linear filters H, G, Q and R, and additional downsampling ($\uparrow 2$) or upsampling ($\downarrow 2$) operations.

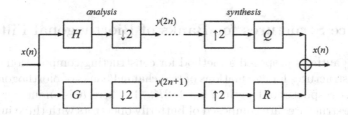

Fig. 1. An analysis and synthesis stages for wavelet transform

In order to ensure perfect reconstruction of data (PR condition) impulse response of filters should satisfy the following relations (see [12]):

$$\sum_{k=0}^{K-1}(-1)^k h_{K-1+2m-k} q_k = \pm\delta(m),\tag{1}$$

where $m = \pm 1, \pm 2, \ldots, \pm(K/2 - 1)$, K is filter order, $\delta(0) = 1$ and $\delta(m) = 0$ for $m \neq 0$, in addition $r_k = (-1)^k h_k$ for $k = 0, 1, \ldots, 2K$ and $q_k = (-1)^k g_k$ for $k = 0, 1, \ldots, 2K - 2$. For the considered class of biorthogonal filters it is assumed an additional symmetry of impulse responses: $h_k = h_{2K-k}$ for $k = 0, 1, \ldots, K-1$ and $g_k = g_{2K-2-k}$ for $k = 0, 1, \ldots, K - 2$ (c.f. [12]). The signal filtering scheme shown in Fig. 1 that is based on banks of biorthogonal filters with the specified properties, can be implemented in a computationally effective way with aid of lattice structures described in the following section. It should be noted that in case of DWT additional restrictions on impulse responses of filters may be required (e.g. h_k and g_k as lowpass and highpass filters, smoothness conditions on impulse responses [1], etc.).

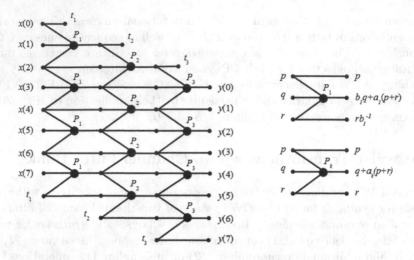

Fig. 2. Forward lattice structure of type I for 7/5 taps DWT ($N = 8$, $K = 3$)

3　Lattice Structures for Banks of Biorthogonal Filters

In paper [7] authors proposed a method for constructing computationally effective lattice structures for calculations of two-channel banks of biorthogonal filters with impulse responses of lengths $2K+1/2K-1$, where K is an integer number. Such lattice structures are composed of butterfly operators with three inputs and three outputs. Starting with this point, we will refer to such structures as type I structures. For an analysis stage a corresponding bank of filters (see Fig. 1) can be implemented as a forward lattice structure of type I depicted in Fig. 2. The first layer of the structure requires P_1 operators parametrized with values a_1 and b_1, $b_1 \neq 0$ and the remaining layers for $k = 2, 3, \ldots, K$ are constructed with uniform operators P_k described by single parameters a_k which can take any values. The definitions of both types of operators can be given in matrix form as:

$$P_1 = \begin{bmatrix} 1 & 0 & 0 \\ a_1 & b_1 & a_1 \\ 0 & 0 & b_1^{-1} \end{bmatrix}, P_k = \begin{bmatrix} 1 & 0 & 0 \\ a_k & 1 & a_k \\ 0 & 0 & 1 \end{bmatrix}. \tag{2}$$

The structure of type I is characterized by a number of $L_{MUL} = \frac{N}{2}(K + 2)$ multiplications and a number of $L_{ADD} = KN$ additions, where N is the size of transformation. The main disadvantage of the structure of type I is an increased number of references to input data resulting from overlapping (by one element) of adjacent operators in layers. In case of GPU implementation it can be important due to specific hardware requirements (memory coalescing) which in order to ensure maximum throughput of calculations require referenced data to be arranged successively in DRAM memory (see [13]).

The synthesis bank of filters from Fig. 1 can be realized in the form of inverse lattice structure of type I shown in Fig. 3. Such a structure can be obtained simply by reversing the order of layers of a forward structure and by replacing P_k operators with inverse ones P_k^{-1} for $k = 1, 2, \ldots, K$. The definitions of inverse

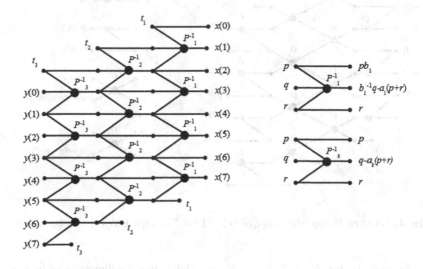

Fig. 3. Inverse lattice structure of type I for 7/5 taps DWT ($N = 8$, $K = 3$)

operators take the following form:

$$P_1^{-1} = \begin{bmatrix} b_1 & 0 & 0 \\ -a_1 & 1 & -a_1 \\ 0 & 0 & 1 \end{bmatrix}, P_k^{-1} = \begin{bmatrix} 1 & 0 & 0 \\ -a_k & 1 & -a_k \\ 0 & 0 & 1 \end{bmatrix}. \tag{3}$$

With an analysis of both structures from Figs. 2 and 3, and formulas (2) and (3) we conclude that the inverse structure is equivalent to the forward one, both in terms of computational complexity and a way of parallel implementation.

In this paper a novel variant of the lattice structure (referred to as type II) that consists of butterfly operators having two inputs and two outputs is proposed. Then such a structure is structurally similar (except the first and last layer) to the known lattice structures used to calculate orthonormal wavelet transforms [6]. The mentioned property is desirable since it allows for a generalization of lattice structures on a wider class of transformations. It gives the possibility to calculate wavelet transforms (orthonormal and biorthogonal) commonly used in practice with almost uniform computational structure. The forward lattice structure of type II is shown in Fig. 4, while the butterfly operators used in this structure can be described in a matrix form as:

$$O_1 = \begin{bmatrix} 1 & 0 \\ a_1 & b_1 \end{bmatrix}, O_2 = \begin{bmatrix} 1 & a_1 \\ a_2 & a_1 a_2 + b_1^{-1} \end{bmatrix}, \tag{4}$$

$$O_k = \begin{bmatrix} 1 & a_{k-1} \\ a_k & a_{k-1} a_k + 1 \end{bmatrix}, O_{K+1} = \begin{bmatrix} 1 & a_K \\ 0 & 1 \end{bmatrix}, \tag{5}$$

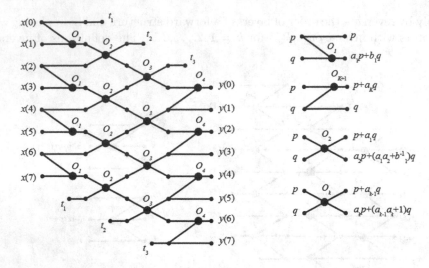

Fig. 4. Forward lattice structure of type II for 7/5 taps DWT ($N = 8$, $K = 3$)

where b_1 and a_k for $k = 1, 2, \ldots, K$ are the same coefficients as in type I structure. The structure of type II is characterized by the number of $L_{MUL} = (3/2)KN$ multiplications and $L_{ADD} = KN$ additions. From these formulas, it follows immediately that such structure requires almost threefold higher number of multiplications for large values of K compared to the structure of type I. The structure of type II is characterized, however, by nearly 30 % smaller number of references to memory buffers that store input data for subsequent stages.

The inverse lattice structure of type II (see Fig. 5) is constructed by reversing layers order and by replacing O_k operators with inverse ones O_k^{-1}, wherein:

$$O_1^{-1} = \begin{bmatrix} b_1 & 0 \\ -a_1 & 1 \end{bmatrix}, O_2^{-1} = \begin{bmatrix} b_1^{-1} + a_1 a_2 & -a_1 \\ -a_2 & 1 \end{bmatrix}, \tag{6}$$

$$O_k^{-1} = \begin{bmatrix} 1 + a_{k-1} a_k & -a_{k-1} \\ -a_k & 1 \end{bmatrix}, O_{K+1}^{-1} = \begin{bmatrix} 1 & -a_K \\ 0 & 1 \end{bmatrix}. \tag{7}$$

It can be easily verified that the reverse lattice structure of type II has exactly the same computational complexity as the forward one.

4   Implementation of Lattice Structures on GPU

During the implementation of the considered lattice structures on GPU cards we assumed regardless of the type of a structure the same approach, which consists in calculating the following layers sequentially by separate kernel functions. Within a single layer butterfly operators are assigned to separate threads which are grouped into blocks in the number of N divided by the assumed number of threads per block (here that number equaled 32 threads). Thus, individual

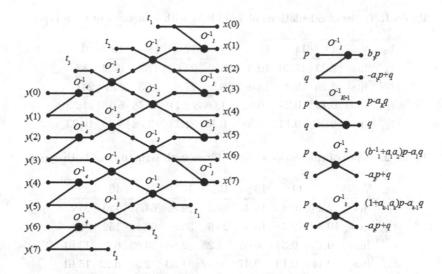

Fig. 5. Inverse lattice structure of type II for 7/5 taps DWT ($N = 8$, $K = 3$)

threads are responsible for the execution of operations required by butterfly operators P_k, P_k^{-1}, O_k or O_k^{-1}.

A shift by one element between successive layers (see Figs. 2, 3, 4 and 5) was compensated by appropriate offset of outputs of butterfly operators (by one index up). This entails the need for the use of two buffers: input and output ones, which are further on switched sequentially between successive layers of the structure. The required implementations of the considered lattice structures were carried out for NVIDIA cards with use of CUDA^{TM} programming library.

5   Experimental Research

As part of the research a series of experiments using NVIDIA GeForce GTX 970 were conducted. Results in the form of time of execution of kernel functions (t_I and t_{II} respectively for the structures of types I and II), which were obtained for forward lattice structures of both types (inverse structures are computationally equivalent) with $N = 8192, 16384, \ldots, 1048576$ and for 7/5 and 9/7 taps filters are summarized in Tables 1 and 2. The corresponding results for CPU implementation on Intel Core^{TM} i7 2.67 GHz processor are described as t_I^{CPU} and t_{II}^{CPU}.

The obtained results show that the structure of type I is more computationally effective when implemented in parallel on GPU. Here, an average increase of efficiency by 20 % (15 %) in comparison to the structure of type II for filters 7/5 (9/7) was observed. Thus, the additional number of calculations in the case of type II structure had a higher impact on the overall time of transform calculation than an increased (by 30 %) number of data references characteristic of the structure of type I. In the case of CPU implementations the situation was

Table 1. Times of calculations of 7/5 filters with structures of both types

$\log_2 N$	13	14	15	16	17	18	19	20
t_I^{GPU}[μs]	7.23	7.26	10.3	13.4	22.6	47.2	94.2	183
t_{II}^{GPU}[μs]	8.26	10.2	13.4	18.5	27.8	54.3	123	241
t_I^{CPU}[ms]	0.13	0.25	0.52	1.06	2.11	3.78	5.33	11.32
t_{II}^{CPU}[ms]	0.08	0.12	0.38	0.75	1.12	2.08	4.80	10.73

Table 2. Times of calculations of 9/7 filters with structures of both types

$\log_2 N$	13	14	15	16	17	18	19	20
t_I^{GPU}[μs]	7.26	10.4	15.4	19.4	29.8	64.5	126	245
t_{II}^{GPU}[μs]	10.3	12.3	16.5	20.6	32.8	69.8	152	300
t_I^{CPU}[ms]	0.15	0.34	0.67	1.28	2.88	3.63	6.42	13.91
t_{II}^{CPU}[ms]	0.14	0.19	0.47	0.62	1.63	2.99	6.22	13.01

different. Here, the structure of type II allowed for an average improvement in 30 % (25 %) for 7/5 (9/7) filters. The comparison of results for the GPU and CPU indicates a significant acceleration of calculations for the GPU, i.e.: for type I structure in average of about 59 times (53 times) for 7/5 (9/7) filters, and for the structure of type II in average of about 31 (33) times for 7/5 (9/7) filters. Such an acceleration is expected in the view of results presented, e.g. in [14].

6   Summary and Conclusions

This paper presents the comparative results in times of calculation of biorthogonal wavelet transform realized in the form of lattice structures of two different types. The mentioned structures have been implemented in parallel on GPUs with CUDA^{TM} library and sequentially on CPU. The experimental results indicate the advantage of a structure of type I in case of the GPU implementation and the structure of type II for CPU. It should be noted that the type I structure requires three times smaller number of multiplications but the number of memory references is increased by about 30 % when compared to the structure of type II. The comparison of results between GPU and CPU indicated significant acceleration of calculations in case of mass parallel implementation, i.e.: above 50 times in case of type I structure and above 30 times for the structure of type II. Hence, it can be concluded that it is advantageous to realize calculations of biorthogonal wavelet transforms with filters of lengths $2K+1/2K-1$ on GPUs. In such a case the lattice structure of type I is computationally more effective.

References

1. Fleet, P.J.: Discrete Wavelet Transformation: An Elementary Approach with Applications. Wiley, Hoboken (2008)

2. Stolarek, J., Lipiński, P.: Improving watermark resistance against removal attacks using orthogonal wavelet adaptation. In: Bieliková, M., Friedrich, G., Gottlob, G., Katzenbeisser, S., Turán, G. (eds.) SOFSEM 2012. LNCS, vol. 7147, pp. 588–599. Springer, Heidelberg (2012)
3. Sheikholeslami, G., Chatterjee, S., Zhang, A.: WaveCluster: a wavelet-based clustering approach for spatial data in very large databases. J. Very Large Databases **8**, 289–304 (2000)
4. Cooklev, T.: An efficient architecture for orthogonal wavelet transforms. IEEE Signal Process. Lett. **13**(2), 77–79 (2006)
5. Olkkonen, J.T., Olkkonen, H.: Discrete lattice wavelet transform. IEEE Trans. Circuits Syst. II: Express Briefs **54**(1), 71–75 (2007)
6. Yatsymirskyy, M., Stokfiszewski, K.: Effectiveness of lattice factorization of two-channel orthogonal filter banks. In: Joint Conference in New Trends in Audio & Video and Signal Processing, pp. 275–279. Lodz, Poland (2012)
7. Yatsymirskyy, M.: A lattice structure for the two-channel bank of symmetrical biorthogonal filters of lengths 2K+1/2K-1. In: 13th International Workshop Computational Problems of Electrical Engineering, Grybów, Poland (2012)
8. Daubechies, I., Sweldens, W.: Factoring wavelet transform into lifting steps. J. Fourier Anal. Appl. **4**(3), 245–267 (1998)
9. Yildrim, A.A., Ozdogan, C.: Parallel wavelet-based clustering algorithm on GPUs using CUDA. Procedia Comput. Sci. **3**, 396–400 (2011)
10. Puchala, D., Szczepaniak, B, Yatsymirskyy, M.: Lattice structure for parallel calculation of orthogonal wavelet transform on GPUs with CUDA architecture. In: Conference on Computational Problems of Electrical Engineering, Terchova, Slovakia (2014)
11. Cohen, A., Daubechies, I., Feauveau, J.C.: Biorthogonal bases of compactly supported wavelets. Comm. Pure Appl. Math. **45**, 485–560 (1992)
12. Yatsymirskyy, M.: New matrix model for two-channel bank of biorthogonal filters. Metody Informatyki Stosowanej **1**, 205–212 (2011). (in Polish)
13. NVIDIA: Whitepaper. NVIDIA's Next Generation CUDA^{TM} Compute Architecture. Fermi^{TM}
14. Hussein, M.M., Mahmoud, A.O.: Performance evaluation of discrete wavelet transform based on image compression technique on both CPU and GPU. In: International Conference on Innovations in Engineering and Technology, pp. 68–7 (2013)

Deterministic Extraction of Compact Sets of Rules for Subgroup Discovery

Juan L. Domínguez-Olmedo[✉], Jacinto Mata Vázquez, and Victoria Pachón

Escuela Técnica Superior de Ingeniería, University of Huelva, Huelva, Spain
{juan.dominguez,mata,vpachon}@dti.uhu.es

Abstract. This work presents a novel deterministic method to obtain rules for Subgroup Discovery tasks. It makes no previous discretization for the numeric attributes, but their conditions are obtained dynamically. To obtain the final rules, the AUC value of a rule has been used for selecting them. An experimental study supported by appropriate statistical tests was performed, showing good results in comparison with the classic deterministic algorithms CN2-SD and APRIORI-SD. The best results were obtained in the number of induced rules, where a significant reduction was achieved. Also, better coverage and less number of attributes were obtained in the comparison with CN2-SD.

Keywords: Data mining · Machine learning · Rule-based systems

1 Introduction

Data mining techniques are usually divided into predictive and descriptive techniques. In predictive techniques, the models are induced from data labelled with a class and the aim is to predict the value of the class for instances in which the class is previously unknown. On the other hand, the main objective of the descriptive techniques is to find understandable and useful patterns, based on unlabelled data.

Another kind of task consists of obtaining descriptive knowledge regarding a property of interest, that is, to discover significant differences between groups that exist in a dataset and which a human can interpret. In recent years, three approaches have aroused the interest of researchers: Contrast Set Mining [1], Emerging Pattern Mining [2] and Subgroup Discovery (SD) [3, 4]. These techniques are known as Supervised Descriptive Rule Discovery [5] whose goal is the discovery of interesting patterns in relation to a specific property of interest.

Differing SD algorithms have been presented throughout the literature in order to solve SD tasks based on beam search such as CN2-SD [6] or APRIORI-SD [7], exhaustive such as SD-Map [8], or genetic fuzzy systems such as NMEEF-SD [9], amongst others. Other studies propose algorithms based on evolutionary computation for extracting rules, as EDER-SD [10]. Recently, an evolutionary fuzzy system based on genetic programming for SD has been described in [11]. The handling of continuous attributes is a huge problem in the discovery of the most valuable set of rules. Most of

© Springer International Publishing Switzerland 2015
K. Jackowski et al. (Eds.): IDEAL 2015, LNCS 9375, pp. 138–145, 2015.
DOI: 10.1007/978-3-319-24834-9_17

the aforementioned algorithms perform a prior discretization of the continuous attributes before carrying out the rule induction process. Consequently, these algorithms may fail to optimise data breakpoints during the discretizing process leading to less accurate results. The replacement of continuous attributes by nominal attributes can result in suboptimal results [12]. For example, for the APRIORI-SD and CN2-SD algorithms, a discretization for the continuous attributes by using the Fayyad discretization method [13] is carried out. In the evolutionary fuzzy systems, a fixed number of linguistic labels is assigned to the continuous attributes.

This paper presents a novel approach to find compact sets of rules for SD: DEQAR-SD (Deterministic Extraction of Quantitative Association Rules). One contribution of DEQAR-SD is that it does not carry out a discretization of numeric attributes before the rule induction process. Conditions for the numeric attributes are obtained in the rules induction process itself. Also, the value of AUC is used to select the rules, instead of the use of the *weighted relative accuracy* measure. An experimental study supported by appropriate statistical tests has been realized to confirm the effectiveness of the DEQAR-SD algorithm, in comparison with the classic algorithms CN2-SD and APRIORI-SD.

This paper is structured as follows: next section gives a short description of SD. In Sect. 3, the proposed approach is explained. In Sect. 4, the experimental study and the results obtained are presented. Finally, in Sect. 5 the major conclusions and further research are outlined.

2 Subgroup Discovery

The SD task can be defined as discovering properties of a population by obtaining simple rules (understandable), highly significant and with high support. An induced subgroup takes the form of a rule $A \rightarrow C$, where A is the rule antecedent (usually as a conjunction of conditions) and C, the consequent, is the property of interest (class). Descriptive measures of rule interestingness evaluate each individual subgroup and are thus appropriate for evaluating the success of subgroup discovery. The proposed quality measures compute the average over the induced set of subgroup descriptions, which enables the comparison between different algorithms [6]. Given a dataset with N examples and a conjunction of conditions X, $n(X)$ represents the number of examples that satisfy X. A description of some of these measures is the following:

- Coverage. Measures the frequency of examples covered by the rule:

$$cov(R) = cov(A \rightarrow C) = \frac{n(A)}{N}$$

- Support. Measures the frequency of correctly classified covered examples. For a ruleset, the overall support counts only once the examples covered by several rules:

$$overall\ support\ (Ruleset) = \frac{1}{N} * \sum_c n\left(c \wedge \bigvee_{A \rightarrow c} A\right)$$

- Significance. Measured in terms of the likelihood ratio statistic of the rule:

$$sig\,(R) = sig\,(A \rightarrow C) = 2 * \sum_i n\,(A \wedge C_i) * \log \frac{n\,(A \wedge C_i)}{n\,(C_i) * n\,(A)\,/N}$$

3   Proposed Approach

In this work, we have used a method based on a deterministic approach to treat the problem of generating rules without a previous discretization of the numerical attributes. In contrast to the typical deterministic fashion of obtaining rules by previously discretizing those attributes, the method employed is based on a dynamic generation of conditions for the attributes. The developed method is an adaptation for SD of the corresponding to association rule mining [14], in which the conditions were formed based on an interval for numerical attributes. In this case, intervals have not been used for numeric attributes. Instead, a condition for a numeric attribute will be restricted to be one of the two following: $X \geq v$ or $X \leq v$, being v a value existing in the dataset for the attribute X. And for categorical attributes, conditions will be in the form $X = v$.

3.1   Method Description

The main characteristics of the proposed method are the following:

- The search of conditions for attributes in the antecedent is done by a depth-first search with backtracking. It scans from left to right the attributes, looking for possible valid conditions, and storing the support reductions due to each condition.
- In the case of a numeric attribute, it checks the existing values in the dataset for that attribute, from the smallest towards the greatest (for conditions of type $\geq$) or from the greatest towards the smallest (for conditions of type $\leq$).
- A typical threshold in association rule mining is the minimum support (*minsup*) that a rule has to satisfy to be considered valid. In this method, *minsup* is calculated from the parameter *minsens* (minimum sensitivity):

$$minsup = \min_c \left(minsens * \frac{n\,(\text{class} = c)}{N} \right)$$

- The other typical threshold is the minimum confidence (*minconf*). In this method *minconf* is not used, because simply the rules with a lift value less than 1 are discarded (their confidence don't reach the probability of the class).
- The exhaustive search for conditions in the antecedent can be very costly, mainly if the number of attributes is not low and the numeric ones have a lot of different values. Trying to reduce this search, the parameters *maxAttr* and *delta* are used. The parameter *maxAttr* limits the number of attributes in the antecedent of the rule, and *delta* is used to control the search of conditions for the numerical attributes, by defining dynamically the minimum decrease in support for the next condition of an attribute. This minimum decrease in support is evaluated at the beginning of the search of

conditions for an attribute, by computing *(sup - minsup) * delta*, where *sup* is the current support of the antecedent.

- The final process is to select a subset of the generated rules. In this method a separate ranking (sorted list) is employed for each class of the dataset. The measure used for sorting the rules has been the AUC value *(Area Under the ROC Curve)*. For a single rule, its AUC value can be simply calculated by:

$$AUC(R) = AUC(A \rightarrow C) = \frac{sensitivity(R) + specificity(R)}{2}$$

$$= \frac{1}{2} * \left(\frac{n(R)}{n(C)} + 1 - \frac{n(A) - n(R)}{N - n(C)} \right)$$

- The parameters *rankingSize* and *mincover* are used to try to select the less number of rules. The parameter *rankingSize* limits the number of rules in each ranking. And *mincover* helps to filter rules in each ranking, when with a subset of the best rules is possible to reach a minimum percentage of covered examples.

3.2 Algorithm

The summary of the algorithm DEQAR-SD is shown in Table 1. In line 10, each ranking is filtered suppressing the lower rules that are similar to any one upper in the ranking. Two rules are considered to be similar if they have the same attributes and equality or overlapping is present in all their values. In line 11, the lower rules that doesn't give additional covering are suppressed, or if the minimum percentage of covered examples has been reached with the rules upper in the ranking.

Table 1. Algorithm DEQAR-SD

Input: Dataset *D*, *minsens*, *delta*, *maxAttr*, *rankingSize*, *mincover*
Output: Ruleset *R*
  1. while (is possible to obtain a new antecedent) do
  2.   *A* = getNextAntecedent( )
  3.   for each class *c* do
  4.   if (rule(*A* → *c*) is valid) do
  5.   insert rule(*A* → *c*) in the ranking[*c*]
  6.   end if
  7.   end for
  8. end while
  9. for each class *c* do
10.   filter ranking[*c*] to suppress similar rules
11.   filter ranking[*c*] to get only the first rules covering a minimum of examples
12.   add the rules of ranking[*c*] to *R*
13. end for
14. return *R*

---

4   Experimental Study

The proposed method has been compared to the classic deterministic algorithms Apriori-SD and CN2-SD, using a variety of datasets from the UCI machine learning repository [15]. The characteristics of the 16 employed datasets are summarized in Table 2.

Table 2. Datasets employed

dataset	# examples	# attributes	# numeric	# classes
appendicitis	106	7	7	2
australian	690	14	6	2
balance	625	4	4	3
bridges	102	7	3	2
cleveland	303	13	13	5
contraceptive	1473	9	9	3
diabetes	768	8	8	2
echo	131	6	5	2
ecoli	336	7	7	8
haberman	306	3	3	2
hayes-roth	160	4	4	3
heart	270	13	7	2
iris	150	4	4	3
led	500	7	7	10
wine	178	13	13	3
wisconsin	699	9	9	2

The implementations of Apriori-SD and CN2-SD algorithms have been the ones present in the Keel software [16]. The Fayyad discretization method was employed due to it was required for treating the numerical attributes. The value for the parameters used in Apriori-SD have been *MinSupport* = 0.03, *MinConfidence* = 0.6, *Number_of_Rules* = 5 and *Post-pruning_type* = SELECT_N_RULES_PER_CLASS. For CN2-SD, the values for the parameters have been *Nu_Value* = 0.5, *Percentage_Examples_To_Cover* = 0.95, *Star_Size* = 5, *Use_multiplicative_weigths* = NO and *Use_Disjunt_Selectors* = NO. The values for the parameters for DEQAR-SD have been *minsens* = 0.7, *delta* = 0.25, *maxAttr* = 2, *rankingSize* = 5 and *mincover* = 0.95.

The quality measures which have been possible to compute in the three algo-rithms are the following five: average coverage, overall support, number of rules, average number of attributes in the antecedent, and average significance. The final value for the quality measures were obtained through a 10-fold stratified cross vali-dation. The results obtained in the experiments are shown in Tables 3 and 4. DEQAR-SD obtained good average results in coverage, support and ruleset size, and its results in number of attributes and significance were not the worst of the three algorithms. Statistical tests have been performed to assess the significance of the comparison, using the Wilcoxon signed-ranks test [17]. The results of the statistical comparison by pairs is shown in Table 5. As it can be seen, with a significance level $\alpha = 0.05$ the null hypothesis (both algorithms perform equally well) is rejected for ruleset size in the comparison with both algorithms. Also, it is rejected for coverage and number of attributes in the comparison with CN2-SD.

Table 3. Results obtained in coverage and overall support.

	Coverage			Support		
	DEQAR-SD	Apriori-SD	CN2-SD	DEQAR-SD	Apriori-SD	CN2-SD
appendicitis	0.620	0.469	0.409	0.899	0.935	0.926
australian	0.500	0.499	0.418	0.855	0.958	0.962
balance	0.600	0.457	0.330	0.899	0.910	0.961
bridges	0.562	0.297	0.411	0.933	0.874	0.943
cleveland	0.425	0.542	0.332	0.858	0.535	0.898
contraceptive	0.615	0.090	0.346	0.937	0.236	0.965
diabetes	0.595	0.263	0.468	0.899	0.848	0.945
echo	0.565	0.735	0.430	0.840	0.703	0.931
ecoli	0.261	0.233	0.207	0.920	0.872	0.917
haberman	0.693	0.848	0.500	0.964	0.735	0.735
hayes-roth	0.674	0.073	0.263	0.938	0.194	0.594
heart	0.482	0.484	0.394	0.893	0.981	0.937
iris	0.365	0.295	0.347	0.947	0.987	0.973
led	0.223	0.110	0.197	0.934	0.858	0.924
wine	0.346	0.301	0.306	0.944	0.977	0.944
wisconsin	0.535	0.453	0.507	0.954	0.987	0.954
	0.504	0.384	0.367	0.913	0.787	0.907

Table 4. Results obtained in number of rules, number of attributes and significance.

	Ruleset Size			# Attributes			Significance		
	DEQAR	Apriori	CN2	DEQAR	Apriori	CN2	DEQAR	Apriori	CN2
appendicitis	3.90	10.00	8.80	1.42	1.20	1.81	1.402	1.864	1.867
australian	2.20	10.00	12.20	1.07	1.60	2.58	20.505	17.398	16.770
balance	4.00	7.80	10.10	1.00	1.13	1.72	4.651	6.102	7.486
bridges	3.60	8.10	7.30	1.38	1.41	2.18	0.740	0.656	0.653
cleveland	14.90	5.00	16.10	1.93	1.36	2.12	6.477	7.198	7.932
contraceptive	11.40	8.00	17.00	1.69	2.25	2.35	4.356	8.161	5.792
diabetes	7.10	10.00	11.10	1.82	1.46	2.15	5.087	4.876	4.667
echo	6.70	5.50	9.10	1.76	1.12	2.22	0.585	0.249	0.532
ecoli	13.10	20.50	16.60	1.99	1.88	3.43	9.194	9.000	12.193
haberman	6.30	2.00	2.00	1.21	1.00	1.00	0.563	0.550	1.543
hayes-roth	6.90	5.00	4.00	1.00	1.40	1.50	1.360	3.769	4.565
heart	7.60	10.00	8.80	1.87	1.40	1.73	4.135	3.873	5.013
iris	3.30	15.00	3.00	1.62	1.38	1.10	8.091	7.310	9.637
led	24.30	40.40	25.70	1.89	3.33	2.49	21.689	16.132	20.390
wine	4.20	15.00	5.00	1.82	1.69	2.46	9.301	8.604	9.896
wisconsin	2.80	10.00	3.30	2.00	1.00	2.97	29.719	24.911	32.278
	7.64	11.39	10.01	1.59	1.54	2.11	7.991	7.541	8.826

For last, Fig. 1 shows a short analysis of how the size of the dataset affects the execution time in DEQAR-SD. Three datasets have been reduced in their number of examples, and the times have been measured using a computer equipped with a 1.8 GHz processor, 2 GB of memory, Windows XP and C++ compiler. It can be seen that the algorithm scales well in the size of the dataset (number of examples).

Table 5. Results for the Wilcoxon tests.

Comparison	Quality measure	R+	R-	p-value	Hypothesis
DEQAR-SD vs. Apriori-SD	Coverage	103.000	33.000	0.07392	Non-rejected
	Support	100.000	36.000	0.10458	Non-rejected
	Ruleset Size	108.000	28.000	0.03864	Rejected
	# Attributes	52.000	84.000	0.40654	Non-rejected
	Significance	83.000	53.000	0.43540	Non-rejected
DEQAR-SD vs. CN2-SD	Coverage	136.000	0.000	0.00003	Rejected
	Support	40.500	95.500	0.16721	Non-rejected
	Ruleset Size	115.000	21.000	0.01309	Rejected
	# Attributes	124.000	12.000	0.00214	Rejected
	Significance	30.000	106.000	0.05066	Non-rejected

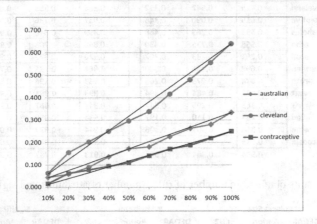

Fig. 1. Execution times (s) varying the number of examples.

5   Conclusions and Future Research

In this work a novel deterministic method to find rules for SD has been described and tested. DEQAR-SD does not carry out a previous discretization of numeric attributes. Instead, conditions for the numeric attributes are obtained dynamically. To select the best rules, the AUC value of a rule is employed.

An experimental study supported by appropriate statistical tests was performed, showing good results in comparison with the classic deterministic algorithms CN2-SD and APRIORI-SD. The best results were obtained in the number of induced rules, where a significant reduction was achieved. Also, in the comparison with CN2-SD, better coverage and less number of attributes were obtained.

As future research, an adaptation of the method for classification tasks should be easily achieved, and comparisons with other techniques, such as evolutionary algorithms, could be of interest.

Acknowledgments. This work was partially funded by the Regional Government of Andalusia (Junta de Andalucía), grant number TIC-7629.

References

1. Bay, S.D., Pazzani, M.J.: Detecting group differences. Mining contrast sets. Data Min. Knowl. Discov. **5**(3), 213–246 (2001)
2. Dong, G., Li, J.: Efficient mining of emerging patterns. Discovering trends and differences. In: Proceedings of the 5th ACM SIGKDD International Conference on Knowledge Discovery and Data Mining, pp. 43–52 (1999)
3. Klösgen, W.: Explora: A multipattern and multistrategy discovery assistant. Advances in Knowledge Discovery and Data Mining, pp. 249–271. American Association for Artificial Intelligence, Cambridge (1996)
4. Wrobel, S.: An algorithm for multi-relational discovery of subgroups. In: Proceedings of the 1st European Conference on Principles of Data Mining and Knowledge Discovery (PKDD-97), pp 78–87 (1997)
5. Novak, P.N., Lavrač, N., Webb, G.: Supervised descriptive rule discovery: a unifying survey of contrast set, emerging pattern and subgroup mining. J. Mach. Learn. Res. **10**, 377–403 (2009)
6. Lavrač, N., Kavsek, B., Flach, P.A., Todorovski, L.: Subgroup discovery with CN2-SD. J. Mach. Learn. Res. **5**, 153–188 (2004)
7. Kavsek, B., Lavrač, N.: APRIORI-SD: adapting association rule learning to subgroup discovery. Appl. Artif. Intell. **20**(7), 543–583 (2006)
8. Atzmüller, M., Puppe, F.: SD-Map – a fast algorithm for exhaustive subgroup discovery. In: Fürnkranz, J., Scheffer, T., Spiliopoulou, M. (eds.) PKDD 2006. LNCS (LNAI), vol. 4213, pp. 6–17. Springer, Heidelberg (2006)
9. Carmona, C.J., González, P., del Jesus, M.J., Herrera, F.: NMEEF-SD: non-dominated multi-objective evolutionary algorithm for extracting fuzzy rules in subgroup discovery. IEEE Trans. Fuzzy Syst. **18**(5), 958–970 (2010)
10. Rodríguez, D., Ruiz, R., Riquelme, J.C., Aguilar-Ruiz, J.S.: Searching for rules to detect defective modules: a subgroup discovery approach. Inf. Sci. **191**, 14–30 (2012)
11. Carmona, C.J., Ruiz-Rodado, V., del Jesus, M.J., Weber, A., Grootveld, M., González, P., Elizondo, D.: A fuzzy genetic programming-based algorithm for subgroup discovery and the application to one problem of pathogenesis of acute sore throat conditions in humans. Inf. Sci. **298**, 180–197 (2015)
12. Grosskreutz, H., Rüping, S.: On subgroup discovery in numerical domains. Data Min. Knowl. Discov. **19**(2), 210–226 (2009)
13. Fayyad, U., Irani, K.B.: Multi-interval discretization of continuous-valued attributes for classification learning. In: 13th International Joint Conference on Artificial Intelligence, pp. 1022–1029 (1999)
14. Domínguez-Olmedo, J.L., Mata, J., Pachón, V., Maña, M.J.: A deterministic approach to association rule mining without attribute discretization. In: Snasel, V., Platos, J., El-Qawasmeh, E. (eds.) ICDIPC 2011, Part I. CCIS, vol. 188, pp. 140–150. Springer, Heidelberg (2011)
15. Lichman, M.: UCI Machine Learning Repository. School of Information and Computer Science, University of California, Irvine, CA (2013). http://archive.ics.uci.edu/ml
16. Alcalá-Fdez, J., Fernandez, A., Luengo, J., Derrac, J., García, S., Sánchez, L., Herrera, F.: KEEL data-mining software tool. J. Multiple-Valued Logic Soft Comput. **17**, 255–287 (2011)
17. Demšar, J.: Statistical comparisons of classifiers over multiple data sets. J. Mach. Learn. Res. **7**, 1–30 (2006)

Variable Transformation for Granularity Change in Hierarchical Databases in Actual Data Mining Solutions

Paulo J.L. Adeodato[✉]

Centro de Informática, Universidade Federal de Pernambuco, Recife, Brazil
pjla@cin.ufpe.br

Abstract. This paper presents a variable transformation strategy for enriching the variables´ information content and defining the project target in actual data mining applications based on relational databases with data at different grains. In an actual solution for assessing the schools´ quality based on official school survey and students tests data, variables at the student and teachers´ grains had to become features of the schools they belonged. The formal problem was how to summarize the relevant information content of the attribute distributions in a few summarizing concepts (features). Instead of the typical lowest order distribution momenta, the proposed transformations based on the distribution histogram produced a weighted score for the input variables. Following the CRISP-DM method, the problem interpretation has been precisely defined as a binary decision problem on a granularly transformed student grade. The proposed granular transformation embedded additional human expert´s knowledge to the input variables at the school level. Logistic regression produced a classification score for good schools and the AUC_ROC and Max_KS assessed that score performance on statistically independent datasets. A 10-fold cross-validation experimental procedure showed that this domain-driven data mining approach produced statistically significant improvement at a 0.99 confidence level over the usual distribution central tendency approach.

Keywords: Granularity transformation · Relational data mining · School quality assessment · Educational decision support system · CRISP-DM · Domain-driven data mining · Logistic regression · Ten-fold cross-validation

1 Introduction

The development of an actual data mining solution for school quality assessment from relational databases faced technical issues that motivated this scientific research.

The quality of and accessibility to education are key aspects focused on by high level offices responsible for educational policies around the world. They define metrics for monitoring the quality of education and goals to be achieved at each of its stages. In recent years, the ease of data gathering and sheer volume has allowed for the application of new methods, particularly, those based on data mining approaches.

Monitoring and assessing the quality of education are often delicate issues but government and regulatory bodies have to define standards to verify the effectiveness

© Springer International Publishing Switzerland 2015
K. Jackowski et al. (Eds.): IDEAL 2015, LNCS 9375, pp. 146–155, 2015.
DOI: 10.1007/978-3-319-24834-9_18

of public policies. In Brazil, the National Institute for Educational Studies (INEP) produces two annual databases [1] used in this research: (1) the School Census which is a survey of the schools for basic education and (2) the National Secondary School Exam (ENEM) that evaluates student performance at end of basic education.

Recently established as the mechanism for students' admission to higher education in public (free) universities, ENEM has become a reliable and rich data source. Added to the technical knowledge, ENEM captures information on a socio-economic-cultural questionnaire from each student. Integrating the School Census with the performance and profile of students in ENEM enables the Federal Government to define and validate public policies for education [2]. These databases can be seen as tables of a relational database once that they use unique keys school, student and teacher.

Nevertheless, not much systematic knowledge has been extracted from these databases, mainly because they are in different grains; students and teachers in one level and the schools in the level above. This was the motivation for this paper that proposes transformations for embedding human expert's knowledge from the students' and teachers' distributions of variables as features of the schools. The quality assessment of the secondary schools also illustrates the proposed approaches following the CRISP-DM (CRoss Industry Standard Process for Data Mining) [3] method.

Even being an apparent straightforward binary decision application, there are two important issues that require scientific consideration; the granularity change from the students and the teacher levels to the school level, and the characterization of a "good school" without entering the sensitive meanders of educational approaches.

The paper is organized in four more sections. Section 2 presents problem characterization and the proposed strategy for the formal definition of the binary goal. Section 3 proposes a systematic granularity change transformation of categorical variables for embedding domain experts knowledge. Section 4 describes the experimental procedure used to validate the proposed approaches, from data preparation to performance evaluation. Section 5 concludes the paper presenting its main results and identifying its limitations and proposing ways to solve them.

2 Problem Characterization and Goal Setting

In business, one of the most common decision strategies for selecting the eligible candidates for an action is ranking them according to a classification score and choosing those above a pre-defined threshold [4]. That is used in applications such as staff selection, fraud detection [5], and resources allocation in public policies, for instance.

This score is computed by either weighing a set of variables based on human defined parameters or by applying a function learned by a classification algorithm from a set of data with binary labels as desired response, according to specific optimization criteria. In some domains of application, such as education, several problems are ill-defined simply because stakeholders do not reach consensus on either methods [6].

This paper proposes a systematic approach for well defining the problem to be solved by machine learning algorithms based on the supervised learning paradigm with a data dependent strategy to label each example as "good" or "bad" for binary decision making. That involves two scientific issues which represent controversial points in the application domain: (1) which metrics should be used as a ranking score for evaluating the quality of the school and (2) which threshold should be adopted as a criterion to define what would be a "good" school in the binary decision.

2.1   Granularity Change for the Goal Variable

The ENEM [2] has been conceived to assess the quality of the Brazilian secondary schools based on their students' evaluation on the test. Despite arguments among experts on education, they have agreed that the performance of the students at the 11th grade would represent the performance of the school and also agreed that the mean student score would be the most relevant indicator of each school [2]. This school score is already used to generate an annual ranking of Brazilian secondary schools.

Technically, the attribute distribution at a finer grain is summarized by a single value at the decision level; a definition that is in accordance with the granularity change recommendation for using the lowest order momenta as features of the distribution of a continuous variable [7]. The central tendency of the distribution could alternatively have been defined as the median of the students' scores which is more robust in cases of asymmetric distributions [8]. This paper preserves the mean of the students' scores as the school main performance indicator for both its technical recommendation and its acceptance by experts in the application domain.

The grade for each student is the arithmetic average of their grades in each test while the school grade is the arithmetic mean of their students' grades considered above filtered only for schools with over 14 students who have attained valid scores.

2.2   Binary Goal Definition ("Good" or "Bad" School)

Once the metrics has been defined (mean), the next controversial definition to be made is the threshold for considering a "good" or a "bad" school for labeling the target dichotomized as a binary classification. Several experts on the domain disagree that this dichotomization is either useful or fair [2].

To circumvent the controversy and bring a higher level of abstraction that enables future comparison across years, regardless of the degree of difficulty of the exams, this study used statistics concepts for setting the threshold. Separatrices (quartiles, deciles, *etc.*) away from the tails of the distributions not only are robust against extreme values (outliers) [9] but also can be a straightforward data dependent dichotomizing criterion of interest for the application domain. The upper quartile has already been successfully used as threshold [10] on a continuous goal variable for creating a binary target-variable thus converting the problem into a binary classification. If the research objective were to estimate the propensity for excellent schools, the top decile would be a more appropriate grade threshold. However, as the research objective is to estimate the propensity for schools being good and using the knowledge extracted for improving bad schools, good schools will be considered those whose grades are in the upper quartile of the grades distribution of the schools included in the research scope.

3   Granularity Transformation

"How can each school have a *school home type* attribute obtained from the distribution of the attribute *student home type* of its students? How can each school its *faculty education level* attribute obtained from the distribution of the attribute *teacher education level* of its teachers?"

The questions above exemplify a broad class of problems of how to summarize concepts of relational databases that represent systems composed of subsystems in various hierarchical levels (with *1:n* relationships between them). That occurs in several application domains (credit, health, education *etc*.) where detailed information belonging to the lower levels entities (transactions, examinations, students, teachers *etc*.) need to be summarized as features of the object in the decision level above (cardholder, patient, school *etc*.) in a process referred to as granularity transformation. The number of hierarchical levels and diversity of entities at each level are factors that increase the complexity of the data mining problems.

In the relational database structure exemplified below, part of the attributes is in the school grain and part in the student and teacher grains. If the decision-making process flows on the school level, it is necessary to transform the attributes (the explanatory variables) from the student and teacher grains to the grain school (Fig. 1).

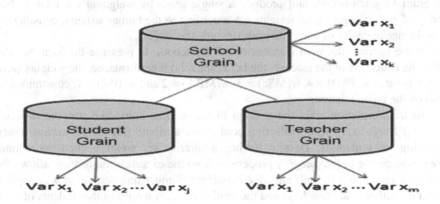

Fig. 1. Example of database entities hierarchical relations and different information grains.

The problem is how to consolidate each attribute from the multitude of students and teachers as a school collective attribute still preserving the conceptual essence of the original individual attribute. Mathematically, the representation of a distribution by a single or a few concepts could be done by its lowest order momenta; an approach similar to the moment invariants approaches used for feature extraction in image processing [7, 11], for example. However, for some applications such as fraud detection [5] and breast cancer detection, extreme values are the most relevant features. In both cases, a data dependent approach combines the use of statistics and knowledge of experts in the application domain to transform the attributes' granularity, in a process coined **D**omain-**D**riven **D**ata **M**ining (D^3M) [12]. In this paper, for comparison of the proposed approach, each distribution from the grains "Students" and "Teachers" was represented in the grain "School" by the mode to represent categorical variables.

3.1   Granularity Change of Categorical Variables with Embedded Knowledge

The second contribution of this paper is the proposal to embed human knowledge on granularity changes of categorical variables which either possess semantics on their categories or, at least, an order relation in terms of propensity towards the binary target from the domain expert's point of view. Specific categorical variables deemed very relevant by educators, such as the level of education of parents and teachers, are the focus of specific transformations on the distribution in the lower grain to produce features for the decision grain instead of the distribution mode.

Just to illustrate, the distribution mode for the variables above would rarely (if ever) be "Ph.D." on a secondary school. However, it is generally accepted that the higher the educational level of the parents and teachers of a school, the better the performance of its students on tests [2] and the distribution mode would be unable to capture this knowledge.

So, for granularity transformation, this paper proposes to construct the variable histogram for each school and produce a single score by weighing the relative frequency of each category with weights set according to the human experts, considering the influence/intensity of each category towards the target.

For the faculty education distribution of the school, to preserve the idea that the higher the education of the teachers the better the school performance, the weights were arbitrarily set as w(PhD) = 4, w(MSc) = 3, w(MBA) = 2 and w(BSc) = 1, constrained to preserve the order relation.

This transformation maps the concept of the original individual attribute (*teacher education level*) in a single collective continuous attribute (*faculty education level*), preserving its semantics. Despite having arbitrarily set weights, their monotonic increase according to the category propensity from the experts point of view allows for non-linear approaches to built effective classifiers. Similar transformation was applied to both the *father education level* and the *mother education level* of the students of each school, with integer weights in a broader range starting from the category "illiterate".

These transformations are not only based on human expert knowledge; their semantics could have been obtained from the years of education of parents and teachers. However, this strategy does not apply for other categorical attributes from the student table (*ethnic group, home type, home location etc.*) and the teacher table (*ethnic group, marital status, contract regime etc.*) when there is no knowledge available.

4   Experimental Project

This section describes all the experimental procedures involved in the validation process of the proposed approach for the weighted score granularity transformation. The procedure was carried out on a database on education, using logistic regression as classifier and using the Area Under the ROC curve and the Maximum Kolmogorov-Smirnov distance as performance metrics, as described below. The goal was to compare if the proposed weighted score was more efficient than the usual mode for granularity transformation of the categorical attributes.

The procedure was decomposed in two stages: (1) learn the logistic regression parameters from the whole dataset in a hold-out process for an initial estimation of the β coefficients and, (2) measure the overall classifier performance in a 10-fold cross-validation process for defining confidence intervals for the hypothesis test.

4.1 Data Sets and Preparation

The research used two Brazilian official public databases with microdata from the National Secondary School Exam 2012 (ENEM 2012) [1] with the students' socio-economic-cultural information and their performance grades on the test, and the School Census 2012 [1]. After integration and filtering, the dataset ended up with 4,400 secondary private schools with over 14 students with valid scores on the exam.

The binary classification target has been defined as described in Sect. 2 by setting the binarizing threshold on the third quartile of the schools' scores distribution.

The granularity change for the school grain followed the procedures described in Sect. 3. The three categorical variables on the parents and teachers' education level were represented by their mode for the usual approach and by the human expert weighed proposed approach.

The missing data in these particular categorical attributes were filled according to the domain interpretation and parents' missing education was coded as illiterate.

4.2 Classifier - Logistic Regression

The technique chosen to generate the classification score for a school being good was the logistic regression which has been successfully applied to binary classification problems in various areas such as credit, among others. Besides the score, the technique identifies and quantifies the key attributes that significantly influence the binary target variable as a set of independent or explanatory variables (attributes).

The behavior of a binary dependent variable based on a set of p independent variables x_p (explanatory features $x_1, \ldots, x_p$) is expressed by the logit function indicated in Eq. (1) below. It presents explicitly the knowledge extracted from data in terms of the coefficients (β) [8].

$$\log\left\{\frac{\pi(x)}{1 - \pi(x)}\right\} = \beta_0 + \beta_1 x_1 + \beta_2 x_2 + \cdots + \beta_p x_p, \tag{1}$$

where:

$$\pi(x) = \frac{\exp(\beta_0 + \beta_1 x_1 + \beta_2 x_2 + \ldots + \beta_p x_p)}{1 + \exp(\beta_0 + \beta_1 x_1 + \beta_2 x_2 + \ldots + \beta_p x_p)}, \tag{2}$$

The output is a continuous score that ranks the chances for a school to be good.

The paired hold-out experiment was run for both the usual mode and the weighed representations of the three categorical attributes on educational levels for the teachers, the fathers and the mothers. Table 1 shows the attributes influence in the models measured by their β coefficients.

Table 1. The three categorical attributes in the logistic regression model.

	Human expert	Mode	
Attributes	β expert weight	Categories	β mode
FatherEducationLevel	0.012	Illiterate	−18.700
		Incomplete primary	−2.506
		Primary	−0.661
		Incomplete secondary	−0.645
		Secondary	−0.331
		Incomplete BSc	0.210
		BSc	1.074
		MBA	1.976
MotherEducationLevel	0.012	Incomplete primary	18.700
		Primary	18.294
		Incomplete secondary	20.646
		Secondary	18.997
		Incomplete BSc	18.914
		BSc	20.298
		MBA	20.427
TeacherEducationLevel	0.008	BSc	−0.105
		MBA	0.168
Constant	−13.870		−21.099

It is important to highlight the difficulty for analyzing the effect of the attribute in the model when decomposed in each of its categories isolate. That is one advantage of the proposed approach.

Referring to the application domain, there is a clear influence of the educational level of the teachers and the fathers on the school performance. However the mothers influence greatly the students education irrespective of their own educational level. That somehow contradicts the human knowledge the approach attempts to embed.

4.3   Performance Metrics - ROC and KS

Considering that one of the purposes of KDD binary solutions is to produce a ranking score for an actual application, the main goal of the performance comparison is to assess if there is significant improvement in the discriminating power of the score with the proposed granularity variable transformation compared to the mode.

The model overall performance has to be assessed on a statistically independent test set. From the 4,400 schools of the data sample, the top quartile was labeled as being "good", as defined in Sect. 2. The hold-out sampling partitioned the dataset stratified by the target label in modeling and test sets containing 3/4 and 1/4 of the examples, respectively. The test set had 1,100 examples with 275 from the target class. The 10-fold cross-validation process partitioned the dataset in 10 sets having 440 examples each with 110 from the target class, since the sampling was stratified.

The ROC Curve [4, 13] is a non parametric performance assessment tool that represents the compromise between the true positive and the false positive example classifications based on a continuous output along all its possible decision threshold values (the score). This paper uses the Area Under the Curve (AUC_ROC) as performance indicator because it is valid for assessing the performance throughout the whole continuous range of scores [13].

The Kolmogorov-Smirnov distribution has been originally conceived as an adherence test for distribution fitting to data [14]. In binary classification problems, it has been used as dissimilarity metrics for assessing the classifier's discriminant power. The maximum distance that its score produces between the cumulative distribution functions (CDFs) of the two data classes (Max_KS2) [15], is the key metrics being invariant to score range and scale, making it suitable for classifiers' comparisons.

4.4 Ten-Fold Cross-Validation and Hypothesis Test

The second stage of the experiments was carried-out by proceeding a stratified 10-fold cross-validation comparison, using 9 folds for training the classifier and the held-out fold for measuring the performance difference with the AUC_ROC and the Max_KS2 metrics described above [16] (Table 2).

Despite some variations in intensity, the difference was systematically favorable to the proposed approach. That can be formally verified in Table 3 below which shows the

Table 2. Performance comparison of the logistic regression model for each fold.

Metrics	Max_KS2		AUC_ROC	
Fold no.	Mode	Human expert	Mode	Human expert
1	0.582	0.640	0.832	0.887
2	0.585	0.610	0.834	0.880
3	0.520	0.547	0.796	0.853
4	0.457	0.544	0.773	0.852
5	0.599	0.629	0.841	0.885
6	0.530	0.560	0.814	0.868
7	0.626	0.602	0.842	0.870
8	0.558	0.615	0.818	0.884
9	0.446	0.536	0.766	0.841
10	0.509	0.586	0.813	0.869
Mean	**0.541**	**0.587**	**0.813**	**0.869**
SD	**0.060**	**0.038**	**0.027**	**0.016**

1-tailed paired t-test. The results have showed that the proposed approach presented a statistically significant improvement in performance for both metrics at a 0.95 confidence level.

Table 3. Performance 1-tailed paired t-test of the logistic regression model.

	Paired differences					t	df	Sign. 1-tailed
	Mean	Std. dev.	Std. err. mean	Conf. int. dif. lower limit				
				$\alpha = 0.05$	$\alpha = 0.01$			
Max_KS_Dif	0.0457	0.0351	0.0111	0.0206	0.0096	4.123	9	0.0013
AUC_ROC_Dif	0.0560	0.0150	0.0047	0.0453	0.0406	11.809	9	0.0000

5   Concluding Remarks

This paper has presented two approaches for developing effective data mining solutions in relational databases with hierarchical relation between entities. One focused on a data-dependent statistical criterion for labeling classes in binary decision problems. The other embedded expert's knowledge in granularity changes for categorical variables based on their distribution histograms and human specified weights.

Thorough experimental procedure proved that the proposed weighted histogram granularity transformation produced statistically significant improvement in performance in a problem for quality assessment of Brazilian secondary schools.

Despite the success verified in this case study, the human knowledge used for setting the weights of the categories on the educational level is not always confirmed. The mode approach has shown that the mother's educational level influences equally the model, contradicting the initial expectation. Also, human expert knowledge is not always available as in the case of other variables such as home type or field of study, for instance, which would still be represented by their distribution mode.

The author's research group is carrying out a research on a granularity change approach which does not depend on the domain expert's knowledge for optimizing the information content of the categorical variables. To generalize the approach in granularity changes, statistical knowledge is embedded by learning the weights through regression on a small labeled sample reserved only for that purpose. That is very much in accordance with the concept of "trainable feature extractors" currently on the fore for its demand in deep learning [17].

Acknowledgments. The author would like to thank Mr. Fábio C. Pereira for running the experiments.

References

1. INEP Databases. <http://portal.inep.gov.br/basica-levantamentos-acessar>. Accessed 15 March 2015. (In Portuguese)

2. Travitzki, R.: ENEM: limites e possibilidades do Exame Nacional do Ensino Médio enquanto indicador de qualidade escolar. Ph.D. thesis, USP, São Paulo (2013). (In Portuguese)
3. Shearer, C.: The CRISP-DM model: the new blueprint for data mining. J. Data Warehouse. **5**(4), 13–22 (2000)
4. Fawcett, T.: An introduction to ROC analysis. Patt. Recognition Lett. **27**, 861–874 (2006)
5. Bolton, R.J., Hand, D.J.: Statistical fraud detection: a review. Statist. Sci. **17**(3), 235–255 (2002)
6. Nordin, F., Kowalkowski, C.: Solutions offerings: a critical review and reconceptualisation. J. Serv. Manage. **21**(4), 441–459 (2010)
7. Hu, M.K.: Visual pattern recognition by moment invariants. IRE Trans Info. Theor. **8**(2), 179–187 (1962)
8. Hair, Jr., J.F., Black, W.C., Babin, B.J., Anderson, R.E., Tatham, R.L.: Multivariate Data Analysis, 6th edn. Pearson Prentice Hall, Upper Saddle River (2006)
9. Johnson, R.A., Wichern, D.W.: Applied Multivariate Statistical Analysis, 6th edn. Pearson Prentice Hall, Upper Saddle River (2007)
10. Sousa, M.U.R.S., Silva, K.P., Adeodato, P.J.L.: Data mining applied to the processes celerity of Pernambuco's state court of accounts. In: Proceedings of CONTECSI 2008 (2008). (In Portuguese)
11. Flusser, J., Suk, T.: Pattern recognition by affine moment invariants. Pattern Recogn. **26**(1), 167–174 (1993)
12. Cao, L.: Introduction to domain driven data mining. In: Cao, L., Yu, P.S., Zhang, C., Zhang, H. (eds.) Data Mining for Business Applications, pp. 3–10. Springer, US (2008)
13. Provost, F., Fawcett, T.: Robust classification for imprecise environments. Mach. Learn. J. **42**(3), 203–231 (2001)
14. Conover, W.J.: Practical Nonparametric Statistics, 3rd edn. Wiley, New York (1999)
15. Adeodato, P.J.L., Vasconcelos, G.C., et al.: The power of sampling and stacking for the PAKDD-2007 cross-selling problem. Int. J. Data Warehouse. Min. **4**(2), 22–31 (2008)
16. Han, J., Kamber, M., Pei, J.: Data Mining: Concepts and Techniques, 3rd edn. Morgan Kaufmann, Waltham (2012)
17. Kavukcuoglu, K.: Learning feature hierarchies for object recognition. Ph.D. thesis, Department Computer Science, New York University, January 2011

OMAIDS: A Multi-agents Intrusion Detection System Based Ontology

Imen Brahmi[✉] and Hanen Brahmi

Computer Science Department, Faculty of Sciences of Tunis,
Campus University, 1060 Tunis, Tunisia
imen.brahmi@gmail.com

Abstract. Nowadays, as a security infrastructure the *Intrusion Detection System* (IDS) have evolved significantly since their inception. Generally, most existing IDSs are plugged with various drawbacks, *e.g.*, excessive generation of false alerts, low efficiency, etc., especially when they face distributed attacks. In this respect, various new intelligent techniques have been used to improve the intrusion detection process. This paper introduces a novel intelligent IDS, which integrates the desirable features provided by the multi-agents methodology with the benefits of semantic relations. Carried out experiments showed the efficiency of our distributed IDS, that sharply outperforms other systems over real traffic and a set of simulated attacks.

Keywords: Intrusion Detection System · Multi-agents · Ontology

1 Introduction

Due to the growing threat of network attacks, the efficient detection as well as the network abuse assessment are becoming a major challenge. In this respect, the Intrusion Detection System (IDS) has been of use to monitor the network traffic thereby detect whether a system is being targeted by network attacks [2]. Even that IDSs have become a standard component in security infrastructures, they still have a number of significant drawbacks. In fact, they suffer from problems of reliability, relevance, disparity and/or incompleteness in the presentation and manipulation of knowledge as well as the complexity of attacks. This fact hampers the detection ability of IDS, since it causes the generation excessive of false alarms and decreases the detection of real intrusions [2,4].

Indeed, needless to remind that the integration of a multi-agents technology within the IDS can effectively improve the detection accuracy and enhance the system's own security. In fact, the use of multi-agents system for intrusion detection offers an appropriate alternative to the IDS with several advantages listed in literature, *e.g.*, independently and continuous running, minimal overhead, distributivity, *etc.*, [2]. Therefore, multi-agents technology makes the resilience of the system strong and thus ensures its safety [3].

In addition, the concept of ontology has emerged as a powerful method that can improve the intrusion detection features. Thus, the ontology has been shown

© Springer International Publishing Switzerland 2015
K. Jackowski et al. (Eds.): IDEAL 2015, LNCS 9375, pp. 156–163, 2015.
DOI: 10.1007/978-3-319-24834-9_19

to be useful in enabling a security analyst to understand, characterize and share a common conceptual understanding threats [6,9]. Besides, it provides semantic checking to design the signature rules using the SWRL (*Semantic Web Rule Language*) [5], that can solve the disparity issue of security knowledge.

In this paper, we introduce a new distributed IDS, called OMAIDS (*Ontology based Multi-Agents Intrusion Detection System*). OMAIDS stands within the crossroads of the multi-agents system and the ontology technique. Through extensive carried out experiments on a real-life network traffic and a set of simulated attacks, we show the effectiveness of our system.

The remaining of the paper is organized as follows. Section 2 sheds light on the related work. We introduce our new distributed intrusion detection system in Sect. 3. We then relate the encouraging results of the carried out experiments in Sect. 4. Finally, Sect. 5 concludes and points out avenues of future work.

2   Scrutiny of the Related Work

Recently, few approaches are dedicated to the use of semantic web within the intrusion detection field. Worth of mention that the first research in this area was done by Undercoffer *et al.* [9] in 2003. The authors developed an ontology DAML-OIL focused on the target (*centric*) and supply it within the format of the logical description language DARPA *DARPA Agent Markup Language + Ontology Inference Layer*. The proposed ontology is based on the traditional taxonomy classification migrated to semantic model. It allows modeling the domain of computer attacks and facilitates the process of reasoning to detect and overcomes the malicious intrusions.

Based on the DAML-OIL ontology [9], Mandujano [7] investigates an attack ontology, called OID (*Outbound Intrusion Detection*). The introduced ontology provides agents with a common interpretation of the environment signatures, through a FORID system. The latter detects the intrusions based on a matching strategy using a data structure based on the internals of the IDS Snort [8]. Similarly to the works done in [7], Djotio *et al.* [4] proposed a multi-agents IDS, MONI, based on an ontology model, called NIM-COM. The agents are then responsible for enabling the analysis of network traffic and the detection of malicious activities, using the Snort signatures [8]. However, the approach does not consider the secure state which is important to judge false positive alerts and successful possibility of attacks [6]. With the same preoccupation, Abdoli and Kahani [1] proposed a multi-agents IDS, called ODIDS. Based on the techniques of the semantic web, they have built an ontology for extracting semantic relationships between intrusions. The criticism of the ODIDS system is time wasting, since the system needs more time to make a connection between the agents on the network and to send and receive messages between them.

Due to its usability and importance, detecting the distributed intrusions still be a thriving and a compelling issue. In this respect, the main thrust of this paper is to propose a distributed IDS, called OMAIDS, which integrates : (*i*) a multi-agents technology and (*ii*) an ontology model. In this respect, it is shown that the use of such architecture reveals conducive to the development of IDSs [1,3,4,6,7]. The main idea behind our approach is to address limitations of centralized IDSs, by taking advantage of the multi-agents paradigm as well as the ontological representation.

3   The OMAIDS System

Agents and multi-agents systems are one of the paradigms that best fit the intrusion detection in distributed networks [2]. In fact, the multi-agents technology distributes the resources and tasks and hence each agent has its own independent functionality, so it makes the system perform work faster [3].

The distributed structure of OMAIDS is composed of different cooperative, communicant and collaborative agents for collecting and analyzing massive amounts of network traffic, called respectively: SNIFFERAGENT, MISUSEAGENT and REPORTERAGENT. Figure 1 sketches at a glance the overall architecture of OMAIDS.

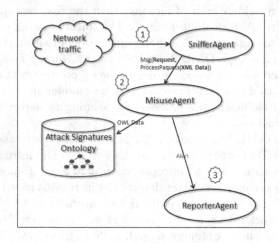

Fig. 1. The architecture of OMAIDS at a glance.

The processing steps of OMAIDS can be summarized as follows:

1. The SNIFFERAGENT captures packets from the network. Indeed, a distributed IDS must undertake to analyze a huge volumes of events collected from different sources around the network. Consequently, the SNIFFERAGENT permits to filter the packets already captured. Besides, it converts them to XML, using the XSTREAM library[1]. Finally, the pre-processed packets will be sent to others agents to be analysed;

---

[1] Available at: http://xstream.codehaus.org/.

2. The MISUSEAGENT receives the packets converted to XML from the SNIFFER-AGENT. It transforms these packets to OWL format in order to be compatible with the SWRL rules stored in the ontology. Now, it is ready to analyze the OWL packets to detect those that correspond to known attacks. Indeed, the MISUSEAGENT searches for attack signatures[2] in these packets, by consulting the ontology ASO (*Attack Signatures Ontology*). Consequently, if there is a similarity between the OWL packets and the SWRL rules that define the attack's signatures, then the agent raises an alert to the REPORTERAGENT;
3. Finally, the REPORTERAGENT generates reports and logs.

OMAIDS detects the attacks through the intelligent agent MISUSEAGENT, which uses an ontology to enrich data intrusions and attack signatures by semantic relationships. In what follows, we present the proposed ontology.

3.1   The Attack Signatures Ontology (ASO)

Ontologies present an extremely promising new paradigm in computer security domain. They can be used as basic components to perform automatic and continuous analysis based on *high-level* policy defined to detect threats and attacks [6]. Moreover, they enable the IDS with improved capacity to reason over and analyze instances of data representing an intrusion [4,9]. Furthermore, the interoperability property of the ontologies is essential to adapt to the problems of the systems distribution, since the cooperation between various information systems is supported [4,6].

Within the OMAIDS system, an ontology, called ASO (*Attack Signatures based Ontology*), is implemented, in order to optimize the knowledge representation and to incorporate more intelligence in the information analysis. The ASO ontology allows the representation of the signatures basis for attacks, used with the agent MISUSEAGENT. Figure 2 depicts a fragment of the ontology ASO, which implements the intrusion detection knowledge. The power and usefulness of ontology, applied to the signature basis issue, provide a simple representation of the attacks expressed by the semantic relationships between intrusion data. We can also infer additional knowledge about intrusion due to the ability of the ontology to infer new behavior by reasoning about data. Therefore, this fact improves the process of decision support for an IDS [1,3,9].

The signature basis incorporates rules provided by the ASO ontology, that allows a semantic mean for reasoning and inferences. In fact, the rules are extracted using the SWRL language (*Semantic Web Rule Language*). The latter extend the ontology and enriches its semantics by the deductive reasoning capabilities [5]. It allows to handle instances with variables ($?x$, $?y$, $?z$). Thus, the SWRL rules are developed according to the scheme: *Antecedent* $\rightarrow$ *Consequent*, where both antecedent and consequent are conjunctions of atoms written $a_1 \wedge$ ...$\wedge a_n$. Variables are indicated using the standard convention of prefixing them

---

[2] An attack signature is a known attack method that exploits the system vulnerabilities and causes security problem [2].

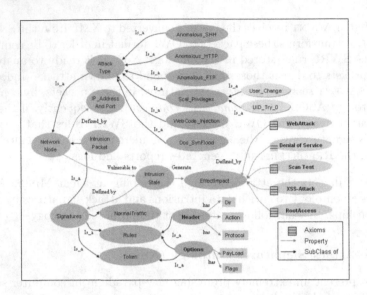

Fig. 2. The Attack Signatures based Ontology ASO.

with a question mark (*i.e.*, "?x"). The following example shows a rule represented with SWRL.

Example. NetworkHost(?z) ∧ IntrusionState(?p) ∧ GeneratedBY(?p,?z) ∧ SQLInjection(?p) ∧ Directd_To(?p,?z) → SystemSQLInjectionState(?p,?z).

Using this syntax, a rule asserting that the composition of the network host(z) and an intrusion state(p) properties implies the attack "*SQL Injection*" property.

4   Experimental Results

In order to assess the overall performance of OMAIDS in a realistic scenario, a prototype of the proposed architecture was implemented using Sun's Java Development Kit 1.4.1, the well known platform JADE[3] 3.7, the Eclipse and the JPCAP[4] 0.7. The ontology ASO is designed using PROTÉGÉ[5].

Through the carried out experiments, we have a twofold aim: (*i*) first, we focus on the assessment of the interaction between agents; (*ii*) Second, we have to stress on evaluating the performance of our system in term of detection ability.

4.1   Interaction Between Agents

Generally, within the existing centralized IDSs, the designed feature of communication and cooperation between their components are badly missing [2]. This

---

[3] Available at: http://jade.tilab.com.

[4] Available at: http://netresearch.ics.uci.edu/kfujii/jpcap/doc/.

[5] Available at: http://protege.stanford.edu/download/download.html.

latter constitutes the main hamper towards efficient detection of attacks [2]. Tacking in the account this issue, the multi-agents architecture of OMAIDS allows to facilitate the communication and the interaction between the agents that operate as the IDS components. In fact, the agents use the ACL (*Agent Communication Language*) language to communicate. Moreover, the information transmitted among agents is sent as text messages and the process complies with the FIPA (*Foundation for Intelligent Physical Agents*)[6] protocols.

The OMAIDS uses several agent's group (sniffing, filtering, analyzing, reporting). Some of these agents need high communication, with rich information, and others just need to share a reduced amount of information. Firstly, the SNIFFERAGENT is responsible of capturing network traffic needed to carry out its task of generating and converting the packets. Figure 3 shows a SNIFFERAGENT whenever a *TCP Connect Scan*[7] is captured and filtered.

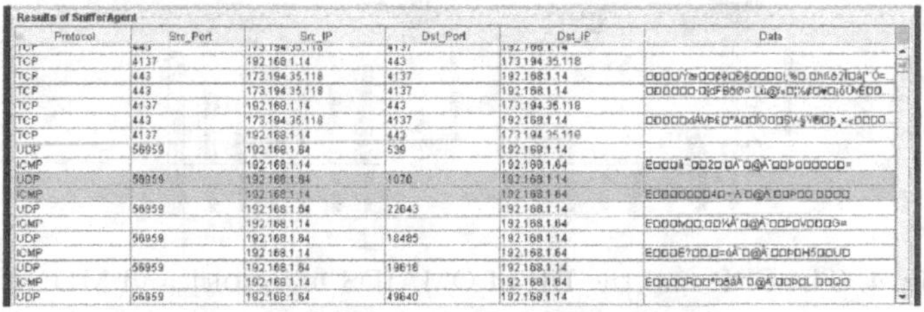

Fig. 3. SNIFFERAGENT within *TCPConnect*.

Once the captured packets are filtered and converted to XML, the SNIFFER-AGENT informs the MISUSEAGENT to analyze these packets. The information includes: (*i*) the protocol; (*ii*) the source IP and port; and (*iii*) the destination IP and port. Based on the signature rules stored in the knowledge base of the ontology ASO, loaded during startup, whenever the MISUSEAGENT perceives a similarity between a packet and a rule, then it detects an attack. Besides, it informs the REPORTERAGENT with the "abnormal" network status. The information includes: (*i*) an alert information indicating that an attack occurs; (*ii*) the date and the time of detection; (*iii*) the IP addresses of both attacker and victim; and (*iv*) the name of the attack.

Finally, we conclude that the agents of our system OMAIDS cooperate by using a reliable communication mechanism. This cooperation is driven by interests expressed by the agents.

4.2   The Detection Ability

In order to evaluate the detection ability of an IDS, two interesting metrics are usually of use [2]: the *Detection Rate* (DR) and the *False Positive Rate*

---

[6] Available at: http://www.fipa.org.

[7] TCP Connect Scan is a scan method used by the operating system to initiate a TCP connection to a remote device. It allows to determine if a port is available.

(FPR). Indeed, the DR is the number of correctly detected intrusions. On the contrary, the FPR is the total number of normal instances that were "incorrectly" considered as attacks. In this respect, the value of the DR is expected to be as large as possible, while the value of the FPR is expected to be as small as possible.

During the evaluations, we compare the results of the OMAIDS system *vs.* that of the IDS SNORT [8] and the multi-agents based ontology one MONI[8] [4]. Moreover, we simulated attacks using the well known tool *Metasploit*[9] version 3.5.1. The simulated eight different attack types are: **attack1**: DoS Smurf; **attack2**: Backdoor Back Office; **attack3**: SPYWARE-PUT Hijacker; **attack4**: Nmap TCP Scan; **attack5**: Finger User; **attack6**: RPC Linux Statd Overflow; **attack7**: DNS Zone Transfer; and **attack8**: HTTP IIS Unicode.

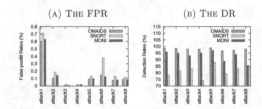

Fig. 4. The FPR and the DR of OMAIDS *vs.* SNORT and MONI.

With respect to Fig. 4(a), we can remark that the FPR of OMAIDS and MONI is significantly lower compared to that of SNORT. This fact is due to the adaptive mechanisms used by the agents, enabling both systems, *i.e.*, OMAIDS and MONI, to better suit the environment. Consequently, the false alarms can be reduced correspondingly. For example, for attack3 the FPR of SNORT can reach values as high as 0.019 % compared to 0.007 % of MONI and 0.005 % of OMAIDS.

Moreover, Fig. 4(b) shows that the DR of OMAIDS is higher than that of MONI. Moreover, among the three investigated IDS, SNORT has the lowest DR. For instance, for attack3, whenever OMAIDS and MONI have the DR 97.9 % and 94.9 %, respectively, SNORT has 74.1 % DR. This is due to his centralized architecture.

Knowing that a main challenge of existing IDSs is to decrease the false alarm rates [2], the main benefit of our system is to lower the false alarm rate, while maintaining a good detection rate.

5   Conclusion

In this paper, we focused on a distributed architecture and multi-agents analysis of intrusions detection system to tackle the mentioned above challenges within

---

[8] We thank Mrs. Djotio *et al.* [4] for providing us with the implementation of MONI system.

[9] Available at: http://www.metasploit.com/.

the IDSs, *i.e.*, the badly communication as well as the low detection ability. Thus, we introduced a multi-agents intrusions detection system called OMAIDS based on an efficient ontology model, called ASO. The carried out experimental results showed the effectiveness of the OMAIDS system and highlighted that our system outperforms the pioneering systems fitting in the same trend.

Worth of mention that the combination of the detection known attacks as well as the unknown ones can lead to improve the performance of the IDS and enhances its detection ability [2]. In this respect, our future work focuses on the integration of data mining techniques within the OMAIDS system.

References

1. Abdoli, F., Kahani, M.: Ontology-based distributed intrusion detection system. In: Proceedings of the 14th International CSI Computer Conference CSICC 2009, Tehran, Iran, pp. 65–70 (2009)
2. Brahmi, I., Ben Yahia, S., Aouadi, H., Poncelet, P.: Towards a multiagent-based distributed intrusion detection system using data mining approaches. In: Cao, L., Bazzan, A.L.C., Symeonidis, A.L., Gorodetsky, V.I., Weiss, G., Yu, P.S. (eds.) ADMI 2011. LNCS, vol. 7103, pp. 173–194. Springer, Heidelberg (2012)
3. Brahmkstri, K., Thomas, D., Sawant, S.T., Jadhav, A., Kshirsagar, D.D.: Ontology based multi-agent intrusion detection system for web service attacks using self learning. In: Meghanathan, N., Nagamalai, D., Rajasekaran, S. (eds.) Networks and Communications (NetCom2013), pp. 265–274. Springer, New York (2014)
4. Djotio, T.N., Tangha, C., Tchangoue, F.N., Batchakui, B.: MONI: Mobile agents ontology based for network intrusions management. Int. J. Adv. Media Commun. **2**(3), 288–307 (2008)
5. Horrocks, I., Patel-Schneider, P.F., Boley, H., Tabet, S., Grosof, B., Dean, M.: SWRL: A Semantic Web Rule Language Combining OWL and RuleML (2004). http://www.w3.org/Submission/SWRL/
6. Isaza, G.A., Castillo, A.G., López, M., Castillo, L.F.: Towards ontology-based intelligent model for intrusion detection and prevention. J. Inf. Assur. Secur. **5**, 376–383 (2010)
7. Mandujano, S., Galvan, A., Nolazco, J.A.: An ontology-based multiagent approach to outbound intrusion detection. In: Proceedings of the International Conference on Computer Systems and Applications, AICCSA 2005, Cairo, Egypt, pp. 94-I (2005)
8. Roesch, M.: Snort - lightweight intrusion detection system for networks. In: Proceedings of the 13th USENIX Conference on System Administration (LISA 1999), Seattle, Washington, pp. 229–238 (1999)
9. Undercoffer, J., Joshi, A., Pinkston, J.: Modeling computer attacks: an ontology for intrusion detection. In: Vigna, G., Kruegel, C., Jonsson, E. (eds.) RAID 2003. LNCS, vol. 2820, pp. 113–135. Springer, Heidelberg (2003)

ICA for Detecting Artifacts
in a Few Channel BCI

Izabela Rejer and Paweł Górski[✉]

Faculty of Computer Science and Information Technology,
West Pomeranian University of Technology Szczecin,
Żołnierska 49, 71-210 Szczecin, Poland
{irejer,pagorski}@wi.zut.edu.pl

Abstract. Eye blinking, body parts movements, power line, and many other internal and external artifacts deteriorate the quality of EEG signal and the whole BCI system. There are some methods for removing artifacts or at least reducing their influence on the BCI system, however, they do not work efficiently when only few channels are used in the system and an automatic artifact elimination is required. The paper presents our approach to deal with artifacts in such a case by adding artificially generated signals to the set of originally recorded signals and to perform Independent Component Analysis on such an enlarged signal set. Our initial experiment, reported in this paper, shows that such an approach results in a better classification precision than when Independent Component Analysis is performed directly on the original signals set.

Keywords: Brain computer interface · BCI · EEG · Preprocessing · ICA · Independent Component Analysis · Artifact removal

1 Introduction

A Brain Computer Interface (BCI) can be viewed as a classification system whose task is to classify EEG (electroencephalographic) signals to one of the predefined classes and trigger commands associated with the recognized class. As any other classification system, the BCI system works in two modes: training mode and executing mode. In the training mode six tasks are performed one after the other: EEG signal acquisition, preprocessing, feature extraction, feature selection, classifier training, and feedback. In the executing mode, two changes in this scheme are made. First, the features selection is not longer needed and hence it is removed. Second, the feedback stage is transformed to the command execution stage.

To obtain a BCI system with a high classification accuracy, all steps from the training mode have to be performed with a great care. However, even the most powerful algorithms for feature selection, extraction or classification do not help when the underlying data are of a poor quality. Since EEG data reflects not only the actual brain activity but also internal and external artifacts like: muscle artifact, eye artifact, power line artifact etc. [1, 2], the signal quality usually is very poor. In order to improve the quality of EEG data, these interfering signals have to be eliminated or at least reduced.

© Springer International Publishing Switzerland 2015
K. Jackowski et al. (Eds.): IDEAL 2015, LNCS 9375, pp. 164–171, 2015.
DOI: 10.1007/978-3-319-24834-9_20

The one strategy for improving the quality of EEG data is to record the artifacts directly with additional electrodes and then subtract them from the signals of interest [3]. This method is simple and reliable, however, it is possible to use only in laboratory conditions. When the brain computer interface is to be used in everyday life, additional electrodes are not welcomed since they prolong the time needed for applying EEG equipment.

Another commonly used method is the manual analysis of succeeding epochs extracted from the original signal. The epochs are inspected one by one by a researcher and those containing artifacts are discarded from the experiment. This procedure is very good at improving the signal quality, but it has two drawbacks. First, it reduces the amount of information gathered in the experiment, second, it needs the assistance of a human being and hence it cannot be used in an automatic mode.

In real-life conditions, when only a few electrodes record the EEG signals from the user scull and the whole system has to work without external management, only one strategy remains - to use a procedure for automatic transformation of the original set of EEG signals and then remove signals which waveforms resemble artifacts (in manual mode) or return the poorest classification results (in automatic mode) [4]. There are some procedures for making a transformation of the set of given signals to another form, such as PCA (Principal Component Analysis) [5], MNF (Maximum Noise Fraction) [6], CSP (Common Spatial Patterns) [7] or ICA (Independent Component Analysis) [8]. Among them ICA is most often applied for artifact detection.

Independent Component Analysis (ICA) is a method for transforming a set of mixed signals into a set of independent components (ICs) [8]. It is often used in electroencephalographic (EEG) signal analysis for detecting ocular, muscle, electro-cardiographic, and power lines artifacts [1, 8–10]. These types of artifacts have a more or less periodic nature and that is why they are recognized by ICA as individual ICs (sometimes one individual IC comprises more than one artifact).

When a large number of signals is recorded, also a large number of ICs is obtained after ICA transformation. If artifacts overweight signals generated by true brain sources, they will be reflected in ICs instead of true brain signals. Hence, when the large number of signals is recorded, it is relatively easy to detect artifacts with ICA. This fact has been confirmed in many research for different numbers of channels, from 16 [10], through 19–20 [1, 9] up to 71 [2] and even many more.

In the case of artificially created signals ICA can find true ICs not only for a large set of mixed signals but also for a very small one. For example if we take three signals, two sinuses of different frequency and random noise of about 10 %, mixed them together and then use ICA over the set of the three mixed signals, we should obtain both original sinuses and the random noise. Now, the question is whether ICA behaves in the same way when applied for real EEG signals. In other words, is it possible for ICA to separate artifacts and true brain sources when only some signal electrodes are applied on the user's head?

The aim of this paper is to report the results of the experiment that was designed to address this question and to find out whether such separation (if successful) can bring any additional benefits. The experiment was very straightforward - we used a bench-mark data set composed of signals recorded from three channels (C3, Cz and C4) and added three artificially created random noise signals. We assumed that when we add some channels with small random noise to the set of EEG signals recorded from the subject, then after applying ICA on this enlarged set of signals, some of the additional channels would absorb the most powerful artifacts and others would absorb the random noise from all channels. In the experiment we used a benchmark data set submitted to the second BCI Competition (data set III – motor imaginary).

2   Methods

The main task of our experiment was to find out whether additional artificially created channels would absorb some of the noise from the original EEG channels. We assumed that if this was true, the ICs obtained after ICA transformation should have higher classification capabilities not only than original EEG signals, but also than ICs obtained without additional channels. In order to verify this assumption we had to built three types of classifiers with: the original EEG channels, ICs obtained after ICA transformation of the set of original channels, and ICs obtained after ICA transformation of the set of original and artificial channels.

In our experiment we used a benchmark data set submitted to the second BCI Competition (data set III – motor imaginary) by Department of Medical Informatics, Institute for Biomedical Engineering, Graz University of Technology [11]. The data set contained EEG signals recorded from one female subject during BCI training session. The subject task was to control the movements of a feedback bar by means of imagery movements of the left and right hand. During the experiment EEG signal was recorded from three bipolar channels (C3, C4, and Cz) with sampling rate 128 Hz. The data was preliminarily filtered between 0.5 and 30 Hz. A single trial lasted 9 s. The whole data set was composed of 140 training trails published with target values (1 - left hand, 2 - right hand) and 140 testing trials. In our experiment we used only data from the training trials.

2.1   Preprocessing

The preprocessing stage was the core of our experiment. It was composed of four steps: removing mean values from the original channels, performing ICA transformation on the original data set, adding artificial channels, and performing ICA transformation on the enlarged set of channels. The first step was a classic one - we just removed the channels mean. In the second step the original data set was transformed with FastICA algorithm symmetric approach shortly described below. We chose this algorithm since it returned the best results in our previous research that were reported in [4]. As a result of ICA transformation 3 ICs were obtained. All of them were inserted to the second data set, named ICA3 data set. Next, we enlarged the original data set by introducing three more channels, unconnected to the brain activity: N1, N2, and N3. Each of these artificially created channels contained random noise from the interval <−0.05, 0.05>. And in the last step ICA transformation was applied once again, this time over the enlarged data set. Six ICs obtained after ICA transformation were inserted to the third file, named ICA6 data set.

2.2   Independent Component Analysis

ICA problem can be stated as follows. Let's assume that there are n linear mixtures $x_1, \ldots x_n$ of n independent components. Vector x (observed signals) can be written as:

$$x = As \tag{1}$$

where A represents a mixing matrix with the size of $n \times n$, and s is the vector of independent components. The aim of ICA is to find a matrix W (i.e. an inverse of the matrix A) to reverse the mixing effect. Then, after computing the matrix W, we can obtain the independent components by [8]:

$$y = wX \cong s \tag{2}$$

The FastICA algorithm, proposed by Oja and Yuan, is an iterative method to find local maxima of a defined cost function [12]. The purpose of this algorithm is to find the matrix of weights w such that the projection $(w^T x)$ maximizes non-Gaussianity [12]. As a measure for non-Gaussianity, simple estimation of negentropy based on the maximum entropy principle is used [12]:

$$J(v) \propto [E\{G(y)\} - E\{G(v)\}]2 \tag{3}$$

where: y – standardized non-Gaussian random variable, v – standardized random variable with Gaussian distribution, $G(.)$- any non-quadratic function.

There are two approaches that can be used to calculate the matrix of weights in FastICA algorithm, symmetric and deflation. The difference between them is that in the deflation approach the weights for each IC are estimate individually and in the symmetric approach the estimation of all components (all weights vectors) proceeds in parallel [12].

2.3   Correlation

Independent Components obtained after ICA transformation are unordered. That means that it is impossible to state directly that the first component corresponds to the first channel in the original data set (C3), the second component corresponds to the second channel (Cz), and the last component corresponds for example to the third artificial component (N3). Hence, while the classification results obtained after ICA transformation were comparable among each other, they could not be directly compared with the classification results obtained for signals from the original data set and also they could not be compared to the ICs obtained after another ICA transformation. Therefore, in order to ensure the comparability between signals from all three sets, the ICs had to be assigned to the corresponding channels (real and artificial) from the original data set.

In order to deal with this task we checked the similarity between signals from all three sets [4]. We used the absolute value of Pearson correlation coefficient as a similarity measure. Since we removed the sign of the Pearson correlation coefficient, our measure took values from 0 (lack of similarity) to 1.0 (ideal correlation). Next, for each independent component we chose the highest correlation coefficient and name the component as the original channel from the chosen correlation pair.

2.4 Feature Extraction

Next, for each data set we calculated features describing our signals. We used classic band power features in our experiment and calculated signal power in 12 frequency bands separately for each channel and each second of the recording. Hence, our feature matrix was composed of 140 rows (140 trials) and 324 columns for the original and ICA3 data sets and 648 columns for ICA6 data set (12 frequency bands × 9 s × (3 or 6) channels). The frequency bands used in the experiment were: alpha band (8–13 Hz), beta band (13–30 Hz), five sub-bands of alpha band (8–9 Hz; 9–10 Hz; 10–11 Hz; 11–12 Hz; 12–13 Hz), five sub-bands of beta band (13–17 Hz; 17–20 Hz; 20–23 Hz; 23–26 Hz; 26–30 Hz).

The features were calculated inside a three-loop structure: the first loop was for 12 bands (the bands order: alpha band, alpha sub-bands from 8–9 to 12–13 Hz, beta band, and beta sub-bands from 13–17 to 26–30 Hz), the second loop was for 3 channels in the case of original data set and ICA3 data set or for 6 channels in the case of ICA6 data set (the channels order: C3, Cz, C4, N1, N2, and N3), and the last one - for 9 s (the seconds order: from 1 to 9). The indexes of features, reported further in the paper, were established inside the most inner loop. Hence, for example the index 1 denotes the band power calculated for band 8–13 Hz, channel C3 and 1s and the index 129 denotes the band power calculated for band 11–12 Hz, channel C4 and 3s.

2.5 Feature Selection and Classification

After feature extraction stage each trial was described by 324 or 648 features. Comparing this large number of features with the number of trials (140), it was clear that a feature selection process should be carried out to improve the generalization capabilities of the future classifiers. According to Raudys and Jain, at least 10 times more training examples per class than the features should be gathered to train the classifier correctly [13]. Taking this recommendation into consideration and additionally assuming 10 % of observations in the validation sets, no more than six features could be used in the classifier in the reported survey. In order to reduce the size of each matrix of features from 324 or 648 to 6 we applied LASSO algorithm (Least Absolute Shrinkage and Selection Operator), proposed by Tibshirani [14]. We used Matlab implementation of LASSO with the following levels of parameters: $Alpha = 1$, $cv = 10$, $DFmax = 6$. Remaining parameters were left on their default levels.

In the classification stage classic linear SVM classifiers [15] were used. Since we wanted to examine the classification results separately for each channel (real and artificial) and for each combination of channels, we built 7 classifiers for original and ICA3 data sets and 63 classifiers for ICA6 data set. Each classifier was built according to the same scheme. First, LASSO algorithm was applied to choose 6 most significant features describing the signal that had been recorded over the given channel or combination of channels. When 6 most important features were chosen, the classifier was trained. The training process was carried out over the whole data set with 10-fold cross-validation. The mean value calculated over the classification accuracy obtained for all 10 validation subsets was used as the final accuracy metrics of the classifier.

3 Results and Discussion

According to the main publications in the motor imagery field, during the left/right hand movement (real or imagery) the alpha rhythm measured over the primary motor cortex contralateral to the hand is desynchronized in the first seconds of the movement [16]. Usually also late beta synchronization (so called beta rebound) over the same cortex areas is reported [16, 17]. Hence, if ICA properly transformed the enlarged set of signals to ICs, only features calculated for two ICs should be marked as important, IC corresponding to C3 channel (left motor cortex) and IC corresponding to C4 channel (right motor cortex). As it can be noticed in Table 1, regardless of the number of channels used for feature extraction, the final set of features chosen in the selection process was always the same. It was composed only of features coming from ICs corresponded to channels C3 and C4. The only exception were feature sets calculated over the single channels, but this was expected since for two-hand classification we needed features from both hemispheres. Moreover, the features that were chosen in the selection process were fully in agreement with theory:

- feature no. 5 was the signal power in 8–13 Hz band in the 5-th second,
- feature no. 24 was the signal power in 8–13 Hz band in the 6-th second.
- feature no. 167 was the signal power in 10–11 Hz band in the 5-th second,
- feature no. 168 was the signal power in 10–11 Hz band in the 6-th second,
- feature no. 185 was the signal power in 10–11 Hz band in the 5-th second,
- feature no. 291 was the signal power in 12–13 Hz band in the 3-th second.

Table 1. The average values of fitness function (classification accuracy) and indexes of features for different combinations of signals from ICA6 data set

Channels	Accuracy [%]	Features indexes					
		1	2	3	4	5	6
C3	0,74	5	112	167	491	546	598
C4	0,83	24	185	187	291	453	561
C3, C4	0,92	5	24	167	168	185	291
C3, C4, and any other channel	0,92	5	24	167	168	185	291
C3, C4, and any other two channels	0,92	5	24	167	168	185	291
C3, C4, and any other three channels	0,92	5	24	167	168	185	291
C3, C4, and any other four channels	0,92	5	24	167	168	185	291
C3, C4, Cz, N1, N2, N3	0,92	5	24	167	168	185	291

Considering the results gathered in Table 1 and the meaning of the selected features, we concluded that ICA correctly separated true brain signals from the noise. Now the question was whether the same features were chosen in the case of original signals and signals after ICA transformation of the original channels. As it can be noticed in Table 2 the collections of features for the best classifiers was exactly the same in the case of two and three channels combinations for signals from all three data sets. That means that the features important for the classification process did not change when we added artificial noise signal. But, what is our most important result, the classification precision changed. According to Table 3 the classification precision was high for signals from all three data sets, however, the most precise classification for each number of channels was obtained for ICs from ICA6 data set. One reasonable explanation of this fact is that the additional artificial channels did absorb some noise from the original channels and at the same time enhanced their discrimination capabilities.

Table 2. The collections of features of the best classifiers obtained for the original data set, ICA3 and ICA6 data sets.

Signal	Original	ICA3	ICA6
C4	185 186 241 291 453 510	185 186 241 291 453 510	24 185 187 291 453 561
C3,C4	5 24 167 168 185 291	5 24 167 168 185 291	5 24 167 168 185 291
C3,Cz,C4	5 24 167 168 185 291	5 24 167 168 185 291	5 24 167 168 185 291

Table 3. The classification accuracy of the best classifiers obtained for each combination of signals corresponding to C3, Cz, and C4 channels for all data sets: original, ICA3, and ICA6.

Signal	Original	ICA3	ICA6
C3	0,74	0,72	0.74
Cz	0,66	0,60	0.65
C4	**0,76**	**0,81**	**0.83**
C3, Cz	0,75	0,74	0.75
Cz,C4	0,87	0,76	0.81
C3,C4	**0,89**	**0,90**	**0.92**
C3,Cz,C4	**0,89**	**0,89**	**0.92**

4   Conclusion

The aim of the experiment whose results were reported in this paper was to find out whether adding artificially created channels to the set of true brain signals can bring any benefits. Since the classification precision calculated for ICs obtained after performing ICA on the enlarged set of channels was higher than for original channels, and for ICs calculated over those channels, we conclude that adding the artificial channels was a good idea. The additional channels apparently absorbed some noise from the original channels improving their discrimination capabilities. Of course this was the preliminary study and hence we cannot formulate any general conclusions.

References

1. Vigario, R.N.: Extraction of ocular artifacts from EEG using independent component analysis. Electroencephalogr. Clin. Neurophysiol. **103**(3), 395–404 (1997)
2. Wallstrom, G.L., Kass, R.E., Miller, A., Cohn, J.F., Fox, N.A.: Automatic correction of ocular artifacts in the EEG: a comparison of regression-based and component-based methods. Int. J. Psychophysiol. **53**(2), 105–119 (2004)
3. Guerrero-Mosquera, C., Vazquez, A.N.: Automatic removal of ocular artifacts from eeg data using adaptive filtering and independent component analysis. In: 17th European Signal Processing Conference (EUSIPCO 2009). Glasgow, Scotland (August 24–28 2009)
4. Rejer, I., Gorski P.: Benefits of ICA in the case of a few channel EEG. In: 37th Annual International Conference of the IEEE Engineering in Medicine and Biology Society of the IEEE Engineering in Medicine and Biology Society in MiCo, Milano (still in-print) (August 25–29 2015)
5. Shlens, J.: A tutorial on principal component analysis derivation. http://arxiv.org/pdf/1404. 1100.pdf. Accessed May 2015
6. Switzer P., Green A.: Min/max autocorrelation factors for multivariate spatial imagery. In: Technical report 6, Department of Statistics, Stanford University (1984)
7. Ramoser, H., Muller-Gerking, J., Pfurtscheller, G.: Optimal spatial filtering of single trial EEG during imagined handmovement. IEEE Trans. Rehabil. Eng. **8**(4), 441–446 (2000)
8. Hyvärinen, A., Oja, E.: Independent component analysis: algorithms and applications. Neural Netw. **13**(4–5), 411–430 (2000)
9. Jung, T.P., Humphries, C., Lee, T.W., Makeig, S., McKeown, M.J., Iragui, V., Sejnowski, T.J.: Extended ICA removes artifacts from electroencephalographic recordings. Adv. Neural Inf. Process. Syst. **10**, 894–900 (1998)
10. Delorme, A., Palmer, J., Onton, J., Oostenveld, R., Makeig, S.: Independent EEG sources are dipolar. PloS ONE **7**(2), e30135 (2012)
11. Data set III, II BCI Competition, motor imaginary. http://bbci.de/competition/ii/index.html
12. Oja, E., Yuan, Z.: The FastICA algorithm revisited: convergence analysis. IEEE Trans. Neural Netw. **17**(6), 1370–1381 (2006)
13. Raudys, S.J., Jain, A.K.: Small sample size effects in statistical pattern recognition: recommendations for practitioners. IEEE Trans. Pattern Anal. Mach. Intell. **13**(3), 252–264 (1991)
14. Tibshirani, R.: Regression shrinkage and selection via the lasso. J. Roy. Stat. Soc. Ser. B (Methodol.) **58**(1), 267–288 (1996)
15. Fung, G.M., Mangasarian, O.L., Shavlik, J.W.: Knowledge-based support vector machine classifiers. Adv. Neural Inf. Process. Syst. **15**, 537–544 (2002)
16. Pfurtschellera, G., Lopes da Silvab, F.H.: Event-related EEG/MEG synchronization and desynchronization: basic principles. Clin. Neurophysiol. **110**, 1842–1857 (1999)
17. McFarland, D.J., Miner, L.A., Vaughan, T.M., Wolpaw, J.R.: Mu and Beta rhythm topographies during motor imagery and actual movements. Brain Topogr. **12**(3), 177–186 (2000)

An Empirical Evaluation of Robust Gaussian Process Models for System Identification

César Lincoln C. Mattos[1]([✉]), José Daniel A. Santos[2],
and Guilherme A. Barreto[1]

[1] Department of Teleinformatics Engineering, Center of Technology,
Federal University of Ceará, Campus of Pici, Fortaleza, CE, Brazil
cesarlincoln@terra.com.br, gbarreto@ufc.br
[2] Department of Industry, Federal Institute of Education,
Science and Technology of Ceará, Maracanaú, CE, Brazil
jdalencars@gmail.com

Abstract. System identification comprises a number of linear and non-linear tools for black-box modeling of dynamical systems, with applications in several areas of engineering, control, biology and economy. However, the usual Gaussian noise assumption is not always satisfied, specially if data is corrupted by impulsive noise or outliers. Bearing this in mind, the present paper aims at evaluating how Gaussian Process (GP) models perform in system identification tasks in the presence of outliers. More specifically, we compare the performances of two existing robust GP-based regression models in experiments involving five benchmarking datasets with controlled outlier inclusion. The results indicate that, although still sensitive in some degree to the presence of outliers, the robust models are indeed able to achieve lower prediction errors in corrupted scenarios when compared to conventional GP-based approach.

Keywords: Robust system identification · Gaussian process · Approximate Bayesian inference

1 Introduction

Gaussian processes (GPs) provide a principled, practical, probabilistic approach to learning in kernel machines [1]. Due to is versatility, GP models is receiving considerable attention from the machine learning community, leading to successful applications to classification and regression [2], visualization of high dimensional data [3] and system identification [4], to mention just a few.

Of particular interest to the present paper is the application of GP models to dynamical system identification, which comprises a number of linear and nonlinear tools for black-box modeling. Contributions to GP-based system identification seem to have started with the work of Murray-Smith et al. [5], who applied it to vehicle dynamics data. Since then, a number of interesting approaches can be found in the literature, such as GP with derivative observations [6], GP

© Springer International Publishing Switzerland 2015
K. Jackowski et al. (Eds.): IDEAL 2015, LNCS 9375, pp. 172–180, 2015.
DOI: 10.1007/978-3-319-24834-9_21

for learning non-stationary systems [7], GP-based local models [8], evolving GP models [9], and GP-based state space models [10].

Nevertheless, these previous GP-based system identification approaches have adopted the Gaussian likelihood function as noise model. However, as a light-tailed distribution, this function is not able to suitably handle impulsive noise (a type of outlier). When such outliers are encountered in the data used to tune the model's hyperparameters, these are not correctly estimated. Besides, as a nonparametric approach, the GP model carries the estimation data along for prediction purpose, i.e. the estimation samples containing outliers and the mis-estimated hyperparameters will be used during out-of-sample prediction stage, a feature that may compromise the model generalization on new data.

In this scenario, heavy-tailed distributions are claimed to be more appropriate as noise models when outliers are present. Such distributions are able to account for, or justify, extreme values, as they have higher probability than in light-tailed distributions. This feature prevents the estimation step from being too affected by outliers. However, while inference by GP models with Gaussian likelihood is tractable, non-Gaussian likelihoods models are not, requiring the use of approximation methods, such as Variational Bayes (VB) [11] and Expectation Propagation (EP) [12].

Robust GP regression started to draw the machine learning community attention more recently. In Faul and Tipping [13], impulsive noise is modeled as being generated by a second Gaussian distribution with larger variance, resulting in a mixture of Gaussian noise models. Inference is done with the VB method. In Kuss et al. [14], a similar noise model is chosen, but the inference makes use of the EP strategy. In Tipping and Lawrence [15], GP models with Student-t likelihood are also considered in a variational context. The same likelihood is used in Jylänki et al. [16], but it is tackled by a Laplace approximation approach. The same approach is used in Berger and Rauscher [17] to calibrate a diesel engine from data containing outliers. In Kuss' thesis [18], besides reviewing some of the approaches for robust GP regression, a Laplacian noise model is detailed and tackled by an EP-based inference strategy.

From the exposed, the goal of this work is to evaluate some of the afore-mentioned robust GP models in nonlinear dynamical system identification in the presence of outliers. More specifically, we apply a Student-t noise model likelihood with VB inference, as in [15], and a Laplace noise model with EP inference, following [18]. Our objective is to assess if such algorithms, originally proposed for robust regression, are able to achieve good performance in dynamical system identification scenarios contaminated with outliers and compare them with standard (i.e. non-robust) GP models that have been used in the system identification literature.

The remainder of the paper is organized as follows. In Sect. 2 we describe the task of nonlinear dynamical system identification with standard GP modeling and the two aforementioned robust variants. In Sect. 3 we report the results of the performance evaluation of the models for five artificial datasets with different levels of contamination with outliers. We conclude the paper in Sect. 4.

2   GP for Nonlinear Dynamical System Identification

Given a dynamical system modeled by a nonlinear autoregressive model with exogenous inputs (NARX), its i-th input vector $\boldsymbol{x}_i \in \mathbb{R}^D$ is comprised of L_y past observed outputs $y_i \in \mathbb{R}$ and L_u past exogenous inputs $u_i \in \mathbb{R}$ [4]:

$$y_i = t_i + \epsilon_i, \qquad t_i = f(\boldsymbol{x}_i), \qquad \epsilon_i \sim \mathcal{N}(\epsilon_i|0, \sigma_n^2), \qquad (1)$$

$$\boldsymbol{x}_i = [y_{i-1}, \cdots, y_{i-L_y}, u_{i-1}, \cdots, u_{i-L_u}]^T \qquad (2)$$

where i is the instant of observation, $t_i \in \mathbb{R}$ is the true (noiseless) output, $f(\cdot)$ is an unknown nonlinear function and ϵ_i is a Gaussian observation noise. After N instants, we have the dataset $\mathcal{D} = \{(\boldsymbol{x}_i, y_i)\}_{i=1}^N = (\boldsymbol{X}, \boldsymbol{y})$, where $\boldsymbol{X} \in \mathbb{R}^{N \times D}$ is the so-called *regressor matrix* and $\boldsymbol{y} \in \mathbb{R}^N$.

An estimated model may be used to simulate the output of the identified system. We use an iterative test procedure where past estimated outputs are used as regressors, which is called *free simulation* or *infinite step ahead* prediction.

2.1   Traditional GP Modeling

In the GP framework, the nonlinear function $f(\cdot)$ is given a multivariate Gaussian prior $\boldsymbol{t} = f(\boldsymbol{X}) \sim \mathcal{GP}(\boldsymbol{t}|\boldsymbol{0}, \boldsymbol{K})$, where a zero mean vector was considered and $\boldsymbol{K} \in \mathbb{R}^{N \times N}$, $K_{ij} = k(\boldsymbol{x}_i, \boldsymbol{x}_j)$, is the covariance matrix, obtained with a *kernel* function $k(\cdot, \cdot)$, which must generate a semidefinite positive matrix $\boldsymbol{K}$. The following function is a common choice and will be used in this paper [19]:

$$k(\boldsymbol{x}_i, \boldsymbol{x}_j) = \sigma_f^2 \exp\left[-\frac{1}{2} \sum_{d=1}^D w_d^2 (x_{id} - x_{jd})^2 \right] + \sigma_l^2 \boldsymbol{x}_i^T \boldsymbol{x}_j + \sigma_c^2. \qquad (3)$$

The vector $\boldsymbol{\theta} = [\sigma_f^2, w_1^2, \ldots, w_D^2, \sigma_l^2, \sigma_c^2]^T$ is comprised of the hyperparameters which characterize the covariance of the model.

Considering a multivariate Gaussian likelihood $p(\boldsymbol{y}|\boldsymbol{t}) = \mathcal{N}(\boldsymbol{y}|\boldsymbol{t}, \sigma_n^2 \boldsymbol{I})$, where $\boldsymbol{I}$ is a $N \times N$ identity matrix, the posterior distribution $p(\boldsymbol{t}|\boldsymbol{y}, \boldsymbol{X})$ is tractable. The inference for a new output t_*, given a new input $\boldsymbol{x}_*$, is also tractable:

$$p(t_*|\boldsymbol{y}, \boldsymbol{X}, \boldsymbol{x}_*) = \mathcal{N}(t_*|\boldsymbol{k}_{*N}(\boldsymbol{K} + \sigma_n^2 \boldsymbol{I})^{-1}\boldsymbol{y}, k_{**} - \boldsymbol{k}_{*N}(\boldsymbol{K} + \sigma_n^2 \boldsymbol{I})^{-1}\boldsymbol{k}_{N*}), \qquad (4)$$

where $\boldsymbol{k}_{*N} = [k(\boldsymbol{x}_*, \boldsymbol{x}_1), \cdots, k(\boldsymbol{x}_*, \boldsymbol{x}_N)]$, $\boldsymbol{k}_{N*} = \boldsymbol{k}_{*N}^T$ and $k_{**} = k(\boldsymbol{x}_*, \boldsymbol{x}_*)$. The predictive distribution of y_* is similar to the one in Eq. (4), but the variance is added by σ_n^2.

The vector of hyperparameters $\boldsymbol{\theta}$ can be extended to include the noise variance σ_n^2 and be determined with the maximization of the marginal log-likelihood $\ln p(\boldsymbol{y}|\boldsymbol{X}, \boldsymbol{\theta})$ of the observed data, the so-called *evidence* of the model:

$$\boldsymbol{\theta}_* = \arg\max\left\{ -\frac{1}{2}\ln|\boldsymbol{K} + \sigma_n^2 \boldsymbol{I}| - \frac{1}{2}\boldsymbol{y}^T(\boldsymbol{K} + \sigma_n^2 \boldsymbol{I})^{-1}\boldsymbol{y} - \frac{N}{2}\ln(2\pi) \right\}. \qquad (5)$$

The optimization process is guided by the gradients of the marginal log-likelihood with respect to each component of the vector $\boldsymbol{\theta}$. It is worth mentioning that the optimization of the hyperparameters can be seen as the model selection step of obtaining a plausible GP model from the estimation data.

2.2 Robust GP with Non-Gaussian Likelihood

The previous GP model with Gaussian likelihood is not robust to outliers, due its light tails. An alternative is to consider a likelihood with heavy tails, such as the Laplace and the Student-t likelihoods, respectively given by

$$p_{\text{Lap}}(\boldsymbol{y}|\boldsymbol{t}) = \prod_{i=1}^{N} \frac{1}{2s} \exp\left(-\frac{|y_i - t_i|}{s}\right), \tag{6}$$

and

$$p_{\text{Stu}}(\boldsymbol{y}|\boldsymbol{t}) = \prod_{i=1}^{N} \frac{\Gamma((\nu+1)/2)}{\Gamma(\nu/2)\sqrt{\pi\nu\sigma^2}} \left(1 + \frac{1}{\nu}\frac{(y_i - t_i)^2}{\sigma^2}\right)^{-(\nu+1)/2}, \tag{7}$$

where s, ν and σ^2 are likelihood hyperparameters and $\Gamma(\cdot)$ is the gamma function.

However, once a non-Gaussian likelihood is chosen, many of the GP expressions become intractable. In the present paper, we apply approximate Bayesian inference methods to overcome those intractabilities. More specifically, we are interested in the Variational Bayes and the Expectation Propagation algorithms, briefly presented below.

Variational Bayes (VB). In the case of applying VB to the Student-t likelihood, it must be rewritten as follows [18]:

$$p(\boldsymbol{y}|\boldsymbol{t}, \boldsymbol{\sigma}^2) = \mathcal{N}(\boldsymbol{y}|\boldsymbol{t}, \text{diag}(\boldsymbol{\sigma}^2)), \qquad p(\boldsymbol{\sigma}^2|\boldsymbol{\alpha}, \boldsymbol{\beta}) = \prod_{i=1}^{N} \text{Inv}\Gamma(\sigma_i^2|\alpha_i, \beta_i), \tag{8}$$

where $\boldsymbol{t}, \boldsymbol{\sigma}^2 \in \mathbb{R}^N$ are latent variables, $\text{diag}(\cdot)$ builds a diagonal matrix from a vector and σ_i^2 has an inverse gamma prior with parameters α_i and β_i.

The joint posterior of $\boldsymbol{t}$ and $\boldsymbol{\sigma}^2$ is considered to be factorizable as

$$p(\boldsymbol{t}, \boldsymbol{\sigma}^2|\boldsymbol{y}, \boldsymbol{X}) \approx q(\boldsymbol{t})q(\boldsymbol{\sigma}^2) = \mathcal{N}(\boldsymbol{t}|m, \boldsymbol{A})\left(\prod_{i=1}^{N} \text{Inv}\Gamma(\sigma_i^2|\tilde{\alpha}_i, \breve{\beta}_i)\right), \tag{9}$$

where $m \in \mathbb{R}^N$, $\boldsymbol{A} \in \mathbb{R}^{N \times N}$ and $\tilde{\boldsymbol{\alpha}}, \breve{\boldsymbol{\beta}} \in \mathbb{R}^N$ are unknown variational parameters.

A lower bound $\mathcal{L}(q(\boldsymbol{t})q(\boldsymbol{\sigma}^2))$ to the log-marginal likelihood can be found relating it to the factorized posterior $q(\boldsymbol{t})q(\boldsymbol{\sigma}^2)$ [15]:

$$\ln p(\boldsymbol{y}|\boldsymbol{X}, \boldsymbol{\theta}) = \mathcal{L}(q(\boldsymbol{t})q(\boldsymbol{\sigma}^2)) + \text{KL}(q(\boldsymbol{t})q(\boldsymbol{\sigma}^2)||p(\boldsymbol{t}, \boldsymbol{\sigma}^2|\boldsymbol{y}, \boldsymbol{X})), \tag{10}$$

where the last term is the Kullback-Leibler divergence between the approximate distribution and the true posterior. The maximization of the bound $\mathcal{L}(q(\boldsymbol{t})q(\boldsymbol{\sigma}^2))$ also minimizes the KL divergence term, improving the approximation [15].

The optimization of the hyperparameters and the latent variables can be done in an Expectation-Maximization (EM) fashion, as detailed in [18]. Then, the moments of the prediction $p(t_*|\boldsymbol{y}, \boldsymbol{X}, \boldsymbol{x}_*) = \mathcal{N}(t_*|\mu_*, \sigma_*^2)$ for a new input $\boldsymbol{x}_*$ are given by

$$\mu_* = \boldsymbol{k}_{*N}(\boldsymbol{K} + \boldsymbol{\Sigma})^{-1}\boldsymbol{y}, \quad \text{and} \quad \sigma_*^2 = k_{**} - \boldsymbol{k}_{*N}(\boldsymbol{K} + \boldsymbol{\Sigma})\boldsymbol{k}_{N*}, \qquad (11)$$

where $\boldsymbol{\Sigma} = \text{diag}(\tilde{\boldsymbol{\beta}}/\tilde{\boldsymbol{\alpha}})$. Although the calculation of the predictive distribution of y_* is intractable, its mean is equal to the previously calculated μ_*.

Expectation Propagation (EP). EP usually works by approximating the true posterior distribution by a Gaussian which follows a factorized structure [12,18]:

$$p(\boldsymbol{t}|\boldsymbol{y}, \boldsymbol{X}) \approx \frac{\mathcal{N}(\boldsymbol{t}|\boldsymbol{0}, \boldsymbol{K})}{q(\boldsymbol{y}|\boldsymbol{X})} \prod_{i=1}^{N} c(t_i, \mu_i, \sigma_i^2, Z_i) = q(\boldsymbol{t}|\boldsymbol{y}, \boldsymbol{X}) = \mathcal{N}(\boldsymbol{t}|\boldsymbol{m}, \boldsymbol{A}), \qquad (12)$$

where $c(t_i, \mu_i, \sigma_i^2, Z_i) = Z_i \mathcal{N}(t_i|\mu_i, \sigma_i^2)$ are called *site functions*. The mean vector $\boldsymbol{m} \in \mathbb{R}^N$ and covariance matrix $\boldsymbol{A} \in \mathbb{R}^{N \times N}$ of the approximate distribution may be computed as $\boldsymbol{m} = \boldsymbol{A}\boldsymbol{\Sigma}^{-1}\boldsymbol{\mu}$ and $\boldsymbol{A} = (\boldsymbol{K}^{-1} + \boldsymbol{\Sigma}^{-1})^{-1}$, where $\boldsymbol{\Sigma} = \text{diag}(\sigma_1^2, \cdots, \sigma_N^2)$ and $\boldsymbol{\mu} = [\mu_1, \cdots, \mu_N]^T$.

The prediction $p(t_*|\boldsymbol{y}, \boldsymbol{X}, \boldsymbol{x}_*) = \mathcal{N}(t_*|\mu_*, \sigma_*^2)$ for a new input $\boldsymbol{x}_*$ is given by

$$\mu_* = \boldsymbol{k}_{*N}\boldsymbol{K}^{-1}\boldsymbol{m}, \quad \text{and} \quad \sigma_*^2 = k_{**} - \boldsymbol{k}_{*N}(\boldsymbol{K}^{-1} - \boldsymbol{K}^{-1}\boldsymbol{A}\boldsymbol{K}^{-1})\boldsymbol{k}_{N*}. \qquad (13)$$

Although the predictive distribution of y_* is intractable, its mean is also μ_*.

The variables μ_i, σ_i^2 and Z_i are obtained by iterative moment match, which simultaneously minimizes the reverse Kullback-Leibler divergence between the true posterior and the approximate distribution. The convergence is not guaranteed, but it has been reported in the literature that EP works well within GP models [1]. The complete algorithm for a Laplace likelihood is detailed in [18].

3 Experiments

In order to verify the performance of the previously described models in the task of nonlinear system identification in the presence of outliers, we performed computational experiments with five artificial datasets, detailed in Table 1. The first four datasets were presented in the seminal work of Narendra et al. [20]. The fifth dataset was generated following Kocijan et al. [4].

Besides the Gaussian noise, indicated in the last column of Table 1, the estimation data of all datasets was also incrementally corrupted with a number of outliers equal to 2.5 %, 5 % and 10 % of the estimation samples. Each randomly chosen sample was added by a uniformly distributed value $\text{U}(-M_y, +M_y)$, where M_y is the maximum absolute output. We emphasize that only the output values were corrupted in this step. Such outlier contamination methodology is similar to the one performed in [21]. The orders L_u and L_y chosen for the regressors were set to their largest delays presented in the second column of Table 1.

Table 1. Details of the five artificial datasets used in the computational experiments. The indicated noise in the last column is added only to the output of the estimation data. Note that U(A, B) is a random number uniformly distributed between A and B.

| # Output | Input/Samples | | Noise |
	Estimation	Test	
1 $y_i = \frac{y_{i-1}y_{i-2}(y_{i-1}+2.5)}{1+y_{i-1}^2+y_{i-2}^2} + u_{i-1}$	$u_i = \mathrm{U}(-2,2)$ 300 samples	$u_i = \sin(2\pi i/25)$ 100 samples	$\mathcal{N}(0, 0.29)$
2 $y_i = \frac{y_{i-1}}{1+y_{i-1}^2} + u_{i-1}^3$	$u_i = \mathrm{U}(-2,2)$ 300 samples	$u_i = \sin(2\pi i/25) + \sin(2\pi i/10)$ 100 samples	$\mathcal{N}(0, 0.65)$
3 $y_i = 0.8y_{i-1} + (u_{i-1}-0.8)u_{i-1}(u_{i-1}+0.5)$	$u_i = \mathrm{U}(-1,1)$ 300 samples	$u_i = \sin(2\pi i/25)$ 100 samples	$\mathcal{N}(0, 0.07)$
4 $y_i = 0.3y_{i-1} + 0.6y_{i-2} + 0.3\sin(3\pi u_{i-1}) + 0.1\sin(5\pi u_{i-1})$	$u_i = U(-1,1)$ 500 samples	$u_i = \sin(2\pi i/250)$ 500 samples	$\mathcal{N}(0, 0.18)$
5 $y_i = y_{i-1} - 0.5\tanh(y_{i-1} + u_{i-1}^3)$	$u_i = \mathcal{N}(u_i\|0,1)$ $-1 \le u_i \le 1$ 150 samples	$u_i = \mathcal{N}(u_i\|0,1)$ $-1 \le u_i \le 1$ 150 samples	$\mathcal{N}(0, 0.0025)$

We compare the performances of the following GP models: conventional GP, GP with Student-t likelihood and VB inference (GP-tVB) and GP with Laplace likelihood and EP inference (GP-LEP). The obtained root mean square errors (RMSE) are presented in Table 2.

In almost all scenarios with outliers both robust variants presented better performances than conventional GP. Only in one case, *Artificial 3* dataset with 10 % of corruption, GP performed better than one of the robust models (GP-tVB). In the scenarios without outliers, i.e., with Gaussian noise only, the GP model achieved the best RMSE for *Artificial 1* and *4* datasets, but it also performed close to the robust models for the other datasets with 0 % of corruption.

A good resilience to outliers was obtained for *Artificial 1* and *2* datasets, with GP-LEP and GP-tVB models being less affected in the cases with outliers. The most impressive performance was the one achieved by the GP-tVB model for all cases of the *Artificial 2* dataset, with little RMSE degradation.

For the *Artificial 3* dataset, only the GP-tVB model with 2.5 % of outliers achieved error values close to the scenario without outliers. In the other cases, both variants, although better than conventional GP model, presented greater RMSE values than their results for 0 % of outliers.

Likewise, in the experiments with *Artificial 4* and *5* datasets, we also observed that all models were affected by the corruption of the estimation data, even with lower quantities of outliers. However, it is important to emphasize that both GP-tVB and GP-LEP models achieved better RMSE values than conventional GP, often by a large margin, as observed in the *Artificial 4* dataset for the GP-tVB model. In such cases, the robust variants can be considered a valid improvement over the conventional GP model.

Table 2. Summary of simulation RMSE without and with outliers in estimation step.

	Artificial 1				Artificial 2			
% of outliers	0%	2.5%	5%	10%	0%	2.5%	5%	10%
GP	**0.2134**	0.3499	0.3874	0.4877	0.3312	0.3724	0.5266	0.4410
GP-tVB	0.2455	0.3037	0.2995	**0.2868**	**0.3189**	**0.3247**	**0.3284**	**0.3306**
GP-LEP	0.2453	**0.2724**	**0.2720**	0.3101	0.3450	0.3352	0.3471	0.3963

	Artificial 3				Artificial 4			
GP	0.1106	0.4411	0.7022	0.6032	**0.6384**	2.1584	2.2935	2.4640
GP-tVB	0.1097	**0.1040**	**0.3344**	0.8691	0.6402	**0.7462**	2.2220	**2.1951**
GP-LEP	**0.0825**	0.3527	0.4481	**0.5738**	0.9188	1.1297	**2.1742**	2.3762

	Artificial 5			
GP	0.0256	0.0751	0.1479	0.1578
GP-tVB	**0.0216**	0.0542	**0.0568**	**0.1006**
GP-LEP	0.0345	**0.0499**	0.0747	0.1222

Finally, we should mention that during the experiments, the variational approach of the GP-tVB model has been consistently more stable than the EP algorithm of the GP-LEP model, even with the incorporation of the numerical safeties suggested by Rasmussen and Williams [1] and Kuss [18], which might be a decisive factor when choosing which model to apply for system identification.

4   Conclusion

In this paper we evaluated robust Gaussian process models in the task of nonlinear dynamical system identification in the presence of outliers. The experiments with five artificial datasets considered a GP model with Student-t likelihood and variational inference (GP-tVB) and a model with Laplace likelihood with EP inference (GP-LEP), besides conventional GP with Gaussian likelihood.

Although the robust variants performed better in the scenarios with outliers, we cannot state categorically that they were insensitive to the corrupted data, for both GP-tVB and GP-LEP models obtained considerable worse RMSE in some cases with outliers. Depending on the task in hand, such degradation may or may not be tolerable. This observation, as well as some numerical issues encountered in the EP algorithm, encourages us to further pursue alternative GP-based models which are more appropriate for robust system identification.

Acknowledgments. The authors thank the financial support of FUNCAP, IFCE, NUTEC and CNPq (grant no. 309841/2012-7).

References

1. Rasmussen, C., Williams, C.: Gaussian Processes for Machine Learning, 1st edn. MIT Press, Cambridge (2006)
2. Williams, C.K.I., Barber, D.: Bayesian classification with Gaussian processes. IEEE Trans. Pattern Anal. **20**(12), 1342–1351 (1998)
3. Lawrence, N.D.: Gaussian process latent variable models for visualisation of high dimensional data. In: Advances in Neural Information Processing Systems, pp. 329–336 (2004)
4. Kocijan, J., Girard, A., Banko, B., Murray-Smith, R.: Dynamic systems identification with Gaussian processes. Math. Comput. Model. Dyn. **11**(4), 411–424 (2005)
5. Murray-Smith, R., Johansen, T.A., Shorten, R.: On transient dynamics, off-equilibrium behaviour and identification in blended multiple model structures. In: European Control Conference (ECC 1999), Karlsruhe, BA-14. Springer (1999)
6. Solak, E., Murray-Smith, R., Leithead, W.E., Leith, D.J., Rasmussen, C.E.: Derivative observations in Gaussian process models of dynamic systems. In: Advances in Neural Information Processing Systems 16 (2003)
7. Rottmann, A., Burgard, W.: Learning non-stationary system dynamics online using Gaussian processes. In: Goesele, M., Roth, S., Kuijper, A., Schiele, B., Schindler, K. (eds.) Pattern Recognition. LNCS, vol. 6376, pp. 192–201. Springer, Heidelberg (2010)
8. Ažman, K., Kocijan, J.: Dynamical systems identification using Gaussian process models with incorporated local models. Eng. Appl. Artif. Intel. **24**(2), 398–408 (2011)
9. Petelin, D., Grancharova, A., Kocijan, J.: Evolving Gaussian process models for prediction of ozone concentration in the air. Simul. Model. Pract. Theory **33**, 68–80 (2013)
10. Frigola, R., Chen, Y., Rasmussen, C.: Variational Gaussian process state-space models. In: Advances in Neural Information Processing Systems (NIPS), vol. 27, pp. 3680–3688 (2014)
11. Jordan, M.I., Ghahramani, Z., Jaakkola, T.S., Saul, L.K.: An introduction to variational methods for graphical models. Mach. Learn. **37**(2), 183–233 (1999)
12. Minka, T.P.: Expectation propagation for approximate Bayesian inference. In: Proceedings of the 17th Conference on Uncertainty in Artificial Intelligence (UAI 2001), pp. 362–369. Morgan Kaufmann (2001)
13. Nakayama, H., Arakawa, M., Sasaki, R.: A computational intelligence approach to optimization with unknown objective functions. In: Dorffner, G., Bischof, H., Hornik, K. (eds.) ICANN 2001. LNCS, vol. 2130, pp. 73–80. Springer, Heidelberg (2001)
14. Kuss, M., Pfingsten, T., Csató, L., Rasmussen, C.E.: Approximate inference for robust Gaussian process regression. Technical report 136, Max Planck Institute for Biological Cybernetics, Tubingen, Germany (2005)
15. Tipping, M.E., Lawrence, N.D.: Variational inference for student-t models: robust Bayesian interpolation and generalised component analysis. Neurocomputing **69**(1), 123–141 (2005)
16. Jylänki, P., Vanhatalo, J., Vehtari, A.: Robust gaussian process regression with a student-t likelihood. J. Mach. Learn. Res. **12**, 3227–3257 (2011)
17. Berger, B., Rauscher, F.: Robust Gaussian process modelling for engine calibration. In: Proceedings of the 7th Vienna International Conference on Mathematical Modelling (MATHMOD 2012), pp. 159–164 (2012)

18. Kuss, M.: Gaussian process models for robust regression, classification, and reinforcement learning. Ph.D. thesis, TU Darmstadt (2006)
19. Rasmussen, C.E.: Evaluation of Gaussian processes and other methods for nonlinear regression. Ph.D. thesis, University of Toronto, Toronto, Canada (1996)
20. Narendra, K.S., Parthasarathy, K.: Identification and control of dynamical systems using neural networks. IEEE Trans. Neural Networks **1**(1), 4–27 (1990)
21. Majhi, B., Panda, G.: Robust identification of nonlinear complex systems using low complexity ANN and particle swarm optimization technique. Expert Syst. Appl. **38**(1), 321–333 (2011)

Throughput Analysis of Automatic Production Lines Based on Simulation Methods

Sławomir Kłos[✉] and Justyna Patalas-Maliszewska

Faculty of Mechanical Engineering,
University of Zielona Góra, ul. Licealna 9,
65-417 Zielona Góra, Poland
{s.klos,j.patalas}@iizp.uz.zgora.pl

Abstract. The effectiveness of manufacturing systems is a very important ratio that decides about the competitiveness of enterprises. There are a lot of factors which affect the efficiency of production processes such as: operation times, setup times, lot sizes, buffer capacities, etc. In this paper, a computer simulation method is used to model the throughput and provide a product life span analysis of a number of automatic production lines. The research was done for two topologies of manufacturing systems and different capacities of intermediate buffers. Due to the fact that an increase of intermediate buffers results in an increase of work in process, a flow index of production is proposed that includes a relationship between the throughput of a system and the average life span of products. The simulation model and experimental research was made using Tecnomatix Plant Simulation software.

Keywords: Computer simulation · Production line · Buffer capacity · Throughput · Life span of products

1 Introduction

Computer simulation is commonly used in manufacturing to analyze the productivity and effectiveness of production processes. Especially for manufacturing system design and the prototyping of new production processes; simulation methods are very useful because they enable us to find those combinations of components (resources, operation and setup times, intermediate buffers, etc.) that guarantee a sufficient level of throughput in the system. In this paper, a new simulative efficiency study for the allocation of buffer capacities in serial production lines is developed and a set of experiments are provided as a tool to determine the optimal buffer size and product life span index. The research problem considered in the paper can be formulated as follows: Given a manufacturing system, including flow production lines with defined operation and setup times, number of resources and production batch sizes; how do different topologies, intermediate buffer allocations and capacities affect the system throughput and the average life span of products? The model of the manufacturing system was created on the basis of part of a real manufacturing system of a company that carries out make-to-order production for the automotive industry. The production system in this study includes three production stages: pre-machining, appropriate machining and

© Springer International Publishing Switzerland 2015
K. Jackowski et al. (Eds.): IDEAL 2015, LNCS 9375, pp. 181–190, 2015.
DOI: 10.1007/978-3-319-24834-9_22

finishing. The number of production resources is the same for each production stage (three machines in each production stage).

The problem of intermediate buffer allocation (BAP) is one of the most important questions to face a designer in the field of serial production. It is an NP-hard combinatorial optimization problem in the design of production lines and the issue is studied by many scientists and theorists around the world. The buffer allocation problem is concerned with the allocation of a certain number of buffers, N, among the K – 1 intermediate buffer locations of a production line to achieve a specific objective [3]. This formulation of the problem expresses the maximization of the throughput rate for a given fixed number of buffers, achieving the desired throughput rate with the minimum total buffer size or the minimization of the average work-in-process inventory which is subject to the total buffer size and the desired throughput rate constraint.

The problem of maximizing the throughput of production lines by changing buffer sizes or locations using simulation methods was studied by Vidalis et al. [1]. A critical literature overview in the area of buffer allocation and production line performance was done by Battini et al. [2]. Demir et al. proposed a classification scheme to review the studies and presented a comprehensive survey on the buffer allocation problem in production systems [3]. Staley and Kim presented results of simulation experiments carried out for buffer allocations in closed serial-production lines [4]. In a production line, a single buffer space is the room and the associated material handling equipment that is needed to store a single job that is a work-in-process, and buffer allocation is the specific placement of a limited number of buffer spaces in a production line. The authors demonstrated a buffer allocation decomposition result for closed production lines, and also provided evidence that optimal buffer allocations in closed lines are less sensitive to bottleneck severity than in open production lines. The placement of buffers in a production line is an old and well-studied problem in industrial engineering research. Vergara and Kim proposed a new buffer placement method for serial production lines [5]. The method is very efficient and uses information generated in a production line simulation whose conceptual representation of job flow and workstation interaction can be described with a network which aims to place buffers in such a way as to maximize throughput. They compared the results of the new method against a method for buffer placement based on a genetic algorithm. Yamashita and Altiok [6] proposed an algorithm for minimizing the total buffer allocation for a desired throughput in production lines with phase-type processing times. They implemented a dynamic programming algorithm that uses a decomposition method to approximate system throughput at every stage. Gurkan used a simulation-based optimization method to find optimal buffer allocations in tandem production lines where machines are subject to random breakdowns and repairs, and the product is fluid-type [7]. He explored some of the functional properties of the throughput of such systems and derived recursive expressions to compute one-sided directional derivatives of throughput from a single simulation run. Fernandes and Carmo-Silva [8] presented a simulation study of the role of sequence-dependent set-up times in decision making at the order-release level of a workload controlled make-to-order flow-shop. They indicated that the local strategy, which has been traditionally adopted in practice and in most of the studies dealing with sequence-dependent set-up times, does not always give the best results. Matta [9] presented mathematical programming representations for a simulation-based optimization of buffer allocation in flow lines. Shi and Gershwin presented an effective algorithm for maximizing profits through buffer size optimization for production lines [10]. They considered both buffer space cost and average inventory cost with distinct cost coefficients for different buffers. To solve the problem, a corresponding unconstrained problem was introduced and a nonlinear programming approach was adopted [11]. In the next chapter, a simulation model of the manufacturing system is presented.

2   A Simulation Model of a Manufacturing System

The simulation experiments were made for two models of manufacturing systems with different topologies. The models of the automated production line were prepared on the basis of a real example of a manufacturing system dedicated to metal tooling in an automotive company. Both the models and simulation experiments were implemented using Tecnomatix PLM simulation software. The first model of a studied manufacturing system includes three technological lines with common input and output storage (Fig. 1).

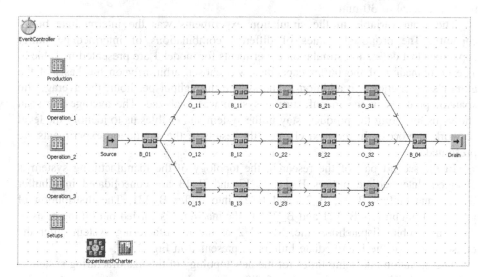

Fig. 1. Simulation model 1 of the manufacturing system

Machines O_11, O_12 and O_13 perform the same technological operations and are identical (the machines O_21, O_22, O_23 and O_31, O_32, O_33 are also identical). The manufacturing system is dedicated to the production of four kinds of products. The batch sizes for every kind of product are presented in Table 1.

The operation times defined for the products on the machines are presented in Table 2.

Table 1. The batch size matrix

Name of product	Batch size
Product A	20
Product B	25
Product C	15
Product D	10

Table 2. The matrix of operation times [minutes]

Machine	Product A	Product B	Product C	Product D
O_11,O_12, O_13,	5:00	8:00	6:00	9:00
O_21, O_22, O_23	14:00	10:00	12:00	11:00
O_31, O_32, O_33	5:00	4:00	3:00	5:00

The set-up times are defined in a set-up matrix but to simplify the model the setup times are equal to 30 min.

The main variable in the simulation experiments was the intermediate buffer capacity. The proposed values of different combinations of intermediate buffer capacities that define the simulation experiments for model 1 are presented in Table 3. The combination of the buffer capacities was chosen arbitrarily on the basis of the author's experiences. In addition to the system throughput per hour, the average life span of products in the manufacturing system was analyzed. The average life span shows us how long the products stay in the system and enables us to identify the level of work-in-process. In the next chapter, the results of the simulation experiments are presented.

In the second part of the research, the topology of the manufacturing system is changed. Intermediate buffers B_11, B_12 and B_13 are consolidated into buffer B_121 and buffers B_21, B_22 and B_23 into buffer B_221. The distribution of products from buffers B_121 and B_221 to the machines is carried out on the basis of the round robin dispatching rule. Model 2 of the manufacturing system, after the consolidation of the intermediate buffers, is presented in Fig. 2.

The operation and setup times for the new topology of the manufacturing system is not changed. The proposed values of different combinations of intermediate buffer capacities that define the simulation experiments for model 2 are presented in Table 4.

The input values of the simulation experiments are the variants of the intermediate buffer capacities presented in Tables 2 and 3. The output values of the experiments were the throughput per hour and average lifespan of the products. The input and output buffers in both models have the same capacity (equal to 5), but the capacity has no influence on the throughput of the investigated manufacturing system. It was assumed that in both models the efficiency of manufacturing resources is approximately 95 %.

Table 3. The matrix of intermediate buffer capacities for simulation experiments of model 1

Experiment	B_01	B_11	B_12	B_13	B_21	B_22	B_23	B_04
Exp 01	5	1	1	1	1	1	1	5
Exp 02	5	1	5	1	1	5	1	5
Exp 03	5	5	5	5	1	1	1	5
Exp 04	5	1	1	1	5	5	5	5
Exp 05	5	1	2	3	3	2	1	5
Exp 06	5	2	1	2	2	1	2	5
Exp 07	5	1	2	3	1	2	3	5
Exp 08	5	4	1	4	5	1	5	5
Exp 09	5	2	2	3	3	2	2	5
Exp 10	5	4	2	4	2	4	2	5
Exp 11	5	2	3	2	2	3	2	5
Exp 12	5	3	4	2	2	4	3	5
Exp 13	5	3	4	3	3	3	2	5
Exp 14	5	5	5	5	5	5	5	5
Exp 15	5	10	10	10	10	10	10	10
Exp 16	5	2	2	2	2	2	2	5
Exp 17	5	3	2	1	3	2	1	5
Exp 18	5	1	3	2	1	3	2	5
Exp 19	5	3	2	1	3	2	1	5
Exp 20	5	3	1	2	1	2	3	5
Exp 21	5	1	3	1	1	3	1	5
Exp 22	5	1	1	1	1	1	1	5

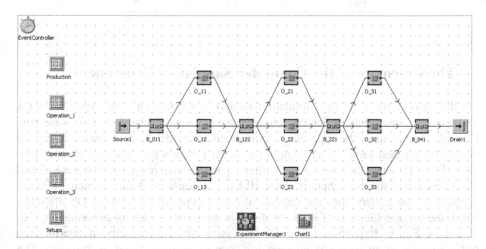

Fig. 2. Simulation model 2 of the manufacturing system after consolidation of intermediate buffers

Table 4. The matrix of intermediate buffer capacities for simulation experiments of model 2

Experiment	B_011	B_121	B_221	B_041
Exp 01	5	1	1	5
Exp 02	5	1	2	5
Exp 03	5	1	3	5
Exp 04	5	1	4	5
Exp 05	5	1	5	5
Exp 06	5	2	1	5
Exp 07	5	2	2	5
Exp 08	5	2	3	5
Exp 09	5	2	4	5
Exp 10	5	2	5	5
Exp 11	5	3	1	5
Exp 12	5	3	2	5
Exp 13	5	3	3	5
Exp 14	5	3	4	5
Exp 15	5	3	5	5
Exp 16	5	4	1	5
Exp 17	5	4	2	5
Exp 18	5	4	3	5
Exp 19	5	4	4	5
Exp 20	5	4	5	5
Exp 21	5	5	5	5
Exp 22	5	8	8	5
Exp 23	5	10	10	5
Exp 24	5	20	20	5

3   The Outcomes of the Computer Simulation Experiments

On the basis of the described simulation models, a set of simulation experiments for variants of intermediate buffer capacity, as presented in Tables 3 and 4, were performed. The results of the experiments performed for model 1 are presented in Figs. 3 and 4. The greatest values of throughput for model 1 were achieved in experiments Exp 14 and Exp 15 where the intermediate buffer capacities were respectively equal to 5 and 10. The value of the throughput is about 10.1 pieces per hour. A further increase in the capacity of the buffers does not increase the throughput of the system. On the other hand, the average life span of products displays minimum values for experiments: Exp 1, Exp 4, Exp 6, Exp 16 and Exp 22. Increasing the total capacity of intermediate buffers results in an increased average life span of products (work in process). The minimum value of the average life span of products (Exp 4) is equal to 1:50:37. The results of the simulation research show that, generally, the throughput of the system increases with increasing buffer capacities, but together with an associated increase in

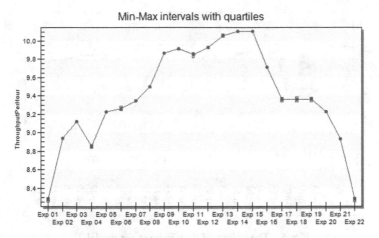

Fig. 3. The throughput per hour for model 1

the average life span of products (work-in-process). To find the best compromise between throughput and average life span; a flow index is proposed.

To calculate the index, the value of the throughput of the system is divided by the average life span. The values of the flow index for model 1 are presented in Fig. 5.

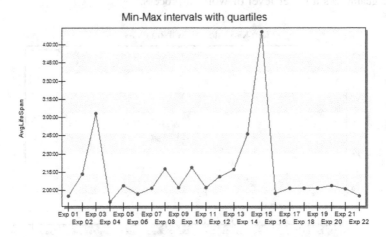

Fig. 4. The average life span for model 1

The greatest value of the flow index is achieved in experiment 16 (Exp 16) for model 1 and is equal to 118.78. It means that the allocation of intermediate buffers defined in experiment 16 guarantee the best compromise between the throughput of the manufacturing system and the average lifespan of products.

In the Figs. 6 and 7, the throughput and life span characteristics for model 2 are presented. The greatest values of the throughput for model 2 are achieved for the large size of intermediate buffers; equal to 8, 10 and 20 (experiments: Exp 22, Exp 23 and

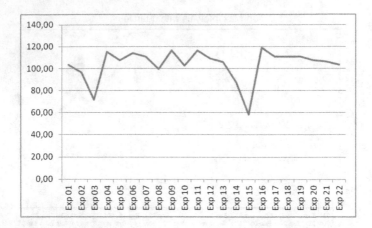

Fig. 5. The values of flow indexes for model 1

Exp 24). The value of throughput for the experiments is equal to 10.1 (the same greatest value of throughput is achieved in model 1). From the chart presented in Fig. 6, it is evident that the single capacity of intermediate buffers results in a decrease in the throughput of the manufacturing system.

The average value of product life span for model 2 is equal to 1:57. It means that model 2 guarantees a lower level of work-in-process.

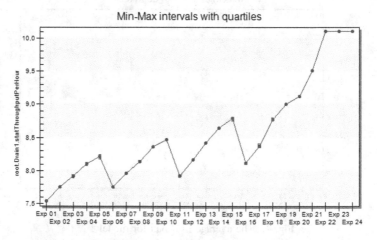

Fig. 6. The throughput per hour model 2

It is interesting to note that if the capacity of buffer B_121 is constant and the capacity of buffer B_221 increases, the throughput of the system increases and the average lifespan of products decreases. The lowest value of the average life span is achieved by the combination of buffer capacities used in Exp 5 (1:45). Increasing the intermediate buffer capacity to 20 units results in a rapid increase of the average life span (more than 3 h).

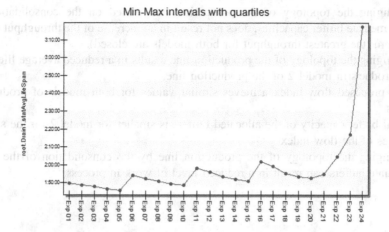

Fig. 7. The average life span for model 2

The flow index for model 2 is presented in Fig. 8. The best value of the index is achieved in Exp 21 in which intermediate buffer capacities equal 5. The maximal flow index for model 2 achieves a similar value to the maximal flow index of model 1 (119.34).

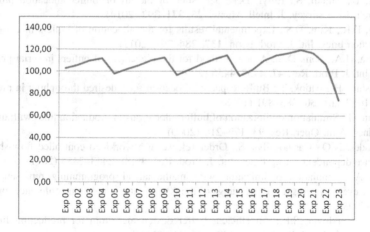

Fig. 8. The values of flow indexes for model 2

4   Conclusions

In this paper, an analysis of the throughput and life span of an automatic production line based on a computer simulation is presented, and the problem of intermediate buffer allocation is addressed. Two topologies of production lines are taken into account. Simulation models are prepared and experiments are carried out for different allocations of intermediate buffers for the two models of production line. The results of the simulation research enabled us to formulate the following conclusions:

1. Changing the topology of the production line, based on the consolidation of intermediate buffer capacities, does not result in an increase of the throughput of the system (the greatest throughput for both models are closed).
2. Changing the topology of the production line results in a reduced average life span of products in model 2 of the production line.
3. The proposed flow index achieves similar values for both models of production lines.
4. Total buffer capacity of the allocated buffers is smaller for model 2 for the similar values of the flow index.
5. Changing the topology of the production line by the consolidation of the intermediate buffers can result in a reduced level of work in process.

References

1. Vidalis, M.I., Papadopoulos, C.T., Heavey, C.: On the workload and 'phase load' allocation problems of short reliable production line with finite buffers. Comput. Ind. Eng. **48**, 825–837 (2005)
2. Battini, D., Persona, A., Regattieri, A.: Buffer size design linked to reliability performance: a simulative study. Comput. Ind. Eng. **56**, 1633–1641 (2009)
3. Demir, L., Tunali, S., Eliiyi, D.T.: The state of the art on buffer allocation problem: a comprehensive survey. J. Intell. Manuf. **25**, 371–392 (2014)
4. Staley, D.R., Kim, D.S.: Experimental results for the allocation of buffers in closed serial production lines. Int. J. Prod. Econ. **137**, 284–291 (2012)
5. Vergara, H.A., Kim, D.S.: A new method for the placement of buffers in serial production lines. Int. J. Prod. Res. **47**, 4437–4456 (2009)
6. Yamashita, H., Altiok, T.: Buffer capacity allocation for a desired throughput in production lines. IIE Trans. **30**, 883–891 (1998)
7. Gurkan, G.: Simulation optimization of buffer allocations in production lines with unreliable machines. Ann. Oper. Res. **93**, 177–216 (2000)
8. Fernandes, N.O., Carmo-Silva, S.: Order release in a workload controlled flow-shop with sequence-dependent set-up times. Int. J. Prod. Res. **49**(8), 2443–2454 (2011)
9. Matta A.: Simulation optimization with mathematical programming representation of discrete event systems. In: Proceedings of the 2008 winter simulation conference, pp. 1393–1400 (2008)
10. Shi, C., Gershwin, S.B.: An efficient buffer design algorithm for production line profit maximization. Int. J. Prod. Econ. **122**, 725–740 (2009)
11. Nahas, N., Ait-Kadi, D., Nourelfath, M.: Selecting machines and buffers in unreliable series-parallel production lines. Int. J. Prod. Res. **47**(14), 3741–3774 (2009)

A Novel Recursive Solution to LS-SVR
for Robust Identification of Dynamical Systems

José Daniel A. Santos[1(✉)] and Guilherme A. Barreto[2]

[1] Department of Industry, Federal Institute of Education,
Science and Technology of Ceará, Maracanaú, CE, Brazil
jdalencars@gmail.com
[2] Department of Teleinformatics Engineering, Center of Technology,
Federal University of Ceará, Campus of Pici, Fortaleza, CE, Brazil
gbarreto@ufc.br

Abstract. Least Squares Support Vector Regression (LS-SVR) is a powerful kernel-based learning tool for regression problems. However, since it is based on the ordinary least squares (OLS) approach for parameter estimation, the standard LS-SVR model is very sensitive to outliers. Robust variants of the LS-SVR model, such as the WLS-SVR and IRLS-SVR models, have been developed aiming at adding robustness to the parameter estimation process, but they still rely on OLS solutions. In this paper we propose a totally different approach to robustify the LS-SVR. Unlike previous models, we maintain the original LS-SVR loss function, while the solution of the resulting linear system for parameter estimation is obtained by means of the Recursive Least M-estimate (RLM) algorithm. We evaluate the proposed approach in nonlinear system identification tasks, using artificial and real-world datasets contaminated with outliers. The obtained results for infinite-steps-ahead prediction shows that proposed model consistently outperforms the WLS-SVR and IRLS-SVR models for all studied scenarios.

Keywords: Nonlinear regression · Outliers · LS-SVR · System identification · M-estimation

1 Introduction

Least Squares Support Vector Regression (LS-SVR) [4,5] is a widely used tool for regression problems, being successfully applied to time series forecasting [6], control systems [7] and system identification [8]. Its standard formulation makes use of equality constraints and a least squares loss function [5]. Then, the solution for the parameters is obtained by solving a system of linear equations, using ordinary Least Squares (OLS) algorithm. However, OLS provides optimal solution only if errors follow a Gaussian distribution. Unfortunately, in several real-world applications of regression it is a too strong assumption because the available data is usually contaminated with impulsive noise or outliers. In these scenarios, the performance of LS-SVR models may decline significantly.

© Springer International Publishing Switzerland 2015
K. Jackowski et al. (Eds.): IDEAL 2015, LNCS 9375, pp. 191–198, 2015.
DOI: 10.1007/978-3-319-24834-9_23

Thereby, some authors have devoted attention to develop robust variants for the LS-SVR models. For example, Suykens *et al.* [1] introduced a weighted version of LS-SVR based on *M*-Estimators [9], and named it *Weighted Least Squares Support Vector Regression* (WLS-SVR). Later, De Brabanter *et al.* [2] also introduced a robust estimate upon the previous LS-SVR solutions using an iteratively reweighting approach, named *Iteratively Reweighted Least Squares Support Vector Regression* (IRLS-SVR).

In essence, while these robust variants modify the original LS-SVR loss function to penalize large error, they still rely on the OLS algorithm to provide a solution to the resulting linear system. In this paper, we introduce a different approach to add outlier robustness to the LS-SVR model. Instead of modifying its loss function, we decide to solve the resulting linear system for parameter estimation by using the Recursive Least *M*-estimate (RLM) algorithm [3]. The RLM rule is itself a robust variant of the standard *recursive least squares* (RLS) algorithm, which was modified by the use of *M*-estimators to handle outliers.

The proposed approach, referred to as the RLM-SVR model, updates recursively the vector of Lagrange multipliers and bias for each input sample. By doing this, the occurrence of outliers in a given input sample may be treated individually, improving the quality of the solution. For the purpose of validation, we use two synthetic datasets, whose outputs are contaminated with different amounts of outliers, and a real world dataset. The performance of the proposed RLM-SVR model is then compared with those provided by the standard LS-SVR, WLS-SVR and IRLS-SVR models in infinite-step-ahead prediction tasks.

The remainder of the paper is organized as follows. In Sect. 2 we describe briefly the LS-SVR, WLS-SVR and IRLS-SVR models. In Sect. 3 we introduce our approach. In Sect. 4 we report the achieved results. Finally, we conclude the paper in Sect. 5.

2   Evaluated Models

Initially, let us consider the estimation dataset $\{(\boldsymbol{x}_n, y_n)\}_{n=1}^{N}$, with the inputs $\boldsymbol{x}_n \in \mathbb{R}^p$ and corresponding outputs $y_n \in \mathbb{R}$. In a regression problem, the goal is to search for a function $f(\cdot)$ that approximates the outputs y_n for all instances of the available data. For the nonlinear case, f usually takes the form

$$f(\boldsymbol{x}) = \langle \boldsymbol{w}, \varphi(\boldsymbol{x}) \rangle + b, \tag{1}$$

where $\varphi(\cdot) : \mathbb{R}^p \to \mathbb{R}^{p_h}$ is a nonlinear map into a higher dimensional feature space, $\langle \cdot, \cdot \rangle$ denotes a dot product in this feature space, $\boldsymbol{w} \in \mathbb{R}^{p_h}$ is a vector of weights and $b \in \mathbb{R}$ is a bias.

The formulation of the parameter estimation problem in LS-SVR leads to the minimization of the following functional [4,5]

$$J(\boldsymbol{w}, e) = \frac{1}{2}\|\boldsymbol{w}\|_2^2 + C\frac{1}{2}\sum_{n=1}^{N} e_n^2, \tag{2}$$

subject to

$$y_n = \langle w, \varphi(x_n) \rangle + b + e_n, \quad n = 1, 2, \ldots, N \tag{3}$$

where $e_n = y_n - f(x_n)$ is the error due to the n-th input pattern and $C > 0$ is a regularization parameter.

The Lagrangian function of the optimization problem in Eqs. (2) and (3) is

$$L(w, b, e, \alpha_0) = \frac{1}{2}\|w\|_2^2 + C\frac{1}{2}\sum_{n=1}^{N} e_n^2 - \sum_{n=1}^{N} \alpha_n[\langle w, \varphi(x_n)\rangle + b + e_n - y_n], \tag{4}$$

where α_n's are the Lagrange multipliers. The optimal dual variables correspond to the solution of the linear system $A\alpha = y$, given by

$$\underbrace{\begin{bmatrix} 0 & \mathbf{1}^T \\ \mathbf{1} & \Omega + C^{-1}\mathbf{I} \end{bmatrix}}_{A} \underbrace{\begin{bmatrix} b \\ \alpha_0 \end{bmatrix}}_{\alpha} = \underbrace{\begin{bmatrix} 0 \\ y_0 \end{bmatrix}}_{y}, \tag{5}$$

where $y_0 = [y_1, \ldots, y_N]^T$, $\mathbf{1} = [1, \ldots, 1]^T$, $\alpha_0 = [\alpha_1, \ldots, \alpha_N]^T$, $\Omega \in \mathbb{R}^{N \times N}$ is the kernel matrix whose entries are $\Omega_{i,j} = k(x_i, x_j) = \langle \varphi(x_i), \varphi(x_j) \rangle$ where $k(\cdot, \cdot)$ is the chosen kernel function. Moreover, $A \in \mathbb{R}^{(N+1) \times (N+1)}$, $\alpha \in \mathbb{R}^{(N+1)}$ and $y \in \mathbb{R}^{(N+1)}$. Thus, the solution for α is provided by the OLS algorithm as

$$\alpha = (A^T A)^{-1} A^T y. \tag{6}$$

Finally, the resulting LS-SVR model for nonlinear regression is given by

$$f(x) = \sum_{n=1}^{N} \alpha_n k(x, x_n) + b. \tag{7}$$

The Gaussian kernel, $k(x, x_n) = \exp\left\{ \frac{\|x - x_n\|_2^2}{2\gamma^2} \right\}$, was adopted in this paper.

2.1   The WLS-SVR Model

The WLS-SVR model [1] is formulated by the minimization of the loss function

$$J(w, e) = \frac{1}{2}\|w\|_2^2 + C\frac{1}{2}\sum_{n=1}^{N} v_n e_n^2, \tag{8}$$

subject to the same constraints in Eq. (3). The error variables from the unweighted LS-SVR are weighted by the vector $v = [v_1, \ldots, v_N]^T$, according to Eq. (11). The optimal solution for WLS-SVR is provided by solving the linear system $A_v \alpha = y$ also by means of the OLS algorithm:

$$\alpha = (A_v^T A_v)^{-1} A_v^T y, \tag{9}$$

where the matrix $A_v \in \mathbb{R}^{(N+1) \times (N+1)}$ is defined as

$$A_v = \begin{bmatrix} 0 & \mathbf{1}^T \\ \mathbf{1} & \Omega + C^{-1}\mathbf{V} \end{bmatrix}, \tag{10}$$

and the diagonal matrix $\mathbf{V} \in \mathbb{R}^{N \times N}$ is given by $\mathbf{V} = \text{diag}\left\{\frac{1}{v_1}, \ldots, \frac{1}{v_N}\right\}$.

Each weight v_n is determined based on the error variables $e_n = \alpha_n/C$ from the original LS-SVR model. In this paper, the robust estimates are obtained from *Hampel* weight function as follows

$$
v_n = \begin{cases} 1 & \text{if } |e_n/\hat{s}| \le c_1, \\ \frac{c_2 - |e_n/\hat{s}|}{c_2 - c_1} & \text{if } c_1 < |e_n/\hat{s}| \le c_2, \\ 10^{-8} & \text{otherwise,} \end{cases} \tag{11}
$$

where $\hat{s} = IQR^1/1.349$ is a robust estimate of the standard deviation of the LS-SVR error variables e_n. Values $c_1 = 2.5$ and $c_2 = 3.0$ are typical ones.

2.2   The IRLS-SVR Model

The weighting procedure for WLS-SVR model can be repeated iteratively giving rise to the IRLS-SVR model [2]. At each iteration i, one can weight the error variables $e_n^{(i)} = \alpha_n^{(i)}/C$ for $n = 1, \ldots, N$. The weights $v_n^{(i)}$ are calculated based upon $e_n^{(i)}/\hat{s}$ and using the weight function in Eq. (11). The next step requires the solution of the linear system $\mathbf{A}_v^{(i)} \boldsymbol{\alpha}^{(i)} = \boldsymbol{y}$, where

$$
\mathbf{A}_v^{(i)} = \begin{bmatrix} 0 & \mathbf{1}^T \\ \mathbf{1} & \boldsymbol{\Omega} + C^{-1}\mathbf{V}_i \end{bmatrix}, \quad \text{and} \quad \mathbf{V}^{(i)} = \text{diag}\left\{ \frac{1}{v_1^{(i)}}, \ldots, \frac{1}{v_N^{(i)}} \right\}. \tag{12}
$$

The resulting model in the i-th iteration is then given by

$$
f^{(i)}(\boldsymbol{x}) = \sum_{n=1}^{N} \alpha_n^{(i)} k(\boldsymbol{x}, \boldsymbol{x}_n) + b^{(i)}. \tag{13}
$$

Then, we set $i = i+1$ and the new weights $v_n^{(i+1)}$ and vector $\boldsymbol{\alpha}_{i+1}$ are calculated. In this paper, this procedure continues until $\max_n(|\alpha_n^{(i)} - \alpha_n^{(i-1)}|) \le 10^{-3}$.

3   The Proposed Approach

Initially, we compute matrix $\boldsymbol{A} = [\boldsymbol{a}_1, \ldots, \boldsymbol{a}_n, \ldots, \boldsymbol{a}_{N+1}]$, where $\boldsymbol{a}_n \in \mathbb{R}^{N+1}$, and vector $\boldsymbol{y}$ in Eq. (5) as in the standard LS-SVR model. The proposed idea involves the application of a robust recursive estimation algorithm to compute at each iteration n the vector $\boldsymbol{\alpha}_n$ considering as the input vectors the columns of matrix $\boldsymbol{A}$, rather than using the batch OLS algorithm. As we will show in the simulations, this simple approach turned out to be very effective in providing robust estimates of the vector $\boldsymbol{\alpha}$ resulting in very good performance for the RLM-SVR model when the data are contaminated with outliers.

---

[1] IQR stands for InterQuantile Range, which is the difference between the 75th percentile and 25th percentile.

The RLM algorithm [3] was chosen as the recursive estimation rule in our approach, and its loss function is given by

$$J_\rho^{(n)} = \sum_{n=1}^{N+1} \lambda^{N+1-n} \rho(e_n), \tag{14}$$

where $\rho(\cdot)$ is an M-estimate function and $0 \ll \lambda \leq 1$ is a exponential forgetting factor, since the information of the distant past has an increasingly negligible effect on the coefficient updating.

In this paper, the Hampel's three-part M-estimate function is considered

$$\rho(e) = \begin{cases} \frac{e^2}{2} & 0 \leq |e| < \xi_1, \\ \xi_1|e| - \frac{\xi_1^2}{2} & \xi_1 \leq |e| < \xi_2, \\ \frac{\xi_1}{2}(\xi_3 + \xi_2) - \frac{\xi_1^2}{2} + \frac{\xi_1}{2}\frac{(|e|-\xi_3)^2}{\xi_2-\xi_3} & \xi_2 \leq |e| < \xi_3, \\ \frac{\xi_1}{2}(\xi_3 + \xi_2) - \frac{\xi_1^2}{2} & \text{otherwise}, \end{cases} \tag{15}$$

where ξ_1, ξ_2 and ξ_3 are threshold parameters which need to be estimated continuously. Let the error e_n have a Gaussian distribution possibly corrupted with some impulsive noise. The error variance σ_n^2 at iteration n is estimated as follows:

$$\hat{\sigma}_n^2 = \lambda_e \hat{\sigma}_{n-1}^2 + c(1 - \lambda_e)\text{med}(F_n), \tag{16}$$

where $0 \ll \lambda_e \leq 1$ is a forgetting factor, $\text{med}(\cdot)$ is the median operator, $F_n = \{e_n^2, e_{n-1}^2, \ldots, e_{n-N_w+1}^2\}$, N_w is the fixed window length for the median operation and $c = 1.483(1 + 5/(N_w - 1))$ is the estimator's correction factor. In this paper, we set $\xi_1 = 1.96\hat{\sigma}_i$, $\xi_2 = 2.24\hat{\sigma}_i$ and $\xi_3 = 2.576\hat{\sigma}_i$ [3]. Moreover, the values $\lambda_e = 0.95$ and $N_w = 14$ were fixed.

The optimal vector $\boldsymbol{\alpha}$ can be obtained by setting the first order partial derivative of $J_\rho^{(n)}$ with respect to $\boldsymbol{\alpha}_n$ to zero. This yields

$$\mathbf{R}_\rho^{(n)} \boldsymbol{\alpha}_n = \mathbf{P}_\rho^{(n)}, \tag{17}$$

where $\mathbf{R}_\rho^{(n)}$ and $\mathbf{P}_\rho^{(n)}$ are called the M-estimate correlation matrix of $\boldsymbol{a}_n$ and the M-estimate cross-correlation vector of $\boldsymbol{a}_n$ and y_n, respectively. Equation (17) can be solved recursively using RLM by means of the following equations:

$$\boldsymbol{S}_n = \lambda^{-1}(\boldsymbol{I} - \boldsymbol{g}_n \boldsymbol{a}_n^T)\boldsymbol{S}_{n-1}, \tag{18}$$

$$\boldsymbol{g}_n = \frac{q(e_n)\boldsymbol{S}_{n-1}\boldsymbol{a}_n}{\lambda + q(e_n)\boldsymbol{a}_n^T\boldsymbol{S}_{n-1}\boldsymbol{a}_n}, \tag{19}$$

$$\boldsymbol{\alpha}_n = \boldsymbol{\alpha}_{n-1} + (y_n - \boldsymbol{a}_n^T\boldsymbol{\alpha}_{n-1})\boldsymbol{g}_n, \tag{20}$$

where $\boldsymbol{S}_n = \boldsymbol{R}_\rho^{-1(n)}$, $q(e_n) = \frac{1}{e_n}\frac{\partial\rho(e_n)}{\partial e_n}$ is the weight function and $\boldsymbol{g}_n$ is the M-estimate gain vector. The maximum number of iterations is $N + 1$; however, it may be necessary to cycle over the data samples $N_e > 1$ times in order to the algorithm to converge. In this paper, we used $N_e = 20$ for all experiments. The RLM-SVR model is summarized in Algorithm 1.

Algorithm 1. Pseudo-code for RLM-SVR algorithm.

Require: $C, \gamma, \delta, N_e, \lambda, \lambda_e, N_w$
  calculate A and y from Eq. (5)
  set $S_0 = \delta I, g_0 = \alpha_0 = 0$
  for $i = 1 : N_e$, **do**
    for $n = 1 : N + 1$, **do**
      $e_n = y_n - \alpha_{n-1}^T a_n$
      calculate $\hat{\sigma}_n^2$ from Eq. (16)
      calculate $q(e_n)$ from the derivative of Eq. (15)
      update S_n from Eq. (18)
      update g_n from Eq. (19)
      update α_n from Eq. (20)
    end for
  end for

Table 1. Details of the artificial datasets used in the computational experiments. The indicated noise in the last column is added only to the output of the estimation data.

		Input/Samples		
#	Output	Estimation	Test	Noise
1	$y_n = y_{n-1} - 0.5\tanh(y_{n-1} + u_{n-1}^3)$	$u_n = \mathcal{N}(u_n\|0,1)$ $-1 \le u_n \le 1$ 150 samples	$u_n = \mathcal{N}(u_n\|0,1)$ $-1 \le u_n \le 1$ 150 samples	$\mathcal{N}(0, 0.0025)$
2	$y_n = \frac{y_{n-1}y_{n-2}(y_{n-1}+2.5)}{1+y_{n-1}^2+y_{n-2}^2}$	$u_n = \mathrm{U}(-2,2)$ 300 samples	$u_n = \sin(2\pi n/25)$ 100 samples	$\mathcal{N}(0, 0.29)$

4   Computer Experiments

In order to assess the performance of the proposed approach in nonlinear system identification for infinite-step-ahead prediction and in the presence of outliers, we carried out computer experiments with two artificial datasets (Table 1) and a real-world benchmarking dataset.

In addition to Gaussian noise, the estimation data of all datasets were also incrementally corrupted with a fixed number of outliers equal to 2.5 %, 5 % and 10 % of the estimation samples. Each randomly chosen sample was added by a uniformly distributed value $\mathrm{U}(-M_y, +M_y)$, where M_y is the maximum absolute output. Only the output values were corrupted. The artificial datasets 1 and 2 were generated following [10,11], respectively. The real-world dataset is called wing flutter and it is available at DaISy repository[2]. This dataset corresponds to a SISO (Single Input Single Output) system with 1024 samples of each sequence u and y, in which 512 samples were used for estimation and the other 512 for test.

---

[2] http://homes.esat.kuleuven.be/smc/daisy/daisydata.html.

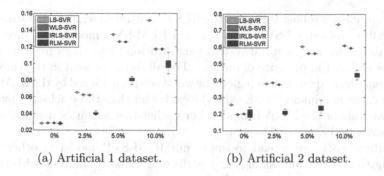

(a) Artificial 1 dataset.   (b) Artificial 2 dataset.

Fig. 1. RMSE with test samples in free simulation.

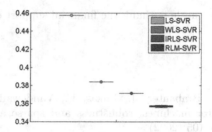

Fig. 2. RMSE with test samples of flutter dataset in free simulation.

The orders L_u and L_y chosen for the regressors in artificial datasets were set to their largest delays, according to Table 1. For wing flutter dataset the orders $L_u, L_y \in \{1, \ldots, 5\}$ were set after a 10-fold cross validation strategy in LS-SVR model. The same strategy was used to set $C \in \{2^0, \ldots, 2^{20}\}$, $\gamma \in \{2^{-10}, \ldots, 2^0\}$ and $\lambda \in \{0.99, 0.999, 0.9999\}$ in the search for their optimal values. Furthermore, we use $\delta = 10^2$ or $\delta = 10^3$ to initialize the matrix S.

We compare the performance of RLM-SVR, standard LS-SVR, WLS-SVR and IRLS-SVR models over 20 independent runs of each algorithm. The Root Mean Square Error (RMSE) distributions for test samples with artificial datasets and flutter wing dataset are shown in Figs. 1 and 2, respectively.

In all scenarios with artificial and real datasets, the LS-SVR, WLS-SVR and IRLS-SVR algorithms obtained the same RMSE value over the 20 runs. Although RLM-SVR algorithm presented some variance for RMSE values over the runs, their maximum values were even lower than those obtained with the other models in all contamination scenarios with artificial datasets and with flutter dataset. Moreover, in the scenarios with artificial datasets without outliers, the RLM-SVR model achieved average value of RMSE closer to the other robust models.

5 Conclusion

In this paper we presented an outlier-robust recursive strategy to solve the parameter estimation problem of the standard LS-SVR model. The chosen recursive estimation rule is a robust variant of the RLS algorithm which uses M-estimators. The resulting robust LS-SVR model, called RLM-SVR model, was succesfully applied to nonlinear dynamical system identification tasks under free simulation scenarios and in the presence of outliers. For all datasets used in the computer experiments carried out in this paper, the worst results achieved by the RLM-SVR model (i.e. the maximum RMSE values) were better than those achieved by other robust variants of the LS-SVR model. For outlier-free scenarios, the RLM-SVR model performed similarly to them.

In future work, we intend to apply our RLM-SVR model in other tasks, including time series forecasting, and perform a stability analysis with respect to the hyperparameters of the method.

Acknowledgments. The authors thank the financial support of IFCE, NUTEC and CNPq (grant no. 309841/2012-7).

References

1. Suykens, J.A.K., De Brabanter, J., Lukas, L., Vandewalle, J.: Weighted least squares support vector machines: robustness and sparse approximation. Neurocomputing **48**(1), 85–105 (2002)
2. De Brabanter, K., Pelckmans, K., De Brabanter, J., Debruyne, M., Suykens, J.A.K., Hubert, M., De Moor, B.: Robustness of Kernel based regression: a comparison of iterative weighting schemes. In: Alippi, C., Polycarpou, M., Panayiotou, C., Ellinas, G. (eds.) ICANN 2009, Part I. LNCS, vol. 5768, pp. 100–110. Springer, Heidelberg (2009)
3. Zou, Y., Chan, S., Ng, T.: A recursive least m-estimate (RLM) adaptive filter for robust filtering in impulse noise. IEEE Signal Proccess. Lett. **7**(11), 324–326 (2000)
4. Saunders, C., Gammerman, A., Vovk, V.: Ridge regression learning algorithm in dual variables. In:Proceedings of the 15th International Conference on Machine Learning (ICML 1998), pp. 515–521. Morgan Kaufmann (1998)
5. Suykens, J.A.K., Van Gestel, T., De Brabanter, J., De Moor, B., Vandewalle, J.: Least Squares Support Vector Machines. World Scientific, Singapore (2002)
6. Van Gestel, T., Suykens, J.A., Baestaens, D.E., Lambrechts, A., Lanckriet, G., Vandaele, B., De Moor, B., Vandewalle, J.: Financial time series prediction using least squares support vector machines within the evidence framework. IEEE Trans. Neural Netw. **12**(4), 809–821 (2001)
7. Khalil, H.M., El-Bardini, M.: Implementation of speed controller for rotary hydraulic motor based on LS-SVM. Expert Syst. Appl. **38**(11), 14249–14256 (2011)
8. Falck, T., Suykens, J.A., De Moor, B.: Robustness analysis for least squares kernel based regression: an optimization approach. In: Proceedings of the 48th IEEE Conference on Decision and Control (CDC 2009), pp. 6774–6779 (2009)
9. Huber, P.J., et al.: Robust estimation of a location parameter. Ann. Math. Stat. **35**(1), 73–101 (1964)
10. Kocijan, J., Girard, A., Banko, B., Murray-Smith, R.: Dynamic systems identification with gaussian processes. Math. Comput. Model. Dyn. Syst. **11**(4), 411–424 (2005)
11. Narendra, K.S., Parthasarathy, K.: Identification and control of dynamical systems using neural networks. IEEE Trans. Neural Netw. **1**(1), 4–27 (1990)

NMF and PCA as Applied to Gearbox Fault Data

Anna M. Bartkowiak[1](✉) and Radoslaw Zimroz[2]

[1] Institute of Computer Science, Wroclaw University, 50-383 Wroclaw, Poland
aba@ii.uni.wroc.pl
[2] Diagnostics and Vibro-Acoustics Science Laboratory,
Wroclaw University of Technology, 50-421 Wroclaw, Poland

Abstract. Both Non-negative matrix factorization (NMF) and Principal component analysis (PCA) are data reduction methods. Both of them act as approximation methods permitting to represent data by lower rank matrices. The two methods differ by their criteria how to obtain the approximation. We show that the main assumption of PCA demanding orthogonal principal components leads to a higher rank approximation as that established by NMF working without that assumption. This can be seen when analyzing a data matrix obtained from vibration signals emitted by a healthy and a faulty gearbox. To our knowledge this fact has not been clearly stated so far and no real example supporting our observation has been shown explicitly.

Keywords: Non-negative matrix factorization · Principal component analysis · Matrix approximation · Optimization with constraints · Gearbox fault detection

1 Introduction

Nowadays, when a huge amount of data is collected anywhere, the problem of extracting the proper information hidden in the data and reduce their amount – without losing the hidden information – becomes crucial. One popular method of finding the intrinsic dimensionality of the gathered data and reduce them to a smaller representative amount is the method known widely as Principal Component Analysis (PCA) [8]. It permits to construct new independent variables called principal components (PCs), which are representative for the data matrix dealt with.

Nonnegative Matrix Factorization (NMF), launched as a bio-inspired method [9], is in its goals similar to PCA, however in its principles quite different. The NMF method was presented as an analogy to recognizing objects by the human brain which learns objects by learning parts of them [9]. The main idea of NMF is that a given observable data matrix $\mathbf{V}$, necessarily with nonnegative elements, is approximated by the product of two smaller rank matrices $\mathbf{W}$ and $\mathbf{H}$, also with nonnegative elements: $\mathbf{V} \approx \mathbf{W} * \mathbf{H}$. The derived approximation matrices $\mathbf{W}$ and

© Springer International Publishing Switzerland 2015
K. Jackowski et al. (Eds.): IDEAL 2015, LNCS 9375, pp. 199–206, 2015.
DOI: 10.1007/978-3-319-24834-9_24

H offer frequently contextual explanation on the structure of the data contained in **V**. The explanation – because of the non-negativeness of the elements of **W** and **H** – is easier to perceive. A recent description of the NMF principles with its extensions and real data examples may be found the book by Cichocki et al. [3], also in Gillis [5] and in references therein.

The publication [9] gave beginning of large amount of discussion and analysis of experimental non-negative data. The method has found interesting applications in image analysis [6,12], technical fault detection [10], analysis of satellite data [13], music data [4], analysis of sound waves, EEG, etc., and became more and more popular.

On the other hand, the classic PCA is still much in use. It is developing by adapting a probabilistic approach [1]. Our question is: What are the advantages and the disadvantages of the two methods? In particular: May NMF give something different as PCA? We elaborate these questions when looking at a particular set of data collected for a healthy and a faulty gearbox. Our main goal will be to compare the approximation of **V** obtained by NMF and PCA.

In the following we introduce in Sect. 2 both methods formally and indicate for similarities and differences between them.

In Sect. 3 we describe shortly the gearbox data (given as matrix $\mathbf{V}_{n \times m}$), which will serve as an test rig for our considerations. We will calculate for out matrix **V** its approximates of lower rank using the NMF and PCA methods. The obtained results for rank r = 2 and r = 3 are shown in Figs. 1 and 2. We consider specifically 3 types of results produces both by NMF and PCA.

Short discussion and closing remarks are presented in Sect. 4.

2   The NMF and PCA Methods: Definitions, Similarities and Dissimilarities

Firstly we describe here briefly the NMF and the PCA methods. Next we present a summarization of their main properties valuable for practitioners.

2.1   The NMF Method

After [9], we consider a data matrix **V** of size $n \times m$, denoted $\mathbf{V}_{n \times m}$), with n rows and m columns. The authors [9] call this matrix 'visible' (hence denotation by **V**), because this should be a real observed data matrix. The elements (v_{ij}) of **V** should be expressed as non-negative real numeric values satisfying the constraints: $v_{ij} >= 0$, $i = 1, \ldots n$, $j = 1, \ldots, m$.

The authors [9] proposed to approximate the recorded data matrix $\mathbf{V}_{n \times r}$ by the product of two lower rank components **W** and **H** with non-negative elements:

$$\mathbf{V}_{n \times m} \simeq \mathbf{W}_{n \times r} * \mathbf{H}_{r \times m}, \tag{1}$$

where $0 < r <= min(m, n)$ is an integer denoting the rank of the sought component matrices. For practical reasons the following inequality should be satisfied: $(n + m) * r << n * m$.

The first derived approximation matrix $\mathbf{W}_{n \times r}$ is built from r columns vectors $\mathbf{w}_k$, $k = 1, \ldots, r$, each of size $n \times 1$, which may be written as

$$\mathbf{W} = [\mathbf{w}_1, \mathbf{w}_2, \ldots, \mathbf{w}_r], \text{ with } r <= min(n, m).$$

The column vectors $\mathbf{w}_k$ $(k = 1, \ldots, r)$ play the role of representatives of all the m columns of $\mathbf{V}$. They are referred to as the *basis vectors* of the entire data [7]. They may be mutually dependent. Each of them has size $n \times 1$, similarly as each column of the data matrix $\mathbf{V}$. We call them *meta-features* (MFs) of the analyzed data. They may be also considered as building blocks for the observed data matrix $\mathbf{V}$ [7].

The second component matrix $\mathbf{H}$ appearing in (1) contains elements called *encoding coefficients*. Matrix $\mathbf{H}$ is frequently viewed as composed from r row vectors denoted as $\mathbf{h}^{(k)}$, $k = 1, \ldots, r$. Obviously: $\mathbf{h}^{(k)} = [H_{k,1}, H_{k,2}, \ldots, H_{k,m}]$.

The original observed data matrix $\mathbf{V}$ may be approximated either directly by the product of $\mathbf{W}_{n \times r} * \mathbf{H}_{r \times m}$, as shown in (1), or in r cumulative steps as:

$$\mathbf{V}_{n \times m} \simeq \sum_{k=1}^{r} \mathbf{w}_k * \mathbf{h}^{(k)}. \tag{2}$$

The above formula (2) says that the approximation of the entire data matrix $\mathbf{V}$ may be done in r steps; each step uses the product matrix $[\mathbf{w}_k * \mathbf{h}^{(k)}]$, which provides a rank one approximation matrix of size $n \times m$.

Algorithms for finding the factorization matrices $\mathbf{W}$ and $\mathbf{H}$.

The authors [9] launched three algorithms for three objective functions yielding the sought factorizing matrices (mean square error, Poissonian likelihood and Kullback-Leibler divergence). All algorithms work as iterative multiplicative algorithms and all they –being easily programmable in Matlab – contributed to the popularity of NMF.

The discussion of these algorithms may be found in [2]. The extensions of NMF went in the direction of using other optimization criteria, in particular, *differently defined divergences*, also in adding additional constraints suitable, for example, for discriminant analysis or cluster analysis [3–5, 7, 11, 12, 16, 17].

2.2   The PCA Method

Let $\mathbf{X}_{n \times m}$ denote the observed data matrix with real numeric values. The matrix $\mathbf{X}$ is centered and usually additionally standardized to have unit variances. The rows of $\mathbf{X}$ denote objects ('individuals'), each of them characterized by m variables, contained in the columns of $\mathbf{X}$. The PCA method seeks for new (constructed) variables which (i) are linear functions of the primary variables contained in $\mathbf{X}$, (ii) are mutually orthogonal, and (iii) explain as much as possible of the total sum of variances evaluated for all m variables included into $\mathbf{X}$. The sought new variables (their realizations $Y_1, Y_2, \ldots, Y_r$ called *Principal Components*) are then constructed as [8]:

$$Y_k = \mathbf{X} * \mathbf{a}_k, \quad k = 1, \ldots, r \ (r \leq m), \tag{3}$$

where $\mathbf{a}_k$ are eigenvectors of the covariance (or correlation) matrix of $\mathbf{X}$.

The reconstruction of $\mathbf{X}$ is done on the basis of the derived new variables (i.e. principal components Y_k) and connected with them eigenvectors $\mathbf{a}_k$:

$$\hat{\mathbf{X}}^{(r)} = \sum_{k=1}^{r} Y_k * \mathbf{a}_k^T, \quad \text{or equivalently} \quad \hat{\mathbf{X}}^{(r)} = [Y_1, \ldots, Y_r] * ([\mathbf{a}_1, \ldots, \mathbf{a}_r]^T). \quad (4)$$

Comparing the two methods
Similarity

- Both methods have the same goal: Reduction of dimensionality by constructing new meaningful features which for $r = 2$ and $r = 3$ may serve for a 2D- and a 3D-visualization of the original data vectors projected to the reduced space.

Dissimilarity

- Both methods use different optimization criteria how to obtain their goals, hence different algorithms for constructing the approximation matrices. NMF approximates the entire data matrix $\mathbf{X}$, PCA approximates the trace of $\mathbf{S} = \frac{1}{n-1} * (\mathbf{X}^T * \mathbf{X})$ of properly standardized data matrix $\mathbf{X}$.
- Different properties of the constructed features: PCA yields uncorrelated features, NMF yields features which may be correlated.
- NMF needs a data matrix with non-negative elements; PCA takes any real numeric values, however the work *in the classic setting* is done using centered data matrix (where the elements are *ex definitione* both positive and negative) and the proper calculations (evaluating eigen-values and eigen-vectors) are carried out using covariance or correlation matrix of the (rescaled) data.
- NMF is iterative with a random start; results may vary in different runs. Algorithm for PCA is stabile, and in a repeated runs one obtains the same solution.

3   The Analyzed Data Matrix and Its Approximation by NMF and PCA

3.1   The Data

We have used data described and analyzed partially in [14,15]. For the present analysis, we have sampled a data matrix $\mathbf{V}$ of size 1000×15. All elements of this 'visible' matrix are positive. In rows (called segments) we have firstly 500 data vectors from the healthy, and next 500 data vectors from the faulty gearbox. Columns in $\mathbf{V}$ contain values of 15 variables called $pp1, \ldots, pp15$ evaluated for each row vector of $\mathbf{V}$. The pp variables are power spectra characteristics obtained from a PSD analysis of subsequent segments of the vibration signal emitted by the machines (their gearboxes).

To identify more easily graphical results, the 500-samples were sorted according to an external variable called ZWE denoting load state of the gearboxes.

3.2 Main Results from NMF and PCA Applied to the Gearbox Data

We have investigated the approximation for $r = 2$ and $r = 3$. For NMF we have used the matlab nmfmse function [7] working with the mean square error (mse) criterion applied to $\mathbf{V}$. For PCA we have used own script in Matlab which needed only the eig function from basic package, without any additional toolboxes.

Applying NMF we obtained for $r = 2$ and $r = 3$ the approximations:
$\mathbf{V}_{1000 \times 15} \simeq \mathbf{W2}_{1000 \times 2} * \mathbf{H2}_{2 \times 15}$ and $\mathbf{V}_{1000 \times 15} \simeq \mathbf{W3}_{1000 \times 3} * \mathbf{H3}_{3 \times 15}$.
The results of our analysis are depicted in Fig. 1.

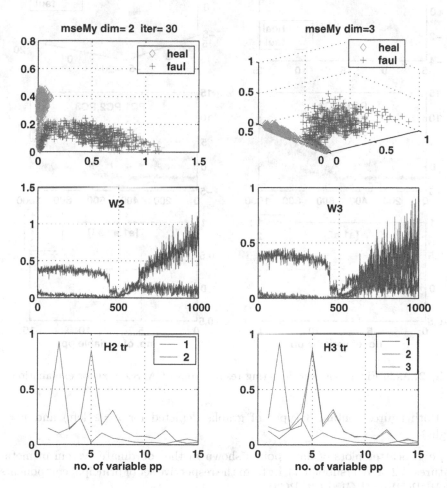

Fig. 1. Main results from NMF when using approximation of rank $dim = r = 2$ (left column of plots) and of rank $dim = r = 3$ (right column of plots). See text for explanation.

When **applying the PCA method** and using the correlation matrix for calculation of the eigenvectors, we obtained for $r = 2$ two eigenvectors $[\mathbf{a}_1, \mathbf{a}_2]$ and two principal components $[PC1, PC2]$. When opting for dimension $r = 3$, we needed additionally the eigenvector $\mathbf{a}_3$ and the additional principal component denoted as PC3. Their profiles are depicted in Fig. 2.

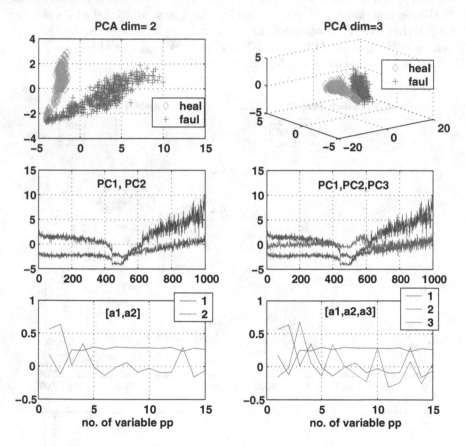

Fig. 2. As in Fig. 1, however depicting results from PCA. See text for explanation.

Both figures contain 3 types of graphs depicted for $r = 2$(left) and $r = 3$ (right):

Type 1. Scatter plots of data points shown in the coordinate system of meta-features W2 or W3 (subplot 1), or from the respective set of principal components (PC1, PC2) or (PC1, PC2, PC3).

Type 2. Profiles of the meta-features W2 and W3 (for NMF), and of the PC-s [PC1, PC2] or [PC1, PC2, PC3] (for PCA).

Type 3. Profiles of the encoding matrices H2 and H3 (for NMF), and of the first two or first three eigenvectors $[\mathbf{a}_1, \mathbf{a}_2]$ or $[\mathbf{a}_1, \mathbf{a}_2, \mathbf{a}_3]$ (for PCA).

3.3 What Is Seen in the 3 Types of Graphs from from NMF and PCA

Graphs of the first type, located in subplots 1 and 2 of both figures, show scatter plots depicting subsequent data vectors (rows) of the analyzed data as represented by the MFs contained in W2 – for NMF, and similarly the representations by the first two or first three PC-s – for PCA. Both methods recognize properly that the data are composed from two different groups of data.

Graphs of the second type, located in subplots 3 and 4 of both figures, show profiles of the derived MFs contained in W2 and W3 (Fig. 1), and of the respective PC-s (Fig. 2).

Concerning the exhibits in subplot 3, based on W2 or [PC1, PC2], one might say that the exhibits show the same pattern for both methods. We may deduce here that the first MF (or PC1) expresses the *faultiness* of the gearbox, and that the second MP (or PC2) expresses the normal functioning of the gearbox. The dip in the middle, around items no.s 500, corresponds to the no/low load state of the gearboxes.

Concerning the exhibits in subplot 4 based on W3 or [PC1, PC2, PC3], it is obvious that the exhibits are essentially different. The exhibit from NMF depicts the three MFs contained in W3. We see here that the first and the third MF are overlapping (with a correlation coefficient $r = 0.92$), moreover they have high variability for the faulty items. One may deduce: Dimension $r = 3$ is too high for the analyzed data. Analogous exhibit from PCA does not rise any suspicion. Segments no.s 1-500 have distinct values for all three PC-s. Only the PC2 for faulty segments is overshadowed by PC3 (however not so much as was observed on the graph obtained by NMF). One may deduce: The dimension $r = 3$ is OK, 3 dimensions are needed, especially when considering the 'healthy' part of the data coded as items 1-500.

Graphs of the third type, located in subplots 5 and 6 of both figures, show profiles of the transposed encoding matrix H2 or H3. The exhibits from NMF and PCA are completely different.

Subplot 5 and 6 obtained from NMF shows very clearly two profiles obtained from the first two columns of $(H2)^T$ and $(H3)^T$ appropriately. They are similar in the left and right exhibit. The third profile in exhibit 6 coincides practically with the 1st profile of that exhibit. Analogous Subplots 5 and 6 obtained from PCA show completely different profiles of $[a_1, a_2]^T$ and $[a_1, a_2, a_3]^T$. The 3 profiles in subplot 6 are different and do not resemble those obtained by NMF, each of them is needed for encoding. The same can be said about profiles exhibited in subplot 5.

4 Discussion and Closing Remarks

We have compared results of NMF and PCA for a set of gearbox data (15 variables characterizing 1000 segments of recorded vibration signals from a healthy and a faulty gearbox). Our goal was to reduce dimensionality of the data. Both methods constructed new features representative for the observed variables. We found that both methods are effective for that purpose. We found also that PCA - due to the requirement of the orthogonality of the constructed new features - is perhaps more stabile, but also comes later to its goal, that is, it chooses higher dimensionality as the NMF does.

References

1. Bartkowiak, A.M., Zimroz, R.: Probabilistic principal components, how this works. In: Saeed, K., Homenda, E. (eds.) CISIM 2015, IFIP LNCS 9339 (in print) (2015)
2. Berry, M.W., et al.: Algorithms and applications for approximate nonnegative matrix factorization. Comput. Stat. Data Anal. **52**, 155–173 (2007)
3. Cichocki, A., Zdunek, R., Phan, A.H., Amari, Sh: Nonnegative Matrix and Tensor Factorizations. Applications to Exploratory Multi-way Data Analysis and Blind Source Separation. Wiley, Chichester (2009)
4. Fevotte, C., Bertin, N., Durrieu, J.-L.: Nonnegative matrix factorization with the Itakura-Saito divergence. With application to music analysis. Neural Comput. **21**(3), 793–830 (2009)
5. Gillis, N.: The why and how of nonnegative matrix factorization. In: Suykens, J.A.K., et al. (eds.) Regularization, Optimization, Kernels and Support Machines, pp. 3–39. Chapman & Hall/CRC, London (2014)
6. Guillamet, D., Vitria, J., Schiele, B.: Introducing a weighted non-negative matrix factorization for image classification. Pattern Recogn. Lett. **24**, 2447–2454 (2003)
7. Hoyer, P.O.: Non-negative matrix factorization with sparseness constraints. J. Mach. Learn. Res. **5**, 1457–1469 (2004)
8. Jolliffe, I.T.: Principal Component Analysis, 2nd edn. Springer, New York (2002)
9. Lee, D.D., Seung, H.S.: Learning the parts of objects by nonnegative matrix factorization. Nature **401**, 788–791 (1999)
10. Li, B., et al.: A new feature extraction and selection scheme for hybrid fault diagnosis of gearbox. Expert Syst. Appl. **38**, 1000–1009 (2011)
11. Li, X., Fukui, K.: Fisher non-negative matrix factorization with pairwise weighting. In: IAPR Conference on Machine Vision Application, MVA 2007, pp. 380–383. Tokyo, Japan, May 16–18 (2007)
12. Li, S.Z., et al.: Learning spatially localized, parts-based representation. In: Proceesings of IEEE International Conference on Computer Vision and Pattern Recognition, vol. 1, pp. 207–2012. Havaii, December 2001
13. Tsagaris, V., Anastassopoulos, V.: Feature extraction of Meris data. In: Proceedings of the 2004 Envisat & ERS Symposium, Salzburg, Austria, 6–10 September 2004 (ESA SP-572, April 2005), pp. 1–5 (2005)
14. Zimroz, R., Bartkowiak, A.: Two simple multivariate procedures for monitoring planetary gearboxes in non-stationary operating conditions. Mech. Syst. Signal Process. **38**, 237–247 (2013)
15. Zimroz, R., Bartkowiak, A.: Investigation on spectral structure of gearbox vibration signals by principal component analysis for condition monitoring purposes. J. Phys. Conf. Ser. **305**(1), 012075 (2011)
16. Zdunek, R.: Trust-region algorithm for nonnegative matrix factorization with alpha- and beta-divergences. In: Pinz, A., Pock, T., Bischof, H., Leberl, F. (eds.) DAGM and OAGM 2012. LNCS, vol. 7476, pp. 226–235. Springer, Heidelberg (2012)
17. Zurada, J.M., Ensari, T., Asi, E.H., Chorowski, J.: Nonnegative matrix factorization and its application to pattern recognition and text mining. In: Proceedings of the 13th FedCSIS Conference on Computer Science and Information Systems, Cracow, pp. 11–16 (2013)

Application of Cascades of Classifiers in the Vehicle Detection Scenario for the 'SM4Public' System

Dariusz Frejlichowski[1]([✉]), Katarzyna Gościewska[1,2], Adam Nowosielski[1],
Paweł Forczmański[1], and Radosław Hofman[2]

[1] Faculty of Computer Science, West Pomeranian University of Technology, Szczecin,
Żołnierska 52, 71-210 Szczecin, Poland
{dfrejlichowski,anowosielski,pforczmanski}@wi.zut.edu.pl
[2] Smart Monitor Sp. z o.o., Cyfrowa 6, 71-441 Szczecin, Poland
{katarzyna.gosciewska,radekh}@smartmonitor.pl

Abstract. In the paper, the use of cascading approaches for vehicle
classification in static images is described. The problem concerns the
selection of algorithms to be implemented in the 'SM4Public' security
system for public spaces and is focused on specific system working scenario: the detection of vehicles in static images. Three feature extractors
were experimentally evaluated using a cascading classification approach
based on AdaBoost. The algorithms selected for feature extraction are
Histogram of Oriented Gradients, Local Binary Patterns and Haar-like
features. The paper contains brief introduction to the system characteristics, the description of the employed algorithms and the presentation
of the experimental results.

1 Introduction

This paper concerns the scientific research on the algorithms to be implemented
in the prototype 'SM4Public' system. The system is now being developed within
the framework of EU cofounded project and is aimed at construction and implementation of innovative video content analysis-based system prototype that will
ensure the safety of various public spaces using real-time solutions and typical computer components. The 'SM4Public' system will be highly customizable
and offer features that enable its adaptation to the actual situation. It means
that different working scenarios could be implemented. The idea of the project
was risen during the development of the previous system entitled 'SmartMonitor' [1–4]—an intelligent security system based on image analysis, created for
individual customers and home use. The analysis of alternative system applications has revealed a need to build other solution for public space video surveillance to effectively detect events threatening public safety, especially in places
where simultaneous movement of large number of people and vehicles is observed.
Therefore, 'SM4Public' system will be able to work under scenarios specific for
public spaces, such as scenarios associated with vehicle traffic (e.g. accident

© Springer International Publishing Switzerland 2015
K. Jackowski et al. (Eds.): IDEAL 2015, LNCS 9375, pp. 207–215, 2015.
DOI: 10.1007/978-3-319-24834-9_25

detection or failing to stop at the red light), infrastructure protection (e.g. theft or devastation detection), breaking the law (e.g. drinking alcohol in public spaces, prohibited in many countries) or threats to life or health (e.g. a fight or a fall).

Various solutions for vehicle detection in static images has been reported in the literature so far. Several examples are given below. In [5] a scheme for detecting vehicles is proposed together with a method that models the unknown distribution of the vehicle images using higher order statistics (HOS) information about 'vehicle class' based on data samples. A HOS-based decision measure is used to classify test patterns. The proposed method was tested on aerial images and gave good results even when an image contained a complicated scene with a cluttered background. The method can also detect non-frontal car views. Another solution to vehicle detection in static images was presented in [6]. The proposed approach uses colour transformation to project all input pixels' colours to a new feature space. Then the Bayesian classifier is applied to distinguish between vehicle and non-vehicle pixels. In the next step the Harris corner detector is used and detected extreme corners enable to identify and mark a vehicle with the rectangular shape. In turn the authors of [7] presented a system for the detection of rear-vehicle views from static images. The proposed method do not require any additional information about road boundary or lane. In the proposed system the region of interest is segmented based on the shadow area underneath the vehicles. Then the region of interest is localized by vehicle features such as edges, shadow information or symmetry. Vehicle detection is performed using a combination of knowledge-based and statistics-based methods. The prior knowledge about vehicles is used to filter some non-vehicles—this step additionally reduces the number of false detections. Then the Support Vector Machine is applied for two-class pattern classification. The authors declare good recognition and performance results of the system.

In the paper, other solutions for the vehicle detection in static scenes scenario are investigated using three feature extractors based on Haar-like features, Histogram of Oriented Gradients (HOG) and Local Binary Patterns (LBP), and one cascading approach for object classification, namely AdaBoost. By employing Visual Content Analysis (VCA) algorithms we aim at automatic detection and differentiation of vehicles in static images. The considered scenario does not require complex calculations and high computational power, and is characterized by a high detection rate and a small error probability. Therefore it is possible to perform static image analysis with a time interval of a few or more seconds. The rest of the paper is organized as follows: the second section describes the algorithms employed for the vehicle detection scenario. The third section discusses the experimental conditions and results, in turn the fourth section concludes the paper.

2   Algorithms Employed for Vehicle Detection

The task of extracting specific objects from a static scene involves the determination of the image part containing the searched object. The main issue concerning the detection of objects from static images is an appropriate selection of characteristic features used to build an object model. The other problematic aspects are the selection of the mechanism for feature matching and the method for scanning the source image. According to the literature, the process of detecting objects in static images assumes the absence of certain information about the object under detection, namely the probable object size and location are unknown. Additional difficulties result from changes in the scene lighting and appearance of objects. Many methods utilize the sliding window approach where the detection of an object refers to an appropriate scanning of the image and matching the selected image parts with the templates from the training set. If there is no information about an object, the detection from static images requires to perform search process in all possible locations and using all probable window (or image) scales, which increases the computational complexity. This strategy is known as the pyramid. In each location of the search window features of a particular image part are calculated. The obtained features are used during the classification process in order to indicate the most probable object location.

In the paper the attention is focused on static scene feature extractors that enable proper object representation for the purpose of vehicle detection. The algorithms were selected based on the computational complexity, simplicity of implementation and the effectiveness declared in the literature. According to that, we have selected three object detectors that use Haar-like features, Histogram of Oriented Gradients and Local Binary Patterns as feature extractors, and a cascading approach for object classification based on AdaBoost. These solutions, in original or modified forms, are popular and constantly applied, which is confirmed by the recently published papers, e.g. [8,9].

Histogram of Oriented Gradients was proposed by Dalal and Triggs in [10] originally as an approach for human detection. It is a typical descriptor which extracts image characteristic features and stores them in a compact form as a feature vector. Since it bases on gradient information it requires greater (compared to other approaches) detector window which provides a significant amount of context that helps detection [10]. The HOG representation takes into account the directional brightness changes in local areas and is derived including five main steps. Firstly, gamma and colour normalization are performed, and secondly— oriented gradients are calculated using directional filters (vertical and horizontal edges are detected). Then, the length and orientation of gradients are calculated. In the third step, an image is divided into cells and frequencies of oriented gradients are calculated for each cell. In turn in the fourth step, cells are grouped into larger overlapping spatial blocks and normalized separately. The overlapping grouping is essential here. One cell is normalized several times, each time with respect to a different block. Such procedure constitutes for local variations in illumination and contrast. Furthermore, it significantly improves the performance [10]. The final HOG representation is obtained in the fifth step by concatenating oriented histograms of all blocks. The representation is invariant to translation within an image plane—if an object changes its position the directional gradients and histogram remain the same.

Local Binary Patterns is a feature used for texture classification, first described in [11], and is a particular case of the Texture Spectrum model proposed in [12]. LBP labels pixels by thresholding the neighbourhood of each pixel, what results in a binary number. Here we consider only a case when a particular pixel has eight neighbouring pixels. In order to derive the LBP representation, several steps must be followed. Firstly, the analysed window is divided into equal cells, and each pixel in a cell is compared with its eight neighbours in a clockwise or counter-clockwise manner (the direction and starting point of the comparison process have to be consistent in all cells). Secondly, the pixel's value is checked and a neighbourhood values are obtained in the following way—if the central pixel's value is equal or greater than the neighbouring pixels' values, then number '1' is written. Otherwise, '0' is put. This gives a binary number, usually converted to decimal. As a third step, the histogram is computed over the cell, based on the occurrence frequency of each number. Ultimately, histograms of all cells are concatenated and a feature vector for the window is produced. The original LBP representation is not invariant to object rotation within an image plane. However, it is robust to monotonic grey-scale changes resulted e.g. due to variations in illumination.

Haar-like features were proposed by Viola and Jones in 2004, and applied together with a variant of AdaBoost in [13] as a solution for face detection. They were also successfully applied to human silhouettes detection [14]. Individual Haar-like features are weak classifiers, therefore they are combined in a cascade using AdaBoost algorithm. During training, the most important image features are selected and only these are used for detection. Features are selected in a way enabling to reject negative regions (without an object of interest) at early stage of recognition, what reduces the number of calculations. During classification subsequent features of an unknown object are calculated only if the answer of the previous feature is equal to the learned value. Otherwise, the examined region is rejected. A simple rectangular Haar-like feature is defined as the difference of the sum of pixels of areas inside the rectangle, which can be characterized by any position and scale in an input image. For instance, using two-rectangle feature, the borders can be indicated based on the pixel intensities under the white and black rectangles of the Haar-like feature. Such features could be calculated fast using the integral image approach.

The derived feature vectors (HOG, LBP or Haar-like features) are further used in the process of classification using e.g. Support Vector Machine or, in our case, AdaBoost. The original AdaBoost (Adaptive Boosting) was proposed in [15]. It is a machine learning meta-algorithm—a method of training boosted classifier—that during the training process selects only those features that would be able to improve the prediction. For each weak classifier from a cascade AdaBoost determines an acceptance threshold, which minimizes the number of false classifications. However, a single weak classifier is not able to classify objects with low error rate. The boosting algorithm used for learning the classifier uses positive and negative samples as an input. The initial weights are equal to $\frac{1}{2}$. In each iteration the weights are normalized and the best weak classifier is selected based on the weighted error value. In the next step weight values are updated and it is determined if an example was classified correctly or not. After all iterations a set of weak classifiers characterized by a specified error rate is selected and a resulting strong learned classifier is obtained. The classification is performed iteratively and its effectiveness depends on the number of learning examples. During classification, an image is analysed using a sliding window approach. Features are calculated in all possible window locations. The window is slid with a varying step, which depends on the required accuracy and speed.

3   Experimental Conditions and Results

Several experiments have been carried out in order to verify the effectiveness of the selected detectors in the task of vehicle detection in static scenes. The detectors are based on Haar-like features, Histogram of Oriented Gradients and Local Binary Patterns. During classification a standard boosted cascade algorithm was applied, namely AdaBoost, and varying values of the following parameters were used: the size of search windows, the scaling step in the pyramid and candidates rejection based on the number of adjacent detections.

Five different cascades of classifiers were built using various learning databases presented in the literature. Cascades no. 1–3 were built using UIUC database [16], containing 1050 learning images (550 car images and 500 images with other objects). Cascade no. 1 used a set of LBP features calculated for 20×20 pixel size windows and frontal car views. Cascade no. 2 used HOG features calculated for 24×48 pixel size windows and images with cars seen form the side. Cascade no. 3 was built of LBP features calculated for 25×50 pixel size windows and images of cars seen from the side. Cascade no. 4 used a set of Haar-like features calculated for 40×40 pixel size windows and images of cars in frontal views taken from the OpenCV database. Cascade no. 5 was prepared for the other set of Haar-like features calculated for 20×20 pixel size windows and frontal car views. The experiments were performed using Matlab 64-bit framework and OpenCV 2.4.10 implementation of Viola-Jones detector. There is no data about execution time of the learning phase—cascades were trained previously. However for the detection phase the execution time equalled less than a second per frame (Core i7, 2nd generation, 64-bit Windows).

The results of vehicle detection for sample video frames taken from the project test database are illustrated on Figs. 1, 2 and 3. These are the results obtained for the cascades no. 1, 2 and 5 for which the least number of false detections have been obtained. White rectangles correspond to the detected regions. Results are given for the same camera view in order to enable the comparison of experimental results based on the number and location of detected regions.

As can be seen from the presented figures, the vehicle detection in static scenes gives acceptable results. The presence of a small number of false detections as well as some missed objects cause that the solution should be modified. One of the possible modifications is to combine the results obtained for various detectors. The other way is to use frames resulted from background modelling

Fig. 1. Exemplary results of the experiment using LBP features (cascade no. 1)

Fig. 2. Exemplary results of the experiment using HOG features (cascade no. 2)

stage to limit the region under detection or to prepare a set of learning samples consisting other vehicle views. Described architecture involves a cascade of classifiers employing different features and a coarse-to-fine strategy. Proposed classifier, in order to be applicable, should focus on one vehicle view only and first detect as many potential vehicles as possible, even if there are false detections. This task could be performed using Haar-like features, since it is the most

Fig. 3. Exemplary results of the experiment using Haar-like features (cascade no. 5)

tolerant approach. Then the LBP features to resulting objects should be applied. The final stage should involve HOG features in order to fine-tune the results.

4   Summary and Conclusions

In the paper, the problem of vehicle detection in static images was analysed. The aim of this analysis was to verify the possibility of implementing various feature extractors and AdaBoost classification in the 'SM4Public' system. The experiments were performed using images extracted from the video sequences recorded for the project test database. Three feature extractors were experimentally tested, namely Histogram of Oriented Gradients, Local Binary Patterns and Haar-like features. It can be concluded from the experimental results that the vehicle detection is possible, however the number of false detections causes the necessity to modify the proposed solution, e.g. based on the combination of individual detectors results. It is also clear, that the position and orientation of the camera strongly influence the detection accuracy. The analysed cascades have been learned on samples presenting vehicles in frontal or side position. In order to make proposed system even more flexible, it is advised to prepare an image sample set depicting vehicles from other orientations, representing expected surveillance camera parameters.

Acknowledgments. The project *"Security system for public spaces — 'SM4Public' prototype construction and implementation"* (original title: *Budowa i wdrożenie prototypu systemu bezpieczeństwa przestrzeni publicznej 'SM4Public'*) is a project co-founded by European Union (EU) (project number PL: POIG.01.04.00-32-244/13, value: 12.936.684, 77 PLN, EU contribution: 6.528.823,81 PLN, realization period: 01.06.2014–31.10.2015).

European Funds—for the development of innovative economy (Fundusze Europejskie—dla rozwoju innowacyjnej gospodarki).

References

1. Frejlichowski, D., Forczmański, P., Nowosielski, A., Gościewska, K., Hofman, R.: SmartMonitor: an approach to simple, intelligent and affordable visual surveillance system. In: Bolc, L., Tadeusiewicz, R., Chmielewski, L.J., Wojciechowski, K. (eds.) ICCVG 2012. LNCS, vol. 7594, pp. 726–734. Springer, Heidelberg (2012)
2. Frejlichowski, D., Gościewska, K., Forczmański, P., Nowosielski, A., Hofman, R.: Extraction of the foreground regions by means of the adaptive background modelling based on various colour components for a visual surveillance system. In: Burduk, R., Jackowski, K., Kurzynski, M., Wozniak, M., Zolnierek, A. (eds.) CORES 2013. AISC, vol. 226, pp. 355–364. Springer, Heidelberg (2013)
3. Frejlichowski, D., Gościewska, K., Forczmański, P., Hofman, R.: 'SmartMonitor'–an intelligent security system for the protection of individuals and small properties with the possibility of home automation. Sensors **14**, 9922–9948 (2014)
4. Frejlichowski, D., Gościewska, K., Forczmański, P., Hofman, R.: Application of foreground object patterns analysis for event detection in an innovative video surveillance system. Pattern Anal. Appl. **18**, 473–484 (2014)
5. Rajagopalan, A.N., Burlina, P., Chellappa, R.: Higher order statistical learning for vehicle detection in images. In: Proceedings of the 7th IEEE International Conference on Computer Vision, vol. 2, pp. 1204–1209 (1999)
6. Aarthi, R., Padmavathi, S., Amudha, J.: Vehicle Detection in static images using color and corner map. In: 2010 International Conference on Recent Trends in Information, Telecommunication and Computing, pp. 244–246 (2010)
7. Wen, X., Zhao, H., Wang, N., Yuan, H.: A Rear-vehicle detection system for static images based on monocular vision. In: 9th International Conference on Control, Automation, Robotics and Vision, pp. 1–4 (2006)
8. Tang, Y., Zhang, C., Gu, R., Li, P., Yang, B.: Vehicle detection and recognition for intelligent traffic surveillance system. Multimed.Tools Appl., 1-16. Springer US (2015)
9. Sun, D., Watada, J.: Detecting pedestrians and vehicles in traffic scene based on boosted HOG features and SVM. In: 2015 IEEE 9th International Symposium on Intelligent Signal Processing (WISP), pp. 1–4, 15–17 May 2015
10. Dalal, N., Triggs, B.: Histograms of oriented gradients for human detection. IEEE Comput. Soc. Conf. Comput. Vis. Pattern Recogn. **1**, 886–893 (2005)
11. Ojala, T., Pietikinen, M., Harwood, D.: Performance evaluation of texture measures with classification based on Kullback discrimination of distributions. In: Proceedings of the 12th International Conference on Pattern Recognition, vol. 1, pp. 582–585 (1994)
12. He, D.C., Wang, L.: Texture unit, texture spectrum, and texture analysis. IEEE Trans. Geosci. Remote **28**, 509–512 (1990)
13. Viola, P., Jones, M.J.: Robust real-time face detection. Int. J. Comput. Vision **57**(2), 137–154 (2004)
14. Forczmański, P., Seweryn, M.: Surveillance video stream analysis using adaptive background model and object recognition. In: Bolc, L., Tadeusiewicz, R., Chmielewski, L.J., Wojciechowski, K. (eds.) ICCVG 2010, Part I. LNCS, vol. 6374, pp. 114–121. Springer, Heidelberg (2010)

15. Freund, Y., Schapire, R.E.: A decision-theoretic generalization of on-line learning and an application to boosting. In: Proceedings of the 2nd European Conference on Computational Learning Theory, pp. 23–37 (1995)
16. Agarwal, S., Awan, A., Roth, D.: Learning to detect objects in images via a sparse, part-based representation. IEEE Trans. Pattern Anal. **26**(11), 1475–1490 (2004)

Assessment of Production System Stability with the Use of the FMEA Analysis and Simulation Models

Anna Burduk[1] and Mieczysław Jagodziński[2(✉)]

[1] Mechanical Department, Wrocław University of Technology, Wrocław, Poland
anna.burduk@pwr.edu.pl
[2] Institute of Automatic Control, Silesian University of Technology, Gliwice, Poland
mieczyslaw.jagodzinski@polsl.pl

Abstract. In order to ensure smooth functioning of a production system, the stability of its processes must be guaranteed, while on the other hand it must be possible to make quick decisions encumbered with the lowest possible risk. The risk results from the uncertainty associated with making decisions as to the future, as well as from the fact that the implementation of innovations is one of the factors that disturb the current manner of operation of the enterprise. The stability of a production system is defined as maintaining the steady state by the system for a certain assumed period of time. The paper describes a method for analysing and assessing the stability in production systems. In order to determine the extent of the impact of individual risk factors on the selected area of the production system, the FMEA analysis was used. When determining the values of the parameters needed for calculating the Risk Priority Number (RPN), defuzzified values of appropriate linguistic variables were used. A process of ore transportation process with the use of a belt conveyor was used as an example.

Keywords: Stability · Production system · Modelling and simulation · FMEA analysis

1 Introduction

The concept of stability is derived from the systems theory. Most definitions found in the literature refer to the concept of the state of balance and define the stability of a system as its ability to return to the state of balance after the disturbances that caused the instability have ceased. The stability of a control system is its most important feature that characterizes the ability to accomplish the tasks, for which it has been built [1].

If the value of the parameter P(ti), which characterizes the production system at the time t_i, is within the predetermined interval $P_1 \leq P(t_i) \leq P_2$, this will indicate a correct course of the process. Otherwise, corrective measures should be taken. Corrective measures usually consist in changing the values of control variables (inputs to the system) in such a way, so that the values of the parameters characterizing the controlled variables (outputs from the system) return to the process course standards established at the planning stage [1]. Production plans and parameters characterizing them usually constitute the standards. A correct decision will cause that the system will return to the steady state.

© Springer International Publishing Switzerland 2015
K. Jackowski et al. (Eds.): IDEAL 2015, LNCS 9375, pp. 216–223, 2015.
DOI: 10.1007/978-3-319-24834-9_26

So it can be said that a production system is in the steady state, if values of the parameters defining it are within the ranges specified in the planning function and recorded in a standard, i.e. a production plan, as schematically shown in Fig. 1.

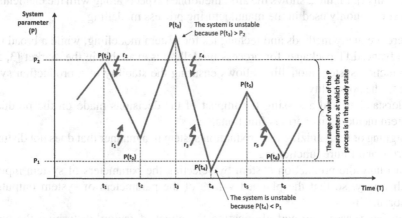

Fig. 1. The variability of the system parameter $P(t_i)$ is caused by the impact of disturbing factors (r_i)

In order to ensure the stability of production systems, on the one hand an appropriate control is needed, while on the other hand it is necessary to analyse, evaluate and eliminate the random factors causing the disturbances (risk factors). Control in the context of production systems means making decisions based on the information or data coming from the controlled system. The impact of single- or multi-criteria decisions on the production system can be verified very well on a model of the production system, which contains important system elements and their parameters as well as the relationships between them.

2 The Role of Production Systems Modelling in Ensuring the Stability

Operations performed on a model instead of the actual production system do not disturb the stability of production processes. Treating a model as a duplicate of the actual system enables, inter alia, the transfer of the conclusions from the studies performed on the computer model to the actual production system. Modelling and computer simulation allow verifying solutions being introduced before their actual implementation, which is not possible in the case of conventional methods of conducting design work [2]. An additional advantage is a reduction in the costs of the changes made on the basis of the simulations carried out at the beginning of the project implementation. The changes, which have been foreseen and planned at the beginning of the project, cost significantly less than in the later stages and they disturb the functioning and stability of the system to a lesser extent [3, 4].

In order to be able to make right decisions based on a model, it must include the company's aspects adequate to the scope of the studies. When modelling production systems, regardless of the purpose of modelling and the optimization criteria adopted, generally six aspect of a company are taken into account: management, structures, resources, production processes, basic manufacturing measures and tasks of the production system [2]. Figure 2 shows the aforementioned aspects along with the elements that are most commonly used in the manufacturing process modelling.

- There are many methods and techniques for system modelling, while a broad range of advanced IT packages for process modelling is available in the market [3, 5–7]. Production system modelling allows ensuring the stability of a production system due to the possibility of:
- understanding and assessing the impact of the decisions made on the production system including its various functional areas,
- designing or reorganizing the production system in a manner that does not disturb its current and future functioning,
- controlling the production system by selecting the parameters of system inputs in such a way, so that the planned values of the parameters of system outputs are ensured,
- identifying, assessing and eliminating the effect of factors disturbing the correct functioning of the production system.

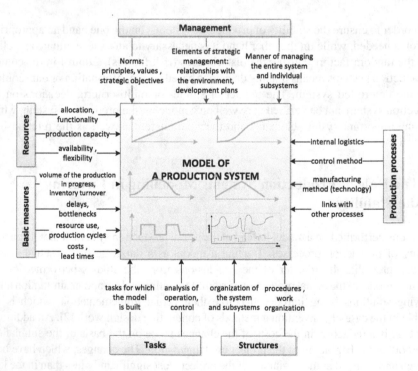

Fig. 2. Production system modelling usually takes account of parameters, an aggregated model of a production company, as well as selected components.

3   Identification and Assessment of Risk in Production Systems

In order to reduce the level of risk in a production system, a series of actions must be taken. The first of them is the risk identification, which determines the threats that might occur during realization of company's goals. Due to a potential possibility that many risk factors may occur, it is important to find the source risk, which is the key cause of the problems. During the identification, it is important to search for the answers to the following questions: in which area of the production system the risk occurs and which area is affected by the highest risk.

The next step in reducing the risk level is measuring the risk and determining the extent of the impact on the production system. Failure Mode and Effect Analysis (FMEA) is one of the methods which allow determining the extent of risk in the designated area of a production process or in a product, as well as the resulting effects. Thanks to this, corrective actions aiming at mitigation of the risk can be found subsequently [8]. *"One of the key factors in proper implementation of the FMEA program is to act before an event occurs and not to gain experience after the event. In order to obtain the best results, FMEA should be performed before a particular type of construction or process defect is "designed" for a given product."* [3].

3.1   Determination of the Risk Priority Number (RPN) in the FMEA Method

When assessing the risk in a production process with the use of the FMEA method, the first step is to detail the operations in the process, then to identify the risk factors present in the process, determine the effects caused by their presence, and to find possible causes. The next step in the analysis is to assign numerical values to the following parameters shown in Table 1.

Table 1. Characteristics of the parameters used in the FMEA method for determining RPN

Parameter symbol	Parameter name	Description
Z	degree of threat	It determines the extent of the effects which arise as a result of the occurrence of a defect during a production process and during the use of a product
P	probability	The probability of the occurrence of a defect
T	detection rate	It determines the probability that a potential defect or its cause will be detected later

Risk Priority Number (RPN), i.e. the extent of the risk, is calculated for each of the selected areas of the production system using the formula [3]:

$$RPN = (Z) \times (P) \times (T) \tag{1}$$

This obtained value allows assessing the estimated risk and is used as a point of reference in relation to the corrective actions taken. The value of RPN may be in the range between 1 and 1000. So a high value of RPN corresponds to a high risk in the process. If the RPN value is high, efforts should be taken to mitigate the risk using corrective actions [3]. The corrective actions shall be taken first in the areas with the highest RPN level.

Figure 3 shows 4 areas representing a area of high losses and risk. These areas are presented together with the parameters described above.

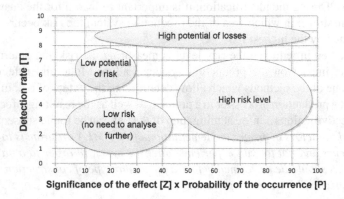

Fig. 3. The results of the RPN analysis depending on values of the parameter

Determination of a general limit for a high RPN value is not easy. Each FMEA analysis is unique and the risk estimation in this method cannot be compared with other analyses. This is caused by some sort of subjectivity, the dependence during the assessment, and the decisions made by the person performing the analysis. Therefore for each FMEA analysis a system of criteria should be developed and it should be determined from which values of RPN the corrective actions should be taken [3]. In the era of dynamic changes in the market environment, the FMEA method proved to be a good alternative solution that enables quick identification of potential risks for a company.

4   Determination of Risk in the Process of Haulage of Excavated Material by Belt Conveyors, Using Linguistic Variables

Belt conveyors are mechanical means of transport with a limited range and continuous movement. Typically they are used for conveying bulk materials. Material is transported on a specific route limited by the distance between the loading and unloading stations. Depending on the construction, material can be transported along a straight line or a curve, at any angle. Belt conveyors are characterized by simple construction, high reliability and safety.

The problem of failures of belt conveyors was subjected to an analysis. This is a very important issue in respect of transportation of excavated material in a mine, because failures lead to unplanned downtimes and thus to stopping the haulage of excavated material for several shifts. On the other hand, the information about a failure may come from production workers only, which results from the conditions occurring in a mine, the length of the transport system and provisions of the mining law. Information about a failure was verbal and depended on individual impressions of workers.

The FMEA analysis was prepared on the basis of the stages of the process of transportation by a belt conveyor and the risk factors. (Table 2) shows the FMEA analysis performed only for first stage of the belt conveyor operation – start-up.

Table 2. FMEA analysis of two stages of the belt conveyor operation

Opera-tion /Process stage	Possible risk factors	Effects caused by the risk factors	Current state			RPN
			Risk factor assessment (P) [rank]	Effect assessment Z [rank]	Hazard assessment T [rank]	
Start-up of the belt conveyor	Transmission system failure	Gearbox failure	5	8	2	80
		Coupling failure	5	8	3	120
		Motor failure	6	7	3	126
		Pulley failure	6	6	5	180
	Belt damage	Belt breakage	5	9	2	90
		Belt slip-off	5	9	7	315

5   Assessing the Stability of Haulage of Excavated Material by Belt Conveyors, Using Linguistic Variables and Simulation Models

An analysis of the stability of the production system for different variants concerning the occurrence of risk factors during one shift is presented below. Due to the random nature of risk factors, the analysis will concern the occurrence of one and two risk factors. Such a combination will also give an answer to the question at which variants of the occurrence of risk factors the system will remain stable and at which not. In addition, it will be possible to assess which risk factors disturb the stability of the production system to the greatest extent. It has been assumed that the system will be stable, if the transportation volume of excavated material W = 2500 ± 10 % tons/week. The analysis of variants of the studies on the extent of the impact of individual risk factors was performed using the iGragix Process for Six Sigma computer modelling and simulation system.

Thanks to the use of IT systems, models were populated with appropriate data from an actual system. Modern production systems are measured and monitored to a higher and higher degree. The experiments conducted on models do not disturb the functioning of the actual system and thus allow predicting the effects as well as selecting an optimal variant of decisions on the parameters and types of inputs to the system.

In order to assess the stability of the analysed production system, a computer simulation model of the system was built. This model did not include any identified risk factors and was named the base model. Values of different types of the risk factors (Table 2) identified in FMEA analysis were introduced to the base model. In this way,

2 simulation models were created to assess the influence of risk factors on the analysed production system (Table 3).

Table 3. Two models used for analysing the stability of the production system depending on the occurrence of a single risk factor

Model name	Model description
Model 1.	impact of the risk of transmission system failure
Model 2.	impact of the risk of belt damage

The results obtained from the experiments conducted with the use of the models presented in Table 3 are shown in Fig. 4.

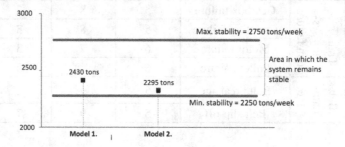

Fig. 4. The impact of a single risk factor on the FMEA analysed production system

As shown in Fig. 4, if risk factors occur one at time, the system will be stability.

The next step involved examining the stability of the production system in the event of the occurrence of two risk factors. As in the case of the analysis of a single risk factor, in the base model were introduced two types of risk factors identified in FMEA analysis. The results obtained from the experiments conducted with the use of the model presented are shown in Fig. 5.

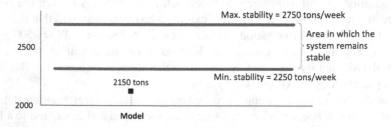

Fig. 5. The impact of a combination of two risk factors on the FMEA analysed production system

As is appears from Fig. 5, if any combination of two risk factors occurs, the analysed transportation system will become out of balance. As can be seen, if the production organization level is not improved, the risk factors present in the system will prevent the accomplishment of the assumed goal.

6   Conclusion

Smooth operation of a production system is a phenomenon that occurs less and less often. It happens more and more frequently that the attention is drawn to the need of detecting the threats early and collecting the information concerning the cause-effect relationships occurring in the system. The FMEA analysis helped to determine the cause-effect relationships associated with the occurrence of risk factors and then minimize their impact on the production system. Proposed in literature quantitative methods of risk analysis and evaluation treat single issues, assuming certain factors and conditions as well as impose constraints. Taking into consideration the complexity of modern production systems as well as a number of influencing them external, random factors, this kind of approach seems to be unsuitable.

Acknowledgements. This work has been partly supported by the Institute of Automatic Control under Grant BK/265/RAU1/2014.

References

1. Bubnicki, Z.: Modern Control Theory. Springer, Berlin (2005)
2. Azadegan, A., Probic, L., Ghazinoory, S., Samouei, P.: Fuzzy logic in manufacturing: a review of literature and a specialized application. Int. J. Prod. Econ. **132**(2), 258–270 (2011)
3. Chrysler Cooperation, Ford Motor Company, General Motors Cooperation, Potential Failure Mode and Effects Analysis (FMEA), First Edition Issued (February 1993)
4. Roux, O., Jamali, M., Kadi, D., Chatelet, E.: Development of simulation and optimization platform to analyse maintenance policies performance for manufacturing systems. Int. J. Comput. Integr. Manuf. **2008**(21), 407–414 (2008)
5. Krenczyk, D., Skolud, B.: Transient states of cyclic production planning and control. Appl. Mech. Mater. **657**, 961–965 (2014)
6. Krenczyk, D., Skolud, B.: Production preparation and order verification systems integration using method based on data transformation and data mapping. In: Corchado, E., Kurzyński, M., Woźniak, M. (eds.) HAIS 2011, Part II. LNCS, vol. 6679, pp. 397–404. Springer, Heidelberg (2011)
7. Rojek, I.: Neural networks as performance improvement models in intelligent CAPP systems. Control Cybern. **39**(1), 55–68 (2010)
8. Sankar, N., Prabhu, B.: Modified approach for prioritization of failures in a system failure mode and effects analysis. Int. J. Qual. Reliab. Manag. **18**(3), 324–336 (2001)

Study of Collective Robotic Tasks Based on the Behavioral Model of the Agent

Fredy Martínez[✉], Edwar Jacinto, and Fernando Martínez

District University Francisco José de Caldas, Bogotá, Colombia
{fhmartinezs,ejacintog,fmartinezs}@udistrital.edu.co
http://www.udistrital.edu.co

Abstract. In recent years, much research has been devoted to the analysis, modeling and design of multi-agent robotic systems. Such systems are composed of a set of simple agents that self-organize to perform a task. Given the parallel structure, also happens to be a very robust solution. This paper focuses on the development of a behavioral model of each agent of the system, from which, defining simple behavior and interaction rules, it is possible to set the emergent behavior of the system. The interaction model takes elements observed in bacteria and establishes a structure at the agent level and the system level. The proposed model is validated through the design of a basic navigation task where the robots form autonomously groups without any external interaction or prior information of the environment or other robots.

Keywords: Bio-inspired model · Parallel system · Path planing · Robotics

1 Introduction

The theory of self-organization is based on the assumption that the system functionality is not a result of the individual performance of its elements; but rather a result of the interaction among the elements. Studying the characteristics of self-organizing dynamic systems [9], it is possible to identify three key aspects of a self-organized system [8]: (1) The existence in the system of a lot of units (agents) [10] that interact among themselves, so that the system evolves from a less organized state to a more organized state dynamically, along time, and maintains some kind of exchange; (2) The organization is evident in the global system behavior as a result of the interaction of agents [1], its functioning is not imposed on the system by any kind of external influence; and (3) the agents, all of them with identical characteristics, have only local information [8,11], which implies that the process of self-organization involves some transfer of local information.

Our research aims at providing a different approach into the application of self organization principles to the digital processing: to devise a reconfigurable logic hardware and an architecture that permits self-organizing processes.

© Springer International Publishing Switzerland 2015
K. Jackowski et al. (Eds.): IDEAL 2015, LNCS 9375, pp. 224–231, 2015.
DOI: 10.1007/978-3-319-24834-9_27

The practical application of our ideas focuses on the self-coordination of a group of robots to perform a task collectively. However, the first step in the design of the application is to develop a mathematical model of interaction between robots. This paper focuses specifically on this problem.

We have assumed as interaction model the bacterial behavior [7]. The multi-agent model adopted is composed of a set of artificial bacteria or agents. This set of agents, and their interactions, reflect the dynamics that will solve the tasks. All agents are identical in design. Nevertheless, to solve tasks, each of them undergoes certain behavior inside the system along the time.

The variables of this system are defined in a continuous space. However, the different agent behaviors are triggered by certain events (event-based), which implies that the appropriate analysis model is a hybrid system (the dynamic changes from one state to another when a condition is present). In this way, our ideas are closely linked to research in behavior-based control systems for autonomous robots [5,13]. In this control scheme, the designer works the group of robots as a system, where each node corresponds to a robot, and to solve the task, each node has a different behavior. This structure is called a hybrid automaton, and allows to use in the system design the capacity for abstraction of hybrid systems [2,14].

In the paper, we want to characterize the behavior using a micro-scale model for each system agent. This model allows to describe the dynamics of the agents, and may even include their physical characteristics. This structure allows the use equations to describe the system state variables, and from them, following a scheme of top-down design, to define the behavior of individual agents.

The paper is organized as follows. In Sect. 2 problem formulation is discussed, Sect. 3 describes the control strategy. Section 4 analyzes the performance of the proposed control scheme. Section 5 shows some results with prototypes. Finally, conclusion and discussion are presented in Sect. 6.

2   Problem Statement

Let $W \subset \mathbb{R}^2$ be the closure of a contractible open set in the plane that has a connected open interior with obstacles that represent inaccessible regions [4]. Let $\mathcal{O}$ be a set of obstacles, in which each $O \subset \mathcal{O}$ is closed with a connected piecewise-analytic boundary that is finite in length. Furthermore, the obstacles in $\mathcal{O}$ are pairwise-disjoint and countably finite in number. Let $E \subset W$ be the free space in the environment, which is the open subset of W with the obstacles removed.

Let us assume a set of n agents in this free space. The agents know the environment E in which they move from observations, using sensors. These observations allow them to build an information space I. A information mapping is of the form:

$$q : E \longrightarrow S \tag{1}$$

where S denote an *observation space*, constructed from sensor readings over time, i.e., through an observation history of the form:

$$\tilde{o} : [0, t] \longrightarrow S \tag{2}$$

The interpretation of this information space, i.e., $I \times S \longrightarrow I$, is that which allows the agent to make decisions [6].

We assume the agents are able to sense the proximity of their team-mates and/or obstacles within the environment, using minimal information. The environment E is unknown to the robot. Furthermore, the robot does not even know its own position and orientation. Our goal is to design the control rules for the n robots in order to independently solve navigation tasks in a dynamic and unknown environment.

The particular equations of motion $\dot{x} = f(x)$ for each robot is unimportant in this approach. We do not explicitly control their motions and do not even attempt to measure their precise state. Instead, we rely on the fact that the robot moves in a uncontrollable way, but the trajectory satisfies the following high-level property: For any region $r \in R$, it is assumed that the robot moves on a trajectory that causes it to strike every open interval in ∂r (the boundary of r) infinitely often, with non-zero, non-tangential velocities.

However, the robots should not move randomly in the environment, they should jointly develop a task. To achieve this, we propose a design model based on the collective behavior of robots. This collective behavior involves a local interaction at agent level, reflecting a system-level behavior. These behaviors in conjunction with some control modes encoded in all agents, enable the coordinated navigation of the robots.

We develop an event-based system. Each robot starts with an initial control mode. During execution, the control mode may change only when receiving an sensor observation event y. Let Y denote the set of all possible observation events for a robot in a particular system, then $y \in Y$. The control modes of robot i are set during execution according to a policy. Since state feedback is not possible, *information feedback* is instead used. A *control policy* is specified as an information-feedback mapping which enables the control mode to be set according to the sensor observation history.

3   Behavioral Model

The motivation, rather than compute the exact trajectories and sensory information of individual robots, we want our agents to be able to build an information space from the local information that they receive from the environment and its near neighbors, and from this information space, define its behavior required for the development of the task.

We start from the assumption that an agent is capable of switching behavior, and develop together with other agents a task if it is able to make and remember local observations of the environment. These two properties are essential for the construction of the information space, which is used in the control hybrid structure for the feedback of each agent.

We consider our system as a set of n agents (differential wheeled robots in our experiments), each with a continuous state space $X_a \subset \mathbb{R}^2$, where the actions or behaviors of the agent belongs to a set L_a of l_a behaviors (Fig. 1). An agent may switch its state according to its control policy when it determines it is appropriate to do so. The activation of each of these behaviors should be done in a way that improves the overall performance of the system (the performance function should be encoded in its genome), in our case this means performing the task in the shortest time possible. Accordingly, each agent is a hybrid automaton whose continuous and discrete dynamics can be represented by:

$$H_a = \{X_a, L_a\} \tag{3}$$

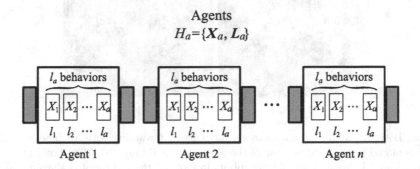

Fig. 1. The agents in the system

The agents have no global information from their environment or other agents. Instead, the agents can sample the environment. They are able to move in the environment and make local observations of its neighbors and the environment. We assume that the agents are able to observe behaviors and discriminate their types.

4   A Basic Task of Testing: Grouping

A first exercise that could be interesting is to coordinate the grouping of a group of robots. To do this, we define two basic behaviors: an initial stage of reproduction and a final stage of activation or cell differentiation.

In the stage of reproduction, robots are activated each other. When a robot is activated (initial robot or robot activated by another robot), it begins its behavior at this stage. The robot looks randomly inactive robots around, and when one is found, it is headed in that way to activate it. Robots differ from the environment due to their physical characteristics.

When the reproduction ends, the second behavior is activated, the cell differentiation. At this stage we program the grouping task. The grouping is achieved optimizing the behavior to favor the robot movement toward to areas of the environment with a high degree of movement. According to this, the following performance function is defined:

$$s = \frac{1}{\|h_{actual} - 15.1\|} \tag{4}$$

This function gives greater performance to weight vectors that direct the robot to areas with other robots in motion. In our prototype, the robot performs a random search in the environment, and when it detects any movement, it goes to that point (Fig. 2).

(a)   (b)

Fig. 2. Basic behavior for autonomous navigation: Grouping. (a) An inactive robot (red) is placed in the field of view of the active robot (blue). (b) After 1 min of exploration without being able to identify robots in motion, the red robot is shaken in front of the blue robot to force its identification and movement (Color figure online).

We have taken the task of grouping as a basic example for observing the performance of the proposed control structure. Based on it, it is possible to think about more complex tasks such as carrying objects, adjusting the algorithm for robots to surround the object, and then moving together to a certain point in the environment.

5   Results

For our experiments, we use the Oomlout Company's open-source design SERB Robot. We modify the design to add a camera as visual sensor. Our robot and all its peripherals, including the camera is handled by a single 8-bit microcontroller (Atmel ATmega328, 8-bit AVR RISC-based microcontroller).

We can control the translational and rotational speeds of the robot. However, we decide to design a software actuator on the displacement of the robot, a high-level control, which generates control states in order to facilitate the construction of state diagrams.

Due to the impossibility of building tens of these robots, we only use the prototype robots to observe the basic behaviors of the algorithm. We have done the analysis of the dynamics of the system with a large number of robots through simulation.

The simulations are performed under Linux (kernel 3.13, 64 bits) on Player/ Stage [3,12], this platform allows for the simulation to run exactly the same code developed in C for the real robot. The simulations consider not only the functionality of the algorithm, but also the physical characteristics of the robot, i.e. shape, size (0.26 m × 0.22 m) and weight (1.2 kg). The camera is installed just above its geometric center both in the real prototype as in the simulation.

The simulation environment is designed in square shape with a total area of $6\,\text{m} \times 6\,\text{m} = 36\,\text{m}^2$. Inside the environment, there are three holes inaccessible

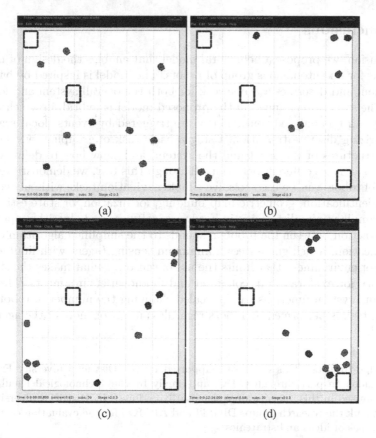

(a)   (b)

(c)   (d)

Fig. 3. Simulation of the autonomous navigation: Grouping. (a) Initial state of the simulation (00:00:26). (b) The two robots in the center (blue and golden) begin a dance that lasts for 38 s (00:06:42). (c) At 9 min. (00:09:00) we observe the formation of small groups of robots. (d) Final state of the simulation, the robots have established a stable group in the lower right side of the environment (00:12:16) (Color figure online).

to the robots as an obstacle for navigation. In the simulations there are 10 independent agents.

We start the simulation with the robots scattered in the environment. The simulation starts at t = 26 s, and the case documented here ends at 12 min and 16 s (00:12:16, time in the simulation, Fig. 3). We see first signs of grouping of 2 or 3 agents throughout the simulation, and a behavior that resembles a dance between two robots at 6 min and 42 s (00:06:42) that lasts a total of 38 s. At the end of this simulation we can observe a fairly stable group of robots in the lower right side of the environment.

The simulation demonstrates the ability of the system to solve a complex task. It also shows that the performance is tied to the interpretation of the local information sensed by the agent (genome), rather than the platform. That is, the behavior can be implemented on other hardware, or emulated by software.

6   Conclusions

In conclusion, we propose a behavioral model that enables the design of navigation tasks for an autonomous group of robots. The model is inspired by bacterial interaction, and describes a framework for both the overall system and for each robot. The structure assumed in the proposed model is hybrid, in which behaviors that can be analyzed continuously are triggered by events (local readings) characterizing discrete transition. Using a basic task of grouping, we show the hybrid structures of the agent and the system, and show how to define control policies to activate different behaviors. Through this test, we demonstrated that a group of robots can be used to solve simple navigation tasks without requiring system identification, geometric map building, localization, or state estimation, using a simple minimalist design in hardware, software and algorithm operation. The interaction between the robots responds to the simplified algorithm of local communication, which guarantees a minimum sensing (agent with limited sensing and/or environment that limits the use of sensors), what makes the strategy promissory for exploration in collapsed and unknown environments. The capabilities and system functions can be scaled changing the number of robots. The grouping task is performed by robots regardless if one or more of the agents are damaged.

Acknowledgments. This work was supported by the District University Francisco Jos de Caldas, in part through CIDC, and partly by the Technological Faculty. The views expressed in this paper are not necessarily endorsed by District University. The authors thank the research groups DIGITI and ARMOS for the evaluation carried out on prototypes of ideas and strategies.

References

1. Camazine, S., Deneubourg, J., Franks, N., Sneyd, J., Theraulaz, G., Bonabeau, E.: Self-organization in Biological Systems. Princeton University Press, Princeton (2001). ISBN 978-0691012117

2. Chaohong, C., Rafal, G., Sanfelice, R.G., Teel, A.R.: Hybrid dynamical systems: Robust stability and control. Proc. Chinese Control Conf. CCC **2007**, 29–36 (2007)
3. Gerkey, B., Vaughan, R.T., Howard, A.: The player/stage project: tools for multi-robot and distributed sensor systems. In: Proceedings IEEE 11th International Conference on Advanced Robotics ICAR 2003, pp. 317–323. IEEE, Coimbra (Portugal), June 2003
4. Gonzalez, A., Ghaffarkhah, A., Mostofi, Y.: An integrated framework for obstacle mapping with see-through capabilities using laser and wireless channel measurements. IEEE Sensors J. **14**(1), 25–38 (2014)
5. Hernandez-Martinez, E.G., Albino, J.M.F.: Hybrid architecture of multi-robot systems based on formation control and som neural networks. In: Proceedings of IEEE International Control Applications (CCA) Conference, pp. 941–946 (2011)
6. LaValle, S.M.: Planning Algorithms. Cambridge University Press, Cambridge (2006). http://planning.cs.uiuc.edu/
7. Martínez, F.H., Delgado, J.A.: Hardware emulation of bacterial quorum sensing. In: Huang, D.-S., Zhao, Z., Bevilacqua, V., Figueroa, J.C. (eds.) ICIC 2010. LNCS, vol. 6215, pp. 329–336. Springer, Heidelberg (2010)
8. Polani, D.: Measuring self-organization via observers. In: Banzhaf, W., Ziegler, J., Christaller, T., Dittrich, P., Kim, J.T. (eds.) ECAL 2003. LNCS (LNAI), vol. 2801, pp. 667–675. Springer, Heidelberg (2003)
9. Prokopenko, M.: Advances in Applied Self-organizing Systems. Advanced Information and Knowledge Processing, 1st edn. Springer, Berlin (2008). ISBN 978-1-84628-981-1
10. Russell, S., Norvig, P.: Artificial Intelligence: A Modern Approach, 2nd edn. Pearson Prentice Hall, Englewood Cliffs (2002). ISBN 0137903952
11. Santini, C., Tyrrell, A.: Investigating the properties of self-organization and synchronization in electronic systems. IEEE Trans. NanoBiosci. **8**(3), 237–251 (2009). ISSN 1536–1241
12. Vaughan, R.: Massively multi-robot simulation in stage. Swarm Intell. **2**(2), 189–208 (2008)
13. Weikersdorfer, D., Conradt, J.: Event-based particle filtering for robot self-localization. In: 2012 IEEE International Conference on Robotics and Biomimetics (ROBIO), pp. 866–870 (2012)
14. Yoon, K.H., Lee, J.K., Kim, K.H., Park, B.S., Yoon, J.S.: Hybrid robust controller design for a two mass system with disturbance compensation. In: Proceedings of International Conference on Control, Automation and Systems, ICCAS 2008, pp. 1367–1372 (2008)

Minimalist Artificial Eye for Autonomous Robots and Path Planning

Omar Espinosa, Luisa Castañeda, and Fredy Martínez[✉]

District University Francisco José de Caldas, Bogotá, Colombia
{oaespinosag,lfcastannedaf}@correo.udistrital.edu.co,
fhmartinezs@udistrital.edu.co
http://www.udistrital.edu.co

Abstract. The visual tracking is a feature of great importance for an artificial autonomous system that interfaces with the environment. It is also an open research problem of great activity in the field of computer vision. In this paper a minimalist tracking system for autonomous robots, designed primarily for navigation tasks, is proposed. The concept of minimalism translates into a processing system of very low range, but still, developing identification and tracking in real time. The proposed scheme is evaluated experimentally with a basic tracking task in which the system identifies a geometric mark on the environment, calculate its three-dimensional position and coordinates the movement of the eye in order to reduce the distance to the target and improve its focus.

Keywords: Autonomous robots · Bio-inspired system · Object detection · Object tracking

1 Introduction

The detection and visual tracking is a dynamic field of research in robotics and systems for human-machine interaction [1,7]. It is a process by which the eyes follow a target through a visual field. The visual system adjusts its resolution for a target, using still images. When the object is in motion, the visual system performs what is called dynamic resolution [3,13].

This skill is needed in an artificial system that requires the interaction between its elements and the environment. In developing control schemes is essential to have reliable sensing information, explicit and in real time according to the application. In robotics, for example, we want basic functionality equivalent to human, so the artificial systems have the possibility of action on human environments.

Several approaches have been proposed to solve the problem of visual tracking [3,8,11,12,14]. The differences between them are generally summarized as follows: (1) characteristics to identify in the image, (2) how to internally represent these features and the processing strategy, y (3) the hardware required to implement the solution. In the first case the solution is to use specific filters on the image, which allows to extract the features of interest [2]. In the second case,

© Springer International Publishing Switzerland 2015
K. Jackowski et al. (Eds.): IDEAL 2015, LNCS 9375, pp. 232–238, 2015.
DOI: 10.1007/978-3-319-24834-9_28

it is quite widespread the use of *machine learning* techniques [4,6]. In terms of hardware, it is common to use customized systems, with a camera as optical sensor and some kind of joint that allows movement in two or three DOF, as well as a control and processing unit.

The main problems associated with visual tracking are listed in [3]. On the one hand there is the problem of target recognition in the image, and on the other hand there is the problem of tracking this object when it is in motion (video). The object tracking is becoming ever more complex as it moves at high speed [8]. Some solutions, for example, involve speed visual estimation [12], process that requires adequate filtering.

In structure is of importance the articulation of movement and the processing system. In the first case there are alternatives inspire biologically, for example mimicking the neural circuitry of the human eye [14]. However, most research propose a movement system according to the final functionality of the prototype [3,5,10], within a pan and tilt platform. Regarding the processing system, most solutions focus on high throughput, running complex algorithms in real time, which generally produces good performance in terms of higher frequency imaging and less tracking error [3,9,11].

The tracking scheme proposed in this research aims to its use in autonomous mobile robots, particularly in navigation applications in dynamic environments. A basic design principle is the minimalism, which is why it was decided to perform the processing tasks in real time with a general-purpose microcontroller of 8-bit, processing the minimum necessary amount of information. As sensor system we use a small analog camera.

2 Problem Formulation

The goal is to implement a system capable of controlling an artificial eye, and capture visual information to develop automatically basic tracking tasks. The system is completely independent, i.e. there are no actions on it produced by some superior control unit, internal or external to the robot. The task begins with the receipt of the command tracking, and the information concerning to the target tracking. From that moment, the system must actively seek the inherent characteristic of the target, and keeping track. If the target is lost sight of, then the system must assume its search proactively.

Trace information is comprised of marks on the navigation environment, *landmarks*, recognizable by its geometric shape. This concept can be extended to any other recognizable trace information, even with other sensors. The position of the *landmark* relative to the eye is determined by a scan in the X-Y plane of each digitized image from the camera installed on the eye. The eye is able to move the camera through three engines that make a three DOF with capacity very close to the human eye movement (without considering the iris). This is achieved by applying sinusoidal movements through linear segments linking the three points that provide the basis for the camera (Fig. 1). The design of this prototype always had present its application to humanoid scale prototypes.

Fig. 1. Prototype of anthropometric eye

Accordingly, let us assume a camera attached to the end of a robotic actuator. This actuator is equipped with three DOF that can simultaneously move the triangular base that supports the camera. This system is capable of monitoring a predefined static object (*landmark*) in the environment. Furthermore, assume that during the mechanical movement of the robotic actuator, the camera can always observe a set of k points belonging to the *landmark*. These points should be identified from image to image over time by the processing and control unit of the system.

To describe the motion of the system, we define a reference coordinate frame located in the camera (Fig. 2). Homogeneous coordinates of a point i belonging to the *landmark* with respect to the camera is denoted by (Eq. 1):

$$\mathbf{x}_i \in \mathbb{R}^{4 \times 1} \tag{1}$$

3   Design and Metodology

The key of the image processing in the proposed scheme is the identification and extraction of regions in binary images. In a binary image, each pixel takes only one of two possible values. This modification of the images, produced by a filter, separates the image background of the foreground of interest. The identification of regions implies, first to determine which pixels in the image belong to the regions of interest, second, to determine the number of regions in the image, and third, locate these regions in the image. This process can be done in different ways. The strategy we use is to do a sweep of the image from top to bottom. The process requires the comparison of values between pixels, and according to their proximity, the new pixel is classified as belonging to the region or not. At the end, each region is assigned a unique label.

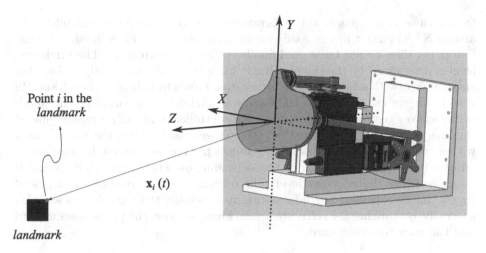

Fig. 2. CAD design and framework for visual tracking system

The requirement of processing of the tracking system was minimized by working at a very low level. We opted for the contrasts of light, making labeling of dark areas. The binary images resulting from the labeling process, called $I(u, v)$ (u and v denote the coordinates of each pixel), use zero values for the background and one for the foreground (regions, Eq. 2).

$$I(u, v) = \begin{cases} 0 \text{ for background pixels} \\ 1 \text{ for foreground pixels} \end{cases} \tag{2}$$

We start from the principle that the tracking system completely understood the characteristics of the *landmarks*. For a first evaluation, we use black squares *landmarks* of 14×14 cm. Both the size in pixels and the asymmetry detected in the image provide to the system with information regarding distance to *landmark* and spatial orientation of the same, i.e., the value of $\mathbf{x}_i$.

The value of $\mathbf{x}_i$ is calculated continuously for each image, and this value for the last nine analyzed images are stored. This matrix is used to calculate the control action on the motors, so that the distortion of the square in the image is reduced, the number of pixels that represent is increased, and focus the landmark in the image is achieved. In case of no keep an eye on the *landmark*, the system enters a query sequence in which the motors make the camera scan the entire field of vision. It is only possible exit from this search sequence when the system positively identifies the *landmark*.

4   Results

We use the idea of minimalist design in terms of information processing to perform the tasks. We develop a simple video pre-processing card to work together with an 8-bit microcontroller. The card interprets an analog NTSC video signal from a camera, and passes this information directly to the microcontroller. The analog NTSC format has an easily manipulable structure. Each line has a negative pulse, then a color burst, and finally the image information. The brightness level is determined by the signal voltage. Lines are fired sequentially to form the entire image, with additional synchronization pulses to indicate a new frame. In the circuit, we use the video sync separator LM1881 for separating sync pulses of the video signal, which allows the microcontroller to identify a specific line of the image. To read the information, the voltage change across the line is sensed with a comparator and at least eight samples per line are achieved. Comparator adjusts its output either high or low depending on whether the video signal is above or below the reference level. The reference level is dynamically adjusted by the microcontroller. For information in grayscale, the same line is sampled successively changing the reference levels. Figure 3 shows the pre-processing card and the microcontroller card.

Fig. 3. Prototype of video pre-processing system and microcontroller

We conducted several tests on laboratory prototype. One of these tests is shown in Fig. 4. The square *landmark* was placed on a wall 1.5 m from the sensor. However, it was further away since it was not located in front of the camera (it was located in the upper left). Because of the distance and size of the *landmark*, this is in itself a point of interest i.

In this test the system immediately identifies the *landmark*, and coordinates the movement of the eye to improve its focus. Figure 4 shows the path of movement used to track and focus the *landmark*. In all the frames the system finds the *landmark*, and the versatility of motion of the mechanical prototype allows faithful following analogous to the human eye. It is also noteworthy that does not require a great camera to solve the problem, a small analog camera without a lens is able to solve this particular type of problem.

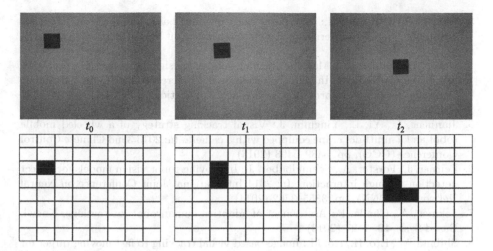

Fig. 4. Results. Top row: pictures corresponding to initial position (left, t_0), middle position (center, t_1) and final position (right, t_2) of a static target tracking. Bottom row: The same three previous cases but digitized by the tracking system.

5 Conclusions

The paper presents a visual tracking scheme for small autonomous robots that does not use a big hardware or complex processing algorithms, but still operates correctly in real time. The methodology can be extended in many different ways, allowing adaptation to different robotic platforms and different types of identification and tracking problems. The system proposes low-level manipulation of digital images of low resolution. To do this, we use an analog camera, a pre-processing system that produces digital frames of 8×8 pixels and a digital processing unit of 8 bits. An artificial eye with capacity equivalent to the human eye movement was used. The move was coordinated with the results of the tracking process, to improve the target focus on the environment.

Acknowledgments. This work was supported by the District University Francisco José de Caldas, in part through CIDC, and partly by the Technological Faculty. The views expressed in this paper are not necessarily endorsed by District University. The authors thank the research groups DIGITI and ARMOS for the evaluation carried out on prototypes of ideas and strategies, and especially Jesús David Borda Guerrero and Carlos Andrés Nieto Ortiz for the support in the development of the prototype.

References

1. Andreopoulosa, A., Tsotsosb, J.K.: 50 years of object recognition: directions forward. Comput. Vis. Image Underst. **117**(8), 827–891 (2013)
2. Bousnina, S., Ammar, B., Baklouti, N., Alimi, A.: Learning system for mobile robot detection and tracking. In: 2012 International Conference on Communications and Information Technology (ICCIT), pp. 384–389 (2012)

3. Chang, O., Olivares, M.: A robotic eye controller based on cooperative neural agents. In: The 2010 International Joint Conference on Neural Networks (IJCNN), pp. 1–6 (2010)
4. Dabre, K., Dholay, S.: Machine learning model for sign language interpretation using webcam images. In: 2014 International Conference on Circuits, Systems, Communication and Information Technology Applications (CSCITA), pp. 317–321 (2014)
5. Jianheng, L., Yi, L., Yingmin, J.: Visual tracking strategy of a wheeled mobile robot and a multi-dof crane equipped with a camera. In: 2013 32nd Chinese Control Conference (CCC), pp. 5489–5493 (2013)
6. Leitner, J., Forster, A., Schmidhuber, J.: Improving robot vision models for object detection through interaction. In: 2014 International Joint Conference on Neural Networks (IJCNN), pp. 3355–3362 (2014)
7. Sekmen, A., Challa, P.: Assessment of adaptive humanrobot interactions. Knowl. Based Syst. **42**, 1–27 (2013)
8. Shibata, M., Eto, H., Ito, M.: Image-based visual tracking to fast moving target for active binocular robot. In: 36th Annual Conference on IEEE Industrial Electronics Society IECON 2010, pp. 2727–2732 (2010)
9. Shibata, M., Eto, H., Ito, M.: Visual tracking control for stereo vision robot with high gain controller and high speed cameras. In: 2011 1st International Symposium on Access Spaces (ISAS), pp. 288–293 (2011)
10. Sorribes, J., Prats, M., Morales, A.: Visual tracking of a jaw gripper based on articulated 3d models for grasping. In: 2010 IEEE International Conference on Robotics and Automation (ICRA), pp. 2302–2307 (2010)
11. Sun, Y., Ding, N., Qian, H., Xu, Y.: Real-time monocular visual self-localization approach using natural circular landmarks for indoor navigation. In: 2012 IEEE International Conference on Robotics and Biomimetics, pp. 495–500 (2012)
12. Wang, H., Liu, Y., Chen, W.: Uncalibrated visual tracking control without visual velocity. IEEE Trans. Control Syst. Technol. **18**, 1359–1370 (2010)
13. Yali, L., Shengjin, W., Qi, T., Xiaoqing, D.: A survey of recent advances in visual feature detection. Neurocomputing **149**, 800–810 (2015)
14. Zhou, Y., Luo, J., Hu, J., Li, H., Xie, S.: Bionic eye system based on fuzzy adaptive PID control. In: 2012 IEEE International Conference on Robotics and Biomimetics, pp. 1268–1272 (2012)

15 DOF Robotic Hand Fuzzy-Sliding Control for Grasping Tasks

Edwar Jacinto, Holman Montiel, and Fredy Martínez[✉]

District University Francisco José de Caldas, Bogotá, Colombia
{ejacintog,hmontiela,fhmartinezs}@udistrital.edu.co
http://www.udistrital.edu.co

Abstract. We present a grasp control scheme for an anthropomorphic multi-fingered robotic hand (kinematic structure similar to the human hand) with five fingers and 15 DOF that use different control surfaces according to the system state and the expected response. The responses of each surface are combined into a single structure using a fuzzy control scheme. We define a general behavior for all fingers, which act independently, it allows that the hand adapts to the shape of the object. To reduce the computational complexity, we established only five internal control blocks, one for each finger; the mechanical actuator distributes the control action along the finger joints. This robot hand can grasp various objects steadily and achieve manipulations with a very simple mechanism. Algorithm effectiveness has been tested on a real robotic prototype.

Keywords: Grasping control · Robotic hand · Sliding mode control

1 Introduction

The ultimate goal of robotic hand is to achieve the functionality of a human hand. Even though grasping has been a central topic in robotics for decades, robots still have great difficulty to pick up objects when they operate in open, unknown environments or under uncontrolled conditions. This continues to be an open unsolved problem in robotics. The task of grasping objects through robotic hands has many levels of complexity. Objects can have a variety of features, they can be hard (like a Rubik cube), soft (like a stuffed toy), heavy (like a brick), light (such as a plastic bottle), robust (like a ball), fragile (like a glass of wine), large or small and with different geometric shapes [4].

The task can be separated into two different problems [10] a first problem related to the hand movement until the object to grab (finger placement), and a second problem corresponding to the grasping process itself. This is the problem of interest in our research. In general terms, the problem comes down to control the force of the fingers in line with the reactions produced during the hand-object interaction. The robotic hand should be able to grab a wide variety of objects firmly without dropping them, and without crushing them.

K. Jackowski et al. (Eds.): IDEAL 2015, LNCS 9375, pp. 239–247, 2015.
DOI: 10.1007/978-3-319-24834-9_29

This problem has been traditionally attacked with two strategies. The first strategy is known as *power grasp*, or *whole hand grasp* [5,9,11–13]. This strategy is the most widely investigated. It, somehow in concept, tries to reduce the complexity of the problem. It involves controlling the closing of the hand (actually fingers) over the object in a way that a balance between the applied force and the force of repulsion of the object is achieved. Under this scheme, the forces applied by the fingers ensure the object immobility, achieving its secure grip. While it is a scheme that significantly reduces the computational cost, it is not possible to generate grasping forces in all directions. In addition, the whole finger is active and in contact with the object (the finger is not adapted to the shape of the object). Another problem with this strategy that significantly reduces the kind of objects to manipulate is that in order to grasp the object firmly, the hand will grasp the object by full force until the object is unable to move. As a result, the object is no longer in the original state (severe target object deformation).

The second strategy is known as *intelligent grasping* [1,7,8,14]. In precision grasping the object is held by tips of the fingers, i.e. the fingers adapt to the shape of the object. Many researchers have focused on this kind of strategy mainly because of the well-established techniques for control of effectors. However, this approach is computationally expensive and require the specification of the object characteristics in mathematical form.

Both strategies are optimized with earlier stages of sensing of the object and estimating the optimal position of the hand prior to the process of grasp. We propose a general strategy of control that can be placed in the second category. This is a fuzzy-sliding control scheme that combines control surfaces, each with different rules of operation which depend on the state identified by sensing each finger. The computational problem is simplified by assuming a single control output by finger.

The paper is organized as follows. In Sect. 2 problem formulation is discussed, Sect. 3 describes the control strategy. Section 4 analyzes the performance of the proposed control scheme. Finally, conclusion and discussion are presented in Sect. 5.

2   Problem Formulation

The problem in a grasping task involves the interaction between the object to be handled and the robotic hand. Consider a fixed object, of proportional and lower size relative to the size of the robotic hand, with known coordinates for each point p_i belonging to the object.

Let $\mathscr{W} \subset \mathbb{R}^3$ be the closure of a contractible open set in the space that has a connected open interior with obstacles that represent inaccessible regions. Let $\mathscr{O}$ be a set of obstacles, in which each $O \subset \mathscr{O}$ is closed with a connected piecewise-analytic boundary that is finite in length. Furthermore, the obstacles in $\mathscr{O}$ are pairwise-disjoint (their intersection is the empty set) and countably finite in number. Let $E \subset \mathscr{W}$ be the free space in the environment, which is the open subset of $\mathscr{W}$ with the obstacles removed.

Let us assume a robotic hand in this free space. The hand know the environment E in which it moves, and the object with which it interacts, from observations, using sensors. These observations allows it to build an information space $\mathscr{I}$. An information mapping is of the form (Eq. 1):

$$g : E \longrightarrow Y \tag{1}$$

where Y denotes an *observation space*, constructed from sensor readings over time, i.e., through an observation history of the form (Eq. 2):

$$\tilde{y} : [0, t] \longrightarrow Y \tag{2}$$

The interpretation of this information space, i.e., $\mathscr{I} \times Y \longrightarrow \mathscr{I}$, is that which allows the hand controller to make decisions [6].

The task of grasping can be separated into two distinct problems. First one is a path planning problem where the hand have to reach the object location. With a given trajectory, the joints position of the fingers are controlled to reach the desired location. As soon as the fingers hit the object, start the second problem.

The second problem is the grasping task itself. During this process a reaction force is generated, which the system controller needs to be dealt with. The reaction force is needed to be limited to a certain optimum value so that it does not cause any slippage or damage of the object while maintaining contact with the hand.

It is assumed that, the object does not produce any motion to the system, i.e. (Eq. 3):

$$\ddot{X}^i_{obj} = \dot{X}^i_{obj} = 0 \tag{3}$$

with (Eq. 4):

$$X^i_{obj} = \left(x^i_1, \ x^i_2, \ x^i_3 \right) \tag{4}$$

where $\left(x^i_1, \ x^i_2, \ x^i_3 \right) \in \mathbb{R}^3$ is the position of the i point belonging to the object.

The goal of control in the grasping process is achieve the desired position X_{hand_d} and the desired force F_d such that (Eq. 5):

$$\lim_{t \longrightarrow \infty} e_X(t) = X_{hand_d}(t) - X_{hand}(t) = 0 \tag{5}$$

and (Eq. 6):

$$\lim_{t \longrightarrow \infty} e_F(t) = f_d(t) - f(t) = 0 \tag{6}$$

where $X(t)$ is the position vector for each point of the hand (we use only a few points of interest), and $f(t)$ is the force vector of the same points of interest. $e_X(t)$ and $e_F(t)$ are the position and force error vectors respectively.

3   Methodology

3.1   Robotic Hand

The robotic hand used in this paper is an anthropomorphic five-fingered hand with the ability to grasp objects of different shapes and sizes. The prototype was built in ABS trying to duplicate the functionality of the human hand. The hand consists of five fingers F_1, F_2, F_3, F_4 and F_5, placed on a palm G. Each of the fingers has a total of three joints, each with one DOF (15 DOF in total); one articulating the finger to the palm, and the other two along the finger. Each finger has three revolute joints (q_1^n, q_2^n and q_3^n, with $n = 1, 2, 3, 4$ or 5, indicating each of the fingers). This structure means that each finger has three links (l_1^n, l_2^n and l_3^n, with $n = 1, 2, 3, 4$ or 5, for each of the fingers). One finger moves in an orthogonal plane to the other four as a thumb (Figs. 1 and 2).

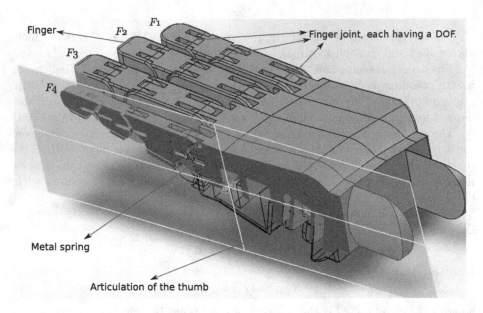

Fig. 1. Robotic hand. Anthropomorphic hand with five fingers (thumb included), each with three DOF. The action is driven by a motor for the grasping process, and a metal springs for the opening.

The angle of rotation for the joint q_1 of F_1, F_2, F_3 and F_4 fingers has a maximum amplitude of $90°$. This angle for the finger F_5 is slightly less, $70°$. The maximum rotation angle for the q_2 joints of the five fingers is equal to $110°$, and for joints q_3 is $75°$.

The actuators and sensors in the hand are handled directly by a embedded microcontroller system [3]. This system communicates with the central control unit, the unit that takes control decisions. Each finger has three strain gauges mounted in the middle of each of the links l_1^n, l_2^n and l_3^n. The value of these three sensors per finger is averaged to obtain a single value of normal force f_N^n during grasping.

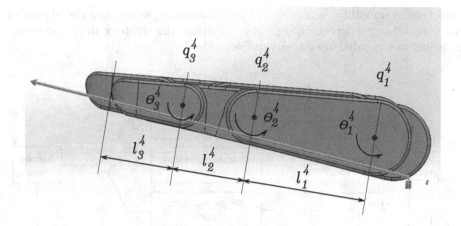

Fig. 2. Robotic hand. Details of the joints, links and angles of rotation for a finger.

3.2 Our Approach of Grasping Control

We define the same control strategy for each of the fingers. Each finger has three joints q_1^n, q_2^n and q_3^n, which generate the three position angles θ_1^n, θ_2^n and θ_3^n, and the three DOF of each finger. The robotic hand has only one actuator (motor) by finger. The mechanical assembly allows simultaneous and proportional to control of three DOF of each finger, i.e. it is not possible to separately control the movement of each of the links of the fingers.

To control each finger we defined three control surfaces, one for each operating condition identified:

1. Steadily grasped object. This is the ideal state of the system. The finger grasps the object without deforming it (ideally) and without slipping.
2. Grasped object slip. This surface is characterized by the existence of slip. We use the strain gauges equipped on the finger to distinguish between slip and non slip events. The desired position corresponds to the finger grasping the object without slipping. The slip increases the position error, so to reduce the position error we should eliminate slipping. You can not move the object in the hand after grasp it (try to grab it better).
3. No grasped object. On this surface the finger does not see repulsive force when it is trying to grasp the object, this may be due to the shape of the object or partial grasp. In this case the finger should return to its initial idle position (open) and does not try to apply force to the object.

Each of these surfaces is characterized by a particular value of the state variables of the finger. According to the value of the variables, the system can move from one surface to another. The control identifies which is the control surface, and applies the control rules defined to the surface.

For a smooth control surface without abrupt jumps between control actions, we require a global control scheme capable of switching between control surfaces correctly and smoothly. According to this requirement, we use as a global control structure a fuzzy inference block, able to define the strategy implementation according to the value of system variables (Fig. 3).

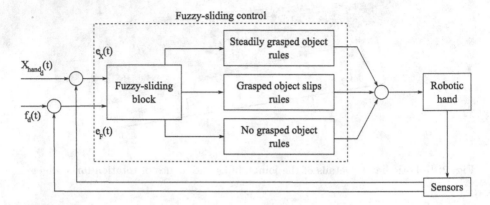

Fig. 3. Diagram of the fuzzy-sliding mode control algorithms proposed for grasping control.

The final control action is interpreted as a force applied by a motor to a pulley system that activates the entire finger. This force is inferred by the control unit in order to minimize errors of position and force, and according to the rules defined in the control surfaces.

3.3   Robotic Grasp Control Rules

The proposed control algorithm was designed to minimize the position and force errors according to the desired values in each control surface. These two errors are the two linguistic input variables of the fuzzy inference, the output variable is the force applied to the finger.

For each of these three linguistic variables we define five (5) possible linguistic values: large negative (GN), small negative (SN), zero (ZE), small positive (SP) and large positive (GP). According to these linguistic variables, it is possible to express the different control laws in a single global control system for the movement of the finger. We use Mamdani for the inference engine.

1. Steadily grasped object: In this surface there is repulsion force and there is no slip. The desired force (reference value) should be reduced up to the border of the sliding surface, this reduces the possibility of damage to the object.
2. Grasped object slip: In this surface there is repulsion force and there is slip. To prevent sliding the control should increase the value of the desired force (reference value) until completely eliminate slipping, even if it means to deform the object. This should take the system to the first control surface.
3. No grasped object: In this surface there is no repulsion force and either sliding. The desired force is set to zero regardless of the presence of slip.

When a grasped object slips, vibrations are produced at the interface between the hand and object. These frequency vibrations (between 20 Hz and 60 Hz) that occur during slip can be filtered and amplified in order to produce a variable which permits to establish the position error. Details on the filter design are available at [2].

4   Performance Evaluation

We conducted many performance tests on the prototype, Fig. 4 shows the results of one of them to a solid object. In Fig. 4(a) the initial grab is observed. At the time when the control detects the presence of reaction force, it increases the reference and tries to reduce the error until a point of equilibrium without the presence of sliding. In Fig. 4(b) we force a sliding, but without removing the object, the finger force is adjusted to maintain the grasp. Finally in Fig. 4(c) we remove the object by force, so that the finger stops applying force.

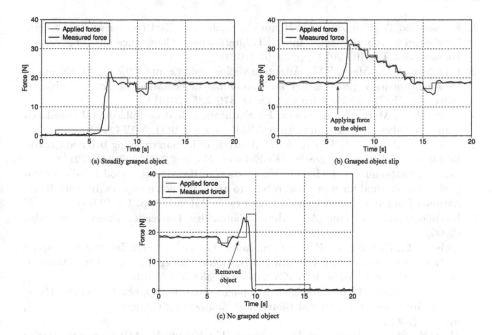

Fig. 4. Laboratory results (index finger). (a) Grasp of the object. (b) Behavior when the object is pushed and slipped from its initial position. (c) Behavior when object is pushed and removed.

5   Conclusions

In this paper a control scheme for achieving object intelligent grasping by an anthropomorphic multi-fingered robotic hand was proposed. The control scheme is based on three control surfaces derived from the behavior of the system variables. We define rules of behavior for each control surface, and combine the output with a fuzzy inference. Control is applied independently and in parallel on each finger, allowing gently adapt to the shape of the object avoiding deforming it. The proposed control method is successful for wide variety of objects, besides, the method does not rely on complex equations of inverse kinematics, or inverse dynamic, and hence is suitable for real-time applications.

Acknowledgments. This work was supported by the District University Francisco Jos de Caldas, in part through CIDC, and partly by the Technological Faculty. The views expressed in this paper are not necessarily endorsed by District University. The authors thank the research groups DIGITI and ARMOS for the evaluation carried out on prototypes of ideas and strategies, and especially Andrés Felipe García Guerrero and Andrés Julian Becerra for the support in the development of the prototype.

References

1. Bernandino, A., Henriques, M., Hendrich, N., Jianwei, Z.: Precision grasp synergies for dexterous robotic hands. In: IEEE International Conference on Robotics and Biomimetics. ROBIO 2013, pp. 62–67 (2013)
2. Engeberg, E.D., Meek, S.G.: Adaptive sliding mode control for prosthetic hands to simultaneously prevent slip and minimize deformation of grasped objects. IEEE/ASME Trans. Mechatronics **18**(1), 376–385 (2013)
3. Esquivel, J., Marin, N., Martínez, F.: Plataforma de desarrollo digital basada en game boy advance y la arquitectura ARM7. Tekhnê **9**(1), 5–12 (2012)
4. Feix, T., Bullock, I.M., Dollar, A.M.: Analysis of human grasping behavior: correlating tasks, objects and grasps. IEEE Trans. Haptics **7**(4), 430–441 (2014)
5. Gori, I., Pattacini, U., Tikhanoff, V., Metta, G.: Ranking the good points: a comprehensive method for humanoid robots to grasp unknown objects. In: 16th International Conference on Advanced Robotics. ICAR 2013, pp. 1–7 (2013)
6. LaValle, S.M.: Planning Algorithms. Cambridge University Press, Cambridge (2006)
7. Lichong, L., Huaping, L., Funchun, S., Meng, G., Wei, X.: The intelligent grasping tactics of dexterous hand. In: 13th International Conference on Control Automation Robotics and Vision. ICARCV 2014, pp. 352–357 (2014)
8. Ma, S., Moussa, M.: An intelligent grasping system for applications in prosthetic hands. In: Cairo International Biomedical Engineering Conference. CIBEC 2008, pp. 1–5 (2008)
9. Man, B., Sin, J., Taewoong, U., Keehoon, K.: Kulex: An ADL power-assistance demonstration. In: 10th International Conference on Ubiquitous Robots and Ambient Intelligence. URAI 2013, pp. 542–544 (2013)
10. Muller, J., Frese, U., Rofer, T.: Grab a mug - object detection and grasp motion planning with the NAO robot. In: 12th IEEE-RAS International Conference on Humanoid Robots Humanoids 2012, pp. 349–356 (2012)
11. Roa, M.A., Argus, M.J., Leidner, D., Borst, C., Hirzinger, G.: Power grasp planning for anthropomorphic robot hands. In: IEEE International Conference on Robotics and Automation. ICRA 2012, pp. 563–569 (2012)

12. Sang-Mun, L., Kyoung-Don, L., Heung-Ki, M., Tae-Sung, N., Jeong-Woo, L.: Kinematics of the robomec robot hand with planar and spherical four bar linkages for power grasping. In: IEEE International Conference on Automation Science and Engineering. CASE 2012, pp. 1120–1125 (2012)
13. Yamaguchi, K., Hirata, Y., Kosuge, K.: Development of robot hand with suction mechanism for robust and dexterous grasping. In: IEEE/RSJ International Conference on Intelligent Robots and Systems. IROS 2013, pp. 5500–5505 (2013)
14. Lin, Y.-C., Wei, S.-T., Fu, L.-C.: Grasping unknown objects using depth gradient feature with eye-in-hand RGB-D sensor. In: IEEE International Conference on Automotion Science and Engineering. CASE 2014, pp. 1258–1263 (2014)

Intelligent Automated Design of Machine Components Using Antipatterns

Wojciech Kacalak, Maciej Majewski$^{(\boxtimes)}$, and Zbigniew Budniak

Faculty of Mechanical Engineering, Koszalin University of Technology,
Raclawicka 15-17, 75-620 Koszalin, Poland
{wojciech.kacalak,maciej.majewski,zbigniew.budniak}@tu.koszalin.pl
http://kmp.wm.tu.koszalin.pl

Abstract. The article presents a methodology for the analysis of similarities between structural features of designed machine elements and the corresponding antipatterns. This methodology allows normalization of selected design solutions' characteristic features. The defined antipatterns are generic definitions of the possible incorrect design solutions. The article also presents antipatterns' attributes, as well as classification based on the root causes of errors in designed solutions. Examples of a step shaft have been used to illustrate the methodology of designed solution and its antipattern correspondence evaluation. Root causes of a design error, its importance, and similarities shared with other design errors have been analyzed for each of the presented antipatterns. Correctly designed shafts - not having the characteristics of an antipattern - have also been presented.

Keywords: Intelligent automated design · Interactive design system · Intelligent interface · Antipatterns · Design automation · Neural networks

1 The Role of Antipatterns in the Machine Design Process

The antipattern concept, which is successfully applied in software engineering, can serve as inspiration to using it in other engineering-related tasks. The most favorable scenario would be a task based on heuristic concept creation, in which the designer wouldn't be limited to using patterns, but instead he would be expected to come up with innovative solutions characterized by a high level of non-obviousness in relation to the established knowledge.

Issues connected with optimal design have been elaborated in many publications on subjects like design method selection [1,4,5]. Methods such as programming tasks, equality constraints with excess variables, dynamic programming, and optimization of variable-topology structures are useful when it comes to accepting decisions regarding the value of a structural feature. When making such decisions, stress and distortions, as well as correspondence of geometries

K. Jackowski et al. (Eds.): IDEAL 2015, LNCS 9375, pp. 248–255, 2015.
DOI: 10.1007/978-3-319-24834-9_30

and topologies from a finite set of combinations with other structural elements are taken into account. However, usefulness of the mentioned methods in concept development is limited.

The set of methods used for solving problems connected with designing machine components comprises many topics related to data analysis and exploration. This set of methods can be expanded with many more details, such as: multidimensional scaling, tree-based classifiers, neural classifiers, search algorithms (including heuristic ones), data-clustering algorithms, and content-based search methods.

The drawback of using design patterns is that not only do they impede innovation, but also they increase the odds of a scenario, where poor application of a pattern produces additional design errors, or contributes to delayed detection of such errors and may result in repeated application of solutions worse than the possible correct ones, coming at the same cost. Antipatterns are generic definitions of possible instances of incorrect solutions to design problems. Work on theoretical and experimental basics of antipattern definition methodology, their features and exceptional matching factor measure evaluation classification ability are important lines of research, important to the automated systems of aided design process, which aspire to being prevalent in the near future.

Antipatterns can comprise multiple attributes: (AFD) the function served by the designed component or structural system, (ASG) the component's structural group, (AIS) the definition of an incorrect solution, (AEC) the error cause, (AER) the error result, (AEI) the error's importance, (AES) other error similarity, and design solution features (AF1), (AF2) ... (AFn). The distinct features of antipatterns are their geometrical and physical properties. Antipatterns can also be classified based on the root causes of flawed designs, the most common of which are presented in Table 1.

Works aiming to develop basics of automation of processes in designing machine elements and assemblies with the use of artificial intelligence in uncertainty and unrepeatability of processes [2] have been started. In automated design systems, which use a natural-language description of structural features and an intelligent interface of natural speech and hand-drawn sketches, application of design antipatterns - especially in combination with artificial-intelligence methods - carries a lot of meaning to effectiveness and development of such systems.

2   Evaluation Measures of Correspondence Between the Designed Structure and Antipatterns

Determination of measures for evaluation of correspondence between the designed structure and antipatterns requires normalization of compared features. Antipatterns' linguistic attributes can be used for selection by (AFD), (ASG), (AIS), (AEC), (AER), (AEI), (AES), (AF1), (AF2) ... (AFn). Input attributes require normalization to fit in the (0,1) range.

Table 1. Root causes of designs with errors

	Symbol	The name of an ERROR's cause
1.	(EPI)	A lack of correct evaluation of PRODUCT'S IMPORTANCE for the future manufacturing program
2.	(EOE)	A lack of project's economical OUTCOME EVALUATION
3.	(ETP)	An underestimation of TECHNOLOGICAL PROBLEMS' impact on the product's quality
4.	(EMR)	A lack of analysis of MAINTENANCE REQUIREMENTS
5.	(ECP)	A lack of analysis of CUSTOMER PREFERENCES in relation to quality, modernity, durability, ease of operation, and visual design
6.	(EDG)	DOMINANCE OF GRAPHICAL tasks over conceptual analysis in the process of structure design
7.	(EDT)	DOMINANCE OF TRUST in results of massive data processing over trust in knowledge
8.	(ESE)	A deferral of SOLUTION EVALUATION, with overemphasized trust in ease of maintenance
9.	(EAF)	An introduction of an ABUNDANT FUNCTIONALITY
10.	(ECF)	Incorrectly COMBINED FUNCTIONS in the designed elements
11.	(EAE)	Existence of ABUNDANT ELEMENTS, which do not serve any functions
12.	(ELH)	A LACK OF HARMONY in functionality, incorrect division of functions into elements of the structure
13.	(ELV)	A LACK OF A VERIFICATION (competent and multicriterial) at each stage, and a lack of quality evaluation, supposed to be carried out with the required procedures
14.	(EDR)	DEADLINE RUSH due to short time
15.	(EAT)	Incorrectly ASSIGNED TASKS within the team
16.	(EDV)	Either a neglect or awareness of DEFERRED VERIFICATION of structural design effects and unclear evaluation of responsibility
17.	(EAS)	Stubborn ADHERENCE TO SOLUTIONS (known and otherwise correct)
18.	(EIM)	INCOMPATIBLE METHODOLOGIES, preferences, and design criteria among team members
19.	(EOC)	Application of OLD COMPONENTS and norms in the designed structure
20.	(ELE)	A LACK OF EVALUATION of a need for scalable, modular and unifiable solutions
21.	(ENE)	An excessive (implying a lack of integration) or an insufficient (implying differentiation) NUMBER OF ELEMENTS
22.	(EDS)	Too easily DEFORMABLE STRUCTURE
23.	(ECM)	Incorrect CHOICE OF MATERIALS
24.	(EDI)	DIFFICULT INSTALLATION, limited or costly parts replacement ability
25.	(EQI)	An insufficient emphasis on the QUALITY OF INTERACTING surfaces and the properties of outer layers
26.	(EAA)	An excessively limited ABILITY TO ADAPT structural features to operating conditions
27.	(EED)	A lack of EVALUATION OF DYNAMICS of operating conditions
28.	(EEO)	A lack of EVALUATION OF OCCURRENCES, which are not characteristic of the typical operation (conditions pacing)

Below are listed examples of attributes used for antipatterns selection:

1. (AFD) - the function served by the designed element or aggregate of components - rotating element, housing, hood, frame, cantilever, adaptor, guide, cylinder, valve, handle, floor, pipe, spring etc.;
2. (ASG) - component's structural groups, for AFD = 'rotating' possible options are: straight or step shaft, spindle, disk, ring, roller, ball, and additionally: crankshaft, axle, toothed wheel etc.;
3. (AIS) - incorrect solution's definition, for AFD = 'rotating' and ASG = 'shaft' such a definition can be: easy deformation, excessive difference between steps' diameters, incorrect diameter tolerance, high level of steps, lack of exit zone, incorrect dimensions of splineway etc.

Evaluation measures of correspondence will depend on the set of parameters typical of the aforementioned attributes, and they should be subjected to normalization, so that they fall within the range of (0,1) using fuzzy sets. Error's importance (AEI) should be - after normalization - attached to the set of attributes describing structural features. Normalization by the shape of the classification function should take into account the meaning of the value to the result of normalization (flow direction).

The developed methodology for design solutions evaluation with antipatterns has been presented in Fig. 1A.

An application example of the design solutions evaluation methodology is a seven-step shaft ($n = 7$), presented in Fig. 1B.

The selected shaft antipattern has been, for the designed element's function AFD = 'rotating', has been chosen from the structural group ASG = 'step shaft'. The incorrect solution AIS_1 = 'easy deformation' is characterized by shaft buckling due to an axial force. The cause of the error AEC_1 = 'high $S_L = L/d_{ekw}$' is such, that a flexible shaft, due to axial compression, may be subject to buckling (bending). In the case of the given example importance AEI_1 is high and it comprises similarity to other errors AES_1: stability loss, which leads to inevitable physical destruction, which usually involves other interacting elements, small stiffness, large deviations, noise, vibrations, unstable operation, etc.

Another incorrect solution is AIS_2 = 'excessive diametral steps', the cause of which AEC_2 = 'high S_d' may lead to excessive stress cumulation caused by indentation. The S_d value is given by the formula:

$$S_d = \sqrt{S_s \cdot S_m}$$
$$\text{where: } S_s = max \in \left\{ \frac{d_j}{d_{j-1}}, \frac{d_k}{d_{k+1}} \right\}, \; j \in \{2, 3, \ldots, i, \ldots, n\},$$
$$k \in \{1, 2, \ldots, i, \ldots, n-1\}, \; d_i \in \{d_1, d_2, \ldots, d_n\}, \; S_m = \frac{d_{max}}{d_{min}}$$
$$d_{max} = max \in \{d_1, d_2, \ldots, d_n\}, \; d_{min} = min \in \{d_1, d_2, \ldots, d_n\} \tag{1}$$

To lessen the indentation's impact it is advised to assume $S_m = d_{max}/d_{min} \leq 1,2$ and also to introduce a possibly large intermediate radius or a conical transition. The importance of the AEI_2 error is considerable and it comprises similarities to other errors AES_2: stress cumulation, fatigue endurance, difficult installation, etc.

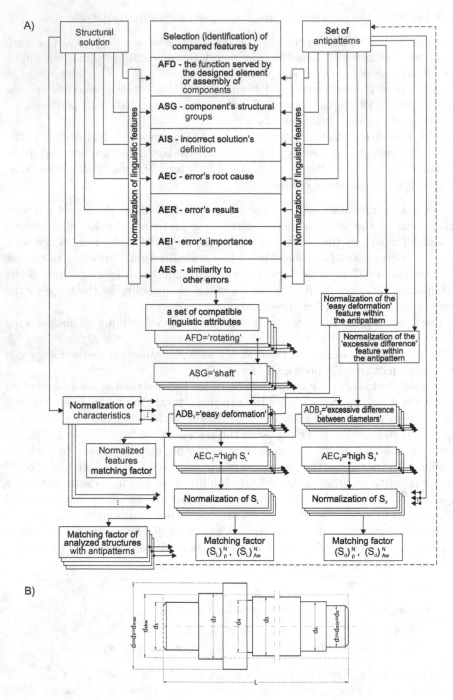

Fig. 1. (A) Design solutions and antipattern correspondence evaluation methodology, (B) a step shaft.

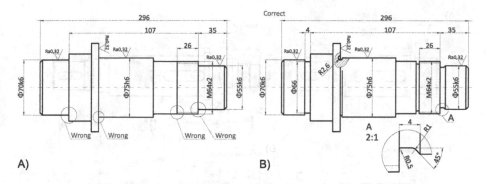

Fig. 2. A design of a shaft manufactured with machine cutting with tool entrance and exit zones: (a) antipattern, (b) correct design.

More examples of applied methodology for design solution and antipattern correspondence evaluation are presented in Figs. 2 and 3ABCD. Selected shaft antipatterns for the designed element's function AFD = 'rotating' have been chosen from the structural group ASG = 'step shaft'.

In the example presented in Fig. 2 AIS = 'tool exit zone' is an incorrect solution, the cause of which AEC = 'tool exit zone's shape' is the fact, that the shaft doesn't provide a zone for the tool to freely move around and exit certain zones during the manufacturing process. In the case of the given example importance of AEI is high and it comprises similarity to other errors AES: manufacturing errors, producibility, poor installation, etc. Fig. 2 presents a shaft designed with tool's reach and exit ability in mind, in accordance with the producibility criterion and not displaying the aforementioned antipattern characteristics.

In the example presented in Fig. 3AB the incorrect solution is AIS_1 = 'installation of fitted elements', the cause of which AEC_1 = 'the L_p fitted surface's length' is such, that the shaft has a mounting surface with the diameter d = F85p6 and length L_p = 203 mm, which is much longer than the width of the element mounted on it, e.g. a toothed wheel. The cause of the error (AEC_1) results in difficult installation - that is forcing the element through a significant length. In the case of the given example the importance of the error AEI_1 is average and it comprises similarity to other errors AES_1, such as: causing damage of the mounting surface, costs of forcing the shaft through, etc. At the same time, the shaft has an error AIS_2 = 'imprecise axial fitting of the mounted components'. The cause of the error AEC_2 = 'filleting radius R' indicates the possible imprecision of components axial mounting (lack of support for the fitted component from the end collar), in which the size of the mounted component's phase F is smaller than the radius of the shaft step's filleting R. The error's importance AEI_2 is average, and its similarities shared with other errors AES_2 are: noise, unstable operation, vibrations, increased wearing out of the fitted elements' surfaces, etc.

An example of an attributes set for the aforementioned cases can be as follows: (AEIN) - error importance normalized value, (LVN) - normalized length value L to the diameter equivalent d_{ekw}, (DVN) - maximal S_m diameter value

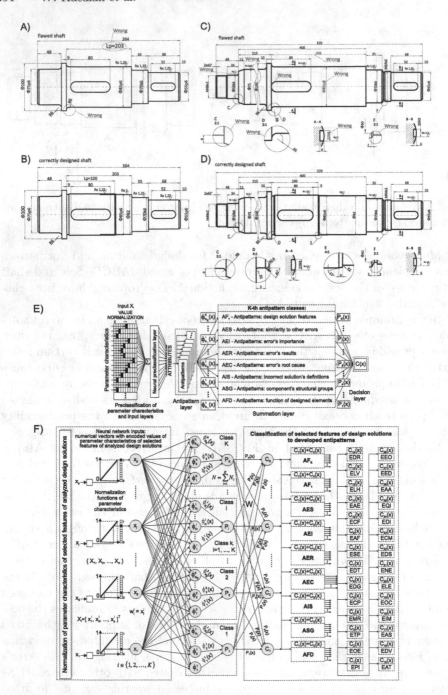

Fig. 3. Design of a shaft with friction fit's mounting surfaces of different lengths: (A) flawed shaft, (B) correctly designed shaft; A machine shaft with highlighted errors: C) an antipattern, (D) correct design; (E) Proposed neural models and the architecture (F).

normalized to the smallest one, (STN) - normalized ratio of surface's roughness parameter S_t to this surface's size tolerance T, etc. The measure of correspondence with an antipattern for the given examples can be the sum of absolute distance values between the parameters of an antipattern and the designed element, the sum of squared distances or their geometrical distance. Before performing the evaluation of correspondence between the structural features of a designed machine elements and antipatterns, it is possible to verify the ability to classify of the used parameters. The requirement would be to have a set of design solutions, for which we want to determine the differentiation ability with each of the parameters.

Figure 3E presents proposed probabilistic neural models for classification of selected features of design solutions to developed antipatterns. The architecture of the developed probabilistic neural networks [6] is shown in Fig. 3F. The inputs of the network consist of parameter characteristics of normalized features of analyzed design solutions. Selected articles [2,3] present innovative solutions in intelligent systems for interactive design of machine elements and assemblies.

3   Conclusions and Perspectives

Defined antipatterns comprise a generic definition of possible incorrect design solutions. The developed methodology for evaluation of correspondence between a design solution and an antipattern, illustrated with step shafts, allows to normalize parameter characteristics of selected features of analyzed design solutions, and, then, to determine the antipattern's and the analyzed structure's matching factor. Work on theoretical and experimental basics of antipattern definition methodology, their features and exceptional matching factor measure evaluation classification ability are important lines of research, important to the automated systems of aided design process, which aspire to being prevalent in the future.

Acknowledgements. This project was financed from the funds of the National Science Centre (Poland) allocated on the basis of the decision number DEC-2012/05/B/ST8/02802.

References

1. Brown, W.J., McCormick, H.W., Thomas, S.W.: AntiPatterns in Project Management. Wiley, New York (2000)
2. Kacalak, W., Majewski, M.: New Intelligent interactive automated systems for design of machine elements and assemblies. In: Huang, T., Zeng, Z., Li, C., Leung, C.S. (eds.) ICONIP 2012, Part IV. LNCS, vol. 7666, pp. 115–122. Springer, Heidelberg (2012)
3. Majewski, M., Zurada, J.M.: Sentence recognition using artificial neural networks. Knowl. Based Syst. **21**(7), 629–635 (2008). Elsevier
4. Piegl, L.A.: Ten challenges in computer-aided design. Comput. Aided Des. **37**(4), 461–470 (2005). Elsevier
5. Riel, A.J.: Object-Oriented Design Heuristics. Addison-Wesley, Reading (1996)
6. Specht, D.F.: Probabilistic neural networks. Neural Netw. **3**(1), 109–118 (1990)

Application of Fuzzy Logic Controller for Machine Load Balancing in Discrete Manufacturing System

Grzegorz Kłosowski[1], Arkadiusz Gola[1(✉)], and Antoni Świć[2]

[1] Faculty of Management, Department of Enterprise Organization,
Lublin University of Technology, Lublin, Poland
{g.klosowski,a.gola}@pollub.pl
[2] Faculty of Mechanical Engineering, Institute of Technological Systems of Information,
Lublin University of Technology, Lublin, Poland
a.swic@pollub.pl

Abstract. The paper presents a concept of control of discrete manufacturing system with the use of fuzzy logic. A controller based on the concept of Mamdani was developed. The primary function realized by the controller was the balancing of machine tool loads taking into account the criteria of minimisation of machining times and costs. Two models of analogous manufacturing systems were developed, differing in the manner of assignment of production tasks to machine tools. Simulation experiments were conducted on both models and the results obtained were compared. In effect of the comparison of the results of both experiments it was demonstrated that better results were obtained in the system utilising the fuzzy inference system.

Keywords: Simulation · Modelling · Control · Fuzzy logic · Manufacturing system

1 Introduction

Most issues related with the organisation of manufacturing processes are characterised by a high level of complexity [1–3]. One of the main reasons for that situation is the large number of factors that have an effect on the decisions which have to be made in that area of business [4]. Moreover, those factors are often hard to define [5–7]. As the description of manufacturing processes by means of classic math functions requires precise determination of both the variables and the manner of transformations, under the conditions of uncertainty better performance is obtained with methods based on artificial intelligence, including also operations based on fuzzy sets.

Fuzzy logic control is used successfully in various aspects of management, especially where decision-making is necessary, as well as process control and optimisation. Examples of application of fuzzy logic can be cases of combining standard techniques of control and fuzzy logic in compound control systems [8]. Another kind of application of fuzzy logic can be exemplified by the selection of a manipulator based on the criteria of adaptation to the kind of object [9]. In practice one can also encounter applications of fuzzy logic control in discrete manufacturing systems. An example here can be the

© Springer International Publishing Switzerland 2015
K. Jackowski et al. (Eds.): IDEAL 2015, LNCS 9375, pp. 256–263, 2015.
DOI: 10.1007/978-3-319-24834-9_31

use of that solution in dynamic organisation of dispatching of parts after machining to various manufacturing cells [10]. Fuzzy logic is also used in issues of production scheduling, e.g. in relation to flexible manufacturing systems (FMS), where planning is made on the basis of diverse efficiency indices [11]. Another kind of application of fuzzy logic is the sphere of human resources management in the aspect of personnel competence management [12]. As can be seen from the above examples, fuzzy inference systems are a highly versatile tool that can be adapted to the solution of a wide spectrum of problems.

2   Problem Description

The objective of the study presented in this paper was the method of solving a problem consisting in ensuring the balancing of machine tool loads in a manufacturing system, taking into account certain limitations.

To realize the above objective a controller was developed, the function of which is adaptive control of the realization of orders in a manufacturing system. In particular, the above problem relates to decisions concerning the assignment of suitable machine tools for the machining of specific parts and operations so as to meet simultaneously three criteria: balancing of machine loads, minimisation of production costs, and minimisation of manufacturing cycle.

The object of the study was a discrete manufacturing system (shop) consisting of a certain number of technological machines. At certain time intervals the next part requiring machining arrived at the shop. In this situation it is first necessary to perform a selection of machine tools which, for technological-design reasons (e.g. kind of operation, dimensions of platen, required accuracy etc.), can be used to perform a specific unit operation. In this manner we obtain a set of non-dominated solutions (machine tools) from among which we need to select one which, in view of the criteria adopted (balancing of loads, minimisation of machining cost and time), is best suited for the machining of a given part.

In view of the level of complexity of the problem and its multi-criterion character, it was decided to use a controller utilising fuzzy logic in accordance with the Mamdani model [13, 14]. The primary advantage of fuzzy logic is its ease of transforming linguistic formulae into logical and mathematical relations. That feature was used during the creation of fuzzification rules taking into account all decision criteria as well as other determinants of the manufacturing system being modelled.

To solve the problem presented in this paper the method of computer simulation was applied. For this purpose a model of the manufacturing system was constructed, containing machine tools equipped with fuzzy logic controllers and a decision element that assigns production tasks to the machine tools. It was assumed that the stage of first selection of machine tools, consisting in the elimination of those machine tools that for technological reasons are not suitable for the realization of specific operations, had been already finished. Therefore a number of machine tools remained that could be used for a specific manufacturing task, differing from one another in terms of load and of the costs and times of unit operations. The model assumes that each next machined part may have different requirements with regard to technology (requires the use of different machines), but for simplification the set of non-dominated machines (that can be used in the machining of a given part) always consists of four machine tools.

3   Modelling of Manufacturing System with Fuzzy Logic Controllers

As mentioned earlier, the model of manufacturing system assumes the use of machine tools from among which we should select one that fulfils the assumed criteria the best. Each machine tool is equipped with a fuzzy logic controller whose function is to transform the input information into an output signal based on which the decision element compares the readiness of the machine tools to accept the next task. The machine tool which generates the highest value of output signal gets assigned to the realization of a given task,

Figure 1 presents the schematic of connections between the machine tool and the fuzzy logic controller. Via the input *Conn1* machine tool Machine1 receives its manufacturing tasks. Machine1 generates two kinds of information: *ServTime* (unit time of unit operation) and *Util* (utilisation). *ServTime* is transmitted to the input port of fuzzy logic controller Fuzzy1 for the calculation of the input value *Utilization ratio*[1]. Apart from that, the value of that signal is directed to the output port *Out1* where it then becomes an element of the sum of signals of that type from all machine tools in the manufacturing system. Signal *Util* from Machine1, via port *ProcTime*, reaches the fuzzy logic controller and exits the subsystem via port *Out3* to the adder, from which it returns, as a sum of unit times, via port *In2*. The quotient of signals *ProcTime* and *SubProcTime* determines the input value *Time ratio*. Output information which is then used by the decision element for pairing the manufacturing tasks with the machine tools exits the subsystem via ports *Fuzzy Out* and *Out2*.

The fuzzy logic controller is composed of three inputs and one output signal (Fig. 2).

All three inputs are real absolute numbers from the range $\langle 0;1 \rangle$. *Utilisation* is the ratio of operation time of a given machine tool to the duration of a work shift, while *utilisation ratio* is the ratio of *utilisation* of a given machine to the sum of loads of all four machine tools (1). *Time ratio* is the ratio of the operation time on a given machine to the sum of unit times of a given operation on all four machine tools (2). *Cost ratio* is the product of unit cost of a given operation on a specific machine and the unit time (3, 4).

$$R_{u_{ij}} = \frac{u_{ij}}{\sum_{i=1}^{N} u_{ij}} \tag{1}$$

where: $R_{u_{ij}}$ – utilisation ratio for process j in machine i (loading of i-th machine tool with j-th operation), u_{ij} – utilisation for process j in machine i,

---

[1] Detailed explanations of all input and output parameters of the fuzzy logic controller are given in further part of the text.

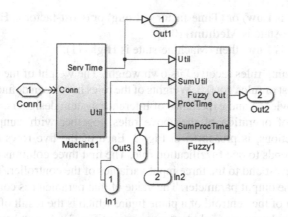

Fig. 1. Schematic of machine-controller subsystem

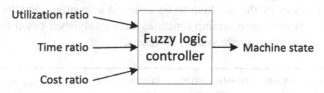

Fig. 2. Fuzzy logic controller

$$R_{t_{ij}} = \frac{t_{ij}}{\sum_{i=1}^{N} t_{ij}} \qquad (2)$$

where: $R_{t_{ij}}$ – time ratio for process j in machine i, t_{ij} – cycle time for process j in machine i,

$$R_{c_{ij}} = c_{ij} t_{ij} \qquad (3)$$

where: $R_{c_{ij}}$ – cost ratio for process j in machine i, c_{ij} – unit cost for process j in machine i,

$$c_{ij} = \frac{c_s}{T_i - k_i T_{ib}}; \ 1 \le k < 2, \ T_i \gg T_{ib} \qquad (4)$$

where: c_s – combined costs of materials and labour resulting from maintenance, overhauls and repairs (including those caused by failures), T_i – time of operation of i-th machine in adopted period (e.g. 12 months), T_{ib} – idle time of i-th machine due to failures in adopted period (e.g. 12 months), k_i – index of idle time costs of i-th machine; the more costly the idle times the higher the index value.

In the controller the following five reasoning rules were applied:

1. If (Util-factor is **Low**) or (Time-factor is **Short**) or (Cost-factor is **Low**) then (Machine-state is **High**) (0.3).
2. If (Util-factor is **High**) or (Time-factor is **Long**) or (Cost-factor is **High**) then (Machine-state is **Low**) (0.3).
3. If (Util-factor is **Medium**) or (Cost-factor is **Medium**) then (Machine-state is **Medium**) (0.3).

4. If (Util-factor is **Low**) or (Time-factor is **Long**) or (Cost-factor is **High**)
   then (Machine-state is **Medium**) (0.3).
5. If (Util-factor is **Low**) then (Machine-state is **High**) (1).

Each of reasoning rules received its own weight. The weight of the most important rule is 1 but the rest of rules got 0.3. Weights of the roles have been defined and assigned based on the knowledge and experience of inference system designers.

The manner of operation of the above rules, together with sample runs of the membership functions, is presented in Fig. 3. Each of the five rows of membership functions corresponds to one fuzzification rule. The first three columns of the membership functions correspond to the three input variables of the controller. The last, fourth, column reflects the output parameter. The value of that parameter is computed through the determination of the centroid of a plane figure which is the result of compilation of several inference rule graphs (right bottom corner of Fig. 3). As can be seen, the higher the values of the input parameter the lower the value of *Machine-state* output, and thus the lower the chances of the machine to be assigned a specific operation. The types (shapes) and numbers of membership functions were established based on the knowledge and experience of inference system designers.

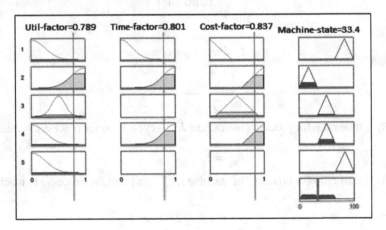

Fig. 3. Visualization of fuzzy reasoning

Figure 4 presents the surface of responses received by the *Machine-state* output of the fuzzy logic controller in relation to the input variables *Cost-factor* and *Util-factor*.

An important issue determining the correct functioning of the model is suitable selection of initial values. The model in question was constructed so that at the moment of starting the simulation all input parameters receive random assigned values from the range of (0; 1). Next, in the course of simulation, the values of the input variables are taken directly from the manufacturing system. Also noteworthy is the input variable concerning the cost aspect. The variable c_{ij} (unit cost for process j in machine i) used in formula (3) depends to a decidedly greater degree on a specific machine and to a lesser extent on the kind of operation or of the part being machined. Therefore, in the model

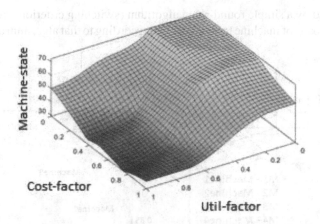

Fig. 4. Machine-state output surface of a fuzzy inference process for *cost* and *utilisation* factors

Table 1. Unit cost per process of the machine tools (Source: own work)

Machine tool	M1	M2	M3	M4
C_{ij} [EUR/s]	0.036	0.4	0.2	0.2

presented here constant values of c_{ij} were assigned for each of the machines. The unit costs for the particular machines are presented in Table 1.

Each machine included in the manufacturing system is equipped with its own fuzzy logic controller which, at one second intervals, receives input signals and generates the output signal. Thanks to this, the moment a next manufacturing tasks appears the decision element is able to select instantaneously a machine tool which is best suited for its realization. The duration of a single simulation experiment was adopted as a single work shift lasting 8 h, which corresponds to 28800 s.

To verify the effectiveness of operation of the controller, apart from the above model of manufacturing system a model of an analogous system was also created, in which the manufacturing tasks are assigned in turn to the particular machine tools. The results of simulations conducted on the basis of both models were compared, which permitted the observation of differences between them.

4   Analysis of Simulation Experiments

In the course of the study simulation experiments were conducted on the basis of two models differing from each other in the manner of assignment of manufacturing tasks to the machine tools.

The first simulation experiment was based on the use of the model without a fuzzy logic controller, due to which none of the criteria of machine tool selection applied. The sequence of assigning operations to machine tools in the manufacturing system modelled was determined by a simple round-robin algorithm (switching criterion – round-robin). Graphs of the loads of machine tools controlled according to that algorithm are presented in Fig. 5a.

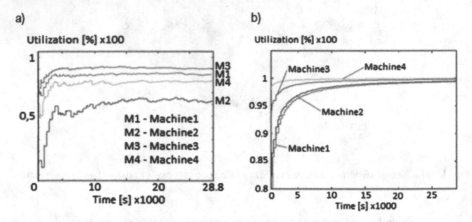

Fig. 5. Load graphs of four machine tools controlled: a) according to the rule of the first free machine (round-robin), b) with the use of fuzzy logic

As can be seen from the graphs, in the initial phase of simulation all machines differ from one another in their load levels. In the course of the simulation experiment the loads of all machine tools increase, but before 3 h elapsed (10800 s) they stabilise at various levels. The least loaded machine tool is M2, while M3 received the highest load. The difference in the loads of the two machines oscillates around 40 %. It is clearly seen that in this case the principle of equal loading of machine tools in a manufacturing system is violated.

In the second experiment the simulation was performed with the use of the model of manufacturing system with machine tools equipped with fuzzy logic controllers, thanks to which each machine generated a signal used by the decision element which assigned manufacturing tasks to specific machines. Load graphs of four machine tools controlled with the use of fuzzy logic are presented in Fig. 5b. As in the preceding experiment, at the beginning of the simulation the loads on the machines were varied. After 3 h the differences in machine loads are already very small, and at the end of the simulation the differences nearly disappear.

5   Summary

The objective of the study was to develop a fuzzy logic controller whose main function was to balance the loads of machine tools included in a manufacturing system. For that purpose two simulation models were created, the first of which reflected a manufacturing system in which manufacturing tasks were distributed among the particular machine tools by a round-robin algorithm in which the tasks were distributed sequentially, beginning again at the start in a circular manner. In the second model the machine tools included in the manufacturing system were equipped with fuzzy logic controllers, one per each machine. The function of the controllers was the transformation of input information into an output signal so as to ensure balancing of loads of the machine tools with simultaneous adherence to the criteria of minimisation of machining time and cost. The development of the models was performed using the simulation environment Matlab and the library SimEvents. The objective of the study has been attained and the results of the experiments confirmed the high effectiveness of the fuzzy logic controller.

References

1. Bzdyra, K., Banaszak, Z., Bocewicz, G.: Multiple project portfolio scheduling subject to mass customized service. In: Szewczyk, R., Zieliński, C., Kaliczyńska, M. (eds.) Progress in Automation, Robotics and Measuring Techniques. AISC, vol. 350, pp. 11–22. Springer, Heidelberg (2015)
2. Kádár, B., Terkaj, W., Sacco, M.: Semantic virtual factory supporting interoperable modelling and evaluation of production systems. CIRP Ann. Manuf. Technol. 62(1), 443–446 (2013)
3. Azadegan, A., Porobic, L., Ghazinoory, S., Samouei, P., Kheirkhah, A.S.: Fuzzy logic in manufacturing: a review of literature and a specialized application. Int. J. Prod. Econ. 132, 258–270 (2011)
4. Sitek, P., Wikarek, J.: A hybrid approach to the optimization of multiechelon systems. Math. Probl. Eng. 2015, 12 (2015)
5. Gola, A., Świć, A.: Computer-aided machine tool selection for focused flexibility manufacturing systems using economical criteria. Actual Probl. Econ. 124(10), 383–389 (2011)
6. Relich, M., Świc, A., Gola, A.: A knowledge-based approach to product concept screening. In: Omatu, S., Malluhi, Q.M., Gonzalez, S.R., Bocewicz, G., Bucciarelli, E., Giulioni, G., Iqba, F. (eds.) Distributed Computing and Artificial Intelligence, 12th International Conference. AISC, vol. 373, pp. 341–348. Springer, Heidelberg (2015)
7. Kłosowski, G., Gola, A., Świć, A.: Human resource selection for manufacturing system using petri nets. Appl. Mech. Mater. 791, 132–140 (2015)
8. Filev, D., Syed, F.: Applied intelligent systems: blending fuzzy logic with conventional control. Int. J. Gen Syst 39(4), 395–414 (2010)
9. Onut, S., Kara, S., Mert, S.: Selecting the suitable material handling equipment in the presence of vagueness. Int. J. Adv. Manuf. Technol. 44(7–8), 818–828 (2009)
10. Naumann, A., Gu, P.: Real-time part dispatching within manufacturing cells using fuzzy logic. Prod. Plann. Control 8(7), 662–669 (1997)
11. Chan, F., Chan, H., Kazerooni, A.: Real time fuzzy scheduling rules in FMS. J. Intell. Manuf. 14(3–4), 341–350 (2003)
12. Karatopa, B., Kubatb, C., Uygunb, Ö.: Talent management in manufacturing system using fuzzy logic approach. Comput. Ind. Eng. 86, 127–136 (2014)
13. Kłosowski, G.: Artificial intelligence techniques in cloud manufacturing. In: Bojanowska, A., Lipski, J., Świć, A. (eds.) Informatics methods as tools to solve industrial problems, pp. 7–19. Lublin University of Technology, Lublin (2012)
14. Nedjah, N., de Macedo Mourelle, L.: Fuzzy Systems Engineering. Springer, Heidelberg (2006)

Data-Driven Simulation Model Generation for ERP and DES Systems Integration

Damian Krenczyk[1]([⊠]) and Grzegorz Bocewicz[2]

[1] Faculty of Mechanical Engineering,
Silesian University of Technology, Gliwice, Poland
damian.krenczyk@polsl.pl
[2] Faculty of Electronics and Computer Science,
Koszalin University of Technology, Koszalin, Poland
bocewicz@ie.tu.koszalin.pl

Abstract. In the paper the concept of data driven automatic simulation model generation method based on hybrid parametric-approach and data mapping and transformation methods in combination with concept of neutral data model is presented. As a key element of the proposed approach, author's own method of data transformation into internal programming languages script code, based on the transformation template is described. Developed simulation model generator is also an effective tool for the integration of ERP and DES systems. A practical implementation of the presented methodology - original software RapidSim is presented as well.

Keywords: ERP · Data-driven · Simulation · Visualization · Data mapping · Data transformation · Automatic model generation

1 Introduction

Today's organizations operate in a very dynamic and complex environment, related among others to the globalization of the market, constant changes of customer requirements and fierce competition. These factors, especially the market globalization, force manufacturers to adapt quickly to new circumstances. Enterprises must be flexible enough so that it can effectively take action to adapt their production systems to produce many variants of products in varying production batches. This, in turn, forces the necessity of investing in an increasingly complex and innovative technologies and more efficient methods of production planning and control [1–5]. Increasing the efficiency of the production planning and scheduling process, which in today's enterprises is one of the main areas where on a large scale computer support applications are used, is made possible by the integration of production planning and control systems. Enterprise Resource Planning (ERP) are the most popular systems of resources planning and management and have crucial impact on the production and business activities of enterprise [6]. Another area of computer aided methods, allowing the increase of efficiency of production planning processes, is computer modelling, simulation and visualization of the production processes flow. The use of computer discrete-event simulation systems (DES), has extensive support capabilities in production planning,

© Springer International Publishing Switzerland 2015
K. Jackowski et al. (Eds.): IDEAL 2015, LNCS 9375, pp. 264–272, 2015.
DOI: 10.1007/978-3-319-24834-9_32

such as verification of the feasibility of production for prepared plans, reducing the workload on the selected schedule, increasing the accuracy of the timing of production orders, transport and storage resources verification, visualization of production processes flow, etc. Nowadays more and more often it is needed to apply DES to assist in solving the problems relating to production planning at the operational level and order management in enterprises [3, 7]. In addition, the need to integrate DES systems with current MRP/MRP II, ERP or PPC (Production Planning and Control) systems has been recognized as an important problem associated with decision support area, inter alia, by [3, 7, 8]. The authors point out the limitations that are caused by the use of MRP/ERP by enterprise in resource planning processes. The most frequently mentioned disadvantages are static models with unlimited storage resources used in these systems and insufficient flexibility.

However, the use of the DES systems is mainly limited to the tactical and strategic levels of production planning. That state results from a relatively large workload and costs associated with simulation model preparation as a highly specialized activity. It is estimated [9] that 30–40 % of the time, associated with conducting of simulation project, is taken by collecting the necessary data, 25–35 % of the time is spent on preparing and testing the simulation model. Carrying out the experiment and analysis of output data represents only 20 % of the entire duration of the project. Increasing interest in the use of computer simulation in supporting production planning at different levels was the cause of the search for effective methods of creating models for DES systems, allowing the elimination of workload and high costs of this process [10].

The main purpose of this study was to develop the concept of the data driven automatic simulation model generation method for DES systems with 3D visualization of processes flow. Developed model generator is also a tool for an integration of ERP and DES systems module. In this paper the concept of integration using data mapping and data transformation methods with the intermediate neutral data model, which is also the effective tool for integration between the systems of production planning and simulation systems is proposed. The following sections present data-driven simulation model generation method, an example of a practical realization of this concept and finally conclusion of the research and directions for future work in presented area.

2 Data-Driven Simulation Model Generation

The concept of data-driven simulation model generation is defined as a method of creating simulation models by users without skill to create the simulation models (or having it at minimal level) and without any programming skills [11–14]. The goal of such approach is set to generate simulation models automatically and directly from the available data sources using algorithms of data analysis and structuring for simultaneous creating and configuring the simulation model. The simulation model is not created manually using the simulation software features and tools, it is rather generated from external data sources using interfaces of the simulator and algorithms for creating the model [13, 14]. This means that the planners/engineers using the systems supporting production planning and having access to simulation systems should be able to prepare simulation models and carry out experiments only by changing or by indication of input data that automatically updates the simulation model.

In particular, a large group of scientific publications on the data-driven generation of simulation models represent the work on generic simulation models. Pidd [15] defines a generic models as models designed for use in the simulation of production systems structurally similar - and this applies both to the production resources, as well as a method for the production flow. This concept implies the possibility of: obtaining data from various systems supporting the enterprise production management, transforming obtained data using modeling languages, data representation, and then automatically generating models for dedicated simulation systems. According to [13, 14] automatic model generation approaches can be classified into three main categories:

(1) Parametric approaches: Models are generated based on existing simulation building blocks (atoms) stored in simulation libraries, which are selected and configured based on parameters obtained from various sources.
(2) Structural approaches: model generation is based on data describing the structure of a system (typically in the form of factory layout data).
(3) Hybrid-knowledge-based approaches: combine AI methods (expert systems, neuronal nets) with two of the above approaches.

This paper present a data-driven approach for automatic model generation based on hybrid parametric-approach, data mapping and transformation methods in combination with concept of neutral data model that makes it possible to increase the versatility of the proposed approach compared to the presently used.

3   ERP and DES Systems Integration

The presented ERP and DES systems integration methodology is intended to enable the use of data obtained from conducted simulation experiments to support the planning area and is designed to form a verification tool. Implementation of the integration process requires the identification of sources and forms of data from ERP systems, and the transformation method into formats (program code) recognized by simulation systems. The achievement of this objective is accomplished in four main stages. The first stage is focused on the acquisition of complete information required for the preparation of functional simulation model. These data, derived from planning support systems (actually their representation), will be saved with the use of a neutral format. It is therefore necessary to develop the definition of structures in neutral data representation formats, using universal formats, independent of the source of data structures as well as hardware and software platforms. In the next step, data exchange processes are performed between the source representation of the data obtained in the first stage and the intermediate neutral data model. In this step data mapping and data transformation methods are used, also it is necessary to develop definition of intermediate data model. In the third step the intermediate model data is transformed using the author's own method of data transformation into program code created in the DES system internal programming languages.

Therefore, it became necessary both to provide a method of generating the script code in DES systems internal programming languages and to choose methods for data transformation. The result of the implementation of this phase are script code fragments, creating specific parts of the simulation model for the particular condition and for the specific structure of the production system. The last, fourth stage is a combination of scripting code of the third stage in one document, containing code responsible for creating resources that make up the manufacturing system, i.e.: machines, inter-operational buffers, warehouses, input and output buffers, elements that generate products that are to be produced in the system and information resources, i.e.: tables containing data about the setup and cycle times of operations on production resources, data on processes routes as a function of carrying out the connection between appropriate atoms in the model to perform simulation experiments. The simulation and visualization results will be useful for the analysis of the production flow, which include, among others, resource utilization, storage capacity optimization, completion date of production orders verification and to manage the quality of the production system (working without deadlocks and starvations) (Fig. 1). In the next chapter data transformation into internal programming languages code according to author's method is presented as a key element of the proposed approach.

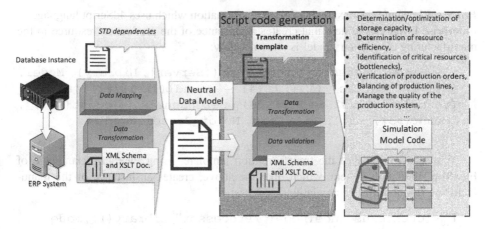

Fig. 1. ERP and DES integration

3.1 Transformation Template

Generating a script code in the internal simulation system language is a sequential process and must be performed in the order of the individual layers to define a simulation model. It is not possible, for example, to generate the parameterization code of the cycle-times for manufacturing resources without previous generating the code to create the appropriate amount of production resources resulting from the input data. Data transformation process to the internal code of the simulation system programming language is implemented based on the defined transformation template - templates containing code with a specific location of the input data (either directly or using a transformation function). The developed functions correspond to individual layers of code of creation of a simulation model.

The transformation template is defined as a three [16, 17]:

$$T_i^S < [tmp(\#1\ldots\#n)]; A_j; m >, \qquad (1)$$

where:

- T_i^S – *i-th* number of the transformation template; this number determines the location of the transformation in the sequence,
- *tmp(#1…#n)* – the template which is a text string containing the lines of the code of internal simulation system programming language with tags *#1…#n*, specifying the place of entry to a data template; tags can be used to implement arithmetical operations on the data, which correspond to (e.g. *#2-#1, #2*#11*),
- A_j – data corresponding to the *j-th* tags in the template, $j = 1 \ldots n$,
- *m* – the number of repetitions in the template code.

The transformations templates are defined at the methodology implementation stage, because the templates must necessarily correspond with syntax of the internal programming language of the chosen simulation system.

Practical Example. An example of the transformation which uses 4dScript language of Enterprise Dynamics system that creates an instance of the production resource in the manufacturing simulation model is shown below:

```
T10S<[Sets(CreateAtom(AtomByName([Server], Library), Model,
[M#1])), setloc(_(4 + 7*(#1 - round((#1-1)/#2))*#2)_, _
(round((#1-1)/#2 + 1)*3)_,0), SetChannels (_(count(#3))_,
(count(#3))_,s)], Id, y_pos, No_of_Processes, 1>
```

While an example of the transformation template, using FlexScript language of FlexSim simulation system, which, like the above, creates an instance of the manufacturing resource in the simulation model is:

```
T10S<[createinstance(node("/Processor",library()),model
()); setname(last(model()),"M#1"); setloc(last(model()),_
(4 + 7*(#1 - round((#1-1)/#2))*#2)_, _(round((#1-1)/#2 + 1)
*3)_,0);], Id, y_pos, No_of_Processes, 1>
```

For prepared transformation template the resulting code in selected by the user data transformation language: XSLT, SQL, XQuery, Java may then be generated. Generated XSLT code for the above example (for Enterprise Dynamics) is as follows:

```
<xsl:for-each select="Production_System/Resources/Resource">
Sets(CreateAtom(AtomByName([Server],   Library),   Model,
[M<xsl:  value-of  select="Id"/>])),   setloc(<xsl:value-of
select="4 + 7 * (Id - ((round((Id - 1) div $rzad)) * $rzad))"/>,
```

```
<xsl: value-of select="(round((Id - 1) div $rzad) + 1) *
3"/>,0),
SetChannels (<xsl:value-of select="count(/Production_
System/Processes/Process)"/>, <xsl:value-of select="count
(/Production_System/Processes/Process)"/>,s),</xsl:for-
each>
```

For the other simulation systems creation of transformation templates is carried out in a similar manner, and its content is dependent on the adopted neutral data model.

Simulation Model Objects Code Generation. By using data, describing the production system and manufacturing processes, stored in the intermediate data model and above transformations templates and generated on the basis of their XSLT code, program code containing instructions for creating a fully functional simulation model can be generated. For this purpose, it can be used on any XSLT processor software, e.g. XMLSpy, Sablotron for C++, XSLT for PHP, etc. The following is a fragment of the generated 4DScript code for creating objects in the simulation model:

```
Sets(CreateAtom(AtomByName([Server], Library), Model,
[M2])),
setloc(18,3,0), SetChannels (5,5,s),
SetEprAtt(1,[czasy(Value(StripString(Name(First(c)),
[P])),Value(StripString(Name(c),[M])))],s),
SetExprAtt(18,[openic(lr(Value(StripString(Name(c),
[M])),(mod (input(c),lr(Value(StripString(Name(c),[M])),
299)) + 1)),c)],s)
```

The full versions of the models and transformation XSLT documents described in this paper, can be found on the website: imms.home.pl/rapidsim.

4   Practical Implementation

The practical implementation of the data driven automatic simulation model generation methodology is conducted in an original software - RapidSim. A schematic flow diagram of the RapidSim has been shown in Fig. 2 [16, 17]. Due to the fact that the

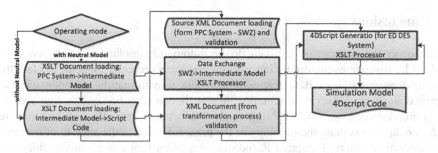

Fig. 2. The algorithms of the RapidSim [17]

data transformation (XSLT documents in RapidSim) can be developed independently, the software can be used to integrate any ERP/DES systems.

Prepared software is a universal tool and can be commercialized in any manufacturing company. The RapidSim includes the following modules: graphical user interface, XML validation and parsing module, XSLT processor module, code of simulation model script generator and module of the automatic parameterization. Depending on whether the data derived from the ERP system is available in accordance with the neural model (NM) or not, the mode of operation is selected and XML document is loaded. In the following steps (only in NM mode) the process of data exchange and data transformation between the data source (originating from ERP system) and the target (neutral intermediate data model) is executed, implemented in XSLT Processor.

Finally, using the XSLT Processor, the code of the simulation model in internal programming language of simulation system is generated. Output code (instructions for creating complete simulation model - including information about the manufacturing system layout, routing of manufacturing processes with setup and cycle times, inter-resources buffers capacity and information about the control rules for each manufacturing resource and processes) is loaded into the DES system to perform simulation experiments. By loading the code directly into the simulation system, the model is being automatically created and ready to carry out simulation experiments (Fig. 3).

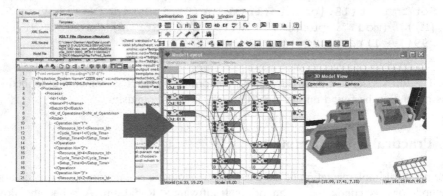

Fig. 3. Simulation model

5   Conclusions

In the paper the concept of the data driven automatic simulation model generation method for discrete-event simulation systems has been presented. Due to the fact that presented approach uses data mapping and data transformation methods with the intermediate neutral data representation model, it can also be an effective tool for integration between the PPC/ERP and simulation systems, regardless of the structures of the production system, the topology of the processes flow or the amount of resources and production orders. Presented RapidSim integration tool can be successfully used with most computer simulation systems, which are characterized by openness. Further research on the presented methodology addresses issues related to the area of the development of the concept of virtual enterprises and dynamic manufacturing networks, by extending the data model with objects related to logistics subsystems.

References

1. Bzdyra, K., Banaszak, Z., Bocewicz, G.: Multiple project portfolio scheduling subject to mass customized service. In: Szewczyk, R., Zieliński, C., Kaliczyńska, M. (eds.) Progress in Automation, Robotics and Measuring Techniques. AISC, vol. 350, pp. 11–22. Springer, Heidelberg (2015)
2. Diering, M., Dyczkowski, K., Hamrol, A.: New method for assessment of raters agreement based on fuzzy similarity. In: Herrero, A., Sedano, J., Baruque, B., Quintián, H., Corchado, E. (eds.) SOCO 2015. ASIC, vol. 368, pp. 415–425. Springer, Heidelberg (2015)
3. Krenczyk, D., Skolud, B.: Transient states of cyclic production planning and control. Appl. Mech. Mater. **657**, 961–965 (2014)
4. Sitek, P., Wikarek J.: A hybrid approach to the optimization of multiechelon systems. Math. Probl. Eng. **2015**, 12, , Article ID 925675 (2015)
5. Wójcik, R., Bzdyra, K., Crisostomo, M.M., Banaszak, Z.: Constraint programming approach to design of deadlock-free schedules in concurrent production systems. In: Proceedings of 10th IEEE International Conference on Emerging Technologies and Factory Automation, ETFA 2005, vol. 1, pp. 135–142 (2005)
6. Wang, C., Liu, X.-B.: Integrated production planning and control: a multi-objective optimization model. J. Ind. Eng. Manage. **6**(4), 815–830 (2013)
7. Lee, S., Son, Y.-J., Wysk, R.A.: Simulation-based planning and control: from shop floor to top floor. J. Manuf. Syst. **26**(2), 85–98 (2007)
8. Heilala, J., et al.: Developing simulation-based decision support systems for customer-driven manufacturing operation planning. In: Proceedings of the 2010 WSC, pp. 3363–3375 (2010)
9. Nordgren, W.B.: Steps for proper simulation project management. In: Proceedings of the 1995 Winter Simulation Conference, pp. 68–73 (1995)
10. Fowler, J.W., Rose, O.: Grand challenges in modeling and simulation of complex manufacturing systems. SIMULATION **80**(9), 469–476 (2004)
11. Chlebus, E., Burduk, A., Kowalski, A.: Concept of a data exchange agent system for automatic construction of simulation models of manufacturing processes. In: Corchado, E., Kurzyński, M., Woźniak, M. (eds.) HAIS 2011, Part II. LNCS, vol. 6679, pp. 381–388. Springer, Heidelberg (2011)
12. Wang, J., et al.: Data driven production modeling and simulation of complex automobile general assembly plant. Comput. Ind. **62**(7), 765–775 (2011)
13. Bergmann, S., Strassburger, S.: Challenges for the automatic generation of simulation models for production systems. In: Proceedings of the 2010 Summer Computer Simulation Conference, SCSC 2010, Ottawa, Canada, pp. 545–549 (2010)
14. Huang, Y., Seck, M.D., Verbraeck, A.: From data to simulation models: component-based model generation with a data-driven approach. In: Proceedings of the Winter Simulation Conference, WSC 2011, pp. 3724–3734 (2011)
15. Pidd, M.: Guidelines for the design of data driven generic simulators for specific domains. Simulation **59**(4), 237–243 (1992)

16. Krenczyk, D., Skolud, B.: Production preparation and order verification systems integration using method based on data transformation and data mapping. In: Corchado, E., Kurzyński, M., Woźniak, M. (eds.) HAIS 2011, Part II. LNCS, vol. 6679, pp. 397–404. Springer, Heidelberg (2011)

17. Krenczyk, D., Zemczak, M.: Practical example of the integration of planning and simulation systems using the RapidSim software. Adv. Mater. Res. **1036**, 1662–8985 (2014)

Improving the NSGA-II Performance
with an External Population

Krzysztof Michalak[✉]

Department of Information Technologies, Institute of Business Informatics,
Wroclaw University of Economics, Wroclaw, Poland
krzysztof.michalak@ue.wroc.pl

Abstract. The NSGA-II algorithm is among the best performing ones
in the area of multiobjective optimization. The classic version of this
algorithm does not utilize any external population. In this work several
techniques of reintroducing specimens from the external population back
to the main one are proposed. These techniques were tested on multi-
objective optimization problems named ZDT-1, ZDT-2, ZDT-3, ZDT-4
and ZDT-6. Algorithm performance was evaluated with the hypervolume measure commonly used in the literature. Experiments show that
reintroducing specimens from the external population improves the performance of the algorithm.

Keywords: Multiobjective optimization · NSGA-II · External population

1 Introduction

This paper deals with multiobjective optimization problems. Such problems can
be formalized as follows:

$$\text{minimize } F(x) = (f_1(x), \dots, f_m(x))^T \\ \text{subject to } x \in \Omega, \tag{1}$$

where:

Ω - decision space,
f_i - i-th optimization criterion.

Most often there exists no single $x_0 \in \Omega$ which minimizes all $f_i, i \in \{1, \dots, m\}$
simultaneously, so there is no possibility of simply choosing "the best" solution.
However, solutions can be compared to each other using Pareto domination
relation. Given two points x_1 and x_2 in the decision space Ω for which:

$$F(x_1) = (f_1(x_1), \dots, f_m(x_1))^T \\ F(x_2) = (f_1(x_2), \dots, f_m(x_2))^T \tag{2}$$

we say that x_1 *dominates* x_2 $(x_1 \succ x_2)$ iff:

$$\forall i \in 1, \dots, m : f_i(x_1) \le f_i(x_2) \\ \exists i \in 1, \dots, m : f_i(x_1) < f_i(x_2) \tag{3}$$

© Springer International Publishing Switzerland 2015
K. Jackowski et al. (Eds.): IDEAL 2015, LNCS 9375, pp. 273–280, 2015.
DOI: 10.1007/978-3-319-24834-9_33

A solution x is said to be *nondominated* iff:

$$\neg \exists x' \in \Omega : x' \succ x. \tag{4}$$

Solving a multiobjective optimization problem usually means finding a *Pareto set* of nondominated solutions or a *Pareto front* of points in the objective space R^m which correspond to solutions from the Pareto set.

There are many optimization techniques that can be used to find Pareto fronts. Evolutionary methods are particularly well-suited for this task because they work with an entire population of solutions among which the Pareto domination relation can be established. One of the most successful multiobjective evolutionary algorithms known in the literature is the NSGA-II algorithm [2], which uses Pareto domination relation to induce ordering among specimens, which in turn is used for selecting winners in a binary tournament procedure.

In the original version, the NSGA-II algorithm does not utilize any external population. However, external populations have been used to improve performance in other algorithms such as SPEA [11] and SPEA-2 [10]. In this paper NSGA-II-EXT algorithm - a modified version of the NSGA-II algorithm - is proposed which stores all the nondominated solutions found so far in an external population. The proposed algorithm includes an additional step in which some or all specimens from the external population are reintroduced to the main population. The NSGA-II-EXT algorithm was tested with various external population merging techniques and compared to the regular NSGA-II algorithm. A comparison with the SPEA-2 algorithm was also performed because the SPEA-2 algorithm also involves merging of the external population with the main population. In experiments performed by other authors, the NSGA-II algorithm was tested on ZDT problems and it was found to outperform PAES and SPEA algorithms [2]. Therefore, these two algorithms have not been included in the tests.

2   External Population Merging

External population is used in some multiobjective genetic algorithms such as SPEA [11] and SPEA-2 [10]. Also, a similar concept of an archive which stores all nondominated specimens found so far by the algorithm is often used to generate a comprehensive overview of possible solutions of the solved problem.

In this paper NSGA-II-EXT - a modified version of the NSGA-II algorithm is proposed in which, after each generation, the current population P is appended to the external population EP and then some specimens from the EP can be reintroduced to the original population P. In the proposed algorithm the external population EP is at the same time involved in the working of the algorithm and plays the role of an archive which stores the nondominated solutions encountered during the entire run of the algorithm.

An overview of the modified algorithm is given in Algorithm 1.

The following procedures are used in the proposed algorithm:

Evaluate - calculates values of the objective functions for specimens from a given population,

Algorithm 1. The NSGA-II-EXT algorithm.

$P = \text{InitPopulation}()$
$EP = \emptyset$
$\text{Evaluate}(P)$
$\text{Rank}(P)$
for $t = 2 \rightarrow N_{Gen}$ **do**
    $P_{parent} = \text{BinaryTournament}(P)$
    $P_{offspring} = \text{GenerateOffspring}(P_{parent})$
    $\text{Evaluate}(P_{offspring})$
    $P = P \cup P_{offspring}$

    $EP = \text{AddToEP}(EP, P)$
    $P = \text{MergeEP}(P, EP)$
    $P = \text{Reduce}(P)$
end for

Rank - assigns nondominated front ranks to specimens based on the domination relation and calculates the crowding distance in each front separately,

BinaryTournament - selects parents for crossover and mutation using a binary tournament procedure based on the nondominated front rank and the crowding distance - the same procedure as in the NSGA-II algorithm,

GenerateOffspring - generates offspring from parents, most often using crossover and mutation operators. In some cases other operators may be used. For example the inver-over operator is often applied in the case of the Travelling Salesman Problem [6]. In the experiments presented in this paper the SBX crossover operator [8] and the polynomial mutation [3] were used because these two operators are dedicated to real-domain problems.

AddToEP - adds the current population P to the external population EP. In this procedure specimens from EP dominated by specimens in P are removed and nondominated specimens from P are added to EP.

MergeEP - reintroduces specimens from the external population EP back to the main population P. Several different reintroduction procedures were tested in this paper:

- **none** - no EP merging. This is equivalent to the regular NSGA-II algorithm,
- **all** - all specimens from EP are added to P,
- **rand** - $q\%$ of specimens from EP selected at random are added to P,
- **cd** - crowding distance (CD) is calculated for specimens in $P \cup EP$, then $q\%$ of specimens from EP with the largest CD values are added to P.

Reduce - performs elitist population reduction used in the NSGA-II algorithm.

Preliminary tests were also performed using a technique based on the Average Linkage clustering method used in the SPEA algorithm [11]. However, due to slow, incremental nature of this clustering procedure this technique resulted in a very long running time and has not been, therefore, extensively tested.

3   Experiments and Results

In the experiments the NSGA-II-EXT algorithm was tested with four external population merging methods: **all**, **cd**, **rand** and **none** (which makes the NSGA-II-EXT algorithm equivalent to the regular NSGA-II). The fraction $q\%$ of EP reintroduced into the main population in the **cd** and **rand** methods was set to 75 %. The proposed value of this parameter has been set so as to differentiate the **cd** and **rand** versions of the algorithm from the **all** version. The results presented in this paper suggest that reintroducing the entire population is the most beneficial, so high values of the $q\%$ parameter should work best. However, a thorough examination of the influence of the $q\%$ parameter on the working of the algorithm has been left as a further work.

The experiments were done on several well known data sets used in the literature for testing multiobjective genetic algorithms, named ZDT-1, ZDT-2, ZDT-3, ZDT-4, ZDT-6 [9,12]. The ZDT test suite also contains the ZDT-5 problem, but this problem was not used for tests in this paper because it uses a different encoding (binary strings as opposed to real-value encoding used in other ZDT suite test problems). Binary encoding can be handled by both the original NSGA-II algorithm and by the version proposed in this paper. However, binary-encoded problems require different operators than real-valued problems. The work on the former class of optimization problems has been left out for another paper.

For each data set and each external population merging scheme 30 iterations of the test were performed. Following [5] the probabilities of crossover and mutation were set to $P_{cross} = 0.9$ and $P_{mut} = 0.033$. Based on parameter settings in [9] the population size was set to $N_{pop} = 100$. The number of generations for all tests was set to $N_{gen} = 250$. Combined with population size N_{pop} this resulted in a total of $N_{eval} = 25000$ objective function evaluations. Following both [2] and [9] the distribution indexes for both the SBX crossover operator and polynomial mutation were set to $\eta_{cross} = 20$ and $\eta_{mut} = 20$.

During the tests the contents of the external population EP was stored after each generation. The quality of results was assessed using the hypervolume measure [4,13] which is often used for comparing the performance of multi-objective evolutionary algorithms. The hypervolume is the Lebesgue measure of the portion of the objective space that is dominated by a set of solutions collectively. In two and three dimensions the hypervolume corresponds to the area and volume respectively. To calculate the hypervolume for a given set of solutions the Hypervolume by Slicing Objectives (HSO) algorithm [7] was used. In order to be able to plot hypervolume changes in time a single nadir point was calculated for each test problem from all generations and test runs of all algorithms. Using this common nadir point as a reference, hypervolume values were calculated for each generation in each test run. Plots of hypervolume values against the generation number for all test problems are presented in Fig. 1. These graphs show hypervolume values starting from the half of the number of generations because quick changes of the hypervolume in the first half of each run extend the scale and make the graph too compressed to be readable. It can be seen in the graphs that in most cases the new algorithm converges faster than the NSGA-II and SPEA2 algorithmns, which could not achieve equally good results in the given 250 generations. Only in the case of the ZDT-4 problem all the algortihms attained an approximately the same hypervolume in 250 generations, but the NSGA-II-EXT worked faster than NSGA-II and SPEA2.

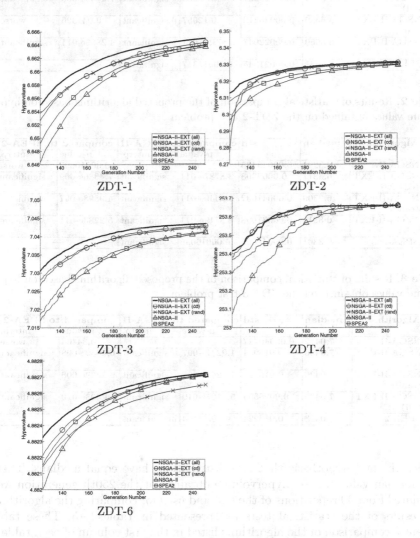

ZDT-1   ZDT-2

ZDT-3   ZDT-4

ZDT-6

Fig. 1. The hypervolume plotted against the generation number.

The performance of different variants of the NSGA-II-EXT algorithm was verified with a statistical test. Since hypervolume values obtained in the tests are not normally distributed a Kruskall-Wallis test [1] was performed which does not make an assumption about distribution normality. The Kruskall-Wallis test

Table 1. Results of statistical comparison of the proposed algorithms based on hypervolume values obtained on the ZDT-1 test problem

Algorithm	median	std	compared to NSGA-II		compared to SPEA-2	
			p-value	interp.	p-value	interp.
NSGA-II	6.6637	0.00027870	—		0.00017317	worse
NSGA-II-EXT (cd)	6.6644	0.00025231	2.7927e-009	significant	7.4755e-006	significant
NSGA-II-EXT (rand)	6.6639	0.00033312	0.12057	insignificant	0.041325	worse
NSGA-II-EXT (all)	6.6650	0.00022047	2.8719e-011	significant	4.2855e-011	significant
SPEA-2	6.6641	0.00021345	0.00017317	significant	—	

Table 2. Results of statistical comparison of the proposed algorithms based on hypervolume values obtained on the ZDT-2 test problem

Algorithm	median	std	compared to NSGA-II		compared to SPEA-2	
			p-value	interp.	p-value	interp.
NSGA-II	6.3298	0.00049425	—		0.00010095	worse
NSGA-II-EXT (cd)	6.3311	0.0004308	3.8787e-011	significant	7.4386e-009	significant
NSGA-II-EXT (rand)	6.3309	0.00041632	9.4449e-011	significant	4.9898e-007	significant
NSGA-II-EXT (all)	6.3316	0.00031698	2.8719e-011	significant	5.2283e-011	significant
SPEA-2	6.3301	0.00041895	0.00010095	significant	—	

Table 3. Results of statistical comparison of the proposed algorithms based on hypervolume values obtained on the ZDT-3 test problem

Algorithm	median	std	compared to NSGA-II		compared to SPEA-2	
			p-value	interp.	p-value	interp.
NSGA-II	7.0412	0.00040727	—		0.34404	worse
NSGA-II-EXT (cd)	7.0430	0.015934	1.0241e-007	significant	6.2639e-008	significant
NSGA-II-EXT (rand)	7.0426	0.015917	1.0241e-007	significant	6.2639e-008	significant
NSGA-II-EXT (all)	7.0435	0.0095554	5.3167e-010	significant	4.403e-010	significant
SPEA-2	7.0413	0.013224	0.34404	insignificant	—	

verifies the null hypothesis that two distributions have equal medians. Therefore, median values of the hypervolume obtained at the 250th generation were calculated from 30 repetitions of the test and used for comparing the algorithms.

Results of the statistical tests are presented in Tables 1−5. These tables present a comparison of the algorithms listed in the first column of each table to the NSGA-II and the SPEA-2 algorithms. P-values given in the tables are calculated for the null hypothesis that hypervolume distributions obtained using the compared algorithms have equal medians. The interpretation of the numerical results was determined as follows. If the median of the hypervolume distribution obtained using the algorithm listed in the left column is lower than the median obtained using the NSGA-II (or the SPEA-2 respectively) then the interpretation is "worse" regardless of the p-value. If the median of the hypervolume

Table 4. Results of statistical comparison of the proposed algorithms based on hyper-volume values obtained on the ZDT-4 test problem

Algorithm	median	std	compared to NSGA-II		compared to SPEA-2	
			p-value	interp.	p-value	interp.
NSGA-II	6.6613	0.97866	—		0.30071	insignificant
NSGA-II-EXT (cd)	6.6614	1.5487	0.56421	insignificant	0.086346	insignificant
NSGA-II-EXT (rand)	6.6618	0.031004	0.28711	insignificant	0.017299	significant
NSGA-II-EXT (all)	6.6613	0.031059	0.47792	worse	0.028663	significant
SPEA-2	6.6590	0.037741	0.30071	worse	—	

Table 5. Results of statistical comparison of the proposed algorithms based on hyper-volume values obtained on the ZDT-6 test problem

Algorithm	median	std	compared to NSGA-II		compared to SPEA-2	
			p-value	interp.	p-value	interp.
NSGA-II	4.8827	1.0912e-005	—		0.0025611	worse
NSGA-II-EXT (cd)	4.8827	9.9921e-006	0.58436	insignificant	0.0084977	worse
NSGA-II-EXT (rand)	4.8827	1.5931e-005	3.5098e-011	worse	2.8719e-011	worse
NSGA-II-EXT (all)	4.8827	8.4028e-006	2.2902e-008	significant	0.0010302	significant
SPEA-2	4.8827	9.4905e-006	0.0025611	significant	—	

distribution obtained using the algorithm listed in the left column is not lower than the median obtained using the NSGA-II (or SPEA-2 respectively) then the interpretation is "significant" if the p-value does not exceed 0.05 or "insignificant" otherwise. If the result obtained by the algorithms listed in the first column is interpreted as significantly better that the result obtained by the NSGA-II or SPEA-2 the corresponding fields in the table are shaded in gray.

4   Conclusions

In this paper the NSGA-II-EXT algorithm - a modified version of the NSGA-II algorithm was proposed which stores all nondominated solutions found so far in an external population EP. Specimens from the external population are reintroduced to the main population at the end of each generation. Several methods of such merging were tested: no merging at all, selection of a fraction of specimens from EP with the highest value of crowding distance, random selection of a fraction of specimens from EP and merging of the entire external population EP with the main population.

The performance of the NSGA-II-EXT algorithm with different merging schemes, the original NSGA-II algorithm and the SPEA-2 algorithm was compared using test problems ZDT-1, ZDT-2, ZDT-3, ZDT-4 and ZDT-6. The comparison was done using hypervolume measure commonly used in the literature. The NSGA-II-EXT (all) algorithm (the variant with merging of the entire external population) is among the best performing algorithms tested in this paper.

For all tests problems except ZDT-4 the NSGA-II-EXT (all) outperformed the NSGA-II algorithm in terms of hypervolume. The SPEA-2 algorithm was outperformed by NSGA-II-EXT (all) in all the tests. Both results were verified using the Kruskall-Wallis test to be statistically significant.

Overall, the NSGA-II-EXT (all) variant of the algorithm proposed in this paper seems to be able to generate many new solutions of the tested multiobjective problems and to produce Pareto fronts with good hypervolume values in the case of most of the problems tested in this paper, compared to algorithms known from the literature and compared to other variants of the NSGA-II-EXT algorithm based on different external population recombination techniques.

References

1. Corder, G.W., Foreman, D.I.: Nonparametric Statistics for Non-Statisticians: A Step-by-Step Approach. Wiley, Hoboken (2009)
2. Deb, K., Pratap, A., Agarwal, S., Meyarivan, T.: A fast and elitist multiobjective genetic algorithm: NSGA-II. IEEE Trans. Evol. Comp. **6**, 182–197 (2002)
3. Hamdan, M.: On the disruption-level of polynomial mutation for evolutionary multi-objective optimisation algorithms. Comput. Inf. **29**(5), 783–800 (2010)
4. Purshouse, R.: On the evolutionary optimisation of many objectives. Ph. D. thesis, The University of Sheffield, UK (2003)
5. Sharma, D., Kumar, A., Deb, K., Sindhya, K.: Hybridization of SBX based NSGA-II and sequential quadratic programming for solving multi-objective optimization problems. In: IEEE Congress on Evolutionary Computation, pp. 3003–3010. IEEE (2007)
6. Tao, G., Michalewicz, Z.: Inver-over operator for the TSP. In: Eiben, A.E., Bäck, T., Schoenauer, M., Schwefel, H.-P. (eds.) PPSN 1998. LNCS, vol. 1498, pp. 803–812. Springer, Heidelberg (1998)
7. While, L., Hingston, P., Barone, L., Huband, S.: A faster algorithm for calculating hypervolume. IEEE Trans. Evol. Comp. **10**(1), 29–38 (2006)
8. Yang, L., Yang, C., Liu, Y.: Particle swarm optimization with simulated binary crossover. In: 2014 Fifth International Conference on Intelligent Systems Design and Engineering Applications (ISDEA), pp. 710–713 (2014)
9. Zhang, Q., Li, H.: MOEA/D: a multiobjective evolutionary algorithm based on decomposition. IEEE Trans. Evol. Comp. **11**(6), 712–731 (2007)
10. Zitzler, E., Laumanns, M., Thiele, L.: SPEA2: Improving the strength pareto evolutionary algorithm for multiobjective optimization. In: Giannakoglou, K., et al. (eds.) Evolutionary Methods for Design, Optimisation and Control with Application to Industrial Problems (EUROGEN 2001), pp. 95–100. International Center for Numerical Methods in Engineering (CIMNE) (2002)
11. Zitzler, E., Thiele, L.: Multiobjective evolutionary algorithms: a comparative case study and the strength pareto approach. IEEE Trans. Evol. Comp. **3**(4), 257–271 (1999)
12. Zitzler, E., Deb, K., Thiele, L.: Comparison of multiobjective evolutionary algorithms: empirical results. Evol. Comput. **8**(2), 173–195 (2000)
13. Zitzler, E., Thiele, L., Laumanns, M., Fonseca, C.M., da Fonseca, V.G.: Performance assessment of multiobjective optimizers: an analysis and review. IEEE Trans. Evol. Comp. **7**, 117–132 (2002)

Local Search Based on a Local Utopia Point for the Multiobjective Travelling Salesman Problem

Krzysztof Michalak[✉]

Department of Information Technologies, Institute of Business Informatics,
Wroclaw University of Economics, Wroclaw, Poland
krzysztof.michalak@ue.wroc.pl

Abstract. Performing a local search around solutions found by an evolutionary algorithm is a common practice. Local search is well known to significantly improve the solutions, in particular in the case of combinatorial problems. In this paper a new local search procedure is proposed that uses a locally established utopia point. In the tests in which several instances of the Travelling Salesman Problem (TSP) were solved using an evolutionary algorithm the proposed local search procedure outperformed a local search procedure based on Pareto dominance. Because the local search is focused on improving individual solutions and the multi-objective evolutionary algorithm can improve diversity, various strategies of sharing computational resources between the evolutionary algorithm and the local search are used in this paper. The results attained by the tested methods are compared with respect to computation time, which allows a fair comparison between strategies that distribute computational resources between the evolutionary optimization and the local search in various proportions.

Keywords: Multiobjective optimization · Combinatorial optimization · Travelling salesman problem · Local search

1 Introduction

Combinatorial problems arise in many practical applications what motivates the need for solving them efficiently. Unfortunately, exact algorithms, while guaranteed to find the optimal solution, often suffer from the effects of combinatorial explosion, which makes them impractical, especially for large instances of optimization problems. To circumvent this limitation approximate methods are often used. Population-based methods are effective in solving such complex optimization problems and are also well suited for multiobjective optimization, because they maintain an entire population of solutions, which, in a multiobjective case, can approximate a Pareto front for a given problem. For combinatorial problems memetic algorithms are often used, in which a population-based algorithm is coupled with a separate individual learning or local search procedure [6]. Among

© Springer International Publishing Switzerland 2015
K. Jackowski et al. (Eds.): IDEAL 2015, LNCS 9375, pp. 281–289, 2015.
DOI: 10.1007/978-3-319-24834-9_34

local search approaches used in multiobjective problems local search procedures based on Pareto dominance are often used [4,8]. In this paper a comparison is performed between a local search procedure based on a locally determined utopia point (ULS) and a local search procedure based on Pareto dominance (PLS). The results produced by these methods are compared using two commonly used indicators: the hypervolume and the Inverse Generational Distance (IGD). In the experiments performed on several instances of the TSP the ULS procedure produced better results than the Pareto-based procedure given equal time for computations using both methods.

This paper is structured as follows. Section 2 presents the local search procedure based on a local utopia point. In Sect. 3 the experiments are described and the results are discussed. Section 4 concludes the paper.

2   Local Search Procedures

In this paper a local search procedure based on a locally determined utopia point (ULS) is proposed. We assume, that the objective space is $\mathbb{R}^d$ and that we minimize all the objectives (application to the maximization case is straightforward). Thus, the optimization problem can be formalized as:

$$\text{minimize } F(x) = (f_1(x), \ldots, f_d(x))$$
$$\text{subject to } x \in \Omega, \tag{1}$$

where:
Ω - the decision space.

The ULS procedure improves a given solution x belonging to a population P as follows. First, a local utopia point $u(x) \in \mathbb{R}^d$ is established by setting each coordinate $u(x)_i$, $i = 1, \ldots, d$ to:

$$u(x)_i = max \left\{ f_i(x') : x' \in P - \{x\} \wedge f_i(x') < f_i(x) \right\} . \tag{2}$$

Thus, the obtained utopia point has all its coordinates set to maximum values of objectives represented in the population and less than those of x. If for a given coordinate i there is no $x' \in P - \{x\}$ satisfying $f_i(x') < f_i(x)$ then a preset special value MIN_VALUE is used (in this paper MIN_VALUE $= -1000000$ is used in this case). As presented in Algorithm 1 the ULS method is a best-improvement search, which replaces the current solution x with a new one x' from a neighbourhood $N(x)$ as long as $d(F(x'), u(x)) < d(F(x), u(x))$, where $d(\cdot)$ denotes the Euclidean distance. The neighbourhood $N(x)$ includes all solutions that can be generated from x using the 2-opt operator [1]. The search is performed until no improvement can be obtained.

To make a fair comparison the Pareto-based method used in this paper is also a best-improvement procedure which works on the same neighbourhoods $N(x)$ as the ULS and also stops when no improvement is obtained, but the acceptance criterion is Pareto dominance rather than the reduction of a distance to a given

---

Algorithm 1. The ULS procedure.

IN: d - the dimensionality of the objective space
       x - the solution to improve
       P - current population, $x \in P$

OUT: x - the solution improved by the local search

for $i = 1, \ldots, d$ **do**   # 1. Establish a local utopia point
    $u(x)_i = \text{MIN\_VALUE}$
    for $x' \in P - \{x\}$ **do**
        if $f_i(x') < f_i(x)$ **and** $f_i(x') > u(x)_i$ **then**
            $u(x)_i = f_i(x')$
        end if
    end for
end for

repeat   # 2. Improve the solution
    improved := **false**
    for $x' \in N(x)$ **do**
        if $d(F(x'), u(x)) < d(F(x), u(x))$ **then**   # 2.1. Check if the new solution
            x := x'   #   is better than the old one
            improved := **true**
        end if
    end for
until (**not** improved)

return x

---

point. Therefore, the entire step 1. in Algorithm 1 is not used in PLS and in step 2.1. the current solution x is replaced if x' dominates x.

Both local search procedures are used in this paper in conjunction with the NSGA-II algorithm [2] for multiobjective optimization of the TSP. In this paper the inver-over operator [11] is used for generating offspring.

Because balancing the amount of computational resources between the evolutionary optimization and the local search can influence the effectiveness of the algorithm [5,10], the Pareto-based method was used with several strategies of allocating computational resources to the evolutionary search and to the local search. Two tested approaches **PLS** (**P** = **1.0**) and **PLS** (**P** = **0.5**) involve applying the PLS with probability 1.0 and 0.5 respectively. Thus, in PLS (P = 1.0) the local search is performed for every specimen in the population and in PLS (P = 0.5) the local search is performed for every specimen with probability 0.5. In the latter approach the local search is performed at half the frequency, which may reduce computational overhead, but also may degrade the quality of the results. Another tested approach is not to perform the local search at all (the **none** strategy). Because the effectiveness of the local search may vary at different stages of the optimization, another two approaches were tested. In the **PLS** (**asc**) approach the probability of applying the local search to each specimen increases linearly from 0 at the beginning of the optimization process to 1 at the end of the optimization. In the **PLS (desc)** approach the probability of applying the local search is 1 at the beginning and it drops linearly to 0 during optimization. Clearly, in the PLS (asc) the local search is mainly applied to solutions from a population that underwent an evolutionary optimization for longer and thus can be expected to be better starting points. In the PLS (desc) the local search is applied most frequently at the beginning of the evolution and thus has to start with solutions that are far from optimal. On the other hand, in the PLS (desc) strategy the local search can positively influence the initial phase of the optimization.

3   Experiments and Results

In the experiments the ULS method was compared to four variants of the PLS using different strategies of allocating computational resources to the evolutionary algorithm and the local search, and to the evolutionary algorithm without the local search. In this paper bi-objective TSP instances kroABnnn made available by Thibaut Lust [12] were used with the number of cities $nnn = 100, 150, 200, 300, 400, 500, 750$ and 1000. The parameters for the algorithm were set as follows. The population size was set to $N_{pop} = 100$ and the random inverse rate $\eta = 0.02$ was used. Algorithms based solely on the evolutionary optimization and those using local search procedures may achieve very different quality of the results in the same number of generations. Also, varying intensity of local search changes both the timing and the attained results in each generation. To allow a fair comparison the quality of the results obtained by different methods was compared after the same running time of $t = 300\,$s rather than after a fixed number of generations. Pareto fronts attained by the tested methods within the available time limit were evaluated using the hypervolume [13] and Inverse Generational Distance (IGD) [7] indicators. The IGD indicator requires a set of reference points from which the distance of the obtained solutions is calculated. There are various techniques for obtaining such points. For example in the case of optimization problems described analytically these points can be calculated directly from the equations describing the problem. In constrained problems one of the possible approaches is to solve a relaxed version of the problem and then use the solutions as idealized reference points. In this paper another approach was followed in which the set of reference points is obtained by taking all the nondominated points from solutions produced by the tested methods for a given problem instance. This approach has the advantage in that it can produce the reference points even for practical problems in which the points cannot be established analytically and no relaxation may be possible because of the complex nature of the problem. For each method and each test set 30 repetitions of the test were performed on a machine with 2.4 GHz Intel Core 2 Quad Q6600 CPU and 4 GB of RAM. From these runs mean, median and standard deviation values of both indicators were calculated and are presented in Tables 1 – 8.

Results obtained in the experiments were processed using the Wilcoxon rank test [9]. This test was chosen because it does not assume the normality of the

Table 1. Hypervolume and IGD values attained for $N = 100$ using different local search methods in 300 s running time.

Algorithm	Hypervolume			IGD		
	Mean	Median	Std. dev.	Mean	Median	Std. dev.
None	$1.6023 \cdot 10^{10}$	$1.6196 \cdot 10^{10}$	$7.6683 \cdot 10^{8}$	$5.1648 \cdot 10^{4}$	$5.1055 \cdot 10^{4}$	$3.3267 \cdot 10^{3}$
PLS $(P = 0.5)$	$2.4334 \cdot 10^{10}$	$2.4342 \cdot 10^{10}$	$4.6754 \cdot 10^{8}$	$9.6821 \cdot 10^{3}$	$9.8788 \cdot 10^{3}$	$1.4756 \cdot 10^{3}$
PLS $(P = 1.0)$	$2.4880 \cdot 10^{10}$	$2.4904 \cdot 10^{10}$	$3.4252 \cdot 10^{8}$	$7.7639 \cdot 10^{3}$	$7.5605 \cdot 10^{3}$	$1.1032 \cdot 10^{3}$
PLS (asc)	$2.2010 \cdot 10^{10}$	$2.2146 \cdot 10^{10}$	$8.0293 \cdot 10^{8}$	$1.7071 \cdot 10^{4}$	$1.6824 \cdot 10^{4}$	$2.1464 \cdot 10^{3}$
PLS (desc)	$2.4837 \cdot 10^{10}$	$2.4878 \cdot 10^{10}$	$2.8748 \cdot 10^{8}$	$7.8428 \cdot 10^{3}$	$7.8020 \cdot 10^{3}$	$9.6257 \cdot 10^{2}$
ULS	$2.5601 \cdot 10^{10}$	$2.5832 \cdot 10^{10}$	$6.7932 \cdot 10^{8}$	$3.9768 \cdot 10^{3}$	$2.8019 \cdot 10^{3}$	$2.9563 \cdot 10^{3}$

Table 2. Hypervolume and IGD values attained for $N = 150$ using different local search methods in 300 s running time.

Algorithm	Hypervolume			IGD		
	Mean	Median	Std. dev.	Mean	Median	Std. dev.
None	$3.2621 \cdot 10^{10}$	$3.2362 \cdot 10^{10}$	$1.7880 \cdot 10^{9}$	$1.0080 \cdot 10^{5}$	$1.0186 \cdot 10^{5}$	$6.1111 \cdot 10^{3}$
PLS $(P = 0.5)$	$5.4549 \cdot 10^{10}$	$5.4433 \cdot 10^{10}$	$9.6264 \cdot 10^{8}$	$1.7823 \cdot 10^{4}$	$1.7760 \cdot 10^{4}$	$1.9285 \cdot 10^{3}$
PLS $(P = 1.0)$	$5.5720 \cdot 10^{10}$	$5.5644 \cdot 10^{10}$	$8.0822 \cdot 10^{8}$	$1.4891 \cdot 10^{4}$	$1.4926 \cdot 10^{4}$	$1.7976 \cdot 10^{3}$
PLS (asc)	$5.0417 \cdot 10^{10}$	$5.0145 \cdot 10^{10}$	$1.2348 \cdot 10^{9}$	$2.8112 \cdot 10^{4}$	$2.8549 \cdot 10^{4}$	$3.0057 \cdot 10^{3}$
PLS (desc)	$5.5453 \cdot 10^{10}$	$5.5454 \cdot 10^{10}$	$6.0025 \cdot 10^{8}$	$1.5162 \cdot 10^{4}$	$1.5086 \cdot 10^{4}$	$1.2078 \cdot 10^{3}$
ULS	$5.8936 \cdot 10^{10}$	$5.9314 \cdot 10^{10}$	$1.4208 \cdot 10^{9}$	$6.2998 \cdot 10^{3}$	$4.7199 \cdot 10^{3}$	$4.9699 \cdot 10^{3}$

Table 3. Hypervolume and IGD values attained for $N = 200$ using different local search methods in 300 s running time.

Algorithm	Hypervolume			IGD		
	Mean	Median	Std. dev.	Mean	Median	Std. dev.
None	$5.2585 \cdot 10^{10}$	$5.2471 \cdot 10^{10}$	$2.4659 \cdot 10^{9}$	$1.4720 \cdot 10^{5}$	$1.4830 \cdot 10^{5}$	$6.5569 \cdot 10^{3}$
PLS $(P = 0.5)$	$9.4544 \cdot 10^{10}$	$9.4447 \cdot 10^{10}$	$1.5599 \cdot 10^{9}$	$2.7332 \cdot 10^{4}$	$2.7231 \cdot 10^{4}$	$2.4114 \cdot 10^{3}$
PLS $(P = 1.0)$	$9.6240 \cdot 10^{10}$	$9.6652 \cdot 10^{10}$	$1.8580 \cdot 10^{9}$	$2.4363 \cdot 10^{4}$	$2.3533 \cdot 10^{4}$	$2.7024 \cdot 10^{3}$
PLS (asc)	$8.7223 \cdot 10^{10}$	$8.7472 \cdot 10^{10}$	$1.9099 \cdot 10^{9}$	$3.9923 \cdot 10^{4}$	$3.9520 \cdot 10^{4}$	$2.9927 \cdot 10^{3}$
PLS (desc)	$9.6239 \cdot 10^{10}$	$9.6039 \cdot 10^{10}$	$1.1415 \cdot 10^{9}$	$2.3962 \cdot 10^{4}$	$2.3868 \cdot 10^{4}$	$1.7321 \cdot 10^{3}$
ULS	$1.0480 \cdot 10^{11}$	$1.0573 \cdot 10^{11}$	$2.7560 \cdot 10^{9}$	$7.5180 \cdot 10^{3}$	$4.5561 \cdot 10^{3}$	$7.8807 \cdot 10^{3}$

distributions, which may be hard to guarantee in case of hypervolume and IGD measurements. It was also recommended in a survey [3] in which methods suitable for evaluating evolutionary and swarm intelligence algorithms were discussed. The results of the statistical tests are presented in Tables 9 and 10 which summarize comparisons of two the best performing methods ULS and PLS (desc) with other algorithms. In Table 9 a sign in a given row and column represents the result of the Wilcoxon test comparing the ULS with a method specified in the column using the results for the test set specified in the row. The null hypothesis of the statistical test is that the median values of a given indicator

Table 4. Hypervolume and IGD values attained for $N = 300$ using different local search methods in 300 s running time.

Algorithm	Hypervolume			IGD		
	Mean	Median	Std. dev.	Mean	Median	Std. dev.
None	$9.8025 \cdot 10^{10}$	$9.9744 \cdot 10^{10}$	$5.7400 \cdot 10^{9}$	$2.5395 \cdot 10^{5}$	$2.5251 \cdot 10^{5}$	$8.3855 \cdot 10^{3}$
PLS ($P = 0.5$)	$1.9927 \cdot 10^{11}$	$1.9914 \cdot 10^{11}$	$3.5869 \cdot 10^{9}$	$4.5034 \cdot 10^{4}$	$4.4860 \cdot 10^{4}$	$3.9594 \cdot 10^{3}$
PLS ($P = 1.0$)	$2.0028 \cdot 10^{11}$	$2.0087 \cdot 10^{11}$	$2.2556 \cdot 10^{9}$	$4.3155 \cdot 10^{4}$	$4.2478 \cdot 10^{4}$	$2.5737 \cdot 10^{3}$
PLS (asc)	$1.8516 \cdot 10^{11}$	$1.8466 \cdot 10^{11}$	$2.8529 \cdot 10^{9}$	$6.1549 \cdot 10^{4}$	$6.2227 \cdot 10^{4}$	$3.5908 \cdot 10^{3}$
PLS (desc)	$2.0187 \cdot 10^{11}$	$2.0224 \cdot 10^{11}$	$1.9251 \cdot 10^{9}$	$4.1816 \cdot 10^{4}$	$4.1515 \cdot 10^{4}$	$1.8804 \cdot 10^{3}$
ULS	$2.2953 \cdot 10^{11}$	$2.3295 \cdot 10^{11}$	$5.5661 \cdot 10^{9}$	$1.0190 \cdot 10^{4}$	$3.5550 \cdot 10^{3}$	$1.0179 \cdot 10^{4}$

Table 5. Hypervolume and IGD values attained for $N = 400$ using different local search methods in 300 s running time.

Algorithm	Hypervolume			IGD		
	Mean	Median	Std. dev.	Mean	Median	Std. dev.
None	$1.4252 \cdot 10^{11}$	$1.4400 \cdot 10^{11}$	$7.2263 \cdot 10^{9}$	$3.5728 \cdot 10^{5}$	$3.5593 \cdot 10^{5}$	$8.9753 \cdot 10^{3}$
PLS ($P = 0.5$)	$3.2235 \cdot 10^{11}$	$3.2327 \cdot 10^{11}$	$4.5732 \cdot 10^{9}$	$7.1950 \cdot 10^{4}$	$7.1490 \cdot 10^{4}$	$4.0573 \cdot 10^{3}$
PLS ($P = 1.0$)	$3.2250 \cdot 10^{11}$	$3.2260 \cdot 10^{11}$	$2.3309 \cdot 10^{9}$	$7.1416 \cdot 10^{4}$	$7.0661 \cdot 10^{4}$	$2.1962 \cdot 10^{3}$
PLS (asc)	$3.0254 \cdot 10^{11}$	$3.0231 \cdot 10^{11}$	$4.4301 \cdot 10^{9}$	$8.9644 \cdot 10^{4}$	$9.0201 \cdot 10^{4}$	$4.0627 \cdot 10^{3}$
PLS (desc)	$3.2774 \cdot 10^{11}$	$3.2797 \cdot 10^{11}$	$3.8239 \cdot 10^{9}$	$6.6905 \cdot 10^{4}$	$6.6328 \cdot 10^{4}$	$3.1366 \cdot 10^{3}$
ULS	$3.8699 \cdot 10^{11}$	$3.9092 \cdot 10^{11}$	$9.4632 \cdot 10^{9}$	$9.6561 \cdot 10^{3}$	$3.7527 \cdot 10^{3}$	$1.4027 \cdot 10^{4}$

Table 6. Hypervolume and IGD values attained for $N = 500$ using different local search methods in 300 s running time.

Algorithm	Hypervolume			IGD		
	Mean	Median	Std. dev.	Mean	Median	Std. dev.
None	$1.9950 \cdot 10^{11}$	$2.0019 \cdot 10^{11}$	$8.5470 \cdot 10^{9}$	$4.7972 \cdot 10^{5}$	$4.7940 \cdot 10^{5}$	$9.7379 \cdot 10^{3}$
PLS ($P = 0.5$)	$4.9077 \cdot 10^{11}$	$4.9095 \cdot 10^{11}$	$3.7994 \cdot 10^{9}$	$8.6907 \cdot 10^{4}$	$8.6609 \cdot 10^{4}$	$2.5111 \cdot 10^{3}$
PLS ($P = 1.0$)	$4.8834 \cdot 10^{11}$	$4.8770 \cdot 10^{11}$	$3.6278 \cdot 10^{9}$	$8.8264 \cdot 10^{4}$	$8.8483 \cdot 10^{4}$	$2.7402 \cdot 10^{3}$
PLS (asc)	$4.6317 \cdot 10^{11}$	$4.6356 \cdot 10^{11}$	$5.5284 \cdot 10^{9}$	$1.0467 \cdot 10^{5}$	$1.0475 \cdot 10^{5}$	$3.6158 \cdot 10^{3}$
PLS (desc)	$4.9502 \cdot 10^{11}$	$4.9455 \cdot 10^{11}$	$3.1035 \cdot 10^{9}$	$8.3402 \cdot 10^{4}$	$8.3579 \cdot 10^{4}$	$1.9820 \cdot 10^{3}$
ULS	$6.0132 \cdot 10^{11}$	$6.0288 \cdot 10^{11}$	$4.0029 \cdot 10^{9}$	$6.9318 \cdot 10^{3}$	$5.8373 \cdot 10^{3}$	$3.4575 \cdot 10^{3}$

are equal for ULS and the compared method. The interpretation of the signs is as follows. The "–" sign is placed in the table when the ULS method attained a worse median value than the compared method. The "+" sign denotes a test in which the ULS method attained a better median value than the compared method and the p-value obtained in the statistical test was not higher than 0.05. If the ULS method attained a better median value than the compared method, but the p-value obtained in the statistical test was higher than 0.05 the "#" sign is inserted, which represents a better performance of the ULS method, but with no confirmed statistical significance. Table 10 summarizes the performance

Table 7. Hypervolume and IGD values attained for $N = 750$ using different local search methods in 300 s running time.

Algorithm	Hypervolume			IGD		
	Mean	Median	Std. dev.	Mean	Median	Std. dev.
None	$3.5962 \cdot 10^{11}$	$3.5960 \cdot 10^{11}$	$1.8593 \cdot 10^{10}$	$7.5376 \cdot 10^{5}$	$7.5480 \cdot 10^{5}$	$1.1816 \cdot 10^{4}$
PLS $(P = 0.5)$	$1.0891 \cdot 10^{12}$	$1.0905 \cdot 10^{12}$	$4.9247 \cdot 10^{9}$	$2.3941 \cdot 10^{5}$	$2.3891 \cdot 10^{5}$	$2.8395 \cdot 10^{3}$
PLS $(P = 1.0)$	$1.0887 \cdot 10^{12}$	$1.0898 \cdot 10^{12}$	$5.5977 \cdot 10^{9}$	$2.3955 \cdot 10^{5}$	$2.4000 \cdot 10^{5}$	$2.9521 \cdot 10^{3}$
PLS (asc)	$1.0529 \cdot 10^{12}$	$1.0507 \cdot 10^{12}$	$1.6065 \cdot 10^{10}$	$2.5560 \cdot 10^{5}$	$2.5590 \cdot 10^{5}$	$7.2508 \cdot 10^{3}$
PLS (desc)	$1.0936 \cdot 10^{12}$	$1.0931 \cdot 10^{12}$	$6.0725 \cdot 10^{9}$	$2.3758 \cdot 10^{5}$	$2.3807 \cdot 10^{5}$	$2.8925 \cdot 10^{3}$
ULS	$1.2795 \cdot 10^{12}$	$1.3003 \cdot 10^{12}$	$6.7021 \cdot 10^{10}$	$5.3185 \cdot 10^{4}$	$3.2315 \cdot 10^{4}$	$5.4241 \cdot 10^{4}$

Table 8. Hypervolume and IGD values attained for $N = 1000$ using different local search methods in 300 s running time.

Algorithm	Hypervolume			IGD		
	Mean	Median	Std. dev.	Mean	Median	Std. dev.
None	$5.2986 \cdot 10^{11}$	$5.2949 \cdot 10^{11}$	$1.4122 \cdot 10^{10}$	$1.0302 \cdot 10^{6}$	$1.0302 \cdot 10^{6}$	$1.2143 \cdot 10^{4}$
PLS $(P = 0.5)$	$1.9127 \cdot 10^{12}$	$1.9131 \cdot 10^{12}$	$9.5281 \cdot 10^{9}$	$4.4548 \cdot 10^{5}$	$4.4488 \cdot 10^{5}$	$4.1141 \cdot 10^{3}$
PLS $(P = 1.0)$	$1.9139 \cdot 10^{12}$	$1.9133 \cdot 10^{12}$	$8.3681 \cdot 10^{9}$	$4.4464 \cdot 10^{5}$	$4.4448 \cdot 10^{5}$	$3.8338 \cdot 10^{3}$
PLS (asc)	$1.8727 \cdot 10^{12}$	$1.8731 \cdot 10^{12}$	$2.0848 \cdot 10^{10}$	$4.5905 \cdot 10^{5}$	$4.5878 \cdot 10^{5}$	$6.7310 \cdot 10^{3}$
PLS (desc)	$1.9204 \cdot 10^{12}$	$1.9185 \cdot 10^{12}$	$7.9682 \cdot 10^{9}$	$4.4170 \cdot 10^{5}$	$4.4259 \cdot 10^{5}$	$4.3471 \cdot 10^{3}$
ULS	$2.0328 \cdot 10^{12}$	$2.0865 \cdot 10^{12}$	$2.2730 \cdot 10^{11}$	$1.2851 \cdot 10^{5}$	$9.6405 \cdot 10^{4}$	$9.3073 \cdot 10^{4}$

Table 9. Results of the statistical analysis comparing hypervolume and IGD values obtained using the ULS method with other tested methods.

N	Hypervolume					IGD				
	none	PLS $(P = 0.5)$	PLS $(P = 1.0)$	PLS (asc)	PLS (desc)	none	PLS $(P = 0.5)$	PLS $(P = 1.0)$	PLS (asc)	PLS (desc)
100	+	+	+	+	+	+	+	+	+	+
150	+	+	+	+	+	+	+	+	+	+
200	+	+	+	+	+	+	+	+	+	+
300	+	+	+	+	+	+	+	+	+	+
400	+	+	+	+	+	+	+	+	+	+
500	+	+	+	+	+	+	+	+	+	+
750	+	+	+	+	+	+	+	+	+	+
1000	+	+	+	+	+	+	+	+	+	+
F-W	+	+	+	+	+	+	+	+	+	+

of the PLS (desc) method in the same way as described above. The last row (F-W) in each of the two tables shows, if the family-wise probability of type I error calculated over all test sets is smaller than 0.05 (the "+" sign) or not (the "#" sign). If the method presented in the table produced a worse result for one or more sets the "−" sign is placed in the last row.

Table 10. Results of the statistical analysis comparing the results obtained using the PLS (desc) method with other tested methods.

N	Hypervolume					IGD				
	none	PLS (P = 0.5)	PLS (P = 1.0)	PLS (asc)	ULS	none	PLS (P = 0.5)	PLS (P = 1.0)	PLS (asc)	ULS
100	+	+	−	+	−	+	+	−	+	−
150	+	+	−	+	−	+	+	−	+	−
200	+	+	−	+	−	+	+	−	+	−
300	+	+	+	+	−	+	+	#	+	−
400	+	+	+	+	−	+	+	+	+	−
500	+	+	+	+	−	+	+	+	+	−
750	+	+	+	+	−	+	#	+	+	−
1000	+	+	+	+	−	+	+	+	+	−
F-W	+	+	−	+	−	+	#	−	+	−

4   Conclusion

In this paper a local search method based on a local utopia point was proposed. This method was compared to a Pareto-based local search on several instances of the multiobjective Travelling Salesman Problem. The proposed method was found to outperform the Pareto-based approach when given equal computational time.

For the Pareto-based method five different strategies of invoking the local search procedure were tested. Out of these strategies the best one performs the local search with probability $P = 1.0$ at the beginning of the optimization and the reduces this probability linearly to $P = 0.0$.

Further work may include using different method of establishing the reference point for the ULS method as well as a research on various strategies of balancing computational resources between the evolutionary optimization and the ULS.

References

1. Chiang, C.W., Lee, W.P., Heh, J.S.: A 2-opt based differential evolution for global optimization. Appl. Soft Comput. **10**(4), 1200–1207 (2010)
2. Deb, K., Pratap, A., Agarwal, S., Meyarivan, T.: A fast and elitist multiobjective genetic algorithm: NSGA-II. IEEE Trans. Evol. Comput. **6**, 182–197 (2002)
3. Derrac, J., Garca, S., Molina, D., Herrera, F.: A practical tutorial on the use of nonparametric statistical tests as a methodology for comparing evolutionary and swarm intelligence algorithms. Swarm Evol. Comput. **1**(1), 3–18 (2011)
4. Dubois-Lacoste, J., et al.: A hybrid TP+PLS algorithm for bi-objective flow-shop scheduling problems. Comput. Oper. Res. **38**(8), 1219–1236 (2011)
5. Goldberg, D.E., Voessner, S.: Optimizing global-local search hybrids. In: Banzhaf, W., Daida, J., Eiben, A.E., Garzon, M.H., Honavar, V., Jakiela, M., Smith, R.E. (eds.) Proceedings of the Genetic and Evolutionary Computation Conference, vol. 1, pp. 220–228. Morgan Kaufmann, Orlando (1999)

6. Ishibuchi, H.: Memetic algorithms for evolutionary multiobjective combinatorial optimization. In: 2010 40th International Conference on Computers and Industrial Engineering (CIE), pp. 1–2 (2010)
7. Li, M., Zheng, J.: Spread assessment for evolutionary multi-objective optimization. In: Ehrgott, M., Fonseca, C.M., Gandibleux, X., Hao, J.-K., Sevaux, M. (eds.) EMO 2009. LNCS, vol. 5467, pp. 216–230. Springer, Heidelberg (2009)
8. Lust, T., Teghem, J.: Two-phase pareto local search for the biobjective traveling salesman problem. J. Heuristics **16**(3), 475–510 (2010)
9. Ott, L., Longnecker, M.: An Introduction to Statistical Methods and Data Analysis. Brooks/Cole Cengage Learning, Boston (2010)
10. Sinha, A., Goldberg, D.E.: Verification and extension of the theory of global-local hybrids. In: Proceedings of GECCO (2001)
11. Tao, G., Michalewicz, Z.: Inver-over operator for the TSP. In: Eiben, A.E., Bäck, T., Schoenauer, M., Schwefel, H.-P. (eds.) PPSN 1998. LNCS, vol. 1498, pp. 803–812. Springer, Heidelberg (1998)
12. Thibaut Lust: Multiobjective TSP (2015). https://sites.google.com/site/thibautlust/research/multiobjective-tsp. Accessed 29 January 2015
13. Zitzler, E., Thiele, L., Laumanns, M., Fonseca, C.M., da Fonseca, V.G.: Performance assessment of multiobjective optimizers: an analysis and review. IEEE Trans. Evol. Comput. **7**, 117–132 (2002)

A Simulated Annealing Heuristic for a Branch and Price-Based Routing and Spectrum Allocation Algorithm in Elastic Optical Networks

Mirosław Klinkowski[1]([✉]) and Krzysztof Walkowiak[2]

[1] Department of Transmission and Optical Technologies,
National Institute of Telecommunications, Warsaw, Poland
m.klinkowski@itl.waw.pl
[2] Department of Systems and Computer Networks,
Wroclaw University of Technology, Wroclaw, Poland
krzysztof.walkowiak@pwr.edu.pl

Abstract. In the paper we focus on developing an efficient algorithm capable of producing optimal solutions to the problem of routing and spectrum allocation (RSA) – a basic optimization problem in elastic optical networks. We formulate the problem as a mixed-integer program and we solve it using a branch and price (BP) algorithm. With the aim to improve the performance of BP, we enhance it with a simulated annealing-based RSA heuristic that is employed in the search for upper bound solutions. The results of numerial experiments show that the heuristic allows to decrease significantly the time required to solve the problem.

Keywords: Simulated annealing · Elastic optical networks · Routing and spectrum allocation · Mixed-integer programming · Branch and price

1 Introduction

The use of advanced transmission and modulation techniques, spectrum-selective switching technologies, and flexible frequency grids (flexgrids), will allow next-generation optical networks to be spectrally efficient and, in terms of optical bandwidth provisioning, scalable and elastic [1,2]. A challenging problem in the design and operation of flexgrid elastic optical networks (EONs) is the problem of routing and spectrum allocation (RSA). RSA consists in establishing optical path (lightpath) connections, tailored to the actual width of the transmitted signal, for a set of end-to-end demands that compete for spectrum resources.

The RSA optimization problem is $\mathcal{NP}$-hard [3]. In the literature, mixed-integer programming (MIP) formulations [4–6], metaheuristics [7–9], and heuristics [3,10], have been proposed to solve it. Both metaheuristics and heuristics can produce locally optimal solutions, however, without guarantees for global optimality. On the contrary, MIP formulations can be solved to optimality. A common approach is to use a standard branch-and-bound (BB) method, which is implemented

© Springer International Publishing Switzerland 2015
K. Jackowski et al. (Eds.): IDEAL 2015, LNCS 9375, pp. 290–299, 2015.
DOI: 10.1007/978-3-319-24834-9_35

in MIP solvers, for instance, in CPLEX [11]. The resolution of MIP using BB can be still difficult and time-consuming due to the processing of a large set of integer variables.

Recently, in [12], we have proposed an RSA optimization algorithm based on a branch-and-price (BP) framework, which is a combination of BB and column generation (CG) methods [13]. In BB, a tree of linear subproblems is generated and the subproblems are gradually solved. By applying CG, the amount of problem variables included into each subproblem is decreased considerably when compared to a standard BB method. In [12], we have shown that the BP algorithm is able to produce optimal solutions and in a vast majority of the considered cases it performs better than the BB method implemented in the CPLEX solver. The effectiveness of BB methods (including BP) depends strongly on the progress in the improvement of upper and lower bounds on the optimal solution during the BB processing. Therefore, efficient procedures estimating such bounds are required.

In this paper, we present and evaluate a heuristic algorithm that produces upper bound solutions in the nodes of the BB search tree in our BP. The heuristic is a combination of a greedy RSA algorithm and a simulated annealing algorithm. Numerical results obtained for a 12-node network show that the use of the heuristic speeds up significantly the BP algorithm.

The remainder of this paper is organized as follows. In Sect. 2, we present an MIP formulation of the considered RSA problem and we briefly describe our BP algorithm. In Sect. 3, we present the node heuristic that aims at improving upper bounds in BP. The optimization algorithm is evaluated in Sect. 4 using the results of numerical experiments. Finally, in Sect. 5, we conclude this work.

2   Optimization Model and Algorithm

2.1   MIP Formulation of RSA

We formulate RSA as an MIP problem and using a link-lightpath (LL) modeling approach that was proposed in [14]. In LL, the spectrum assignment-related constraints are removed from the MIP by using a set of pre-computed lightpaths. At the same time, the LL constraints assure that for each demand a lightpath is selected from the pre-computed set and the selected lightpaths are not in conflict with each other, i.e., their spectra do not overlap on the network links. In this work, we aim at minimizing the spectrum width required to allocate a given set of demands. Such an optimization objective has been frequently used in previous works on RSA [3].

The considered EON network is represented by a graph $\mathcal{G} = (\mathcal{V}, \mathcal{E})$ where $\mathcal{V}$ is the set of optical nodes and $\mathcal{E}$ is the set of fiber links. In each link $e \in \mathcal{E}$, the same bandwidth (i.e., optical frequency spectrum) is available and it is divided into a set $\mathcal{S} = \{s_1, s_2, \ldots, s_{|\mathcal{S}|}\}$ of frequency slices of a fixed width. The set of node-to-node (traffic) demands to be realized in the network is denoted by $\mathcal{D}$.

In the MIP model a notion of a lightpath is used. A lightpath is understood as a pair (p, c), where p is a route and c is a channel. The route is a path through a network from a source node to a termination node of a demand $(p \subseteq \mathcal{E})$, while the channel is a set of contiguous slices assigned to the lightpath $(c \subseteq \mathcal{S})$. Note that channel c should be wide enough to carry the bit-rate of demand d, if it is supposed to satisfy this demand. Channel c is the same for each link belonging to the routing path, what is called the spectrum continuity (SC) constraint. It is assumed that sets of allowable lightpaths $\mathcal{L}(d)$ for each demand are given, thus the problem simplifies to selecting one of those lightpaths for each demand in such a way that there are no two demands that use the same slice on the same link. Let $\mathcal{L}$ be the set of all allowable lightpaths, i.e., $\mathcal{L} = \bigcup_{d \in \mathcal{D}} \mathcal{L}(d)$.

Each lightpath $l \in \mathcal{L}(d)$ is assigned a binary variable $x_{dl}, d \in \mathcal{D}, l \in \mathcal{L}(d)$, where $x_{dl} = 1$ indicates that lightpath l is actually used to realize the traffic (bit-rate) of demand d. Besides, a binary variable $x_{es}, e \in \mathcal{E}, s \in \mathcal{S}$, indicates if there is a used lightpath allocated on slice s of link e. Eventually, the use of slice s in the network is indicated by a binary variable $x_s, s \in \mathcal{S}$. The MIP formulation is the following:

$$\text{minimize } z = \sum_{s \in \mathcal{S}} x_s \tag{1a}$$

$$[\lambda_d] \quad \sum_{l \in \mathcal{L}(d)} x_{dl} = 1 \qquad\qquad d \in \mathcal{D} \tag{1b}$$

$$[\pi_{es} \geq 0] \quad \sum_{d \in \mathcal{D}} \sum_{l \in \mathcal{Q}(d,e,s)} x_{dl} = x_{es} \quad e \in \mathcal{E}, s \in \mathcal{S} \tag{1c}$$

$$x_{es} \leq x_s \qquad\qquad e \in \mathcal{E}, s \in \mathcal{S} \tag{1d}$$

where $\mathcal{Q}(d, e, s)$ is the set of lightpaths of demand d routed through link e and slice s. Optimization objective (1a) minimizes the number of used slices in the network, which is obtained by summing up variables x_s. Constraint (1b) assures that each demand will use one and only one lightpath from a set of allowable lightpaths. Constraint (1c) assures that there are no collisions of the assigned resources, i.e., there are no two lightpaths in the network that use the same slice on the same link. Finally, constraint (1d) defines variables x_s that indicate whether slice s is used on any link.

In the following, the linear programming (LP) relaxation of (1) is called the master problem and the variables in brackets, i.e., λ_d and π_{es}, are its dual variables.

2.2   Optimization Algorithm

For solving problem (1), we use a branch-and-price (BP) algorithm developed in [12]. BP combines the branch and bound (BB) and column generation (CG) methods [13]. In a BB method, a tree of linear subproblems, called restricted

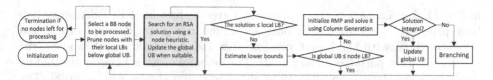

Fig. 1. A branch-and-price algorithm.

master problems (RMPs), related to the master problem is generated through a branching process. At each BB node, a subset of variables is bounded by means of extra constraints in the RMP. Each RMP is solved using a CG procedure in BP. Namely, BP is initiated with a limited set of problem variables (columns) and at each node of the BB search tree, additional variables are generated and included into RMP. For a minimization problem (such as problem (1)), the solution of each RMP provides a lower bound (LB) that is used either to discard certain BB nodes from the search for an optimal solution or to set an upper bound (UB), whenever this solution is also feasible for MIP. To improve the algorithm performance, additional procedures aiming at the estimation of tighter lower and upper bounds can be included. The BB search is terminated whenever there are no nodes left for processing. A block diagram of BP is presented in Fig. 1.

In our implementation of BP, we initialize the algorithm with a small set of allowable lightpaths $\mathcal{L}$ and in BB nodes we generate and include into $\mathcal{L}$ additional lightpaths. For details on lightpath generation, refer to [15]. In the branching step, two child nodes (denoted as Ω_0 and Ω_1) of the currently processed (parent) node are created. The columns generated at the parent node (i.e., set $\mathcal{L}$) are passed to the child nodes. Next, a subset of lightpaths from $\mathcal{L}$ (referred to as restricted) is selected and imposed to be either used or not permitted, respectively, in Ω_1 and Ω_0 as well as in their child nodes. We allow two kinds of branching, namely, for a selected demand we impose/prohibit either (a) a routing path or (b) a lightpath that it may use. For the lightpaths that are not permitted, their corresponding variables x_{dl} are set to 0 in the RMP. Note that these variables may be still regenerated by CG. To mitigate this problem, we assume that a (large) set of candidate lightpaths is given, only lightpaths from this set are processed by CG, and a lightpath can be included into $\mathcal{L}$ only if it is permitted. In the evaluation section, we assume that for each demand d a set of candidate routing paths $\mathcal{P}_d$ is given and the set of candidate lightpaths consists of all possible lightpaths established on these paths and allocating any, appropriate for given demand, segment of spectrum on the flexgrid.

For more details on our BP algorithm, refer to [12].

3   Heuristic for Effective Generation of Upper Bound Solutions in BB nodes

In this section, we develop a heuristic for generating upper bound solutions in the nodes of the BB search tree in our BP algorithm (see the highlighted block in Fig. 1). A main part of the heuristic is a greedy, first-fit (FF) RSA procedure that processes demands one-by-one, according to a given demand order, and allocates them with the lowest possible slice index (primary goal) and on the shortest routing path (secondary goal). The demand order is being optimized by applying a simulated annealing (SA) algorithm, in a similar way as in [8]. The obtained solutions provide UBs on the solution of problem (1). The heuristic takes into account the restrictions imposed on routing paths and lightpaths in BB nodes and they are respected when performing RSA, what makes our implementation different than the one in [8]. In the following, we describe the heuristic and we evaluate it in Sect. 4.

Let $\mathcal{P}^0(d)$ and $\mathcal{L}^0(d)$ denote, respectively, the sets of routes and lightpaths that are not permitted and let $\mathcal{P}^1(d)$ and $\mathcal{L}^1(d)$ denote, respectively, the sets of routes and lightpaths that are allowed for demand d in a certain BB node. Let $\mathcal{P}^0$, $\mathcal{L}^0$, $\mathcal{P}^1$, and $\mathcal{L}^1$ be respective sums of those sets. Let o denote an order of demands and $\mathcal{D}(o)$ be the set of demands ordered according to o.

Procedure 1 specifies a pseudo-code of FF. At the beginning, in lines 2–5, FF allocates the demands (preserving order o) for which the set of allowed light-paths is restricted (i.e., with $\left| \mathcal{L}^1(d) \right| > 0$). If a lightpath cannot be allocated for some demand, the number of unserved demands (denoted as *sol.unserved*) is incremented. Then, in lines 6–13, the rest of demands is processed. Here, either the set of restricted routes (in lines 8–10) or all candidate routes (in lines 11–13) is considered and the routes and lightpaths that are not permitted (i.e., in sets $\mathcal{P}^0$ and $\mathcal{L}^0$, respectively) are excluded from processing. Eventually, in lines 14–17, the objective value of the solution is calculated. If all demands are allocated, the objective represents the maximum allocated slice index in the network. Other-wise, the solution is infeasible and it takes a value exceeding $|\mathcal{S}|$, in particular, the more the demands unserved, the higher value of the objective.

In Algorithm 1, we present a pseudo-code of SA. We implement a standard simulated annealing approach [16] which, starting from an initial RSA solution (in line 1), proceeds in several iterations (lines 4–15). The algorithm control parameters are *temperature* T, which determines the probability of accepting nonimproving solutions, and *cooling rate* R, where $0 < R < 1$, which determines the rate at which the temperature is decreasing. The initial solution is obtained in line 1 using the FF procedure and for the best order o^* found in previously processed BB nodes. In line 3, T is initiated using the value of $|\mathcal{S}|$ and a temper-ature coefficient parameter (denoted as α), similarly as in [8]. At each iteration, a random neighbour is generated (in line 5) by swapping the order of two ran-domly selected demands. Moves that improve the cost function (evaluated in line 6) are always accepted (in lines 8–11) and both the best solution and the best order are updated if they are worse than the generated neighboring solution. Otherwise, in lines 13–14, the neighbour is selected with a given probability that depends on T and the amount of degradation of the objective function (denoted as Δ). Δ is calculated as a difference in the objective value between the current solution and the neighboring solution. As the algorithm progresses, T decreases (line 15) and the probability that such moves are accepted decreases as well. The

Procedure 1. First-Fit (FF) RSA procedure.

Input: Network $\mathcal{G}$, Demands $\mathcal{D}$, Order o, Restricted routes $\mathcal{P}^0$, $\mathcal{P}^1$ and lightpaths $\mathcal{L}^0$, $\mathcal{L}^1$

1: $sol.unserved \leftarrow 0$, $sol.\mathcal{L} \leftarrow \emptyset$
2: **for all** demand $d \in \mathcal{D}(o)$ **do**
3:     **if** $|\mathcal{L}^1(d)| > 0$ **then**
4:        Select lightpath $l \in \mathcal{L}^1(d)$ such that it can be allocated with the lowest slice index in network $\mathcal{G}$
5:        **if** l exists **then** $sol.\mathcal{L}(d) \leftarrow l$ and allocate l in $\mathcal{G}$ **else** increment $sol.unserved$
6: **for all** demand $d \in \mathcal{D}(o)$ **do**
7:     **if** $|\mathcal{L}^1(d)| = 0$ **then**
8:        **if** $|\mathcal{P}^1(d)| > 0$ **then**
9:           Select route $p \in \mathcal{P}^1(d)$ and lightpath $l \in \mathcal{L}(d,p) \setminus \mathcal{L}^0(d)$ such that l can be allocated with the lowest slice index in network $\mathcal{G}$ (first goal) and p is shortest (second goal)
10:           **if** l exists **then** $sol.\mathcal{L}(d) \leftarrow l$ and allocate l in $\mathcal{G}$ **else** increment $sol.unserved$
11:        **else**
12:           Select route $p \in \mathcal{P}(d) \setminus \mathcal{P}^0(d)$ and lightpath $l \in \mathcal{L}(d,p) \setminus \mathcal{L}^0(d)$ such that l can be allocated with the lowest slice index in network $\mathcal{G}$ (first goal) and p is shortest (second goal)
13:           **if** l exists **then** $sol.\mathcal{L}(d) \leftarrow l$ and allocate l in $\mathcal{G}$ **else** increment $sol.unserved$
14: **if** $sol.unserved = 0$ **then**
15:     $sol.objective \leftarrow$ maximum allocated slice index in network $\mathcal{G}$
16: **else**
17:     $sol.objective \leftarrow |\mathcal{S}| + 10 \cdot bestSol.unserved$
Output: RSA solution sol

---

algorithm terminates whenever either the number of performed iterations or the temperature reach their limits (denoted by I_{lim} and T_{lim}, respectively). In the evaluation section, we assume $I_{lim} = 10000$, $T_{lim} = 0.01$, $\alpha = 0.05$, as suggested in [8].

The computation complexity of SA is bounded by $O(|\mathcal{D}| |\mathcal{P}_d| |\mathcal{E}| |\mathcal{S}| I_{lim})$, where $|\mathcal{D}|$ is the number od demands to be processed, $|\mathcal{P}_d|$ is an upper bound on the number of paths analysed, $|\mathcal{E}| |\mathcal{S}|$ corresponds to the (worst-case) complexity of the search for free spectrum in $\mathcal{G}$, and I_{lim} restricts the number of iterations in SA.

4 Numerical Results

In this section, we evaluate the effectiveness of SA in improving upper bound solutions in our BP algorithm. SA is applied in each BB node and with the cooling rate $R = 0.99$ (a value suggested in [8]). As reference scenarios, we consider BP running without a node heuristic and BP with FF (with random demand orders). Our main focus is on processing times of the BP algorithm (t). Also, we report lower bounds obtained in the master node of BP (z^{UB}), the overall number of generated nodes, best (upper bound) solutions found (z^{UB}), and the

---

Algorithm 1. Simulated Annealing (SA) algorithm.

Input: Network $\mathcal{G}$, Demands $\mathcal{D}$, Best order o^*, Restricted routes $\mathcal{P}^0$, $\mathcal{P}^1$ and lightpaths $\mathcal{L}^0$, $\mathcal{L}^1$

   Cooling rate R, Temperature coefficient α, Iteration limit I_{lim}, Temperature limit T_{lim}

1: $bestSol \leftarrow \text{FF}(\mathcal{G}, \mathcal{D}, o^*, \mathcal{P}^0, \mathcal{P}^1, \mathcal{L}^0, \mathcal{L}^1)$
2: $currentSol \leftarrow bestSol$, $o \leftarrow o^*$
3: $T \leftarrow |\mathcal{S}| \cdot \alpha$, $i \leftarrow 0$
4: **while** $i < I_{lim} \wedge T > T_{lim}$ **do**
5:    $\hat{o} \leftarrow o$ and change the order of two randomly selected demands in $\hat{o}$
6:    $sol \leftarrow \text{FF}(\mathcal{G}, \mathcal{D}, \hat{o}, \mathcal{P}^0, \mathcal{P}^1, \mathcal{L}^0, \mathcal{L}^1)$
7:    $\Delta \leftarrow sol.objective - currentSol.objective$
8:    **if** $\Delta \leq 0$ **then**
9:       $currentSol \leftarrow sol$, $o \leftarrow \hat{o}$
10:       **if** $currentSol.objective < bestSol.objective$ **then**
11:          $bestSol \leftarrow currentSol$, $o^* \leftarrow \hat{o}$
12:    **else**
13:       **if** $rand(0,1) < \exp(-\Delta/T)$ **then**
14:          $currentSol \leftarrow sol$, $o \leftarrow \hat{o}$
15:    $T \leftarrow T \cdot R$, increment i
Output: Best solution $bestSol$, Best order o^*

---

status of solutions (either *optimal* or *feasible*). All the results are presented in Table 1 and a detailed histogram of t is presented in Fig. 3.

The evaluation is performed for a generic German network of 12 nodes and 20 links, denoted as DT12 and presented in Fig. 2. We assume the flexgrid of 12.5 GHz granularity and the spectral efficiency of 2 bit/s/Hz. We consider symmetric demands with randomly generated end nodes and uniformly distributed bit-rate requests between 10 and 400 Gbit/s. The number of demands is $|\mathcal{D}| \in \{10, 20, ..., 60\}$ and for each $|\mathcal{D}|$ we generate 10 demand sets, i.e., 60 traffic instances in total. The number of candidate routing paths is $|\mathcal{P}_d| = 10$.

We use CPLEX v.12.5.1 as an LP solver in the column generation phase of BP and as an MIP solver in the search for lower bounds (see [12] for details). CPLEX is run with its default settings and in a parallel mode using 8 threads. The rest of procedures of BP, such as processing of BB nodes and heuristics, are run in a sequential way (1 thread). The BP algorithm and its node heuristics are implemented in C++. Numerical experiments are performed on a 2.7 GHz i7-class machine with 8 GB RAM. We set a 3-hour run-time limit for the algorithm processing. Note that whenever BP terminates its processing within this time limit, the obtained solution is optimal.

At the beginning of our optimization procedure, yet before starting BP, and for each traffic instance we run the SA heuristic (here, with the cooling rate $R = 0.999$) in order to determine an upper bound on $|\mathcal{S}|$, which is required for formulating and solving our RMPs. The computation time of this step is negligible (up to a few seconds for the largest problem instances) and included into the overall reported computation time.

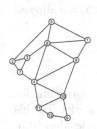

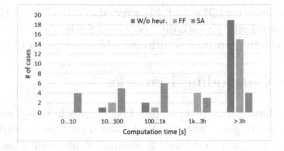

Fig. 2. DT12 network.

Fig. 3. A histogram of computation times.

For 38 traffic instances (out of 60), the initial solutions obtained with SA had the same values as the LBs produced by solving the relaxed problem in the master node of BP. Hence, the solutions were optimal and the algorithm could terminate its processing.

For the rest of traffic instances, BP performed the search for optimal solutions within its BB tree. Here, the BP algorithm without node heuristics was able to solve the RSA problem only for $|\mathcal{D}| = 20$. The use of the FF heuristic in the BB nodes improved the algorithm performance and, in particular, some of traffic

Table 1. Performance of BP without and with node heuristics for the traffic instances not solved in the master node of BP (shorter times marked in bold).

$\mathcal{D}$	z^{LB}	BP w/o node heuristic				BP with FF				BP with SA			
		$Nodes$	z^{UB}	$t[s]$	$Status$	$Nodes$	z^{UB}	$t[s]$	$Status$	$Nodes$	z^{UB}	$t[s]$	$cStatus$
20	33	1103	33	109	optimal	337	33	33	optimal	47	33	**6.5**	optimal
	35	141	35	12	optimal	1109	35	68	optimal	7	35	**2.0**	optimal
	31	8609	31	699	optimal	5359	31	469	optimal	47	31	**6.1**	optimal
30	51	36737	52	>3 h	feasible	22529	51	4816	optimal	1293	51	**332**	optimal
	50	42173	51	>3 h	feasible	9921	50	1974	optimal	63	50	**13**	optimal
	40	65260	41	> h	feasible	14387	40	>3 h	feasible	73	40	**13**	optimal
40	51	51228	53	>3 h	feasible	42110	53	>3 h	feasible	143	51	**41**	optimal
	53	25789	54	>3 h	feasible	34109	53	7972	optimal	687	53	**212**	optimal
	60	19283	63	>3 h	feasible	18673	63	>3 h	feasible	1655	60	**743**	optimal
	69	23137	70	>3 h	feasible	21566	70	>3 h	feasible	5	69	**5.1**	optimal
	73	11581	75	>3 h	feasible	10820	75	>3 h	feasible	1157	73	**625**	optimal
50	89	12515	90	>3 f	feasible	13686	90	>3 h	feasible	9440	90	>3 h	feasible
	79	17470	80	>3 h	feasible	15859	80	>3 h	feasible	407	79	**209**	optimal
	64	13139	66	>3 h	feasible	13464	66	>3 h	feasible	14382	65	>3 h	feasible
	84	8433	86	>3 h	feasible	8538	86	>3 h	feasible	6525	86	>3 h	feasible
	74	11227	76	>3 h	feasible	12033	76	>3 h	feasible	23361	74	**10288**	optimal
60	94	15748	95	>3 h	feasible	15431	95	>3 h	feasible	13	94	**15**	optimal
	86	14855	87	>3 h	feasible	15213	87	>3 h	feasible	13417	87	>3 h	feasible
	91	16129	92	>3 h	feasible	13553	92	>3 h	feasible	215	91	**126**	optimal
	91	16210	92	>3 h	feasible	13742	92	>3 h	feasible	157	91	**99**	optimal
	92	13354	93	>3 h	feasible	14273	93	>3 h	feasible	9319	92	**3558**	optimal
	69	13479	74	>3 h	feasible	10027	74	>3 h	feasible	19887	69	**8061**	optimal

instances for $|\mathcal{D}| \in \{30, 40\}$ were also solved. Still, BP had difficulties with larger demand sets. Eventually, the use of SA in the nodes of the BB search tree allowed to produce optimal solutions in 18 cases (out of 22).

5   Concluding Remarks

We have focused on improving a branch-and-price optimization algorithm for the routing and spectrum allocation problem in elastic optical networks. To this end, we have developed a simulated annealing-based RSA heuristic algorithm for producing upper bound solutions in the nodes of the BB search tree. The results of numerical experiments have shown that the use of the heuristic allows to decrease considerably the time required to produce an optimal solution to the RSA problem. The performance of the heuristic, as well as of BP, might be further improved by implementing parallel processing.

The presented in this paper MIP model and optimization algorithms are appropriate for EONs with a fixed, independent on the transmission distance, transmission format and with unicast traffic demands. Nevertheless, presented solutions could be also adapted to support distance-adaptive transmission as well as anycast and multicast traffic demands. The required modifications concern, among others, improved branching strategies and adequate lightpath/lighttree generation procedures for anycast/multicast demands. Such extensions are left for future work.

Acknowledgment. This work was supported by the Polish National Science Centre (NCN) under Grant DEC-2012/07/B/ST7/01215 and by the FP7 project IDEALIST (grant no. 317999)

References

1. Gerstel, O., et al.: Elastic optical networking: a new dawn for the optical layer? IEEE Commun. Mag. **50**(2), 12–20 (2012)
2. Klinkowski, M., Walkowiak, K.: On the advantages of elastic optical networks for provisioning of cloud computing traffic. IEEE Network **27**(6), 44–51 (2013)
3. Klinkowski, M., Walkowiak, K.: Routing and spectrum assignment in spectrum sliced elastic optical path network. IEEE Commun. Lett. **15**(8), 884–886 (2011)
4. Christodoulopoulos, K., et al.: Elastic bandwidth allocation in flexible OFDM based optical networks. IEEE J. Lightwave Technol. **29**(9), 1354–1366 (2011)
5. Żotkiewicz, M., et al.: Optimization models for flexgrid elastic optical networks. In: Proceedings of IEEE ICTON, Cartagena, Spain (2013)
6. Velasco, L., et al.: Solving routing and spectrum allocation related optimization problems: from off-line to in-operation flexgrid network planning. IEEE J. Lightwave Technol. **32**(16), 2780–2795 (2014)
7. Klinkowski, M.: An evolutionary algorithm approach for dedicated path protection problem in elastic optical networks. Cybern. Syst. **44**(6–7), 589–605 (2013)
8. Aibin, M., Walkowiak, K.: Simulated annealing algorithm for optimization of elastic optical networks with unicast and anycast traffic. In: Proceedings of IEEE ICTON, Graz, Austria (2014)

9. Goścień, R., et al.: Tabu search algorithm for routing, modulation and spectrum allocation in elastic optical network with anycast and unicast traffic. Comput. Network **79**, 148–165 (2015)
10. Walkowiak, K., et al.: Routing and spectrum allocation algorithms for elastic optical networks with dedicated path protection. Opt. Switch Network. **13**, 63–75 (2014)
11. IBM: ILOG CPLEX optimizer (2012). http://www.ibm.com
12. Klinkowski, M., et al.: Spectrum allocation problem in elastic optical networks - a branch-and-price approach. In: Proceedings of IEEE ICTON, Budapest, Hungary (2015)
13. Barnhart, C., et al.: Branch-and-price: column generation for solving huge integer programs. Oper. Res. **46**(3), 316–329 (1998)
14. Velasco, L., et al.: Modeling the routing and spectrum allocation problem for flexgrid optical networks. Photonic Network Commun. **24**(3), 177–186 (2012)
15. Ruiz, M., et al.: Column generation algorithm for RSA problems in flexgrid optical networks. Photonic Network Commun. **26**(2–3), 53–64 (2013)
16. Talbi, E.G.: Metaheuristics: from Design to Implementation. Wiley, New York (2009)

A Hybrid Programming Framework for Resource-Constrained Scheduling Problems

Paweł Sitek[✉] and Jarosław Wikarek

Control and Management Systems Section, Technical University of Kielce,
Al. Tysiąclecia Państwa Polskiego 7, 25-314 Kielce, Poland
{sitek, j.wikarek}@tu.kielce.pl

Abstract. Resource-constrained scheduling problems appear frequently at different levels of decisions in manufacturing, logistics, computer networks, software engineering etc. They are usually characterized by many types of constraints, which often make them unstructured and difficult to solve (NP-complete). Traditional mathematical programming (MP) approaches are deficient because their representation of allocation constraints is artificial (using 0–1 variables). Unlike traditional approaches, declarative constraint logic programming (CLP) provides for a natural representation of heterogeneous constraints. In CLP we state the problem requirements by constraints; we do not need to specify how to meet these requirements. CLP approach is very effective for binary constraints (binding at most two variables). If there are more variables in the constraints and the problem requires further optimization, the efficiency decreases dramatically. This paper presents a hybrid programming framework for constrained scheduling problems where two environments (mathematical programming and constraint logic programming) were integrated. This integration, hybridization as well as a transformation of the problem helped reduce the combinatorial problem substantially.

In order to compare the effectiveness of the proposed framework, also made implementation of illustrative example separately for the two environments MP and CLP.

Keywords: Constraint logic programming · Mathematical programming · Scheduling · Decision support · Hybrid approach

1 Introduction

Today's highly competitive environment makes it an absolute requirement on behalf of the decision makers to continuously make the best decisions in the shortest possible time. That is, there is no room for mistake in making decisions in this global environment. Success depends on quickly allocating the organizational resources towards meeting the actual needs and wants of the customer at the lowest possible cost. Very often, the whole or parts of the decision-making problems are brought to the allocation of resources over time, which must satisfy a set of constraints. This is a generalized form of resource-constrained scheduling problem. These problems appear frequently at different levels of decisions in manufacturing, distribution, logistic, computer networks,

© Springer International Publishing Switzerland 2015
K. Jackowski et al. (Eds.): IDEAL 2015, LNCS 9375, pp. 300–308, 2015.
DOI: 10.1007/978-3-319-24834-9_36

software engineering etc. The resource-constrained scheduling problems involve various numerical, logical and other constraints, some of which are in conflict with each other. Therefore effective and natural ways of modeling different constraint types and structures is a key issue. The most efficient methods for the modeling of linear constraints are OR (Operation Research) methods, mathematical programming (MP) techniques in particular [1]. For non-linear, logical, etc. constraints, MP techniques have proved to be inefficient. Nevertheless, years of using MP methods resulted in the development of ways to cope with such constraints as for example type 0–1 constraints for modeling of allocation problems, etc. Unfortunately, the introduction of this type of constraints complicates the structure of the problem and increases the combinatorial search space. Unlike traditional approaches, declarative constraint logic programming (CLP) provides for a natural representation of heterogeneous constraints. Constraint logic programming (CLP) is a form of constraint programming (CP), in which logic programming (LP) is extended to include concepts from constraint satisfaction problem (CSP) [2]. Constraint satisfaction problems on finite domains are typically solved using a form of search. The most widely used techniques include variants of backtracking, constraint propagation, and local search.

Constraint propagation embeds any reasoning that consists in explicitly forbidding values or combinations of values for some variables of a problem because a given subset of its constraints cannot be satisfied otherwise.

CLP approach is very effective for binary constraints (binding at most two variables). If there are more variables in the constraints the efficiency decreases dramatically. In addition, discrete optimization is not a strong suit of CP-based environments.

Based on [3, 4] and previous works on hybridization [5, 6, 20] some advantages and disadvantages of these environments have been observed. The hybrid approach of constraint logic programming and mathematical programming can help solve decision, search and optimization problems that are intractable with either of the two methods alone [7, 8].

The motivation and contribution behind this work was to apply a hybrid approach as a hybrid programming framework for resource-scheduling problems. The proposed framework allows easy modeling and effective solving of decision-making models as well as enables the decision-makers to ask all kinds of questions.

2 Resource-Constrained Scheduling Problems

In its most general form, the resource-constrained scheduling problem [9–11] asks the following questions.
    Given:

- a set of activities (tasks, machine operations, development software modules, services etc.) that must be executed;
- a set of resources (machines, processors, workers, tools, materials, finances etc.) with which to perform activities;
- a set of constraints (precedence, capacity, time, availability, allocation etc.) must be satisfied;
- a set of objectives with which to evaluate a schedule's performance;

what is the best way to assign the resources to the activities at specific times such that all of the constraints are satisfied (decision and search problems) and the best objective measures are made (optimization problems)?

Constraints and objectives are defined during the modeling of the problem. Constraints define the "feasibility" of the schedule and their solution is sufficient for the decision and search problems. Objectives define the "optimality" of the schedule and require solutions for optimization problems.

Constraints appear in many forms. Precedence constraints define the order in which activities can be performed. Temporal constraints limit the times during which resources may be used and/or activities may be executed, etc.

The general problem includes many variations such as the job-shop and flow-shop problems, production scheduling, the resource-constrained project scheduling problem, and the project scheduling problem for software development etc.

3   A Hybrid Programming Framework

The hybrid approach to modeling and solving of resource-constrained scheduling problems is able to bridge the gaps and eliminate the drawbacks that occur in both MP and CLP approaches. To support this concept, a hybrid programming framework is proposed (Fig. 1), where:

- knowledge related to the problem can be expressed as linear, logical and symbolic constraints;
- two environments, i.e. CLP and MP are integrated;
- the transformation of the problem [12] is an integral part of this framework;
- domain solution obtained by the CLP is used for the transformation of the problem and to build the MP model;
- all types of questions can be asked: general questions for decision problems: *Is it possible ...? Is it feasible ...?* And specific questions for optimization and search problems: *What is the minimum ...? What is the number ...? What is the configuration ... for ..?* (the list of example question in this framework version for illustrative examples is shown in Table 1);
- general questions are implemented as CLP predicates;
- specific questions and optimization are performed by the MP environment;
- the properties of the problem, its parameters, input data are stored as sets of facts.

4   Implementation and Illustrative Example

From a variety of tools for the implementation of the CLP environment in the framework, the ECL^{i}PS^e software [13] was selected. ECLiPSe is an open-source software system for the cost–effective development and deployment of constraint

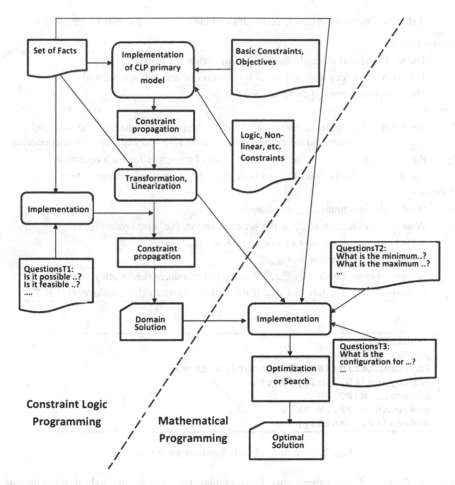

Fig. 1. The schema of hybrid programming framework

programming applications. Environment for the implementation of MP was LINGO by LINDO Systems. LINGO Optimization Modelling Software is a powerful tool for building and solving mathematical optimization models [14].

The illustrative example relates to scheduling in job-shop environment with additional resources (employees).

In order to evaluate the proposed framework, a number of computational experiments were performed for the illustrative example. Variables, the main constraints and questions of the illustrative example are shown in Table 1, while the structure of the facts in Fig. 2. The experiments were performed for the instance consisting of 6 machines (M1... M6), 6 products (A, B, C, D, E, F, G) and different sets of employees (O).

The set of facts for illustrative example is shown in Appendix A. The experimental results for each of the questions are presented in Table 2 and the corresponding schedules for questions Q1, Q2, in Figs. 3a, b and 4a, b.

Table 1. Decision variables, constraint and questions for presented framework

Decision variables	
V_1	The start time and the end time of each operation for activity
V_2	The set of resources assigned to each operation (if this set is not fixed)
V_3	The processing times (if there are not fixed)
Constraints	
C_1	Temporal and precedence constraints - define the possible values for the start and end times of operations and the relations between the start and end time of two operations
C_2	Resource constraints - define the possible set of resources for each operation
C_3	Capacity constraints -limit the available capacity of each resource over time
Questions	
Q_1	What is the minimum C_{max} (makespan)?
Q_2	What is the minimum C_{max}' if the set of resources (i.e. employees) is N?
Q_3	What is the minimum cost of resources K at C_{max}'?
Q_4	Is it possible to schedule in C_{max}'?
Q_5	Is it possible to schedule in C_{max}' if the set of resources (i.e. employees) is N?
Q_6	Is it possible to schedule in C_{max}' if the cost of resources (i.e. employees) is K'?

```
machines(#M).
products(#P).
implementation(#P,#M,execution_time).
employees(#O,limit,cost).
allocation(#O,#M).
precedence(#P,#M,#M).
orders(#P,quantity).
```

Fig. 2. Structure of facts for illustrative example

Table 2. Results of the experiments for questions Q1...Q6 in the hybrid programming framework

Q	parameters	result	Description
Q_1	-------	$C_{max}=14$	
Q_2	N=3	$C_{max}=18$	C_{max}-optimal makspan
Q_3	$C_{max}'=15$	K=1000	C_{max}'-given makspan
Q_4	$C_{max}'=12$	No	K-cost of employees O
Q_5	$C_{max}'=20$, N=2	No	N-the number of employees O
Q_6	$C_{max}'=20$, K=1000	Yes	

This stage of the research showed a very high potential of the framework in supporting various types of decision by questions Q1...Q8. The ease of modeling questions stems from declarativeness of the CLP environment.

In the second stage of the research, effectiveness and efficiency of the proposed framework was evaluated in relation to the MP and CLP. For this purpose, question Q1 (with the largest computational requirements) was implemented in three environments,

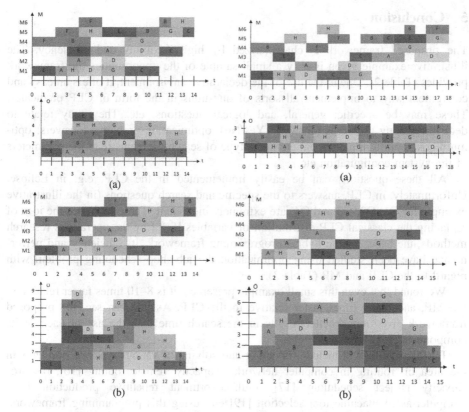

Fig. 3. (a) Schedules for machines and employees corresponding to the question Q1 (only one employee to the machine). (b) Schedules for machines and employees for questions Q1

Fig. 4. (a) Schedules for machines and employees corresponding to the question Q2 for N = 3 (only one employee to the machine). (b) Schedules for machines and employees for questions Q2 for N = 7

CLP, MP, and in the proposed framework. Numerous experiments were performed with varied parameter O. The results are shown in Table 3.

Table 3. Results of the experiments for questions Q1 at different values of set O (employees) in three environments: Framework, MP, the CLP.

N	Framework		MP		CLP		Description
	C_{max}	T	C_{max}	T	C_{max}	T	
1	50	35	50	285	52*	600**	C_{max}-optimal makespan
2	25	34	25	254	27*	600**	T- time of finding solution
3	17	15	17	166	18*	600**	*- feasible solution
4	14	12	14	13	14	131	**-calculation was stopped after 600s
5	14	10	14	12	14	10	N-the number of employees O
6	14	6	14	8	14	10	

5 Conclusion

The proposed framework is characterized by high flexibility and efficiency. The illustrative example given is only a small sample of the opportunities the framework provides (Table 2). Due to the use of the declarative environment, it offers an easy and convenient way to implement all sorts of questions in the form of CLP predicates. These may be specific, general, and logical questions, etc. They may relate to decision-making problems (answers: Yes/No), optimization problems (answers: optimum cost, optimum makespan, etc.) or a type of search problems (e.g. the parameters of the system configuration etc.).

All these questions can be easily implemented in the CLP, e.g. in Eclipse. Unfortunately, in CLP, answers to the specific and search questions (in the illustrative example for Q1, Q2, Q3 and Q4) are extremely inefficient (Table 3), hence the idea of replacing the classical CLP distribution of variables [5, 6] by the operations research methods, the basis for the hybrid programming framework. Its efficiency and performance have been studied using optimization the job-shop scheduling problem with regard to the CLP and MP (Table 3).

We found that even this small example presented, it is 8–10 times faster relative to the MP, and many times faster relative to the CLP. As you can see, the proposed framework gives much better results in the search time than the use of each of its components alone (Table 3).

Further studies will include modeling and solving of various types of problems in the area of: routing in computer networks, capacity vehicle routing [16], resource capacity project scheduling [17], multi-assortment repetitive production [18], computer-aided machine tool selection [19] etc. using this programming framework. Due to the declarativeness and high efficiency of the framework, it is planned to be used for knowledge acquisition from ERP databases for solving the problems above [15].

Appendix A Sets of Facts for Illustrative Example

```
%machines(#M).
machines (m1). machines (m2). machines (m3).
machines (m4). machines (m5). machines (m6).
%products(#P).
products (A). products (B). products (C). products (D).
products (E).
products (F). products (G). products (H). products (I).
products (J).
%implementation(#P,#M,execution_time).
implementation(A,m1,1). ... implementation(J,m6,2).
%employees(#O,limit,cost).
employees(o1,1,40).    employees    (o2,1,30).    employees
(o3,1,30).
```

```
employees(o4,1,20).    employees    (o5,1,20).    employees
(o6,1,20).
%allocation(#O,#M).
allocation (o1,m1). allocation (o1,m2). ... allocation(o6,
m6).
%precedence(#P,#M,#M).
precedence (A,m1,m2). precedence(A,m2,m3). ... precedence
(J,m5,m6).
orders(#P,quantity).
orders(A,1). orders(B,1). ... orders(J,0).
```

References

1. Schrijver, A.: Theory of Linear and Integer Programming. Wiley, New York (1998)
2. Rossi, F., Van Beek, P., Walsh, T.: Handbook of Constraint Programming (Foundations of Artificial Intelligence). Elsevier, New York (2006)
3. Apt, K., Wallace, M.: Constraint Logic Programming using Eclipse. Cambridge University Press, Cambridge (2006)
4. Bocewicz, G., Banaszak, Z.: Declarative approach to cyclic steady states space refinement: periodic processes scheduling. Int. J. Adv. Manuf. Technol. **67**(1–4), 137–155 (2013)
5. Sitek, P., Wikarek, J.: A hybrid approach to the optimization of multiechelon systems. Math. Probl. Eng. Article ID 925675 (2014). doi:10.1155/2014/925675
6. Sitek, P., Nielsen, I.E., Wikarek, J.: A hybrid multi-agent approach to the solving supply chain problems. Procedia Comput. Sci. KES **35**, 1557–1566 (2014)
7. Milano, M., Wallace, M.: Integrating operations research in constraint programming. Ann. Oper. Res. **175**(1), 37–76 (2010)
8. Achterberg, T., Berthold, T., Koch, T., Wolter, K.: Constraint integer programming: a new approach to integrate CP and MIP. In: Trick, M.A. (ed.) CPAIOR 2008. LNCS, vol. 5015, pp. 6–20. Springer, Heidelberg (2008)
9. Guyon, O., Lemaire, P., Pinson, Ă., Rivreau, D.: Solving an integrated job-shop problem with human resource constraints. Ann. Oper. Res. **213**(1), 147–171 (2014)
10. Blazewicz, J., Lenstra, J.K., Rinnooy Kan, A.H.G.: Scheduling subject to resource constraints: classification and complexity. Discrete Appl. Math. **5**, 11–24 (1983)
11. Lawrence, S.R., Morton, T.E.: Resource-constrained multi-project scheduling with tardy costs: comparing myopic, bottleneck, and resource pricing heuristics. Eur. J. Oper. Res. **64**(2), 168–187 (1993)
12. Sitek, P.: A hybrid CP/MP approach to supply chain modelling, optimization and analysis. In: Proceedings of the 2014 Federated Conference on Computer Science and Information Systems, pp. 1345–1352 (2014). doi:10.15439/2014F89
13. Lindo Systems INC: LINDO™ software for integer programming, linear programming, nonlinear programming, stochastic programming, global optimization. www.lindo.com. Accessed 4 May 2015
14. Eclipse: Eclipse - the eclipse foundation open source community website. www.eclipse.org. Accessed 4 May 2015
15. Relich, M.: Using ERP database for knowledge acquisition: a project management perspective. In: Proceedings of International Scientific Conference on Knowledge for Market Practice Use, Olomouc, Czech Republic, pp. 263–269 (2013)

16. Toth, P., Vigo, D.: Models, relaxations and exact approaches for the capacitated vehicle routing problem. Discrete Appl. Math. **123**(1–3), 487–512 (2002)
17. Coelho, J., Vanhoucke, M.: Multi-mode resource-constrained project scheduling using RCPSP and SAT solvers. Eur. J. Oper. Res. **213**, 73–82 (2011)
18. Krenczyk, D., Skolud, B.: Transient states of cyclic production planning and control. Appl. Mech. Mater. **657**, 961–965 (2014). doi:10.4028/www.scientific.net/AMM.657.961
19. Gola, A., Świeć, A.: Computer-aided machine tool selection for focused flexibility manufacturing systems using economical criteria. Actual Probl. Econ. **10**(124), 383–389 (2011)
20. Sitek, P., Wikarek, J.: A hybrid framework for the modelling and optimisation of decision problems in sustainable supply chain management. Int. J. Prod. Res. **1–18**, 2015 (2015). doi:10.1080/00207543.2015.1005762

Tabu Search Algorithm for Routing and Spectrum Allocation of Multicast Demands in Elastic Optical Networks

Róża Goścień[✉]

Department of Systems and Computer Networks,
Wrocław University of Technology, Wrocław, Poland
roza.goscien@pwr.edu.pl

Abstract. In this paper, the problem of routing multicast demands in elastic optical network (EON) is studied. In order to solve the problem in an efficient way, the tabu search algorithm with move prioritization mechanism is proposed. The algorithm is considered in 4 versions, which differ in the move generation process. Next, the numerical experiments are conducted in order to evaluate the algorithm performance and compare its different versions. According to the results, the proposed TS method achieves very good results and significantly outperforms the reference algorithms.

1 Introduction

Elastic optical networks (EONs) are suspected to be a future of optical transport networks, since they combine benefits of flexible frequency grids and advance modulation and transmission techniques [1]. Concurrently, the multicast transmission gains more and more popularity due to efficient realization of such popular services as video on demand, video streaming, content distribution (e.g., software distribution), etc. [2]. The multicast transmission is defined as a data exchange between a source node (root node) and a set of receivers (destination nodes). In this paper, we focus on both these facilities—we study optimization problem of routing multicast demands in elastic optical networks.

In EONs, the entire available frequency resources are divided into narrow, same-size segments, called slices. The width of a particular slice is 6.25 GHz [3]. By grouping an even number of slices, different-width channels can be created and used for data transmission. The basic optimization problem in EONs is called routing and spectrum allocation (RSA). The aim of this problem is to assign for each traffic demand a light-structure, i.e., a routing structure and a channel allocated on this structure. For the considered multicast demands, a routing structure is a tree, which connects source of data and all defined receivers.

Since, the RSA problem is a challenging task, the possibility to solve any problem instance in an efficient way (find optimal or suboptimal solution in reasonable time) is very important. Many different methods were proposed to solve the RSA problem, including integer linear programming (ILP) models

© Springer International Publishing Switzerland 2015
K. Jackowski et al. (Eds.): IDEAL 2015, LNCS 9375, pp. 309–317, 2015.
DOI: 10.1007/978-3-319-24834-9_37

[1,4–7] column generation technique [8], and heuristic algorithms [5–7]. In this paper, we propose to apply tabu search (TS) algorithm for the problem. The TS method was studied in the literature for EONs, however only for unicast and anycast flows [1,6]. Moreover, the problem of routing multicast flows in EONs is a subject of few papers [5,9–11]. Therefore, in order to fill these gaps, we propose a TS-based algorithm for the RSA problem in EONs with multicast demands. To the best of our knowledge, this is the first paper that proposes a TS method for optimizing multicast flows in EONs and one of the few focused on the multicasting in EONs.

2   Problem Formulation

The elastic optical network is presented as a directed graph $G = (V, E)$, where V is a set of network nodes and E is a set of network fiber links. On each network link the frequency resources are divided into slices, wherein each link offers the same number of $|S|$ slices.

The set of multicast demands $d \in D$ is given. Each demand is characterized by a root node, set of receivers, and demand volume (in Gbps). In the modelling we focus only on the transmission from a root node to receivers, due to the fact that the amount of data transmitted in the opposite direction is very small.

We apply link-path modelling approach [12] and assume that for each demand d a set of k candidate trees is given ($t \in T_d$). The trees are calculated in pre-processing stage with respect to demand root node and set of receivers.

In order to model spectrum usage, we use the concept of frequency channels [4]. For each multicast demand and its candidate tree a set of candidate frequency channels is calculated with respect to demand volume and tree length (the distance between root node and the most distant receiver). To calculate the width of a channel required for a particular demand and its tree, we apply the procedure presented in [5].

The aim of the RSA problem is to assign for each demand a routing tree and a channel allocated on this tree. The problem objective to minimize maximum spectrum usage, i.e., number of slices that are required in the network to support all traffic demands.

objective

$$\min z = \sum_s x_s \tag{1}$$

constraints

$$\sum_{t \in T_d} \sum_{c \in C_{dt}} x_{dtc} = 1; d \in D \tag{2}$$

$$\sum_{d \in D} \sum_{t \in T_d} \sum_{c \in C_{dt}} \alpha_{dtcs} \beta_{edt} x_{dtc} = x_{ex}; e \in E, s \in S \tag{3}$$

$$x_{es} \leq x_s; e \in E, s \in S \tag{4}$$

indices

$e \in E$ network links

$s \in S$ frequency slices

$d \in D$ multicast demands

$t \in T_d$ candidate routing trees for demand d

$c \in C_{dt}$ candidate frequency channels for demand d on candidate tree t

constants

α_{dtcs} = 1, if channel c allocated on tree t for a demand d uses slice s; 0, otherwise

β_{edt} = 1, if candidate tree t for a demand d uses link e; 0, otherwise

variables

x_{dtc} = 1, if demand d is realized using tree t and channel c; 0, otherwise

x_{es} = 1, if slice s is used on link e; 0, otherwise

x_s = 1, if slice s id used on any network link; 0, otherwise

The formula (1) defines problem objective (the number of slices used in the network), wherein the Eqs. (2)-(4) describe its constraints. Equation (2) guarantees that for each demand exactly one candidate tree and one candidate channel are assigned. The formula (3) guarantees that a slice on a particular link can be used by at most one traffic demand. Eventually, the Eq. (4) defines variable x_s, which indicates if slice$s$ is used on any network link.

3 Tabu Search Algorithm

The algorithm proposed for the optimization problem is based on the standard tabu search (TS) mechanism introduced initially by Glover [13], with modification proposed in [6]. The idea of the method is presented in Fig. 1.

In the proposed TS notation, a solution is a demands allocation order and a tree selection for each demand. According to this information, traffic demands are allocated one by one to the selected trees with respect to first-fit spectrum assignment method. Next, a move operation is a simple modification of a solution. In the paper, we consider two types of move operation: tree swap and demands swap. The former type changes selected tree for one demand, wherein the latter type swaps two demands in the demand allocation order.

At the beginning of the TS process, an initial solution is calculated by the algorithm Adaptive Frequency Assignment (AFA) [5]. This solution is stored as a current and the best found solution. Next, the TS iteratively search through the solution space in order to find a better solution. In each iteration, a candidate solution is obtained by applying a non-tabu move operation to the current solution. The obtained candidate solution is compared with the current and best found solution. If it is better than the current solution, it is stored as a new current solution. Similarly, if it is better than the already best found solution, it is saved as a new best solution. If the candidate solution is not better than the best found solution, no improvement is reached in the iteration. The TS method tries to improve solution until the maximum number of iterations is reached.

The TS search process is controlled by a diversification process, which is run when the number of iterations without improvement exceeds a predefined threshold. The value of this threshold is defined as a ratio of the demands number and is an algorithm input parameter. The aim of the diversification process is to move search process into another part of the solution space. It can be achieved by changing current solution. To this end, the diversification process multiplies by coefficient $\delta > 1$ value of the objective function related to the current solution. The value of δ is an algorithm input parameter.

The move generation process is supported by a move prioritization and two lists of forbidden moves: tabu list (TL) and used moves (UM) list. Note, that the TS method can also run without move prioritization process. Then, in each iteration a non-tabu move operation is chosen randomly.

The move prioritization process assigns a priority (high, middle, low) to each feasible move operation. The higher move priority, the higher probability of applying this operation in the current iteration. By default, all moves have low priority. Then, moves that are likely to improve current solution are assigned with a higher priority value. In particular, for each predefined tree (taking into account all candidate trees for all demands) a special metric is calculated, which refers to the number of free slices on the tree links. We consider three metric definitions: *Avg*, *Min*, and *Sum*. The first metric is an average number of free slices, wherein the second one is a minimum number of free slices (over all tree links). The last metric sums number of free slices on all tree links. Then, the metric is used to order the trees according to its increasing value. Note, that the ordered set includes all $|T| = kD$ candidate trees (both selected and non-selected for demands). Then, if $\gamma|T|$ first trees from the ordered set are selected for demands, the move operations that change these trees into candidate trees with relatively high metric value are assigned with high priority. Similar process is also applied for the next $\gamma|T|$ trees from the order set, wherein the related move operations are assigned with middle priority. The value of the γ coefficient is an algorithm input parameter.

Concurrently, the TL and UM lists store moves that can not be applied in the current iteration, since they can leed to solutions already considered. The TL stores moves, which led to improvement in previous iterations. It is implemented as a FIFO queue, wherein its size is defined as a ratio of the demands number and is an algorithm input parameter. The UM list contains move operations applied in the previous iterations without improvement. It is cleaned when an improvement is reached and after application of a diversification process.

4   Results

In this section we present results of the investigation focused on the efficiency analysis of the proposed TS method.

We compare 7 different TS versions in terms of the metric used in the prioritization process and value of the γ parameter. We refer to them as $TS_{metric(\gamma)}$ or briefly $metric(\gamma)$, where $metric$ is a metric used in a prioritization process (avg,

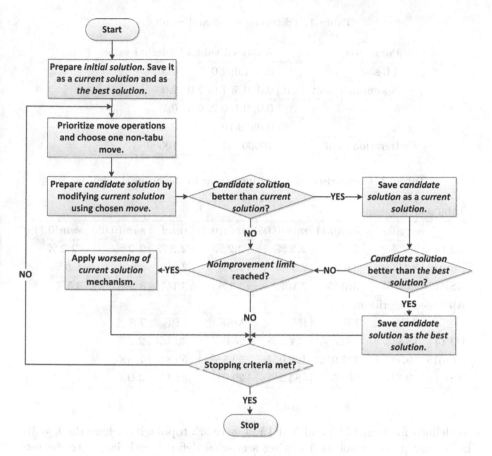

Fig. 1. The tabu search algorithm

min, or sum) and γ refers to the value of γ parameter. Note, that TS method without move prioritization process is indicated as TS_{rand} or *rand*. Moreover, in the experiments we use 5 reference methods, which were already proposed in the literature for optimization problems in EONs and adapted for the considered problem formulation. The reference algorithms are First Fit (FF) [14], Most Subcarrier First (MSF) [7], Longer Path First (LPF) [7], Adaptive Frequency Assignment (AFA) [5], and Random Assignment (RA), where the demands allocation order and routing trees are selected at random. To obtain optimal results we use CPLEX 12.5 solver.

In order to determine values of the TS input parameters that provide very good results, we performed initial experiments. In Table 1 we present analyzed values of the TS parameters and values selected for further experiments.

In the investigation, we apply the same EON assumptions as in [5]. Moreover, we use two groups of testing scenarios with multicast demands. The first group includes 30 relatively small problem instances with total traffic volume 5 Tbps.

Table 1. TS tuning setup and results

Parameter	Analyzed values	Selected value
TL size	0.5; 1.0; 2.0	0.5
No-improvement limit	0.0; 0,5; 1.0; 2.0	2.0
δ	0.0; 0.1; 0.2; 0.3	0.1
γ	0.05; 0.10	0.10
Iteration limit	10000	10000

Table 2. Comparison of algorithms—gap to the optimal solution

Tabu search versions							
	$avg(0.05)$	$avg(0.1)$	$min(0.05)$	$min(0.1)$	$rand$	$sum(0.05)$	$sum(0.1)$
DT14	3.0%	2.7%	**2.2%**	**2.2%**	**2.2%**	**2.2%**	**2.2%**
Euro16	3.1%	2.0%	2.3%	2.3%	2.8%	2.0%	**1.6%**
NSF15	**3.0%**	**3.0%**	3.1%	3.3%	3.1%	3.3%	3.5%
All tested algorithms							
	AFA	FF	RA	MSF	LPF	TS_{best}	
DT14	13.1%	82.5%	77.8%	26.0%	30.1%	**2.2%**	
Euro16	9.4%	137.0%	129.0%	59.5%	59.8%	**1.6%**	
NSF15	9.7%	67.5%	84.2%	29.0%	35.1%	**3.0%**	

It is defined for Euro16 [6] and NSF15 [6] network topologies, where the $k = 10$. The second group involves 5 testing scenarios defined with larger traffic sets (up to 20 Tbps) for US26 [1] and Euro28 [6] network topologies. Moreover, the number of available candidate trees is $k = 10, 30, 100, 200$.

In case of small problem instances we present results as an optimality gaps, i.e., how much more resources are necessary solving problem with considered method compared to the optimal solution. Since the optimal solutions are not available for larger problem instances (the problem is too complex to reach optimal results in reasonable time [5]), as a comparison metric we use gap to the best found solution (among all tested methods).

The presented results are averaged over a number of testing scenarios. Additionally, the TS and RA results are also averaged over 10 algorithm repetitions.

In Table 2, we present optimality gaps for all tested methods (all TS versions and the reference algorithms). Moreover, as TS_{best} we indicate results of the best TS version (for a particular network topology). As we can see, all TS versions reached very good results with small optimality gaps (<4%). However, the performance of the TS implementations strongly depends on the network topology and therefore it is impossible to determine the best TS version (regardless of the considered network topology). However, it is clearly visible that the proposed the TS outperforms other methods.

In case of the small testing scenarios, the processing time of all heuristic algorithms was less than 1 s, wherein the TS method solved a problem instance in less than 10 s. However, the CPLEX solver needed on average up to 600 s to find the optimal solution for a testing scenario.

The results for larger problem instances are presented in Table 3. The results show that the efficiency of the TS versions depends strongly on the problem instance characteristics (network topology and number of candidate trees). However, the pure TS_{rand} and $TS_{avg(0.05)}$ reached best results. What is more, the results once again prove that the proposed TS algorithm outperforms the reference methods.

Table 3. Comparison of algorithms—gap to the best found solution

Euro28

Tabu search versions

	avg(0.05)	avg(0.1)	min(0.05)	min(0.1)	rand	sum(0.05)	sum(0.1)
$k = 10$	**1.0 %**	2.7 %	3.0 %	2.3 %	**1.0 %**	2.0 %	1.4 %
$k = 30$	2.2 %	1.5 %	1.8 %	2.2 %	**0.7 %**	3.6 %	2.6 %
$k = 100$	**0.8 %**	1.2 %	2.3 %	2.7 %	1.2 %	1.9 %	2.3 %
$k = 200$	2.0 %	1.2 %	2.3 %	1.2 %	1.8 %	2.7 %	**1.1 %**

All compared methods

	AFA	FF	RA	MSF	LPF	TS_{best}	
$k = 10$	12.9 %	41.5 %	49.8 %	25.2 %	31.4 %	**1.0%**	
$k = 30$	8.3 %	51.0 %	65.0 %	18.9 %	27.3 %	**0.7%**	
$k = 100$	7.0 %	68.1 %	83.9 %	19.3 %	28.5 %	**0.8%**	
$k = 200$	6.5 %	71.5 %	94.5 %	21.0 %	24.1 %	**1.1%**	

US26

Tabu search versions

	avg(0.05)	avg(0.1)	min(0.05)	min(0.1)	rand	sum(0.05)	sum(0.1)
$k = 10$	**1.5 %**	3.3 %	2.2 %	1.8 %	2.6 %	4.3 %	4.4 %
$k = 30$	3.0 %	3.2 %	4.4 %	3.3 %	2.9 %	**1.8 %**	4.2 %
$k = 100$	3.2 %	2.4 %	4.8 %	4.0 %	**1.6 %**	2.9 %	4.5 %
$k = 200$	2.9 %	2.9 %	6.7 %	**1.7 %**	2.5 %	3.8 %	3.4 %

All compared methods

	AFA	FF	RA	MSF	LPF	TS_{best}	
$k = 10$	10.0 %	67.5 %	68.4 %	24.8 %	42.1 %	**1.5%**	
$k = 30$	13.9 %	79.2 %	86.5 %	25.1 %	41.2 %	**1.8%**	
$k = 100$	5.2 %	93.5 %	98.2 %	27.8 %	35.7 %	**1.6%**	
$k = 200$	8.7 %	96.8 %	103.3 %	24.9 %	35.0 %	**1.7%**	

Table 4. Algorithms processing time [s] for Euro28 topology

	FF, RA, MSF, LPF	AFA	$TS_{avg(0.05)}$	$TS_{min(0.05)}$	TS_{rand}	$TS_{sum(0.05)}$
$k = 10$	< 0.100	0.046	22.664	55.289	36.513	51.513
$k = 30$	< 0.100	0.133	60.446	133.308	19.159	166.466
$k = 100$	< 0.100	0.531	191.459	171.099	30.870	338.258
$k = 200$	< 0.100	0.510	207.884	242.763	52.337	388.920

Additionally, in Table 4 we present average processing time for selected methods and Euro28 topology. As we can see, the processing time increases with increasing number of candidate trees. In case of the TS method, the processing time is the shortest for pure TS_{rand} implementation and increases for implementations with move prioritization process.

5   Conclusion

In this paper, we have proposed an efficient TS algorithm for the optimization problem in EONs. We have implemented different algorithm versions in terms of the metric used in the move prioritization process. Next, we have performed numerical experiments to evaluate efficiency of the proposed method based on the comparison with reference algorithms. The results prove that the proposed TS algorithm is a very efficient tool and significantly outperforms other methods.

In the future work, we plan to extend proposed method to the optimization problem in survivable EONs, which support also unicast and anycast flows.

Acknowledgement. This work was supported by the Polish National Science Centre (NCN) under Grant DEC-2012/07/B/ST7/01215 and statutory funds of the Department of Systems and Computer Networks, Wroclaw University of Technology.

References

1. Goścień, R., Walkowiak, K., Klinkowski, M.: Distance-adaptive transmission in cloud-ready elastic optical networks. J. Opt. Commun. Netw. **6**(10), 816–828 (2014)
2. Kmiecik, W., Goścień, R., Walkowiak, K., Klinkowski, M.: Two-layer optimization of survivable overlay multicasting in elastic optical networks. Opt. Switch. Netw. **14**, 164–178 (2014)
3. ITU-T Recommendation G.694.1 (ed. 2.0), Spectral grids for WDM applications: DWDM frequency grid, February 2012
4. Velasco, L., Klinkowski, M., Ruiz, M., Comellas, J.: Modeling the routing and spectrum allocation problem for flexgrid optical networks. Photonic Netw. Commun. **24**(3), 177–186 (2012)
5. Walkowiak, K., Goścień, R., Klinkowski, M., Woźniak, M.: Optimization of multicast traffic in elastic optical networks with distance-adaptive transmission. IEEE Comm. Lett. **18**(12), 2117–2120 (2014)

6. Goścień, R., Walkowiak, K., Klinkowski, M.: Tabu search algorithm for routing, modulation and spectrum allocation in elastic optical network with anycast and unicast traffic. Comput. Netw. **79**, 148–165 (2015)
7. Christodoulopoulos, K., Tomkos, I., Varvarigos, E.A.: Elastic bandwidth allocation in flexible OFDM-based optical networks. IEEE/OSA J. Lightwave Technol. **29**(9), 1354–1366 (2011)
8. Klinkowski, M., Pióro, M., Żotkiewicz, M., Ruiz, M., Velasco, L.: Valid inequalities for the routing and spectrum allocation problem in elastic optical networks. In: Proceedings of ICTON (2014)
9. Gong, L., Zhou, X., Liu, X., Wenwen, Z., Wei, L., Zuqing, Z.: Efficient resource allocation for all-optical multicasting over spectrum-sliced elastic optical networks. J. Opt. Commun. Netw **5**(80), 836–847 (2013)
10. Wang, Q., Chen, L. K.: Performance analysis of multicast traffic over spectrum elastic optical networks. In: Proceedings of OFC (2012)
11. Yu, Z., Zhao, Y., Zhang, J., Yu, X., Chen, B., Lin, X.: Multicast routing and spectrum assignment in elastic optical networks. In: Proceedings of ACP (2012)
12. Pióro, M., Medhi, D.: Routing, flow and capacity design in communication and computer networks (2014)
13. Glover, F.: Tabu Search Fundamentals and Uses. University of Colorado, June 1995
14. Jinno, M., Kozicki, B., Takara, H., Watanabe, A., Sone, Y., Tanaka, T., Hirano, A.: Distance-adaptive spectrum resource allocation in spectrum-sliced elastic optical path network. IEEE Commun. Mag. **48**(8), 138–145 (2010)

Simulated Annealing Algorithm for Minimization of Bandwidth Fragmentation in Elastic Optical Networks with Multicast and Unicast Flows

Piotr Nagły and Krzysztof Walkowiak[✉]

Department of Systems and Computer Networks,
Wroclaw University of Technology, Wroclaw, Poland
Krzysztof.walkowiak@pwr.edu.pl

Abstract. The recently proposed idea of Elastic Optical Networks (EON) is expected to largely improve performance of optical networks in terms of resource utilization. However, the possibility to divide the optical spectrum into relatively narrow segments called slices in EONs triggers the problem of bandwidth fragmentation. In this paper, we propose a novel metaheuristic algorithm based on the Simulated Annealing approach to solve the Routing and Spectrum Assignment (RSA) problem with the objective function denoting the bandwidth fragmentation. It is assumed that two types of network flows are optimized, namely, multicast and unicast. Using two representative network topologies, we run extensive numerical experiments to tune the SA algorithm and verify its performance in various scenarios. The main conclusion is that the application of our SA algorithm can significantly improve the bandwidth fragmentation in EONs.

Keywords: Simulated Annealing · Elastic Optical Networks · Bandwidth fragmentation · Multicast · Optimization

1 Introduction

In recent years, the behavior of ordinary user of the Internet has been changed. One of the most popular services is sharing of the video through the Internet including services like IPTV, Video on Demand. It is estimated that in the next 3 years global IP traffic will increase by 60 % [1], on the other hand a six-fold increase in overall mobile data traffic till 2019 is expected [2]. This triggers the demand to develop scalable and efficient optical transport platform which can handle demands beyond 100 Gb/s. Moreover, by the introduction of new bandwidth-demanding services, such as Content Delivery Networks (CDNs), IP television or cloud computing can be provisioned in a cost-effective way using multicasting defined as a *one-to-many* transmission.

Elastic Optical Network (EON) is a recent approach proposed for optical networks [3] that meets the above objectives. In EONs, there is a bandwidth fragmentation problem [4]. In more detail, the whole bandwidth (spectrum) is divided into single and relatively narrow frequency slots called *slices* [5]. The channel required to

© Springer International Publishing Switzerland 2015
K. Jackowski et al. (Eds.): IDEAL 2015, LNCS 9375, pp. 318–327, 2015.
DOI: 10.1007/978-3-319-24834-9_38

provision a demand consist of a particular amount of adjacent frequency slices with 12.5 GHz width selected accordingly to serve a requested bit-rate. Thus, the spectrum resources can be utilized in a quite efficient way, while in traditional WDM (Wavelength Division Multiplexing) networks, a fixed grid of spectrum with 50 GHz slots are used each serving up to 100 Gb/s [3], what means that if the required bit-rate is lower than 100 Gb/s, some part of spectrum resources is wasted. Elastic frequency grid in EONs, defined in [5] enables such benefits as [3, 6, 7]:

- Segmentation. On the contrary to WDM, EONs could save some part of spectrum if bitrate of demand is lower than e.g. 100 Gb/s. Such a channel can be adjusted to serve this demands without loss of spectrum.
- Aggregation. EONs allows to aggregate spectrum resources to create channel that could serve big demands above 100 Gb/s. In WDM such a demand is divided and provisioned in two or more channels.
- Possibility of use multiple modulation formats, which can reduce usage of spectrum resources at short transmission range.
- There is no need to rebuild current optical infrastructure to introduce EONs.

The main contribution of this paper is a novel Simulated Annealing (SA) algorithm proposed for the fragmentation problem in EONs with unicast and multicast traffic. Moreover, a detailed tuning of the SA algorithm is performed to find the best values of the algorithm's parameters as well as results various experiments are presented including a comparison of the SA performance against other heuristics.

The rest of the paper is organized in the following way. In Sect. 2, the optimization problem is formulated. Section 3 describes the algorithm. In Sect. 4, we report results of numerical experiments. Finally, the last section concludes this work.

2 Optimization Problem

In EONs, a new optimization problem occurs, known as Routing and Spectrum Assignment (RSA) [8, 10]. It is an extension of the classical Routing and Wavelength Assignment (RWA) problem defined for WDM networks [9]. The RSA problem is to allocate demands on individual links, having regard to following constraints:

- Frequency slices that create a channel need to be contiguous – an assurance of continuity of spectrum for each demand.
- Channel that supports a specific demand must be the same on whole length of optical path – all slices of each link that belong to channel must be allocated in the same part of the spectrum.
- Frequency slice could serve only one demand – each slice could be part of only one optical path.

Above mentioned requirements trigger fragmentation of spectrum resources in network links, which leads to a worse spectral efficiency in the EON [4]. Therefore, there is a need to optimize EONs in terms of spectrum fragmentation metrics. The considered EON is denoted as a directed graph $G = (V,E)$, where V is a set of nodes and E is a set of directed links. The unicast demand is described by a source node and one client node, however the multicast demand is described by a source node and set of client nodes. For each demand d we are given the demand volume h_d expressed in Gb/s and a set $P(d)$ of candidate structures. If d is a unicast demand, structure $p \in P(d)$ is a path connecting end nodes of the demand. If d is a multicast demand, structure $p \in P$ (d) is a tree with the root in the source node and including all client nodes.

Presented below is the optimization model which is based on [11] and describes optimization of unicast and multicast flows in EONs using the candidate tree approach. The objective is to minimize the bandwidth fragmentation. Set $C(d,p)$ includes available spectrum channels for demand d using structure p. Each channel $c \in C(d,p)$ contains sufficient amount of slices to realize demand d on structure p. Authors of [4] raised the issue of bandwidth fragmentation in EONs. They proposed a FR coefficient as a metric in RSA, which can be used to reduce spectrum fragmentation. In general, FR describes level of fragmentation of link e. In [4], two new algorithms that controls the FR growth are proposed: *Minimum Network Fragmentation Ratio RSA* (MNFR-RSA) and *Maximum Local Utilization RSA* (MLU-RSA). In this paper, as objective function we take the MNFR-RSA approach.

Sets

D	demands (unicast and multicast)
$P(d)$	candidate structures for demand d
E	directed network links
S	slices
$C(d,p)$	candidate channels for demand d allocated on structure p

Constants

$\delta_{edp} = 1$,   if link e belongs to structure p realizing demand d; 0, otherwise

$\gamma_{dpcs} = 1$,   if channel c associated with demand d on candidate structure p uses slice s; 0, otherwise

Variables

$x_{dpc} = 1$,   if channel c on candidate structure p is used to realize demand d; 0, otherwise (binary)

$z_{es} = 1$,   if slice s is occupied on link e; 0, otherwise (binary)

$z_s = 1$,   if slice s is occupied on any network link; 0, otherwise (binary)

f_{ei} size of ith available slice block (block of contiguous available slices)

FR_e fragmentation ratio coefficient in link e

Objective

$$\min NFR = \frac{\sum_{e \in E} FR_e}{|E|} \quad (1)$$

Constrains

$$\sum_p \sum_c x_{dpc} = 1, \quad d \in D \quad (2)$$

$$\sum_d \sum_p \sum_c \gamma_{dpcs} \delta_{edp} x_{dpc} \leq z_{es}, \quad e \in E, \quad s \in S \tag{3}$$

$$\sum_e z_{es} \leq |E| z_s, \quad s \in S \tag{4}$$

$$FR_e = 1 - \frac{\sum_i f_{ei}^2}{\left(\sum_i f_{ei}\right)^2} \tag{5}$$

The objective function (1) defines level of fragmentation of whole network. It uses FR of each link defined in (5). In more detail, giving the allocation of demands denoted by x_{dpc}, for each link e the spectrum usage is analyzed and values of f_{ei} are obtained. The smaller value of FR, the lower level of fragmentation of links in the network. Equation (2) assures that for each demand d exactly one candidate structure is used. The next constraint (3) guarantees that each slice on each link can be assigned to only one candidate structure p and the designated channel. Inequality (4) defines that slice s is used in the network only if there is at least one link on which s is allocated.

3 Simulated Annealing Algorithm

In this section, we present a Simulated Annealing (SA) algorithm proposed to solve the RSA problem with joint unicast and multicast traffic defined as (1)–(5). SA is a generic probabilistic heuristic for the global optimization of a given function [12].

In our SA algorithm, a solution of the optimization problem is defined a sequence (ordering) of demands and candidate structure for each demand defined as [(seq_1, p_1), (seq_2, p_2),..., $(seq_{|D|}, p_{|D|})$], where seq_d denotes the position of demand d in the sequence (e.g, if demand d is the first demand in the sequence, then $seq_d = 1$) and p_d denotes the index of candidate path used to allocated demand d. To calculate the objective function of a particular solution, we allocate the demands using a following procedure. For each demand d in the sequence, we allocate the demand on candidate structure p_d using the first fit procedure [3]. The amount of slices required to allocate demand d is determined according to required bit-rate and transmission range of demand d. When all demands are allocated, for each link, the FR coefficient (5) is calculated and after that the objective function (1) is obtained.

We implemented 5 methods of generating the initial solution for the SA method. Each of the methods is a greedy algorithm that orders the demands in a specific sequence and next one-by-one allocates each demand in the EON. Optimization affects both types of flows, namely, unicast and multicast. Therefore, each algorithm processes the demands following a certain order:

- UM – first unicast demands, next multicast demands.
- MU – first multicast demands, next unicast demands.
- RND – random order.

Using some preliminary experiments, we tested each algorithm to find the order than provides the best results. Table 1 presents all implemented algorithms with specific type order, which gave the best results.

Table 1. Initial solution method.

Algorithm	Ordering	Description
FF (First Fit)	UM	Demands are allocated in pursuance to order in files on **first** possible candidate path / tree
BDF (Bigger Demand First)	RND	Demands are allocated in pursuance to decreasing bitrates on **first** possible candidate path / tree
RA (Random Assignment)	MU	Demands are allocated in random order (having regard to types order) on **random** candidate path / tree
MNFR (Minimum Network Fragmentation Ratio)	RND	Method proposed in [4]
BDFMNFR (Bigger Demand First Minimum Network Fragmentation Ratio)	RND	Method that combines BDF and MNFR algorithms. First we sort all demands in pursuance to decreasing bitrates. Next we analyse each demand and choose the candidate structure that gives the lowest NFR

Our SA algorithm uses 2 methods to determine a neighborhood solution:

- Swapping two randomly selected demands. In the case of UM or MU ordering, SA selects at random first demand, checks its type and next selects at random the next demand of the same type. Next, both demands are swapped, their position in the processing sequence is changed accordingly. Otherwise, SA chooses two random demands no matter of the type. After that the algorithm swaps chosen demands in a sequence.
- Changing the path/tree for randomly selected demand. The algorithm selects at random a demand, and simply changes the candidate structure to the next one in the set of all candidate structures available for a particular demand.

Moreover, the SA algorithm uses 4 main tuning parameters:

- Number of iterations.
- Initial temperature.
- Number of iterations without progress – after this number of iterations the method of finding the neighborhood solution is changed.
- Cooling rate parameter.

Figure 1 shows the pseudocode of the algorithm. In line 1, we calculate the initial solution *Sol* denoted with the value of the NFR function denoted as *nfrCurrent* and initial temperature *T* by multiply *nfrCurrent* and initial temperature coefficient *cTemper* (see Table 1). The solution included in *Sol* is modified throughout the algorithm. After that, simulation begins until condition in line 2 is fulfilled (number of iterations *i* is smaller than the maximum number of iterations *iMax*). In lines 4–7, the method of finding a neighborhood solution is chosen. If the amount of iterations without progress *n* is less than *nMax*, chosen demand will be swapped and a swap flag *fSwap* will be *true*. Otherwise, a path of the selected demands will be changed at random and flag *fSwap* will be equal to *false*. Next, value of objective function of a new solution

i	number of iterations
$iMax$	maximum number of iterations
n	number of iterations without progress
$nMax$	maximum number of iterations without progress
crp	cooling rate parameter
T	temperature
$cTemper$	initial temperature coefficient
$SolCurrent$	current solution (selected structures and channels)
$SolTemp$	temporary solution (selected structures and channels)
$nfrCurrent$	current best solution – NFR value
$nfrTemp$	temporary obtained solution – NFR value
$d1, d2$	demands
$fSwap$	logical flag

```
1.    SolCurrent=InitMethod(); nfrCurrent=FindNFR(SolCurrent); T=nfrCurrent*cTemper;
2.    while (i < iMax):
3.          d1 = rand(Demands); d2 = rand(Demands);
4.          if (n < nMax):
5.                SolTemp = Demands.Swap(d1,d2); fSwap = true;
6.          else:
7.                SolTemp = Demands.ChangePath(d1); fSwap = false;
8.          nfrTemp =FindNFR(SolTemp);
9.          if (nfrTemp < nfrCurrent):
10.               nfrCurrent = nfrTemp; SolCurrent = SolTemp; n=0; T=T*crp;
11.         else:
12.               x = rand();
13.               if (x < exp(-delta/T)):
14.                     nfrCurrent = nfrTemp; SolCurrent = SolTemp; n=0; T=T*crp;
15.               else:
16.                     T=T*crp; n++;
17.         i++;
18.   return nfrCurrent;
```

Fig. 1. SA pseudocode

$nfrTemp$ is calculated in line 8, and next $nfrTemp$ is compared with $nfrCurrent$ (line 9). If $nfrTemp$ is better, then it becomes currently the best one. We reset the n number and reduce the temperature T with the s parameter. The new iteration begins. When a new solution is not better than current one, then SA decides with probability equal to the Boltzmann function, if worse solution should be assigned to $nfrCurrent$ (lines 11–14). In case of rejection of the worse solution, the SA brings back to the previous solution, increases n and reduces temperature (lines 15–16). When condition from line 2 is fulfilled then SA stops and $nfrCurrent$ (the best solution) is returned.

4 Results

In this section, we present results of computational experiments. The goal of experiments was threefold. First, we discuss the results of tuning the SA. Second, we compare efficiency of all initial solution methods and the presented SA algorithm.

Third, we present results of the network fragmentation (NFR) as a function of the number of candidate structures (paths and trees).

Two representative network topologies are used: US backbone network called US26 (26 nodes, 84 directed links) and a pan-European network called Euro28 (28 nodes, 82 directed links). All assumptions regarding the EON (physical model, modulation formats) are the same as in [11]. 75 traffic sets including both unicast and multicast demands were generated at random. The bit-rate of an unicast demand is in range 10–400 Gb/s, while the bit-rate of a multicast session is in range 10–200 Gb/s. There are three types of sets different in terms of the overall bit-rate, namely, 60 Tb/s, 52 Tb/s and 40 Tb/s. The multicast flows always contribute 20 Tb/s, while the rest of traffic belongs to unicast flows. Various combinations of candidate structures sets are used in the experiments starting from $k = 2$ candidate paths for unicast demands and $t = 10$ candidate trees for multicast demands denoted as $(2,10)$ up to $k = 30$ and $t = 1000$ denoted as $(30,1000)$. To find the candidate paths the shortest path algorithm with link distance as the metric was applied. The candidate trees were generated using a method described in [11].

The first goal of numerical experiments is to tune the SA algorithm. Each simulation was repeated 10 times. Table 2 describes all tested parameters with the finally selected values. Moreover, Fig. 2 shows performance of SA for US26 network as a function of number of iterations and different values of cooling parameter. As we can suppose, lower values of the iteration number leads to a greater value of the objective function NFR. Cooling rate parameter equals to 0.99 and 0.995 has a specific shape. The reason of this is that the temperature reaches almost immediately a near zero value and algorithm stuck in local minimum, because probability of finding a worse solution is very low. On the base of Fig. 2, we choose 0.999 for the cooling parameter and 14000 iterations, as the best trade-off, since higher numbers of iterations causes minimal progress of optimization and much longer simulations. Performance for the Euro28 network was comparable to US26.

Table 2. Results of tuning process.

Parameter	Tested values	Selected for Euro28	Selected for US26
Number of iterations	5000–25000 with 1000 step	14 000	14 000
Initial temperature	2500, 3000, 3500, 4000, 4500	3500	2500
Number of iterations without progress	25, 50, 75, 100	50	50
Cooling rate parameter	0.99, 0.995, 0.999, 0.9995	0.999	0.999
Initial solution method	FF, BDF, RA, MNFR, BDFMNFR	FF	FF

Figure 3 presents a comparison of all tested methods (including SA) in terms of the objective (1) as a function of overall network traffic. The results are obtained for $k = 15$

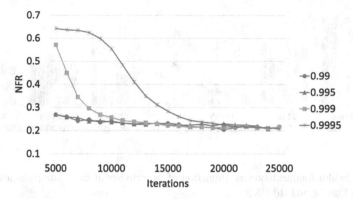

Fig. 2. Convergence of algorithm SA for US26 network as a function of cooling rate

candidate paths and $t = 300$ candidate trees. Each bar is an average result over 25 different traffic sets. We can easily notice that our SA method significantly outperforms other methods. However, the average execution time of the SA algorithm is about 100 s, while other heuristics require 1–20 s. However, it should be mentioned that the reported execution time of SA does not include the time required to tune the algorithm. Nevertheless, the tuning process is done once and next the selected parameters are used for all tests. It is interesting to point out that the FF method selected as the initial method for the SA algorithm does not provide the best results among all methods considered for initial solution calculation. According to our analysis, in cases when a relatively good starting solution was used to initiate the SA method (e.g., MNFR), the SA algorithm was not able to improve this solution and finally offered result worse than the SA with starting solutions yielded by the FF algorithm.

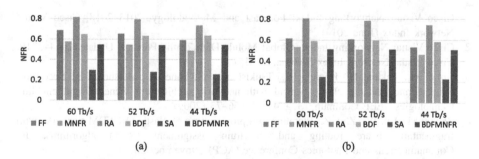

Fig. 3. Algorithms comparison for networks (a) Euro28 and (b) US26.

Finally, we focus on the results showing how the fragmentation expressed as the NFR function depends on the number of candidate structures. For this purpose, we use the SA algorithm tuned as presented above. Figure 4 reports the obtained results. The main observation is that increasing the number of candidate structures does not significantly change the performance of the SA method in terms in the network fragmentation.

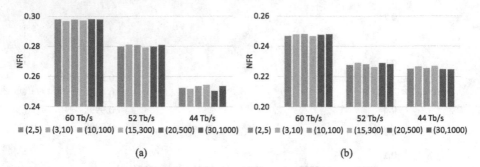

(a)   (b)

Fig. 4. Bandwidth fragmentation as a function of the number of candidate paths and trees for networks (a) Euro28 and (b) US26.

5 Concluding Remarks

In this paper, we focused on the bandwidth fragmentation problem in EONs with unicast and multicast traffic. To solve this problem, we proposed a metaheuristic SA algorithm. To assess the SA performance, we compared its results with results generated by other reference algorithms. The numerical results show that SA outperforms other heuristics.

Acknowledgments. This work was supported by the statutory funds of the Department of Systems and Computer Networks, Wroclaw University of Technology.

References

1. Cisco Visual Networking Index: Forecast and Methodology, 2013–2018, Cisco Visual Network Index (June 2014)
2. Cisco Visual Networking Index: Global Mobile Data Traffic Forecast Update, 2014–2019, Cisco Visual Network Index (February 2015)
3. Jinno, M., Takara, H., Kozicki, B., Tsukishima, Y., Sone, Y., Matsuoka, S.: Spectrum-efficient and scalable elastic optical path network: architecture, benefits, and enabling technologies. IEEE Commun. Mag. **47**(11), 66–73 (2009)
4. Zhang, M., Lu, W., Zhu, Z.: Planning and provisioning of elastic O-OFDM networks with fragmentation-aware routing and spectrum assignment (RSA) algorithms. In: Communications and Photonics Conference (ACP) (November 2012)
5. ITU-T Recommendation G.694.1 (ed. 2.0), Spectral grids for WDM applications: DWDM frequency grid (February 2012)
6. Jinno, M., Takara, H., Yonenaga, K.:Why do we need elastic optical path networking in the 1 Tb/s era?. Quantum Electronics Conference and Lasers and Electro-Optics, (CLEO/IQEC/PACIFIC RIM), pp. 466–468 (2011)
7. Gerstel, O., Jinno, M.: Elastic optical networking: a new dawn for the optical layer? IEEE Commun. Mag. **50**(2), 12–20 (2012)
8. Jinno, M., et al.: Distance-adaptive spectrum resource allocation in spectrum-sliced elastic optical path network. IEEE Commun. Mag. **48**(8), 138–145 (2010)

9. Pioro, M., Medhi, D.: Routing, Flow and Capacity Design in Communication and Computer Networks. Morgan Kaufmann, Los Altos (2004)
10. Klinkowski, M., Walkowiak, K.: Routing and spectrum assignment in spectrum sliced elastic optical path network. IEEE Commun. Lett. **15**(8), 884–886 (2011)
11. Walkowiak, K., Goścień, R., Klinkowski, M., Woźniak, M.: Optimization of multicast traffic in elastic optical networks with distance-adaptive transmission. IEEE Commun. Lett. **18**(12), 2117–2120 (2014)
12. Kirkpatrick, S., Gelatt, C.D., Vecchi, M.P.: Optimization by simulated annealing. Science **220**, 671–680 (1983)

Multi Population Pattern Searching Algorithm for Solving Routing Spectrum Allocation with Joint Unicast and Anycast Problem in Elastic Optical Networks

Michal Przewozniczek[✉]

Department of Computational Intelligence,
Wroclaw University of Technology,
Wybrzeze Wyspianskiego 27, 50-370 Wroclaw, Poland
michal.przewozniczek@pwr.edu.pl

Abstract. Exponentially growing network traffic is triggered by different services including cloud computing, Content Delivery Networks and other. To satisfy these needs, new network technologies like Elastic Optical Networks (EON) are proposed. The EONs bring new, hard optimization problems, like Routing and Spectrum Allocation problem considered in this paper. The well-known and common tools to solve the NP-hard problems are Evolutionary Algorithms (EA). A number of EA-based methods were proposed to solve the RSA problem. However, the papers concerning the RSA problem omit the relatively new propositions of linkage learning methods. Therefore this paper proposes a new effective method that includes linkage learning, local optimization and is based on a novel, effective work schema to fill this gap.

Keywords: Linkage learning · MuPPetS · Local optimization · Hybrid methods · Elastic optical networks · Routing and spectrum allocation

1 Introduction

The recent decades were the time of fast information technologies development. The increasing amount of information that needs to be exchanged causes the constant need for faster and more reliable network technologies. The example of such technology is the concept of Elastic Optical Networks (EON) [1, 7]. EON is the technological answer to the exponentially growing network traffic triggered by the services like cloud computing, Content Delivery Networks (CDNs), IP TV, video streaming, Internet of Things [16]. In EONs, the frequency spectrum that is available for transmitting optical signals in an optical fiber link is divided into narrow frequency segments that are called *slices*. The slices must be properly grouped to create a channel. The issue of proper slices grouping leads to the Routing and Spectrum Allocation (RSA) problem considered in this paper. The RSA is an NP-complete problem and can be solved in optimal way only for relatively small problem instances.

Evolutionary Algorithms (EA) are well known tools that allow to solve hard practical problems also in the field of computer networks. A number of methods were

© Springer International Publishing Switzerland 2015
K. Jackowski et al. (Eds.): IDEAL 2015, LNCS 9375, pp. 328–339, 2015.
DOI: 10.1007/978-3-319-24834-9_39

proposed to solve the RSA problem (see Sect. 2), but to our best knowledge the papers concerning this topic do not take into consideration the more evolved Genetic Algorithm (GA) work schemas that were shown effective. Therefore, the main contribution of this paper is a dedicated method based on the Multi Population Pattern Searching Algorithm (MuPPetS) [11].

This paper is organized as follows: Sect. 2 presents the related work, third section presents the considered problem, the proposed method is described in Sect. 4, the experiments setup, competing methods choice, results of numerical tests and their analysis are presented in Sect. 5, finally the last section concludes this work.

2 Related Work

In the recent years a number of methods to solve EON kind of problems were proposed. Most of these methods are based on evolutionary approach and, more specifically, on genetic algorithms. The already proposed methods are designed to solve static network planning problems [2, 5, 12, 14, 15] and dynamic scenarios [8, 19]. These papers consider the optimization of optical networks with respect to the spectrum usage and request blocking probability for static and dynamic cases, respectively. Different kinds of EA variants are considered, for example multi-population GA [12], problem decomposition into two subproblems [14]. In [2, 15, 19] the multi-objective optimization is discussed.

Also other than GA-based evolutionary methods are considered to solve EON problems. In [7, 18] the dedicated Tabu Search (TS) is proposed to solve the RSA-problem test cases that include unicast and anycast demands. Similar problems are considered in [1, 14], where Simulated Annealing (SA) based method is proposed. Finally in [16] the algorithm called Adaptive Frequency Assignment (AFA) is proposed. AFA allows to find valuable solutions in a very short time, when compared with EA methods. The other algorithms to solve RSA problems which may be found in the literature are: First Fit (FF) [10], Longest Path First (LPF) [4] and Most Subcarriers First (MSF) [4].

Recently, in the GA field a number of papers that consider the linkage learning [3, 11] were proposed. The linkage learning methods use the genetic-based problem encoding and try to discover the possible gene dependencies during their run. The linkage information is usually updated during all the method run. It is used to improve the method effectiveness. For example the crossover operator may exchange only those genes that are considered to be linked together [11]. An interesting method called MuPPetS was proposed [11]. MuPPetS is a linkage learning method that uses messy coding and a flexible number of coevolving subpopulations. It was shown effective against both practical [17] and theoretical problems [11].

To the best of our knowledge the papers [1, 7, 14, 18] are the only that consider metaheuristics to solve the RSA problems with joint anycast and unicast demands. All of them miss the application of modern GA that includes linkage learning. Therefore this paper tries to fill this gap and proposes MuPPetS-based method adjusted to solve the RSA problem.

3 Problem Formulation

The problem of Routing and Spectrum Allocation with Joint Anycast and Unicast (RSA-JAU), considered in this paper is formulated in this section. The notation used is similar to [16]. The RSA-JAU is NP-complete. The objective function value is equal to the spectrum width required to serve all traffic demands. The considered EON is represented by a directed graph $G = (V, E)$, where V is a set of network nodes and E is a set of fiber links that connect pairs of network nodes. On each network link $e = 1,$ $2, ..., E$, available spectrum resources are divided into frequency slices labeled with $s = 1, 2, ..., S$. Different-width optical channels can be created by grouping an even number of frequency slices. In R network nodes data centers (DC) are located. All DCs can offer the same content/service and can serve any number of clients, thus each anycast demand can be assigned to any of the available DC nodes.

A set of two different kind of requests is given: unicast and anycast. Each anycast request is realized by two demands: upstream (from client node to DC node) and downstream (oppositely). All anycast downstream demands are labeled with $d = 1,$ $2, ..., A$ while all upstream demands are labeled $d = A + 1, ..., 2A$. The unicast requests are described by a source node, destination node and a traffic volume. Therefore, each unicast request is realized by a single demand. Unicast demands are labeled with $d = 2A + 1, ..., D$.

For every ordered node pair $|P_d| = k$ different possible paths are available. Since an anycast demand can be served by any available DC, there are $|P_d| = k \times R$ different candidate paths for each anycast demand d. The number of frequency slices required to realize demand d on its candidate path p is calculated with the use of the half distance law [16]. The number of slices required by each path is constant.

To model the spectrum usage the *channel* approach is used. For each traffic demand $(d = 1, 2, ..., D)$ and its predefined routing path $(p = 1, 2, ..., P_d)$ a set of candidate frequency channels $(c = 1, 2, ..., C_{dp})$ is given. Each candidate channel consists of a subset of n_{dp} adjacent frequency slices that are necessary to realize this demand in respect to its volume, path length and applied modulation format. The channel modeling assures that each demand uses the same contiguous fraction of frequency spectrum on network links that belong to the path selected to provide the demand. Thanks to this, frequency slices cannot be overlapped by different demands.

Indices

$e = 1, 2, ..., E$ network links
$s = 1, 2, ..., S$ slices
$d = 1, 2, ..., A$ anycast downstream demands
$d = A + 1, ..., 2A$ anycast upstream demands
$d = 2A + 1, ..., D$ unicast demands
$c = 1, 2, ..., C_{dp}$ candidate channels for demand d allocated on path p
$p = 1, 2, ..., P_d$ candidate paths for flows realizing unicast and anycast demand d.

Constants

$\delta_{edp} = 1$ if link e belongs to path p realizing demand d; 0, otherwise

h_d	volume (requested bit-rate) of demand d (Gb/s)
γ_{dpcs}	=1, if channel c associated with demand d on path p uses slice s; 0, otherwise
$\tau(d)$	index of a demand associated with demand d
$o(p)$	origin (source) node of path p
$t(p)$	destination node of path p

Variables

x_{dpc}	=1, if channel c on path p is used to realize demand d; 0, otherwise (binary)
y_{es}	=1, if slice s is used on network link e; 0, otherwise (binary)
y_s	=1, if slice s is used on any network link; 0, otherwise (binary)

Objective

$$\text{Min } \Phi = \sum_s y_s \tag{1}$$

Subject to

$$\sum_p \sum_c x_{dpc} = 1, \ d = 1, 2, \ldots, D \tag{2}$$

$$\sum_d \sum_p \sum_c g_{dpcs} \delta_{edp} x_{dpc} \leq y_{es}, \ e = 1, 2, \ldots, E \ s = 1, 2, \ldots, S \tag{3}$$

$$\sum_e y_{es} \leq E y_s, \ s = 1, 2, \ldots, S \tag{4}$$

$$\sum_p \sum_c x_{dpc} o(p) = \sum_p \sum_c x_{\tau(d)pc} t(p), \ d = 1, 2, \ldots, A \tag{5}$$

$$y_{s+1} \leq y_s, \ s \in S, s < |S| \tag{6}$$

The objective (1) is to minimize the number of slices that are necessary in the network, i.e., the required spectrum width. The constraint (2) guarantees that for each traffic demand exactly one routing path and one frequency channel are allocated. Equation (3) assures that a particular slice on a particular link can be used to realize only one demand. Constraint (4) is a definition of variable y_e and indicates if slices s is used on any network link. Equation (5) assures that both associated demands, that realize the same anycast request, are related to the same DC. Finally, constraint (6) assures explicitly the contiguity of spectrum allocation in the network.

4 The Proposed Method

In this section the Multi Population Pattern Searching Algorithm for solving Routing Spectrum Allocation with Joint Unicast and Anycast problem in Elastic Optical Networks (MuPPetS-RSA) is presented. MuPPetS-RSA is based on the novel Multi Population

Pattern Searching Algorithm (MuPPetS) proposed in [11]. MuPPetS may be classified as an Evolutionary Algorithm (EA) using many coevolving populations. MuPPetS-RSA combines messy coding [6, 11], linkage learning and the flexible number of coevolving populations. MuPPetS-RSA is a linkage learning method [3, 11] - it gathers an information about gene dependencies during its run and use it to improve its effectiveness. MuPPetS was shown effective when used against both: theoretical [11] and practical [17] problems. Therefore it seems a good starting point for proposition of effective method for solving the RSA-JAU problem.

4.1 Full Solution Encoding

In MuPPetS-RSA each gene in individual's genotype is a list of pairs: $[d, p]$, where d is the demand number p is the candidate path proposed for the demand. On the base of genotype, the final solution, represented by an individual is built. To build the solution, a local optimization algorithm is used. The local optimization algorithm used to build the solution works as follows. The demand d, represented by the first gene, is allocated with the use of path p coded by the gene. The allocation of channel starts from the lowest possible spectrum slice without breaking the spectrum usage constraints. Note that such genotype to solution decoding procedure makes all possible individuals feasible since they will never break any constraints.

4.2 Messy Coding and Operators

MuPPetS-RSA is built from a number of coevolving virus populations. Virus is a new type messy coded individual that was proposed in [11]. Virus is encoded by a chromosome which is a list of pairs: $[(pos, val), (pos, val), (pos, val)]$, where pos is a position of gene in a typical GA-individual genotype, while val is the value of it. In the case of MuPPetS-RSA the demand number d is the gene position, while path p is the gene value. Note that when messy coding is used some of the genes may be specified many times (overspecification), while some of them might be not represented in the messy genotype at all (underspecification). The example of messy individual for the problem of 6 demands may be:

- Genotype: [(**1, 15**) (**6, 4**) (1, 5) (1, 1) (**2, 11**)]
- Phenotype: (1, 15) (2, 11) (3, ?) (4, ?) (5, ?) (6, 4)

In the above example gene on position number 1 is represented 3 times (ovespecification), while genes on positions 3, 4 and 5 are not specified (underspecification). To handle ovespecification it is assumed that the dominating gene value is the one that is closest to the beginning of the genotype. Therefore demand number 1 (the value of gene 1) in the phenotype is served by path number 15 and not by paths 5 and 1. As shown in the example above, some of the genes may be underspecified, to handle this problem all viruses (messy coded individuals) are assigned to classically coded GA individuals called Competitive Templates (CT). The CT supports all missing genes to viruses assigned to it. For example: if virus from the example above was assigned to the

CT with genotype: (1, 8) (2, 11) (3, 6) (4, 24) (5, 22) (6, 33) then the phenotype of the virus presented above would be: (1, 15) (2, 11) (3, 6) (4, 24) (5, 22) (6, 4). Instead of typical crossover operator the cut and splice operators are used. More information about cut and splice operators, under- and overspecification handling, the differences between viruses and typically used messy coded individuals may be found in [6, 11].

4.3   Gene Patterns

The important feature of MuPPetS-RSA inherited from MuPPetS is the linkage information gathering that is performed during the method run. The linkage is represented in the structure called *gene pattern*. Gene patterns are collected during the method run and may be represented as a list of genotype positions without values: [(*d*), (*d*), (*d*), (*d*), (*d*)], where *d* is the demand number (gene position). For example gene pattern: [(5) (2) (2)] marks genes 5 and 2. As shown the gene pattern may be also overspecified.

4.4   General Algorithm Data Structure and Work Schema

MuPPetS-RSA is built from a population of CTs (the typical GA-like individuals with standard GA-like encoding). In the typical method run, the number of CTs is usually in range <1;10>. To each CT the population of viruses is assigned. The number of CTs may increase or decrease during the method run, depending on the results obtained by a method. If MuPPetS-RSA is stuck in the local optima the number of CTs will increase, but any two CTs have the same genotype – one of them will be removed from the population. Thanks this all coevolving virus populations should inspect a different part of the solution space size. The general method overview is presented in pseudocode given in Fig. 1.

At the MuPPetS-RSA start the CT population is empty, therefore to make the method work the first CT is added to the CT population. Any time a new CT is added to the CT population it is randomly generated and it is processed with the virus population run. In the virus population run, the viruses are processed like in the normal GA with respect to the fact that viruses are messy coded and instead of crossover operator, cut and splice are used. The number of iterations of virus population run is set by a user. If during virus population run the virus with fitness better than CT is found then the virus "infects" the CT. The infection means that gene values from the virus phenotype replace appropriate values in the CT genotype. For example: if fitness of CT with genotype: (1, 8) (2, 11) (3, 6) (4, 24) (5, 22) (6, 33) is worse than a virus with phenotype: (**1, 15**) (**2, 11**) (3, ?) (4, ?) (5, ?) (**6, 4**) then after infection the CT genotype wil be: (**1, 15**) (**2, 11**) (3, 6) (4, 24) (5, 22) (**6, 4**). In the main method the genotype of all CTs is first stored, after that every CT is crossed with other CT (if there are more than two CTs in the population) with the use of randomly chosen gene pattern (if no gene patterns have been generated yet then the CTs are not crossed). In the CT crossover only genes marked by a gene pattern are exchanged. For example: if CT_0 (1, 8) (2, 11) (3, 6) (4, 24) (5, 22) (6, 33) is crossed with CT_1 (1, 15) (2, 11) (3, 2) (4, 5) (5, 6)

```
CTPop = empty;
CTPop.push_back(new CT);
CTPop[0].InitializeVirusPop();
CTPop[0].VirusPopRun ();
AvrCTFitness = CTPop.GetAverageFitness();

While(!stop_condition)
{
  If (AvrCTFitness  <  CTPop.GetAverageFitness())
  {
    Remove duplicated CTs from CTPop;
    CTPop.push_back(new CT);
    CTPop[CTPop.size - 1].InitializeVirusPop();
    CTPop[CTPop.size - 1].VirusPopRun();
  }//If (AvrCTFitness  <  CTPop.GetAverageFitness())
  AvrCTFitness = CTPop.GetAverageFitness();

  For (int  ii = 0; ii < CTPop.size; ii++)
  {
    CTFitness = CTPop[ii].GetFitness();
    CTPop[ii].StoreCurrentGenotype();
    CTPop[ii].CrossWithOtherCT();
    If  (CTFitness >= CTPop[ii].GetFitness())
    {
      CTPop[ii].InitializeVirusPop();
      CTPop[ii].VirusPopRun();
      CTPop[ii].GetGenePatternsOnVirusPopRunBase();
      If (CTFitness > CTPop[ii].GetFitness())
        CTPop[ii].RestoreGenotype();
    }//If  (CTFitness >= CTPop[ii].GetFitness())
  }// For (int  ii = 0; ii < CTPop.size; ii++)
}// While(!stop_condition)
```

Fig. 1. MuPPetS-RSA pseudocode

(6, 7), and the gene pattern is [(4) (3) (4) (5)] then the genotype of CT_0 after this operation will be: (1, 8) (2, 11) **(3, 2) (4, 5) (5, 6)** (6, 33). If processed CT fitness increases after crossing with other CT then it is not processed in this iteration any more. If the CT fitness after crossing remains the same or drops down then the virus population run is executed. On the base of the virus population run new gene pattern is generated. If the number of patterns in the gene pattern pool is above the limit (parameter set by a user) then the proper number of gene pattern is randomly deleted. If after the CT crossing and virus population run the CT fitness is lower than it was before, then the CT genotype from before these operations is restored. Note, that after each MuPPetS-RSA run the CT fitness may stay the same or increase, it will never drop down. If none of the CTs in the CT population was able to increase its fitness then it is assumed that the method is stuck and new CT is added to the CT population in order to break out from the local optima.

5 Results

In this section the results of experiments are presented. As the competing methods the standard GA (sGA), Random Search (RS) and AFA, LPF, MSF algorithms were chosen [4, 10, 16]. The sGA and RS use the same problem encoding as MuPPetS-RSA.

Both methods were tuned before tests on the same set of randomly chosen 10 problems. The tuning procedure for all methods was as follows. Each method was given the start configuration on a base of author experience and available literature. Then the key parameters were adjusted one after one in a greedy way: the parameter value was increased or decreased as long as this operation was improving the results. The final method configurations are given in Table 1.

Table 1. Methods configuration

MuPPetS-RSA		simple GA	
Parameter	Value	Parameter	Value
Pattern pool size	500	Population size	12000
Virus generation number	3	Prob. Cross	0.7
Virus population size	30	Prob. Mut	0.4
Prob. Cut	0.12		
Prob. Splice	0.2		
Prob. Mut	0.05		
Prob. Mut add gene	0.05		
Prob. Mut rem gene	0.05		

The UBN24 network representing US national backbone network (24 nodes, 86 links) was used for the experiments [13]. The experiments were performed for $k = 2$. The number of test cases was 96. For every method each test case was executed once on the HP Elite Desk800 TWR server with Intel I7-4770 4.4 GHz processor. All methods were executed in a separate process with no other resource consuming processes running. The computation time was 2400 s for MuPPetS-RSA, sGA and RS. The AFA, LPF and MSF execution times were below 1 ms. Due to the fact that AFA, LPF and MSF are deterministic algorithms and it was not possible to give them more computation time. All methods were coded in C ++. The complete sourcecodes, complete results, may be downloaded from: http://www.mp2.pl/download/ai/ 20150525_mupets_fiber_flow_simple.zip.

Depending on the experiment the number of genes necessary to encode the full solution was in range from 576 up to 741. The solution space size approximation is given in Eq. (7).

$$SolutionSpaceSize = k^{D-2A} * (k * R)^A * k^A = k^{D-A} * (k * R)^A \qquad (7)$$

The approximated solution space size is in range of 10^{100} to 10^{221}. Note, that for the encoding used by MuPPetS-RSA, sGA and RS the solution space could be much wider if the gene order was taken into account. Such solution space limitation may exclude some of the potentially promising search space regions. On the other hand limiting the search space makes the problem more tractable.

5.1   Optimality Gap

To compare the proposed evolutionary methods with optimal results, the tests for NSF15 network (15 nodes and 46 links) where executed. The number of scenarios was 216, the solution space size was in range 10^3 to 10^{10}, the number of genes used was in range from 30 up to 56. All experiments were executed with the same method settings as in the main results section, but the time limit was decreased to 1200 s. The optimal results were generated with CPLEX 11.1 solver [9].

Table 2. Optimality gap results

	RS	sGA	MuPPetS-RSA
Optimality gap	3.84%	4.00%	**3.06%**
Number of optimal solutions	137	137	**145**

As presented in Table 2, MuPPetS-RSA outperformed all competing methods. It was also able to find the optimal result in over 67 % of the test cases. The surprising fact may be that the Random Search, which is quite primitive method, was able to slightly outperform sGA. This observation leads to conclusion that sGA was unable to keep proper level of population diversity. Note, that according to the presented results, it is always important to compare more evolved methods with simple ones to see the gap (or the lack of gap) between them.

5.2   Main Results

In Table 3 the results of performed experiments are presented. The first table part gives average highest used slice (the objective function). In order to reliably compare the results reported by competing methods the *competitive ratio* (*CR*) is used. The CR parameter of a particular algorithm Alg is calculated as $CR(Alg) = (\Phi(Alg) - \Phi(Min))/\Phi(Alg)$, where $\Phi(Alg)$ denotes the value of the objective function yielded by Alg and $\Phi(Min)$ denotes the minimum value obtained for a particular scenario concerning all

4 methods. Note that the lower the CR indicator is the better solution it refers to. The third performance metric is a ranking – the best method for a particular scenario receives first place points, the second method receives second place, etc. The last metric is the number of first places (best results) of a particular method.

Table 3. Experiments results

	RS	AFA	FF	MSF	LPF	sGA	MuPPetS-RSA
Avr. highest slice	537.56	438.02	910.69	455.23	456.79	447.52	**413.15**
CR	40.75%	8.49%	176.18%	12.08%	176.18%	12.60%	**1.13%**
CR Confidence interval	5.63%	1.79%	35.27%	1.92%	35.27%	1.77%	**0.51%**
Ranking	5.94	2.88	6.81	3.42	3.77	3.54	**1.39**
Wins	0	20	0	6	0	4	**71**

The results obtained show that MuPPetS-RSA outperforms all other methods, the advantage is significant. Note that CR indicator is very close to 0 % as well as its confidence interval. The obtained results also show that MuPPetS-RSA mechanisms that help to maintain population diversity and allow the method to jump out from the local optima in which the method may stuck, are proper. The sGA which uses the same problem encoding as MuPPetS-RSA is significantly less effective. As shown in [16], AFA is the most effective among all competing algorithms, but since it is significantly outperformed by MuPPetS-RSA its only, but important, advantage is much lower computation time needed.

6 Summary

In this paper the proposition of MuPPetS-RSA method for solving the RSA-JAU problem was proposed. MuPPetS-RSA is based on MuPPetS method which is a novel work schema for a GA-based methods. MuPPetS were already successfully applied to practical problems, but here MuPPetS-RSA proposes the use of a local optimization algorithm to improve its effectiveness and significantly reduce the solution search space size. The proposed method outperformed all other competing methods including the dedicated TS.

The solution encoding used in MuPPetS-RSA has a unmodifiable order of finding proper channels for the demands (channels are found always in the same manner – from the first demand up to the last one). On one hand such encoding limits the solution space and makes the problem more tractable, but on the other it may exclude from the search procedure some of the promising search space regions. Therefore the main directions of future work shall be the introduction of ordering into the problem encoding and proposing new problem dedicated mechanisms.

Acknowledgements. The author of this work would like to express his sincere appreciation to his advisor professor Krzysztof Walkowiak for his valuable suggestions during the paper redaction process.

This work was supported in by the Polish National Science Centre (NCN) under Grant DEC-2012/07/B/ST7/01215 and by the European Commission under the 7th Framework Programme, Coordination and Support Action, Grant Agreement Number 316097, ENGINE – European research centre of Network intelliGence for INnovation Enhancement.

References

1. Aibin, M., Walkowiak, K.: Simulated Annealing algorithm for optimization of elastic optical networks with unicast and anycast traffic. In: Proceedings of 16th International Conference on Transparent Optical Networks ICTON (2014)
2. Cerutti, I., Martinelli, F., Sambo, N., Cugini, F., Castoldi, P.: Trading regeneration and spectrum utilization in code-rate adaptive flexi-grid networks. J. Lightwave Technol. **32**(23), 4496–4503 (2014)
3. Chen, Y., Sastry, K., Goldberg, D.E.: A Survey of Linkage Learning Techniques in Genetic and Evolutionary Algorithms. In: IlliGAL Report No. 2007014, Illinois Genetic Algorithms Laboratory (2007)
4. Christodoulopoulos, K., Tomkos, I., Varvarigos, E.A.: Elastic bandwidth allocation in flexible OFDM based optical networks. IEEE J. Lightwave Technol. **29**(9), 1354–1366 (2011)
5. Eira, A., Santos, J., Pedro, J., Pires, J.: Multi-objective design of survivable flexible-grid DWDM networks. IEEE/OSA J. Opt. Commun. Netw. **6**(3), 326–339 (2014)
6. Goldberg, D.E., et. al.: Rapid, accurate optimization of difficult problems using fast messy genetic algorithms. In: Proceedings of 5th International Conference on Genetic Algorithms (1993)
7. Goscien, R., Walkowiak, K., Klinkowski, M.: Tabu search algorithm for routing, modulation and spectrum allocation in elastic optical network with anycast and unicast traffic. Comput. Netw. **79**, 148–165 (2015)
8. Huang, T., Li, B.: A genetic algorithm using priority-based encoding for routing and spectrum assignment in elastic optical networks. In: Proceedings of ICICTA (2014)
9. ILOG AMPL/CPLEX software. ILOG website: www.ilog.com/products/cplex/
10. Jinno, M., et al.: Distance-adaptive spectrum resource allocation in spectrum sliced elastic optical path network. IEEE Commun. Mag. **48**(8), 138–145 (2010)
11. Kwasnicka, H., Przewozniczek, M.: Multi population pattern searching algorithm: a new evolutionary method based on the idea of messy genetic algorithm. IEEE Trans. Evol. Comput. **15**(5), 715–734 (2011)
12. Long, G., Xiang, Z., Wei, L., Zuqing, Z.: A two-population based evolutionary approach for optimizing routing, modulation and spectrum assignment (RMSA) in O-OFDM networks. IEEE Commun. Lett. **16**(9), 1520–1523 (2012)
13. Orlowski, S., Wessaly, R., Pioro, M., Tomaszewski, A.: SNDlib 1.0—survivable network design library. Networks **55**(3), 276–286 (2010)
14. Patel, A.N., Ji, P.N., Jue, J.P., Ting, W.: A naturally-inspired algorithm for routing, wavelength assignment, and spectrum allocation in flexible grid WDM networks. In: Proceedings of IEEE GC Workshops (2012)

15. Velasco, L., Wright, P., Lord, A., Junyent, G.: Saving CAPEX by extending flexgrid-based core optical networks toward the edges. IEEE/OSA J. Opt. Commun. Netw. **5**(10), A171–A183 (2013)
16. Walkowiak, K., Klinkowski, M.: Joint anycast and unicast routing for elastic Optical networks: modeling and optimization. In: Proceedings of IEEE ICC (2013)
17. Walkowiak, K., Przewozniczek, M., Pajak, K.: Heuristic algorithms for survivable P2P multicasting. Appl. Artif. Intell. **27**(4), 278–303 (2013)
18. Walkowiak, K., Klinkowski, M., Rabiega, B., Goscien, R.: Routing and spectrum allocation algorithms for elastic optical networks with dedicated path protection. Opt. Switching Netw. **13**, 63–75 (2014). doi:10.1016/j.osn.2014.02.002
19. Xiang, Z., Wei, L., Long, G., Zuqing, Z.: Dynamic RMSA in elastic optical networks with an adaptive genetic algorithm. In: Proceedings of IEEE GLOBECOM (2012)

Hybrid Evolutionary Algorithm with Adaptive Crossover, Mutation and Simulated Annealing Processes to Project Scheduling

Virginia Yannibelli[1,2(✉)] and Analía Amandi[1,2]

[1] ISISTAN Research Institute, UNCPBA University,
Campus Universitario, Paraje Arroyo Seco, 7000 Tandil, Argentina
{vyannibe,amandi}@exa.unicen.edu.ar
[2] CONICET, National Council of Scientific and Technological Research,
Buenos Aires, Argentina

Abstract. In this paper, we address a project scheduling problem that considers a priority optimization objective for project managers. This objective involves assigning the most effective set of human resources to each project activity. To solve the problem, we propose a hybrid evolutionary algorithm. This algorithm uses adaptive crossover, mutation and simulated annealing processes in order to improve the performance of the evolutionary search. These processes adapt their behavior based on the diversity of the evolutionary algorithm population. We compare the performance of the hybrid evolutionary algorithm with those of the algorithms previously proposed in the literature for solving the addressed problem. The obtained results indicate that the hybrid evolutionary algorithm significantly outperforms the previous algorithms.

Keywords: Project scheduling · Human resource assignment · Multi-skilled resources · Hybrid evolutionary algorithms · Evolutionary algorithms · Simulated annealing algorithms

1 Introduction

Project scheduling involves defining feasible start times and feasible human resource assignments for project activities in such a way that a predefined optimization objective is reached. To define human resource assignments, it is essential to have knowledge of the effectiveness of the available human resources in respect of the project activities. This is because the development and also the results of an activity mainly depend on the effectiveness of the human resources assigned to it [1, 2].

In the literature, many different kinds of project scheduling problems have been formally described and addressed. Nevertheless, to the best of our knowledge, only few project scheduling problems have considered human resources with different levels of effectiveness [3–6, 10], a fundamental aspect in real project scheduling problems. These project scheduling problems state different assumptions about the effectiveness of the human resources.

The project scheduling problem presented in [6] considers that the effectiveness of a human resource depends on various factors inherent to its work context (i.e., the

© Springer International Publishing Switzerland 2015
K. Jackowski et al. (Eds.): IDEAL 2015, LNCS 9375, pp. 340–351, 2015.
DOI: 10.1007/978-3-319-24834-9_40

activity to which the resource is assigned, the skill to which the resource is assigned within the activity, the set of human resources that has been assigned to the activity, and the attributes of the resource). This is a really significant aspect of the project scheduling problem presented in [6]. This is because, in real project scheduling problems, the human resources usually have different effectiveness levels in respect of different work contexts [1, 2] and, thus, the effectiveness of a human resource needs to be considered in respect of its work context. To the best of our knowledge, the influence of the work context on the effectiveness of the human resources has not been considered in other project scheduling problems. Based on the above-mentioned, we consider that the project scheduling problem presented in [6] states valuable and novel assumptions about the effectiveness of the human resources in the context of project scheduling problems. Besides, this problem considers a priority optimization objective for managers at the early stage of project scheduling. This objective implies assigning the most effective set of human resources to each project activity.

The project scheduling problem presented in [6] is considered as a special case of the RCPSP (Resource Constrained Project Scheduling Problem) [9] and, thus, is an NP-Hard optimization problem. Because of this, exhaustive search and optimization algorithms only can solve very small instances of the problem in a reasonable period of time. Therefore, heuristic search and optimization algorithms have been proposed in the literature to solve the problem: an evolutionary algorithm was proposed in [6], a memetic algorithm was proposed in [7] that incorporates a hill-climbing algorithm into the framework of an evolutionary algorithm, and a hybrid evolutionary algorithm was proposed in [8] that integrates an adaptive simulated annealing algorithm into the framework of an evolutionary algorithm. These three algorithms use non-adaptive crossover and mutation processes to develop the evolutionary search.

In this paper, we address the project scheduling problem presented in [6] with the aim of proposing a better heuristic search and optimization algorithm to solve it. In this regards, we propose a hybrid evolutionary algorithm that uses adaptive crossover, mutation and simulated annealing processes. The behavior of these processes is adaptive according to the diversity of the evolutionary algorithm population. The utilization of adaptive crossover, mutation and simulated annealing processes is meant to improve the performance of the evolutionary search [18–20].

We propose the above-mentioned hybrid evolutionary algorithm because of the following reason. Evolutionary algorithms with adaptive crossover and mutation processes have been proven to be more effective than evolutionary algorithms with non-adaptive crossover and mutation processes in the resolution of a wide variety of NP-Hard optimization problems [18–20]. Therefore, we consider that the proposed hybrid evolutionary algorithm could outperform the heuristic algorithms previously proposed to solve the problem.

The remainder of the paper is organized as follows. In Sect. 2, we give a brief review of published works that describe project scheduling problems in which the effectiveness of human resources is considered. In Sect. 3, we describe the problem addressed in this paper. In Sect. 4, we present the proposed hybrid evolutionary algorithm. In Sect. 5, we present the computational experiments carried out to evaluate the performance of the hybrid evolutionary algorithm and an analysis of the results obtained. Finally, in Sect. 6 we present the conclusions of the present work.

2 Related Works

In the literature, different project scheduling problems that consider the effectiveness of human resources have been described. Nevertheless, these project scheduling problems state different assumptions about the effectiveness of human resources. In this regards, only few project scheduling problems consider human resources with different levels of effectiveness [3–6, 10], a fundamental aspect in real project scheduling problems. In this section, we focus the attention on analyzing the way in which the effectiveness of human resources is considered in project scheduling problems described in the literature.

In [12–17], multi-skill project scheduling problems are described. In these problems, each project activity requires specific skills and a given number of human resources (employees) for each required skill. Each available employee masters one or several skills, and all the employees that master a given skill have the same effectiveness level in relation to the skill (homogeneous levels of effectiveness in relation to each skill).

In [3], a multi-skill project scheduling problem with hierarchical levels of skills is described. In this problem, given a skill, for each employee that masters the skill, an effectiveness level is defined in relation to the skill. Therefore, the employees that master a given skill have different levels of effectiveness in relation to the skill (heterogeneous levels of effectiveness in relation to each skill). Then, each project activity requires one or several skills, a minimum effectiveness level for each skill, and a number of employees for each pair skill-level. This work considers that all sets of employees that can be assigned to a given activity have the same effectiveness on the development of the activity. Specifically, with respect to effectiveness, such sets are merely treated as unary resources with homogeneous levels of effectiveness.

In [4, 5], multi-skill project scheduling problems are described. In these problems, most activities require only one employee with a particular skill, and each available employee masters different skills. Besides, the employees that master a given skill have different levels of effectiveness in relation to the skill. Then, the effectiveness of an employee in a given activity is defined by considering only the effectiveness level of the employee in relation to the skill required for the activity.

Unlike the above-mentioned problems, the project scheduling problem presented in [6] considers that the effectiveness of a human resource depends on various factors inherent to its work context. Thus, for each human resource, it is possible to define different effectiveness levels in relation to different work contexts. This is a really significant aspect of the project scheduling problem presented in [6]. This is because, in real project scheduling problems, the human resources have different effectiveness levels in respect of different work contexts [1, 2] and, thus, the effectiveness of a human resource needs to be considered in respect of its work context. Taking into account the above-mentioned, we consider that the project scheduling problem presented in [6] states valuable assumptions regarding the effectiveness of human resources in the context of project scheduling problems.

3   Problem Description

In this paper, we address the project scheduling problem described in [6]. We present below a description of this problem.

A project contains a set A of N activities, $A = \{1, ..., N\}$, to be scheduled (i.e., the starting time and the human resources of each activity have to be defined). The duration, precedence relations and resource requirements of each activity are known.

The duration of each activity j is notated as d_j. Moreover, it is considered that pre-emption of activities is not allowed (i.e., the d_j periods of time must be consecutive).

Among some project activities, there are precedence relations. The precedence relations establish that each activity j cannot start until all its immediate predecessors, given by the set P_j, have completely finished.

Project activities require human resources – employees – skilled in different knowledge areas. Specifically, each activity requires one or several skills as well as a given number of employees for each skill.

It is considered that organizations and companies have a qualified workforce to develop their projects. This workforce is made up of a number of employees, and each employee masters one or several skills.

Considering a given project, set SK represents the K skills required to develop the project, $SK = \{1, ..., K\}$, and set AR_k represents the available employees with skill k. Then, the term $r_{j,k}$ represents the number of employees with skill k required for activity j of the project. The values of the terms $r_{j,k}$ are known for each project activity.

It is considered that an employee cannot take over more than one skill within a given activity. In addition, an employee cannot be assigned more than one activity at the same time.

Based on the previous assumptions, an employee can be assigned different activities but not at the same time, can take over different skills required for an activity but not simultaneously, and can belong to different possible sets of employees for each activity.

As a result, it is possible to define different work contexts for each available employee. It is considered that the work context of an employee r, denoted as $C_{r,j,k,g}$, is made up of four main components. The first component refers to the activity j which r is assigned (i.e., the complexity of j, its domain, etc.). The second component refers to the skill k which r is assigned within activity j (i.e., the tasks associated to k within j). The third component is the set of employees g that has been assigned j and that includes r (i.e., r must work in collaboration with the other employees assigned to j). The fourth component refers to the attributes of r (i.e., his or her experience level in relation to different tasks and domains, the kind of labor relation between r and the other employees of g, his or her educational level in relation to different knowledge areas, his or her level with respect to different skills, etc.). It is considered that the attributes of r could be quantified from available information about r (e.g., curriculum vitae of r, results of evaluations made to r, information about the participation of r in already executed projects, etc.).

The four components described above are considered the main factors that determine the effectiveness level of an employee. For this reason, it is assumed that the effectiveness of an employee depends on all the components of his or her work context. Then, for each employee, it is possible to consider different effectiveness levels in relation to different work contexts.

The effectiveness level of an employee r, in relation to a possible context $C_{r,j,k,g}$ for r, is notated as $e_{rCr,j,k,g}$. The term $e_{rCr,j,k,g}$ represents how well r can handle, within activity j, the tasks associated to skill k, considering that r must work in collaboration with the other employees of set g. The mentioned term $e_{rCr,j,k,g}$ takes a real value over the range $[0, 1]$. The values of the terms $e_{rCr,j,k,g}$ inherent to each employee available for the project are known. It is considered that these values could be obtained from available information about the participation of the employees in already executed projects.

The problem of scheduling a project entails defining feasible start times (i.e., the precedence relations between the activities must not be violated) and feasible human resource assignments (i.e., the human resource requirements must be met) for project activities in such a way that the optimization objective is reached. In this sense, a priority objective is considered for project managers at the early stage of the project schedule design. The objective is that the most effective set of employees be assigned each project activity. This objective is modeled by Formulas (1) and (2).

Formula (1) maximizes the effectiveness of the sets of employees assigned to the N activities of a given project. In this formula, set S contains all the feasible schedules for the project in question. The term $e(s)$ represents the effectiveness level of the sets of employees assigned to project activities by schedule s. Then, $R(j,s)$ is the set of employees assigned to activity j by schedule s, and the term $e_{R(j,s)}$ represents the effectiveness level corresponding to $R(j,s)$.

Formula (2) estimates the effectiveness level of the set of employees $R(j,s)$. This effectiveness level is estimated calculating the mean effectiveness level of the employees belonging to $R(j,s)$.

For a more detailed discussion of Formulas (1) and (2), we refer to [6].

$$\max_{\forall s \in S} \left(e(s) = \sum_{j=1}^{N} e_{R(j,s)} \right) \qquad (1)$$

$$e_{R(j,s)} = \frac{\sum_{r=1}^{|R(j,s)|} e_{rCr,j,k(r,j,s),R(j,s)}}{|R(j,s)|} \qquad (2)$$

4   Hybrid Evolutionary Algorithm

To solve the problem, we propose a hybrid evolutionary algorithm. This algorithm uses adaptive crossover, mutation and simulated annealing processes. The behavior of these processes is adaptive according to the diversity of the evolutionary algorithm population. The utilization of adaptive crossover, mutation and simulated annealing processes is meant to improve the performance of the evolutionary search in both exploration and exploitation [18–20].

The general behavior of the hybrid evolutionary algorithm is described as follows. Considering a given project to be scheduled, the algorithm starts the evolution from a random initial population of solutions in which each solution codifies a feasible project schedule. Then, each solution of the population is decoded (i.e., the related schedule is built), and evaluated according to the optimization objective of the problem by a fitness function. As explained in Sect. 3, the objective is to maximize the effectiveness of the sets of employees assigned to project activities. In respect of this objective, the fitness function evaluates the assignments of each solution based on knowledge about the effectiveness of the employees involved in the solution.

Once the solutions of the population are evaluated, a parent selection process is used to decide which solutions of the population will compose the mating pool. The solutions with the highest fitness values will have more probability of being selected. After the mating pool is composed, the solutions in the mating pool are paired. Then, a crossover process is applied to each pair of solutions with an adaptive probability P_c to generate new feasible ones. Then, a mutation process is applied to each solution generated by the crossover process, with an adaptive probability P_m. Then, a survival selection process is applied in order to define which solutions from the solutions in the population and the solutions generated from the mating pool will compose the new population. Finally, an adaptive simulated annealing algorithm is applied to the solutions of the new population.

This process is repeated until a predetermined number of iterations is reached.

4.1 Encoding of Solutions and Fitness Function

To encode the solutions, we used the representation proposed in [6]. Each solution is represented by two lists having as many positions as activities in the project. The first list is a standard activity list. This list is a feasible precedence list of the activities involved in the project (i.e., each activity j can appear on the list in any position higher than the positions of all its predecessors). The activity list describes the order in which activities shall be added to the schedule.

The second list is an assigned resources list. This list contains information about the employees assigned to each activity of the project. Specifically, position j on this list details the employees of every skill k assigned to activity j.

To build the schedule related to the representation, we used the serial schedule generation method proposed in [6]. In this method, each activity j is scheduled at the earliest possible time.

In order to evaluate a given encoded solution, we used a fitness function. This function decodes the schedule s related to the solution by using the serial method above-mentioned. Then, the function calculates the value of the term $e(s)$ corresponding to s (Formulas (1) and (2)). This value determines the fitness level of the solution. The term $e(s)$ takes a real value over $[0, \ldots, N]$.

To calculate the term $e(s)$, the function uses the values of the terms $e_{rCr,j,k,g}$ inherent to s (Formula 2). As mentioned in Sect. 3, the values of the terms $e_{rCr,j,k,g}$ inherent to each available employee r are known.

4.2 Parent Selection, Adaptive Crossover, Adaptive Mutation, and Survival Selection

In order to develop the parent selection, we applied the traditional roulette wheel selection process [18].

To develop the crossover and the mutation, we applied feasible processes for the representation of the solutions. The crossover process contains a feasible crossover operation for activity lists and a feasible crossover operation for assigned resources lists. For activity lists, we used the two-point crossover proposed by Hartmann [21]. For assigned resources lists, we used the traditional uniform crossover [18].

The mutation process contains a feasible mutation operation for activity lists and a feasible mutation operation for assigned resources lists. For activity lists, we used the adjacent pairwise interchange operator described in [21]. For assigned resources lists, we used the traditional random resetting [18].

The crossover and mutation processes are applied with adaptive probabilities P_c and P_m, respectively. In this regards, we used the well-known adaptive probabilities proposed by Srinivas [11, 18]. These probabilities are calculated as detailed in Formulas (3) and (4), where f_{max} is the maximal fitness into the population, f_{avg} is the average fitness of the population, and $(f_{max} - f_{avg})$ is used as a measure of the diversity of the population. In Formula (3), f' is the higher fitness of the two solutions to be crossed, and P_{c1} and P_{c2} are predetermined values for the crossover probability, considering $0 \leq P_{c1}, P_{c2} \leq 1$. In Formula (4), f'' is the fitness of the solution to be mutated, and P_{m1} and P_{m2} are predetermined values for the mutation probability, considering $0 \leq P_{m1}, P_{m2} \leq 1$.

By Formulas (3) and (4), when the diversity of the population decreases, P_c and P_m are increased to promote the exploration of unvisited regions of the search space and thus to prevent the premature convergence of the evolutionary search. When the population is diverse, P_c and P_m are decreased to promote the exploitation of visited regions of the search space. Thus, probabilities P_c and P_m are adaptive to promote either the exploration or exploitation according to the diversity of the population.

$$P_c = \begin{cases} \frac{P_{c2}(f_{max}-f')}{(f_{max}-f_{avg})} & f' \geq f_{avg} \\ P_{c1} & f' < f_{avg} \end{cases} \tag{3}$$

$$P_m = \begin{cases} \frac{P_{m2}(f_{max}-f'')}{(f_{max}-f_{avg})} & f'' \geq f_{avg} \\ P_{m1} & f'' < f_{avg} \end{cases} \tag{4}$$

In order to develop the survival selection, we applied the traditional fitness-based steady-state selection scheme [18]. In this scheme, the worst λ solutions of the current population are replaced by the best λ solutions generated from the mating pool. This scheme preserves the best solutions found by the hybrid evolutionary algorithm [18].

4.3 Adaptive Simulated Annealing Algorithm

Once obtained a new population by the survival selection process, we applied an adaptive simulated annealing algorithm to each solution of this population, except to the best solution of this population which is maintained. This adaptive simulated annealing algorithm is a variant of the one proposed in [8], and is described below.

The adaptive simulated annealing algorithm is an iterative process which starts from a given encoded solution s for the problem, considering a given initial value T_0 for a parameter called temperature. In each iteration, a new solution s' is generated from the current solution s by a move operator. When the new solution s' is better than the current solution s (i.e., the fitness value of s' is higher than the fitness value of s), the current solution s is replaced by s'. Otherwise, when the new solution s' is worse than the current solution s, the current solution s is replaced by s' with a probability equal to $exp(-delta/T)$, where T is the current temperature value and $delta$ is the difference between the fitness value of s and the fitness value of s'. Thus, the probability of accepting a new solution s' that is worse than the current solution s mainly depends on the temperature value. If the temperature is high, the acceptance probability is also high, and vice versa. The temperature value is decreased by a cooling factor at the end of each iteration. The described process is repeated until a predefined number of iterations is reached.

The initial temperature value T_0 is defined before applying the simulated annealing algorithm to the solutions of the population. In this case, T_0 is inversely proportional to the diversity of the population, and is calculated as follows: $T_0 = 1/\left(f_{max} - f_{avg}\right)$, where $\left(f_{max} - f_{avg}\right)$ is used as a measure of the diversity of the population. Therefore, when the population is diverse, the value of T_0 is low, and thus the simulated annealing algorithm behaves like an exploitation process, fine-tuning the solutions of the population. When the diversity of the population decreases, the value of T_0 rises, and thus, the simulated annealing algorithm changes its behavior from exploitation to exploration in order to introduce diversity in the population and therefore to prevent the premature convergence of the evolutionary search. Thus, the behavior of the simulated annealing algorithm is adaptive to either an exploitation or exploration behavior according to the diversity of the population.

In respect of the move operator applied by the simulated annealing algorithm to generate a new solution from the current solution, we used a feasible move operator for the representation of the solutions. The move operator contains a feasible move operation for activity lists and a feasible move operation for assigned resources lists. For activity lists, we used a move operator called simple shift [21]. For assigned resources lists, we used a move operator which is a variant of the traditional random resetting [18]. This variant modifies only one randomly selected position of the list.

5   Computational Experiments

In order to develop the experiments, we utilized the six instance sets presented in [7]. The main characteristics of these instance sets are shown in Table 1. Each instance of these instance sets contains information about a number of activities to be scheduled, and information about a number of available employees to develop the activities. For a detailed description of these six instance sets, we refer to [7].

Each instance of these instance sets has a known optimal solution with a fitness level $e(s)$ equal to N (N refers to the number of activities in the instance). The known optimal solutions of the instances are considered here as references.

Table 1. Characteristics of instance sets

Instance set	Activities per instance	Possible sets of employees per activity	Instances
j30_5	30	1 to 5	40
j30_10	30	1 to 10	40
j60_5	60	1 to 5	40
j60_10	60	1 to 10	40
j120_5	120	1 to 5	40
j120_10	120	1 to 10	40

The hybrid evolutionary algorithm was run 30 times on each instance of the six instance sets. To carry out these runs, the algorithm parameters were set with the values detailed in Table 2. The algorithm parameters were set based on preliminary experiments that showed that these values led to the best and most stable results.

Table 3 presents the results obtained by the experiments. Column 2 presents the average percentage deviation from the optimal solution (Dev. (%)) for each instance set. Column 3 presents the percentage of instances for which the value of the optimal solution is achieved at least once among the 30 generated solutions (Opt. (%)).

The results obtained by the algorithm for j30_5, j30_10, j60_5 and j60_10 indicate that the algorithm has found an optimal solution in each of the 30 runs carried out on each instance of these sets.

The Dev. (%) obtained by the algorithm for j120_5 and j120_10 is greater than 0 %. Considering that the instances of j120_5 and j120_10 have known optimal solutions with a fitness level $e(s)$ equal to 120, we analyzed the meaning of the average deviation obtained for each one of these sets. In the case of j120_5 and j120_10, average deviations equal to 0.1 % and 0.36 % indicate that the average value of the solutions obtained by the algorithm is 119.88 and 119.57 respectively. Thus, we may state that the algorithm has obtained very high quality solutions for the instances of j120_5 and j120_10. Besides, the Opt. (%) obtained by the algorithm for j120_5 and j120_10 is 100 %. These results indicate that the algorithm has found an optimal solution in at least one of the 30 runs carried out on each instance of the sets.

Table 2. Parameter values of the hybrid evolutionary algorithm

Parameter	Value
Population size	90
Number of generations	300
Crossover process	
P_{c1}	0.9
P_{c2}	0.6
Mutation process	
P_{m1}	0.1
P_{m2}	0.05
Survival selection process	
λ	45
Simulated annealing algorithm	
Number of iterations	25
Cooling factor	0.9

Table 3. Results obtained by the computational experiments

Instance set	Dev. (%)	Opt. (%)
j30_5	0	100
j30_10	0	100
j60_5	0	100
j60_10	0	100
j120_5	0.1	100
j120_10	0.36	100

5.1   Comparison with a Competing Algorithm

To the best of our knowledge, three algorithms have been previously proposed to solve the addressed problem: a classical evolutionary algorithm [6], a classical memetic algorithm [7] that incorporates a hill-climbing algorithm into the framework of an evolutionary algorithm, and a hybrid evolutionary algorithm [8] that integrates an adaptive simulated annealing algorithm into the framework of an evolutionary algorithm. These three algorithms use non-adaptive crossover and mutation processes to develop the evolutionary search.

According to the experiments reported in [7, 8], the three algorithms have been evaluated on the six instance sets presented in Table 1 and have obtained the results that are shown in Table 4. Based on these results, the algorithm proposed in [8] is the best of the three algorithms. Below, we compare the performance of this algorithm with that of the hybrid evolutionary algorithm proposed here. For simplicity, we will refer to the algorithm proposed in [8] as algorithm H.

The results in Tables 3 and 4 indicate that the hybrid evolutionary algorithm proposed here and the algorithm H have reached the same effectiveness level (i.e., an optimal effectiveness level) on the first four instance sets (i.e., the less complex sets).

However, the effectiveness level reached by the hybrid evolutionary algorithm on the last two instance sets (i.e., the more complex sets) is higher than the effectiveness level reached by the algorithm H on these sets. Thus, the performance of the hybrid evolutionary algorithm on the two more complex instance sets is better than that of the algorithm H. The main reason for this is that, in contrast with the algorithm H, the hybrid evolutionary algorithm uses adaptive crossover and mutation processes, and these processes adapt their behavior to promote either exploration or exploitation of the search space according to the diversity of the population. Thus, the hybrid evolutionary algorithm can reach better solutions than algorithm H on the more complex instance sets.

Table 4. Results obtained by the algorithms previously proposed for the addressed problem.

Instance set	Evolutionary algorithm [6]		Memetic algorithm [7]		Hybrid algorithm [8]	
	Dev. (%)	Opt. (%)	Dev. (%)	Opt. (%)	Dev. (%)	Opt. (%)
j30_5	0	100	0	100	0	100
j30_10	0	100	0	100	0	100
j60_5	0.42	100	0	100	0	100
j60_10	0.59	100	0.1	100	0	100
j120_5	1.1	100	0.75	100	0.64	100
j120_10	1.29	100	0.91	100	0.8	100

6   Conclusions

In this paper, we have proposed a hybrid evolutionary algorithm to solve the project scheduling problem presented in [6]. This algorithm uses adaptive crossover, mutation and simulated annealing processes in order to improve the performance of the evolutionary search. These processes adapt their behavior to promote either exploration or exploitation of the search space according to the diversity of the evolutionary algorithm population. The computational experiments developed indicate that the performance of the hybrid evolutionary algorithm on the used instance sets is better than those of the algorithms previously proposed for solving the problem.

In future works, we will evaluate other adaptive crossover and mutation processes, and other selection processes. Moreover, we will evaluate the incorporation of other search and optimization techniques into the framework of the evolutionary algorithm.

References

1. Heerkens, G.R.: Project Management. McGraw-Hill, New York (2002)
2. Wysocki, R.K.: Effective Project Management, 3rd edn. Wiley Publishing, Hoboken (2003)

3. Bellenguez, O., Néron, E.: Lower bounds for the multi-skill project scheduling problem with hierarchical levels of skills. In: Burke, E.K., Trick, M.A. (eds.) PATAT 2004. LNCS, vol. 3616, pp. 229–243. Springer, Heidelberg (2005)
4. Hanne, T., Nickel, S.: A multiobjective evolutionary algorithm for scheduling and inspection planning in software development projects. Eur. J. Oper. Res. **167**, 663–678 (2005)
5. Gutjahr, W.J., Katzensteiner, S., Reiter, P., Stummer, Ch., Denk, M.: Competence-driven project portfolio selection, scheduling and staff assignment. CEJOR **16**(3), 281–306 (2008)
6. Yannibelli, V., Amandi, A.: A knowledge-based evolutionary assistant to software development project scheduling. Expert Syst. Appl. **38**(7), 8403–8413 (2011)
7. Yannibelli, V., Amandi, A.: A memetic approach to project scheduling that maximizes the effectiveness of the human resources assigned to project activities. In: Corchado, E., Snášel, V., Abraham, A., Woźniak, M., Graña, M., Cho, S.-B. (eds.) HAIS 2012, Part I. LNCS, vol. 7208, pp. 159–173. Springer, Heidelberg (2012)
8. Yannibelli, V., Amandi, A.: A diversity-adaptive hybrid evolutionary algorithm to solve a project scheduling problem. In: Corchado, E., Lozano, J.A., Quintián, H., Yin, H. (eds.) IDEAL 2014. LNCS, vol. 8669, pp. 412–423. Springer, Heidelberg (2014)
9. Blazewicz, J., Lenstra, J., Rinnooy Kan, A.: Scheduling subject to resource constraints: classification and complexity. Discrete Appl. Math. **5**, 11–24 (1983)
10. Yannibelli, V., Amandi, A.: Project scheduling: a multi-objective evolutionary algorithm that optimizes the effectiveness of human resources and the project makespan. Eng. Optim. **45**(1), 45–65 (2013)
11. Srinivas, M., Patnaik, L.M.: Adaptive probabilities of crossover and mutation in genetic algorithms. IEEE Trans. Syst. Man Cybern. **24**(4), 656–667 (1994)
12. Bellenguez, O., Néron, E.: A branch-and-bound method for solving multi-skill project scheduling problem. RAIRO – Oper. Res. **41**(2), 155–170 (2007)
13. Drezet, L.E., Billaut, J.C.: A project scheduling problem with labour constraints and time-dependent activities requirements. Int. J. Prod. Econ. **112**, 217–225 (2008)
14. Li, H., Womer, K.: Scheduling projects with multi-skilled personnel by a hybrid MILP/CP benders decomposition algorithm. J. Sched. **12**, 281–298 (2009)
15. Valls, V., Pérez, A., Quintanilla, S.: Skilled workforce scheduling in service centers. Eur. J. Oper. Res. **193**(3), 791–804 (2009)
16. Aickelin, U., Burke, E., Li, J.: An evolutionary squeaky wheel optimization approach to personnel scheduling. IEEE Trans. Evol. Comput. **13**(2), 433–443 (2009)
17. Heimerl, C., Kolisch, R.: Scheduling and staffing multiple projects with a multi-skilled workforce. OR Spectr. **32**(4), 343–368 (2010)
18. Eiben, A.E., Smith, J.E.: Introduction to Evolutionary Computing, 2nd edn. Springer, Berlin (2007)
19. Rodriguez, F.J., García-Martínez, C., Lozano, M.: Hybrid metaheuristics based on evolutionary algorithms and simulated annealing: taxonomy, comparison, and synergy test. IEEE Trans. Evol. Comput. **16**(6), 787–800 (2012)
20. Talbi, E. (ed.): Hybrid Metaheuristics. SCI 434. Springer, Berlin, Heidelberg (2013)
21. Kolisch, R., Hartmann, S.: Experimental investigation of heuristics for resource-constrained project scheduling: an update. Eur. J. Oper. Res. **174**, 23–37 (2006)

Building an Efficient Evolutionary Algorithm for Forex Market Predictions

Rafal Moscinski and Danuta Zakrzewska[✉]

Institute of Information Technology, Lodz University of Technology, Lodz, Poland
moscinski.rafal@gmail.com, danuta.zakrzewska@p.lodz.pl

Abstract. Foreign Exchange Market is one of the biggest financial markets in the world. In the paper an efficient algorithm for generating profitable strategies on this market is presented. The proposed technique is based on an evolutionary algorithm and uses the combination of technical indicators, which are to enable obtaining the highest profit depending on training and testing data. The algorithm allows to avoid risky strategies and is enhanced in developed mutation and crossover operators for assuring the effectiveness of obtained Forex trade strategies. The performance of the proposed technique was verified by experiments conducted on real data sets.

Keywords: Evolutionary algorithm · Trend predictions · Forex market

1 Introduction

Financial market predictions have become an object of scientific studies for several years. Foreign Exchange Market is mentioned among the biggest ones in the world. It is the market in which participants are able to buy, sell, exchange and speculate on currencies [1]. In the paper an algorithm which aims at generating profitable strategies for this market is proposed. The presented technique is based on an evolutionary algorithm which uses closing prices of specified time interval and computes them together with technical analysis indicators. The goal of the algorithm is to create such combination of various technical indicators which generates the highest profit on selected training and testing data. Additionally the algorithm detects strategies which generate too high loss during training process and automatically rejects them. Similarly to [2], as a data structure, a pair of decision binary trees is used, what allows to choose effectively combination of technical indicators. However, in the presented approach evolutionary technique has been enhanced in several improvements including input parameters optimization as well as developed mutation and crossover operators, which enable generating new solutions effectively. The performance of the proposed technique has been verified on the real data sets. Experiments showed effectiveness of the presented algorithm in indicating profitable Forex trade strategies.

The remainder of the paper is organized as follows. In the next section related work concerning application of evolutionary and genetic algorithms as well as

© Springer International Publishing Switzerland 2015
K. Jackowski et al. (Eds.): IDEAL 2015, LNCS 9375, pp. 352–360, 2015.
DOI: 10.1007/978-3-319-24834-9_41

artificial intelligence methods in Forex trade prediction is described. In the following section the proposed methodology is depicted with details. Section 4 is devoted to experiments and their results. Finally concluding remarks and future research are presented.

2   Related Work

Many researchers considered application of computational intelligence methods for financial market predictions. Research concerning Forex trade market has taken the significant place in the investigations. As the most commonly used methods one should mention artificial neural networks, genetic programming and evolutionary algorithms. The first method has been broadly investigated by many authors. Ni and Yin considered a hybrid model consisting of various regressive neural networks [3]. Lee and Wong combined an adaptive ANN with multi-value fuzzy logic for supporting foreign currency risk management [4]. Neuro-fuzzy approach for Forex portfolio management was proposed by Yao et al. [5]. A survey of applications of ANN in financial markets can be found in [6].

Li and Suohai [7] considered Forex predictions by combining support vector regression machine (SVR) with artificial fish swarm optimization (AFSA). They proved that SVR optimized by AFSA can be successfully used in Forex predictions. They showed the advantages of their method comparing to the techniques based on combination of SVR and cross validation, genetic algorithm or particle swarm optimization. SVR models were investigated by de Brito and Oliveira ([8,9]). In their papers they considered SVR+GHSOM (growing hierarchical self organizing maps) as well as GA (genetic algorithm) based systems. They concluded that GA model outperforms the one of SVR+GHSOM [9]. Very good performance of GA approach was stated in [10]. But, the authors indicated that when transaction costs are included, GA fails in getting positive results.

Evolutionary approach in Forex trade strategy generation has been investigated by Myszkowski and Bicz in [2]. They built the Evolutionary Algorithm based on BUY/SELL decision trees, that aimed at generating trade strategy, which consisted in technical analysis indicators connected by logical operators.

3   Evolutionary Based Trend Predictions

3.1   Improved Evolutionary Approach

To generate profitable strategies on Forex financial market, we propose application of an evolutionary algorithm, which uses closing prices of specified time interval and computes them with technical analysis indicators. Application of evolutionary algorithm allows to use any data structure. In the current paper, similarly to [2] a pair of decision binary trees has been applied. Evolutionary algorithm creates specified number of initial solutions and then modifies them with two different operators. The solutions are evaluated by fitness function to choose the best ones. The process is repeated until specified iteration number is reached. The mutation operator is responsible for generating new solutions which differ from the ones obtained in the current iteration by random modification of their parameters. Crossover operator is responsible for generating better solutions by combining two of them from the current iteration. Both of the operators have been built to assure that the new combinations of technical indicators are efficiently generated. The impact of algorithm parameters on final solution as well as the entry cost of transaction are also taken into considerations. To exclude too risky strategies maximum loss and profit are specified as input parameters. To generate new solutions effectively, mutation and crossover operators are assumed to have many different behaviors. Wide range of technical indicators such as Exponential, Simple and Weighted Moving Averages, Momentum Indicator, Commodity Channel Index, Relative Strength Index and Stochastic Oscillator [11] is considered. Finally, the impact of all parameters is measured to choose the most effective ones.

3.2   Solutions

The solutions of evolutionary algorithm take the form of pairs of binary trees where one of them is responsible for generating opening transaction signals while the other for closing transaction signals. The tree nodes can be either logical operators AND, OR or rules based on technical indicators and greater/smaller operators. All the parameters such as type of technical indicator, input parameter of the indicator, greater/smaller operator and value of the indicator can be modified randomly. The proposed evolutionary approach is presented on Fig. 1.

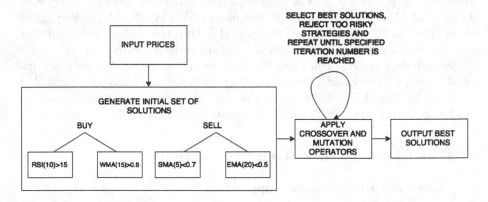

Fig. 1. Proposed evolutionary approach

3.3   Binary Trees

Binary trees are built according to the following rules. The root is one of two possible logical operators AND, OR. Children nodes can be logical operators or technical indicator based rules. After creating the root both of the children are checked, whether for any closing price from the training data set the respective conditions are satisfied. If they are, an opening or closing signals are generated. The next children are added to the tree as long as they are technical indicator based rules or specified maximum depth of the tree is reached. The tree example is presented on Fig. 2, with technical indicators SMA, WMA, EMA, RSI.

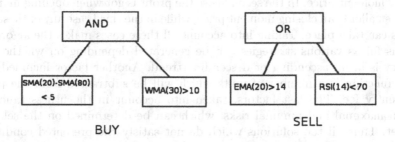

Fig. 2. Exemplary binary tree

In the algorithm such parameters as minimal and maximal tree depth or input ones for technical indicators are determined. The indicators are picked from previously created set. Their parameters are initially randomly determined and then modified by evolutionary operators. The possible value range for every indicator is calculated as a first step of the algorithm according to training and testing data sets. This ensures that mutation operator will select random values within that range. The profit is firstly calculated on the training data set, then the testing one is used. Initialisation function randomly picks tree node types, technical indicators as well as considered rules.

3.4 Evolutionary Operators

There are two evolutionary operators. One of them is a crossover operator. It takes random nodes of the same type from one tree and swaps them with nodes from another tree. The tree nodes are taken from these solutions which generated the best profit on the training data. Another way of modifying solutions by crossover operator consists in taking the whole sub-trees and swapping them so that maximum and minimum tree depths are not exceeded. The third possibility is to take sub-trees from current solution and to treat them as a new one. The fact that nodes are taken from the most profitable trees makes it more probable that next solutions will be better than the previous ones. Mutation operator picks random nodes from the tree and replaces its parameters according to initialization function. That way the obtained solution satisfies all required conditions and gives a chance for creating new, completely different solution. Additionally, a solution selection operator is used. It takes a specified number of best solutions from the previous iteration and creates new solutions using crossover and mutation operators.

3.5   Fitness Function

The choice of a fitness function should ensure indicating the most profitable solutions on specified data set. The profit is calculated by subtracting closing price of the opening moment from price of the closing moment. There are three types of transactions defined in the algorithm: SELL, BUY and ANY. In the first case the profit is obtained when the closing moment price is smaller than the opening moment price. In the second case the profit is got when opening moment price is smaller than closing moment price, while in the third case any of these two options can take place. Taking into account all these cases makes the algorithm more useful as various strategies can be generated depending on whether the currency is in an ascending or descending trend. Another factor included into fitness function is a number of units which will be subtracted from the profit as an entry fee. The last factors, taken into account in the fitness function, concern maximal and minimal risks, which can be determined on the selected data set. Thus all the solutions which do not satisfy the presented conditions can be omitted. Depending on the strategy fitness function may take the form:

$$F(A) = \begin{cases} \sum_{i=1}^{m}(p_k(i) - p_o(i) - P)) & \text{for BUY strategy} \\ \sum_{i=1}^{m}(p_o(i) - p_k(i) - P)) & \text{for SELL strategy} \\ \sum_{i=1}^{m}(|p_k(i) - p_o(i)| - P)) & \text{for ANY strategy} \end{cases} \quad (1)$$

where $A = (p_o(i), p_k(i))|i = 1, 2, ..., m$, $p_o(i)$ is a price of opening transaction moment indexed with i; $p_k(i)$ is a price of closing transaction moment; m means the number of all performed transactions and P is an entry fee.

In the first step of evolutionary algorithm the first set of solutions is initialized. The number of them is specified at the beginning and is fixed. Next, the most and the least profitable solutions are indicated by using fitness function. The copies of the best solutions are used as an input for crossover and mutation operators. The new solutions substitute the worst ones indicated by fitness function. The process is repeated until specified number of iterations is reached.

4   Experiment Results and Discussion

The experiments aimed at evaluating the performance of the proposed algorithm. There were considered two real historical data sets: the first one of four hour interval set of GBPJPY currency pair and the second one consisting of ten minute intervals on EURUSD currency pair [12]. The first data set was used to compare the algorithm results with simple buy-and-hold strategy being a control point for profitability evaluation. The second one was also used to compare obtained results with the evolutionary algorithm (EA) described in [2]. The considered data sets contained ascending as well as descending price trends. Each of them consisted of 1417 instances. The detailed information concerning the data sets is presented in Table 1.

To obtain the best possible results, the procedure of tunning parameters has been carried out. There were examined such parameters as number of iterations,

Table 1. Datasets information

Data type	Currency name	Time interval	Start date	End date
TRAINING	EURUSD	10 min	2009.09.24, 08:00	2009.10.07, 23:56
TESTING	EURUSD	10 min	2009.08.30, 22:01	2009.09.11, 20:00
TRAINING	GBPJPY	4 h	2014.01.01, 23:01	2014.07.10, 21:40
TESTING	GBPJPY	4 h	2014.07.11, 14:20	2014.12.30, 22:35

population size, numbers of mutations and crossovers, minimum and maximum tree depths, as well as minimum and maximum intervals for technical indicators. Table 2 shows the parameter values, for which the final profits for the both of the tested data sets were the highest and that have been used during experiments.

Table 2. Best result parameters

Tested parameter	Value
Number of iterations	100
Population size	80
Number of mutations	30
Number of crossovers	30
Minimum tree depth	5
Maximum tree depth	50
Minimum interval for technical indicator	5
Maximum interval for technical indicator	50

The data sets were equally divided into training and testing parts. Obtained profits for different data types, currency names and time intervals are shown in Table 3.

Table 3. Profits for the best parameters

Data type	Currency name	Time interval	Profit
TRAINING	EURUSD	10 min	2490
TESTING	EURUSD	10 min	2326
TRAINING	GBPJPY	4 h	2495
TESTING	GBPJPY	4 h	2038

The performance of the proposed technique is examined by comparing the resulted profits with the ones obtained by buy-and-hold strategy for the both

Table 4. Exemplary results

Data type	Currency name	Time interval	Profit	EA strategy	Buy-and-hold strat.
TRAINING	EURUSD	10 min	2490	x	1310
TESTING	EURUSD	10 min	2326	2062	1770
TRAINING	GBPJPY	4 h	2495	x	1600
TESTING	GBPJPY	4 h	2038	x	1400

Fig. 3. Best results entry moments

of the data sets. Additionally, in the case of the EURUSD data set the profits have been compared with the ones obtained by evolutionary strategy (EA) [2]. Exemplary results for time intervals of 10 min and 4 h, together with the profits of buy-and-hold strategy as well as EA strategy are presented in Table 4. The first three columns contain data sets features. The three last ones show profits obtained respectively by the proposed approach, EA ([2]) and buy-and-hold strategies.

The effect analysis showed that the results got by the presented algorithm were better than the ones received by buy-and-hold and EA ([2]) strategies. The profit for the EURUSD training data set was even 90.1 % better than the one obtained by buy-and-hold strategy, 31.4 % better on the test data set and 12.8 % better than the ones received by EA [2]. Such amelioration is driven by main advantages of the presented algorithm over the EA approach, that consist in improving evolutionary operators as well as input parameters optimization. The main advantage of the algorithm over buy-and-hold strategy is the ability of finding local minimum and maximum of fitness function while in buy-and-hold strategy they are omitted. Analysis of entry moments showed that the proposed algorithm takes into account more minimum and maximum of price functions on time intervals, what additionally allows to obtain profits higher than the ones obtained by buy-and-hold strategy. The best entry moments for the EURUSD data sets are presented on Fig. 3.

5   Concluding Remarks

In the paper, the new evolutionary algorithm for generating profitable strategies on Forex financial trade market is presented. In spite of the EA approaches investigated so far, several enhancements which are to assure good performance of the evolutionary technique are introduced. Developing of the crossover as well as mutation operators improved effectiveness of the proposed algorithm. Using of the proposed form of mutation operator enables finding local extremes for different price levels. Optimal parameters tunning additionally improves effectiveness of the obtained strategies. Thanks to using technical indicators, strategies which generate too high loss during training process are detected and automatically rejected. The experiments carried out on historical data sets showed good performance of the proposed technique comparing to EA ([2]) and buy-and-hold strategies.

Future research will consist in further experiments carried out on data of different characteristics as well as in development of the presented algorithm, by including fundamental data. It will allow to take into account factors, which are unpredictable by technical analysis. Worth considering is also the problem concerning connection of correlated exchange pairs, as including their growths and falls may significantly ameliorate the performance of the algorithm.

References

1. http://www.investopedia.com/terms/forex/f/foreign-exchange-markets.asp, 30 May 2015
2. Myszkowski, P., Bicz, A.: Evolutionary algorithm in forex trade strategy generation. In: Proceedings of the 2010 International Multiconference on Computer Science and Information Technology (IMCSIT), pp. 81–88 (2010)
3. Ni, H., Yin, H.: Exchange rate prediction using hybrid neural networks and trading indicators. Expert Syst. Appl. **72**, 2815–2823 (2009)
4. Lee, W., Wong, H.T.: A multivariate neuro-fuzzy system for foreign currency risk management decision making. Neurocomputing **70**, 942–951 (2007)
5. Yao, S., Pasquier, M., Quek, C.: A foreign exchange portfolio management mechanism based on fuzzy neural networks. In: 2007 IEEE Congress on Evolutionary Computation, pp. 2576–2583 (2007)
6. Li, Y., Ma, W.: Applications of artificial neural networks in financial economics: a survey. In: 2010 International Symposium of Computational Intelligence and Design, pp. 211–214 (2010)
7. Li, M., Suohai, F.: Forex prediction based on SVR optimized by artificial fish swarm algorithm. In: 2013 Fourth Global Congress on Intelligent Systems, pp. 47–52 (2013)
8. de Brito, R.F.B., Oliveira, A.L.I.: Comparative study of FOREX trading systems built with SVR+GHSOM and genetic algorithms optimization of technical indicators. In: 2012 IEEE 24th International Conference on Tools with Artificial Intelligence, pp. 351–358 (2012)

9. de Brito, R.F.B., Oliveira, A.L.I.: Sliding window-based analysis of multiple foreign exchange trading systems by using soft computing techniques. In: 2014 International Conference on Neural Networks (IJCNN), pp. 4251–4258 (2014)
10. Mendes, L., Godinho, P., Dias, J.: A forex trading system based on genetic algorithm. J. Heuristics **18**, 627–656 (2012)
11. Thiel, S.: Rynek kapitalowy i terminowy (2007) (in polish)
12. http://www.forextester.com/data/datasources, 30 May 2015

Tabu-Search Algorithm for Optimization of Elastic Optical Network Based Distributed Computing Systems

Marcin Markowski[✉]

Department of Systems and Computer Networks,
Wroclaw University of Technology, Wybrzeze Wyspianskiego 27,
50-370 Wroclaw, Poland
marcin.markowski@pwr.edu.pl

Abstract. In the paper we consider expanded routing and spectrum assignment problem in distributed computing system based on elastic optical network. Computational tasks generated by clients at network nodes are divided into subtasks and sent to data centers to provide computations. The problem consists in simultaneously dividing computational tasks into subtasks and for each subtask assigning data center, route and spectrum in elastic optical network. The considered traffic type is an extension of classical anycast traffic, since one on few data centers may be the target (destination) of each demand. In the paper, we formulate a new optimization problem as an ILP model and propose meta-heuristic algorithm based on tabu search method. Then, we present computational experiments for exemplifying network, reporting the quality of heuristic solutions an characteristics of considered optimization problem.

Keywords: Distributed computing · Elastic optical network · Anycast · Tabu search · RSA problem

1 Introduction

In the last years we observe the trend to move the execution of computational and storage tasks to distributed and cloud environment. The main drivers for using cloud in place of traditional computing model are the better agility, nearly unbounded elasticity and high availability. Cloud services and cloud-available resources are the good answer for changeable and unpredictable user demands, they allow to provision IT resources faster and at reduced cost.

Dynamic growth of cloud and distributed services implies in necessity of ensuring stable and high capacity transmission between data centers. Optical networks are often considered as transport layer for inter-data center communication, and elastic optical network (EON) architecture is being found as an promising solution for future optical networks [1]. EON (or SLICE) architecture assumes division of optical spectrum into slots, called also slices, of 6,25 GHz width [2]. Optical channels may be built as continuous blocks of even number of such slots. Division of spectrum into slices is possible with use of bandwidth-variable transponders and wavelength cross connects (WXCs) [3]. The main advantage of EONs is the elasticity in building channel of

© Springer International Publishing Switzerland 2015
K. Jackowski et al. (Eds.): IDEAL 2015, LNCS 9375, pp. 361–369, 2015.
DOI: 10.1007/978-3-319-24834-9_42

demanded bandwidth, what minimize the wasting of spectrum and allows to optimize usage of optical resources [4].

The main optimization problem considered in the scope of elastic optical networks consist in finding continuous unoccupied spectrum (adequate number of slices) in order to establish a lightpaths for traffic demands and it is called Routing and Spectrum Assignment (RSA) problem [5]. Depending on the nature of traffic demands we may distinguish offline and online RSA problems. In offline problem set of demands is known in advance and demands have long-term, permanent nature [6]. The aim of optimization is to minimize amount of consumed spectrum. In online problem requests appear dynamically and lightpaths must be established immediately, what may be impossible due to lack of optical resources. Optimization algorithms must run in real time and the goal is to minimize the blocking probability [7] of requests. RSA problem is solved for unicast [5], anycast [8] and multicast [9] traffic demands.

In this paper, we propose the model of distributed computation system, based on elastic optical transport network. Users of system generate computational demands, which must be computed in date centers. Demands have permanent nature and are characterized by the volume of demand (Gb/s) and the number of instruction required to process data of demand. For each demand one data center may be chosen or demand may be divided into sub-demands (of different size and computational effort) and next computed on few (or all) data centers. The demand traffic is limited with following constraints. Each sub-demand generates the traffic to chosen data center and backward traffic to client, we denote that volume of backward traffic is small and require optical channel with minimal possible bandwidth. Only one route (path) in the EON network and only one range of spectrum may be chosen for each sub-demand (and also for each backward demand). We also denote that the computational power of each data center is limited. Additionally, we consider also unicast traffic in the network. Due to novel traffic demands formulation (one, few or all data centers may be chosen for processing demand) and data center power constraint, the considered problem is original and has not been investigated yet.

2 Problem Formulation

We consider an elastic optical network modeled as the directed graph with V nodes and E links. R data centers, with given processing power are located in some nodes of the network. Demands that must be routed in the network are labeled with d, D is the total number of demands. Unicast traffic demands between clients' systems are defined with source node, destination node and the volume of demand (in Gb/s). Computational tasks are represented by the client node, the volume of demand that must be uploaded to data centers, the number of instructions required to process uploaded demand and the volume of computational results, that must be downloaded from data center to client system. Each computational tasks may result in one or more traffic streams from client to data centers (called *anycast upstream streams*) and the same number of feedback traffic demands (*anycast downstream streams*). The number of demand streams is limited by the number of data centers. Upstream and downstream demands realizing the same anycast request are called associated.

Proposed approach assume using pre-calculated paths between the nodes of the network, we have prepared k candidate paths for each pair of nodes. For each anycast upstream and downstream stream exactly one path must be chosen. Different modulation levels may be used in the network, depending on the length of optical paths. Each optical link in EON is divided into S slices and optical channels built of even number of slices may be used. We consider the average width of spectrum (in number of slices) used on individual network links as an optimization criterion.

In order to formulate the ILP model we use notation proposed in [5]:

indices

$s = 1, 2, \ldots, S$	slices
$m = 2, 4, \ldots, M$	sizes of candidate channels (in number of slices)
$c = 1, 2, \ldots, C_m$	candidate channels of size m (for anycast)
$g = 1, 2, \ldots, G_{dp}$	candidate channels for unicast demand d on path p
$v = 1, 2, \ldots, V$	network nodes
$d = 1, 2, \ldots, D$	demands (both anycast and unicast)
$d = 1, 2, \ldots, A$	upstream anycast demands (from client to data center)
$d = A + 1, 2, \ldots, 2A$	downstream feedback (from data center to client)
$d = 2A + 1, 2, \ldots, D$	unicast demands
$p = 1, 2, \ldots, P_d$	candidate paths for flows realizing demand d
$e = 1, 2, \ldots, E$	network links

constants

$\delta_{edp} = 1,$	if link e belongs to path p realizing demand d; 0, otherwise
h_d	volume of demand d (Gbps)
$\gamma_{mcs} = 1,$	if channel c of size m uses slice s; 0, otherwise
$\gamma_{dpgs} = 1,$	if channel g associated with unicast demand d on path p uses slice s; 0, otherwise
β_{dpm}	maximum capacity of path p of anycast demand d with width of m slices, according to modulation formats and ITU standards (Gbps)
$t_{dpv} = 1,$	if node v is the destination node of path p for demand d; 0, otherwise
$u_{dpv} = 1,$	if node v is the source node of path p for demand d; 0, otherwise
c_v	processing power of data center located at node v (GIPS)
λ_d	number of instructions required to process data of demand d (anycast upload) transmitted with streaming rate 1 Gbps
$\tau(d)$	index of an anycast demand associated with demand d

variables

$x_{dpmc} = 1,$	if channel c of size m on candidate path p is used to realize anycast demand d; 0, otherwise (binary)
x_{dpv}	flow of anycast demand d and assigned to path p sent to data center located at node v, given in Gbps (continuous)
$z_{dpg} = 1,$	if channel g on candidate path p is used to realize unicast demand d; 0, otherwise (binary)
$y_{es} = 1,$	if slice s is occupied on link e; 0, otherwise (binary)
y_e	the largest index of allocated slices in link e (integer)

objective

$$\text{minimize } Q = \left(\sum_e y_e\right)/E \tag{1}$$

subject to

$$x_{dpv} \leq \sum_m \sum_c t_{dpv}\beta_{dpm}x_{dpmc}, \; d = 1, 2, \ldots, 2A, \; p = 1, 2, \ldots, P(d), \; v = 1, 2, \ldots, V \tag{2}$$

$$\sum_p \sum_v x_{dpv} = h_d, \; d = 1, 2, \ldots, D \tag{3}$$

$$\sum_{d=1..2A} \sum_p \sum_m \sum_c \gamma_{mcs}\delta_{edp}x_{dpmc} + \sum_{d=2A+1..D} \sum_p \sum_g \gamma_{dpgs}\delta_{edp}z_{dpg} \leq y_{es},$$
$$e = 1, 2, \ldots, E \; s = 1, 2, \ldots, S \tag{4}$$

$$sy_{es} \leq y_e, \; e = 1, 2, \ldots, E \quad s = 1, 2, \ldots, S \tag{5}$$

$$\sum_{d=1..A} \sum_p \lambda_d x_{dpv} \leq c_v, v = 1, 2, \ldots, V \tag{6}$$

$$\sum_p \sum_m \sum_c t_{dpv}x_{dpmc} \leq 1, d = 1, 2, \ldots, 2A, \; v = 1, 2, \ldots, V \tag{7}$$

$$\sum_p \sum_m \sum_c t_{dpv}x_{dpmc} = \sum_p \sum_m \sum_c u_{\tau(d)pv}x_{\tau(d)pmc}, \; d = 1, 2, \ldots, A \quad v = 1, 2, \ldots, V \tag{8}$$

$$\sum_p \sum_g z_{dpg} = 1, \; d = 2A + 1, 2, \ldots, D \tag{9}$$

The objective (1) is to minimize the width of spectrum, defined as the average number of slices required in each network link. The width of spectrum used in each link is represent with the largest index of the occupied slices. Constraint (2) ensures that the flow volume in each channel selected for demand d does not exceed the capacity of that channel. Equation (3) guarantees that whole demand is realized. Each slice in each link may be used only ones (only in one of assigned channels), which is warranted with constraint (4). Constraint (5) defines the largest index of allocated slices in link e. Constraint (6) assures us that the computational power of any data centers server is not exceeded. For each anycast demand maximally one lightpath to and from respective data centers may be used, which is ensured in (7). Each anycast upstream stream must be associated with corresponding anycast downstream stream (8). Finally, exactly one lightpath (channel) must be assigned for each unicast demand, which is ensured in Eq. (9). Formulated optimization problem is NP-hard as more general than classical RSA problem [5, 6].

3 Tabu Search Algorithm

Tabu search algorithm starts from the initial solution which may be calculated with simple heuristics or generated randomly. Then, in each step of the algorithm neighbor solutions are generated. When the objective value of the neighbor is better than objective value of current solution, then the neighbor becomes the current solution (such operation is called 'a move'). All moves are added to long-term tabu list and are forbidden during next search. Tabu list has limited capacity, what means that tabu moves leave tabu lists after certain number of steps. All tested (but not used) solutions are added to short-term tabu list, which is flushed after each move. Detailed description of tabu algorithm may be found in [10]. In the proposed algorithm solution is defined as the sequence of demands and demands' streams allocation and the path selection for each demand (or demand stream). Demands (streams) are allocated in the established order, at the established path and channel for each demand (stream) starts from the lowest possible slice on path. Then, the decision variables for proposed tabu search algorithm are:

- the position of demand (stream) on allocation list
- the path for demand (stream)
- the volume of the stream for anycast demands

According to the problem formulation and solution format, there are three kinds of moves possible:

1. changing the position of demand or stream in allocation sequence. Move is defined with the current demand position, position change in allocation sequence (positive or negative value) and stream (data center) number (only for anycast demands).
2. changing the path for demand or stream. Move is defined with the demand number, current and new path number and stream (data center) number (for anycast demands).
3. changing the volume of two anycast upstream streams corresponding to one computational task – move possible only for anycast demands. Part of flow may be reallocated from one of the streams existing for demand d (source stream) to another stream of this demand (destination stream). In result the new stream (to previously unused data center) may be created or existing stream may be eliminated. The corresponding downstream streams are updated accordingly. Move is defined with the demand number, stream number and the volume of reallocated flow.

In tabu method, diversification is the process that lets the algorithm to change the search area after reaching the local optima. In proposed algorithm, after certain number (called diversification threshold – DT) of iterations without solution improvement we decide that the local optima has been found. Then tabu algorithm starts to accept the neighboring solutions worse than current one in order to escape from local optima. The linear parameter DR (diversification ratio, $DR > 1$) defines the level of acceptation of worse solutions.

Algorithm terminates after given time limit or when optimal solution is found (optionally – when the optimal solution is known).

Let Ω be the current solution, Ω^* be the best already found solution, Ω_0 - the initial solution and Ω' - candidate (neighbor) solution. Let $Q(\Omega)$ be the value of criterion for solution Ω. The pseudo-code of proposed tabu-search algorithm is presented below.

```
Ω = Ω₀, Q = Q(Ω₀)
repeat
    select possible non-tabu neighbor Ω' (Ω' = Ω + move)
    if ( Q(Ω') < Q ) then {
        Ω = Ω', Q = Q(Ω)
        if ( Q(Ω) < Q(Ω*) then Ω* = Ω
        add move to tabu-list
        flush short-term-tabu-list
    }
    otherwise
        add move to short-term-tabu-list
    if (diversification threshold) then Q = Q * DR
until stopping criteria satisfied
return Ω*
```

4   Computational Results

The main goals of experiments were to evaluate the quality of solutions obtained with the proposed tabu search algorithm and to analyze the properties of considered optimization problem. Experiments were conducted for network topology NFS15, presented in the Fig. 1. To find the optimal solutions for considered problem CPLEX Optimization Studio 12.5 [11] and CPLEX solver were used.

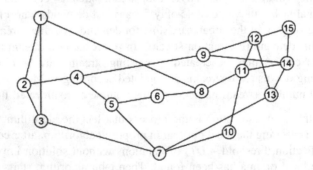

Fig. 1. Examined network topology NFS15

All demand sets were generated randomly: volume of anycast upstream demands in range 40–400 Gbps, volume of anycast downstream (feedback) demands was always set to 10 Gbps, and the volume of unicast demands in range 10–100 Gbps. Demands sets were generated for different anycast ratio (*AR*) defined as the proportion of volume of all anycast demands to volume of all unicast demands. Demands sets with six values of *AR* were examined: 0 % (no anycast traffic), 20 %, 40 %, 60 %, 80 % and 100 % (no unicast traffic).

Two important parameters of tabu algorithm: size of long-term tabu list (TL) and diversification threshold DT (number of iterations without improvement) were selected during tuning process. To perform tuning we have generated demand sets with 5 Tb/s overall demand (sum of all demands), one for each value of AR. For each demand set we have examined scenarios with three sets of data center allocation (with 1, 2 and 3 data centers) and three sets of candidate paths (3, 10 and 30 candidate paths for each pair of nodes). Experiments were conducted ten times for each parameter set, in total 34560 experiments were performed during tuning. Values of examined parameters TL and DT and results of tuning are shown in Table 1. Best metaheuristic solution for each scenario was taken into account as 'best solution' in order to calculate the average gap.

Table 1. Results of tuning (average distance from best solution)

		TL							
		3	5	8	11	14	17	20	23
	160	1,97%	2,01%	1,87%	1,99%	2,16%	1,81%	1,87%	2,13%
	180	1,95%	2,08%	1,81%	1,98%	1,79%	1,80%	1,89%	1,88%
	200	2,06%	1,83%	2,00%	1,87%	1,99%	1,94%	2,06%	1,88%
DT	220	1,79%	1,87%	1,83%	1,98%	1,78%	1,86%	2,02%	1,84%
	240	1,73%	1,92%	1,90%	**1,58%**	1,59%	1,90%	1,77%	1,85%
	260	1,76%	1,74%	1,65%	1,90%	1,74%	1,69%	1,75%	1,82%
	280	1,89%	1,72%	1,73%	1,85%	2,04%	1,78%	1,62%	1,93%
	300	1,87%	1,76%	1,82%	1,81%	1,83%	1,79%	1,72%	1,71%

The best results for tabu search algorithm may be obtained for parameters $LT = 11$ and $DT = 240$. Those values were used in next experiments.

Next goal of experiments was to examine the quality of approximate solutions, basing on optimal solutions yield from CPLEX solver. Optimal solutions were obtained for overall demand in the network equal to 1, 2 Tb/s. Number of candidate paths was set to 2. Experiment were conducted for 6 values of *AR*, 3 sets of demands for each value of *AR*, 3 different number of data centers and 4 sets of nodes of data centers allocation for each number of data centers (total 216 scenarios). Experiments for tabu search algorithm were repeated five times for each scenario. Time limit equal to 300 s and reaching the optimal solution were used as the stopping criteria. In most cases tabu search algorithm was able to find the optimal solution, and the average gap between optimal and tabu solutions was 0, 13 %. The average execution time of tabu search algorithm was 25 s, while average time required for CPLEX to obtain optimal solution was 223 s.

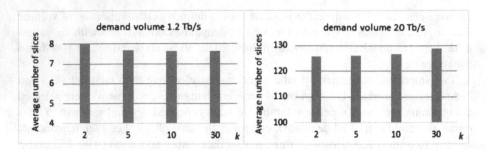

Fig. 2. Influence of number of candidate paths on average spectrum consumption

In the Fig. 2 the influence of number of candidate paths (k) on the solution is presented for two different values of overall demand volume.

We may observe, that increasing the number of candidate paths over $k = 5$ does not effect in improving the quality of obtained solution, and for higher overall demand volume (20 Tb/s) quality of solution degrades with increasing number of candidate paths. For high value of k algorithm is able to search only small fraction of solutions area in limited time (300 s) set for experiments. We may conclude that for relatively small networks (like NFS15) using large sets of candidate paths is not beneficial.

We have analyzed the impact of the value of AR coefficient and number of data centers on solution. Results obtained for network NFS15 are presented in the Fig. 3. Results of experiments show that the average amount of used spectrum decreasing with increasing value of anycast ratio. Also the number of data centers has the significant impact on spectrum consumption. The biggest difference was observed between scenarios with one and two data centers, the slighter improvement may be obtained while increasing number of data centers from 2 to 3. Observed dependences may be helpful for making up decisions about the optimal number of data centers that should be place in the EON based distributed computing system in order to achieve the most spectrum efficient and cost efficient solution.

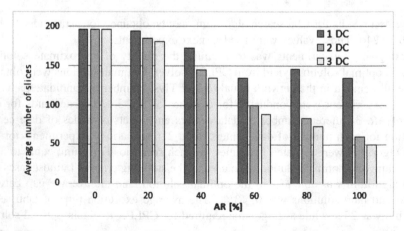

Fig. 3. Influence of anycast ratio (AR) and the number of data centers on average spectrum consumption for overall demand volume equal to 20 Tb/s

5 Conclusion

The problem of optimization of unicast and anycast traffic in elastic optical network based distributed computing systems was considered. Original optimization problem that consists in computational tasks allocation among data centers, route selection and spectrum allocation in EON network was formulated. The optimization problem is new as it takes into account the possibility of dividing task into subtask and execute them on many data centers, we also limit the computational power of data centers. We have proposed the ILP formulation of the problem, allowing to find optimal solution with CPLEX solver for less complex scenarios and meta heuristic algorithm based on tabu search method. Results of experiments, focused on testing the quality of metaheuristic solutions and discovering some properties on considered problem, were analyzed and reported in the paper.

Acknowledgments. This work was supported by the Polish National Science Centre (NCN) under Grant DEC-2012/07/B/ST7/01215 and statutory funds of the Department of Systems and Computer Networks, Wroclaw University of Technology.

References

1. Yoo, S.J.B., Yin, Y., Proietti, R.: Elastic optical networking and low-latency high-radix optical switches for future cloud computing. In: Proceedings of IEEE International Conference on Networking and Communications, pp. 1097–1101 (2013)
2. ITU-T Recommendation G.694.1 (ed. 2.0): Spectral Grids for WDM Applications: DWDM frequency grid (2012)
3. Jinno, M., Takara, H., Kozicki, B., Tsukishima, Y., Sone, Y., Matsuoka, S.: Spectrum-efficient and scalable elastic optical path network: architecture, benefits, and enabling technologies. IEEE Commun. Mag. **47**(11), 66–73 (2009)
4. Gerstel, O., Jinno, M., Lord, A., Yoo, S.J.B.: Elastic optical networking: a new dawn for the optical layer? IEEE Commun. Mag. **50**, S12–S20 (2012)
5. Klinkowski, M., Walkowiak, K.: Routing and spectrum assignment in spectrum sliced elastic optical path network. IEEE Commun. Lett. **15**(8), 884–886 (2011)
6. Talebi, S., et al.: Spectrum management techniques for elastic optical networks: a survey. Opt. Switching Netw. **13**, 34–48 (2014)
7. Walkowiak, K., Kasprzak, A., Klinkowski, M.: Dynamic routing of anycast and unicast traffic in elastic optical networks. In: 2014 IEEE International Conference on Communications, pp. 3313–3318 (2014)
8. Zhang, L., Zhu, Z.: Dynamic anycast in inter-datacenter networks over elastic optical infrastructure. In: 2014 International Conference on Computing, Networking and Communications, pp. 491–495 (2014)
9. Walkowiak, K., Goścień, R., Klinkowski, M., Woźniak, M.: Optimization of multicast traffic in elastic optical networks with distance-adaptive transmission. IEEE Commun. Lett. **18**(12), 2117–2120 (2014)
10. Talbi, E.G.: Metaheuristics: From Design to Implementation. Wiley, Hoboken (2009)
11. IBM ILOG CPLEX Documentation. www-01.ibm.com

Multi-manifold Approach to Multi-view Face Recognition

Shireen Mohd Zaki and Hujun Yin[✉]

School of Electrical and Electronic Engineering, The University of Manchester,
Manchester, M13 9PL, UK
shireen.mohdzaki@postgrad.manchester.ac.uk,
hujun.yin@manchester.ac.uk

Abstract. In this paper a multi-manifold approach is proposed for dealing with facial variations and limited sample availability in face recognition. Face recognition has long been an active topic in computer vision. There have been many methods proposed and most only consider frontal images with controlled lighting. However, variations in environment, pose, expression, constitute huge problems and often make practical implementation unreliable. Although these variations have been dealt with separately with varying degrees of success, a coherent approach is lacking. Here practical face recognition is regarded as a multi-view learning problem where different variations are treated as different views, which in turn are modeled by a multi-manifold. The manifold can be trained to capture and integrate various variations and hence to render a novel variation that can match the probe image. The experimental results show significant improvement over standard approach and many existing methods.

Keywords: Multi-manifolds · Face recognition · Multi-view learning · Invariant features

1 Introduction

Many face recognition algorithms have been proposed and shown to be efficient and to produce some remarkable results [1, 2]. However, many challenges remain for extreme conditions such as uncontrolled illumination and angle of the head. This is added with the complexity of overall appearance of the face image either from the face expression, aging and the background image. Limited variability of training images in a face database is also a common problem, where the images are often in a controlled environment setting.

Examples of early approaches to face recognition are Eigenfaces [3], Fisherfaces [4] and 2-dimensional PCA (2DPCA) [5]. They are known as the appearance-based method, which is based on the whole appearance of a two-dimensional image as opposed to methods that use the shape or key features of a face. These methods need the training set to include face images in all possible variations, and a low dimensional representation of the face images is produced as a linear subspace. Although proved to be simple and efficient, linear subspace techniques suffer some drawbacks, such as huge number of training images is needed to produce a good or complete representation. This is not

© Springer International Publishing Switzerland 2015
K. Jackowski et al. (Eds.): IDEAL 2015, LNCS 9375, pp. 370–377, 2015.
DOI: 10.1007/978-3-319-24834-9_43

always feasible for a practical face recognition system as it often have limited gallery images.

The proposed multi-view face recognition approach differs from previous methods where different variations are handled individually with a dimensionality reduction and manifold learning technique. Upon learning the underlying patterns from a specific variation, parameters that are responsible in controlling the variations are identified and exploited to synthesize novel variations for novel images. Furthermore, this approach only relies on 2-dimensional images and this is advantageous for computation and storage.

2   Related Works

Dimensionality reduction techniques are crucial in face recognition to minimise cost of computation and storage demands and at the same time extract essential face features. Beyond its ability to represent a compact low dimensional visualisation from a very high dimensional data, dimensionality reduction techniques are capable of discovering the intrinsic information of a dataset that may have different dimensionalities and lie on different subspaces. They are also capable of finding underlying structures and patterns that are useful in many applications.

PCA has been the basis of many dimensionality reduction and face recognition algorithms. It finds the orthogonal principal direction and produces a reduced linear subspace that attempts to capture the maximum variability of the data by eliminating the minor components. However, linear methods such as PCA are usually insufficient in interpreting and detangling complex and nonlinear high-dimensional data. Nonlinear methods such as Isomap [6] and Locally Linear Embedding (LLE) [7] are eigen-decomposition based methods with PCA roots. ISOMAP applies a global approach based on geodesic distance instead of Euclidean distance; meanwhile LLE is based on a linear combination of a local neighbourhood of points to learn a manifold structure. Both methods solve an eigen value problem to find high to low-dimensional embedding of data points.

Eigen-decomposition based methods have the advantage to provide unique solution [8] but becomes computationally expensive when being feed with new points. Adaptive manifold learning technique presents an alternative to these methods where they process input data iteratively and suitable for online learning. Self-organising maps (SOM) [9] is an unsupervised learning algorithm that transforms input patterns with arbitrary dimension into 1 or 2-dimensional lattice in a topologically ordered fashion. It plots input patterns with similar features together in the output map, which is useful for high to low dimensional data visualisation and clustering.

However, SOM has limitations in producing flexible map and does not preserve distance relationships which might not produce an accurate low dimensional representation [10]. Visualisation-induced SOM (ViSOM) [11] offers better control on SOM output mapping where it preserves both structure and topology of the manifold. It maintains local distances between the neurons by performing a lateral contraction force when moving the winner's neighbours toward the input. Unlike

SOM, ViSOM enables not only direct representation of the original data but a quantitative measure of the output map.

Most practical data sets such as face images consist of multiple manifolds to represent the variations of illumination, pose or face expression as well as subjects/identity. Hence single manifold method such as PCA, LLE and ISOMAP will not be sufficient to represent these datasets. Multi-manifolds methods have been proposed to address the limitation of single manifold techniques. K-manifold [12], mumCluster [13] and Spectral Multi-manifold Clustering (SMMC) [14] presented multi-manifolds approaches that are based on neighbourhood graphs and clustering analysis. Other techniques directly deal with multi-manifold problem through face recognition point of view. Discriminative Multi Manifold Analysis (DMMA) [15] uses patches of face images that represents key facial features to learn each subject's feature space and represent them as a manifold. For recognition, they applied a manifold-manifold distance mechanism to determine similarities between probe and gallery images.

3   Face Recognition as Multi-view Learning

3.1   Multi-view Learning

Multi-view learning is concerned with learning from diverse data sets or different facets of objects. Many practical data sets or problems are diverse or heterogeneous and/or represented by multiple feature sets or "views", especially in the digital era. Multi-view learning is thus becoming increasingly important as compared to single view based learning in the past. Fusion, association, manifold, clustering and sparse-coding based methods can be used for multi-view learning [16–18].

In this paper, the multi-view learning is formulated as learning and integration from multiple facial variations such as pose and illumination. Every view represents a distinct feature set and reveals certain aspect of the subject or distorted by certain aspect of the environment. A multi-manifold is used to capture these views in a functional manner and thus to characterise these variations. It can synergise these different views coherently and thus render them jointly.

3.2   Problem Formation

Given a set of gallery images with only one sample image per subject, these images were taken with strictly neutral/frontal illumination with frontal pose. The probe images that need to be identified against these images can be significantly different from the gallery images in terms of the illumination and pose condition. As a result, any approach to classifying these images by modelling these variations in a single low-dimensional subspace would fail and degrades recognition performance.

Therefore, in our method, each variation is represented independently through linear subspace, which enables us to discover the underlying patterns in a training set, and the parameters responsible in controlling the variations. These parameters can be exploited to render new variations for a novel image, which in this case the single gallery image for each subject. Consequently, each subject is extended to contain large variations in illumination and pose from the rendered image which benefits the recognition process. However, note that currently in our approach, we do not attempt to handle other variations in face images such as face expressions, aging or other facial details.

3.3 Synthesis of Pose and Illumination

The core approach for the pose and illumination synthesis is to learn the most important modes in controlling the shapes and illumination variation by applying multi-manifold on training sets. As different variations are treated as different views, illumination and pose manifolds are learned from an illumination and shape training set. Each training set needs to contain images with different pose and illumination conditions (the more the better) in order to capture the different modes of both variations and the novel images can be synthesized from them.

An image, X_i from an illumination training set X that contains an adequate sample of lighting variations can be reconstructed by

$$X_i = \psi + Pb \tag{1}$$

where ψ is the mean image of the training set, $P = \left[\phi_1 \,|\phi_2| \,\dots\, |\phi_t\right]$ is the eigenvectors calculated from the covariance matrix of the data and jth component of b (where $j = 1, 2, \dots t$) weighs the jth eigenvector Φ_j. The values of b define the illumination direction on the face in the training images. As the columns of P is orthogonal, b is

$$b = P^T \left(X_i - \psi\right) \tag{2}$$

Hence it is possible to synthesize a novel illumination condition for an image based on the training set by reconstructing the image with a value of b such as,

$$X_i' = P_j \left(b_j + \psi\right) \tag{3}$$

where X_i' is the synthesized image.

Following the principle of the Point Distribution Model (PDM) [19] and a pose invariant method proposed in [20], the parameter that controls the pose changes is defined from the landmarks of a set of images with non-frontal poses. These landmarks are obtained by annotating the key facial features that defines the shapes of the face and other features. The landmark points of a face are stored as rows in an n-by-2 matrix, and reshaped into a vector to create the training shape set such as,

$$Y_i = \left(y_{1i}, y_{2i}, \dots, y_{Ni}, z_{1i}, z_{2i}, \dots, z_{Ni}\right)^T \tag{4}$$

A novel shape is synthesized similar to the principle of illumination synthesis. The pose parameter that explains the pose changes is calculated through (3) while a novel shape can be reconstructed with (4). A novel image with a novel pose can be generated by warping the pixels of an image to the synthesized shape. In our method, we applied Piecewise Affine Warping (PAW) as the warping technique. Figure 1 demonstrates the illumination and pose synthesis process.

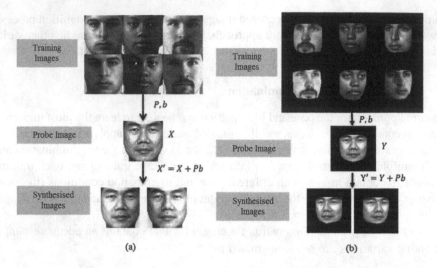

Fig. 1. (a) Illumination synthesis and (b) pose synthesis process.

4   Proposed Method

Figure 2 demonstrates the proposed multi-manifold based multi-view face recognition method. It utilises the simple approach of synthesizing novel illumination and pose condition described in the previous section. In multi-view face recognition, image synthesis is applied to extend the variability of gallery images so that the difference between a gallery and probe image can be minimised. As described in the previous section, the illumination and pose parameters need to be calculated from an illumination and shape training set. These training sets can be developed from available images as long as the illumination and pose variability are present in the training images. Novel images with multiple variations of illumination and pose are synthesized to extend the gallery set.

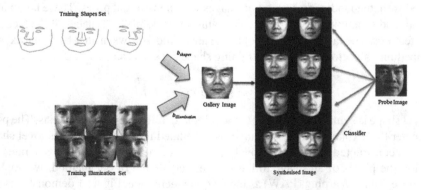

Fig. 2. Multi-manifold based multi-view face recognition.

The recognition engine or classifier for the proposed method is based on Gabor filters [21] and Gaussian kernel function (GKF). We used 40 Gabor filters with 5 frequencies and 8 orientations and 15×15 image patch as a convolution mask for a specific pixel on each image. Each pixel will acquire a vector of 40 coefficients, or known as a jet, $\mathcal{J}$ and we applied this filter on every landmark that we manually annotated. For a pair of images from probe and gallery set, their similarities are calculated based on their jets similarity, S at each point, taking account only the magnitude of the jet coefficients. Labels are assigned to the probe images based on the maximum average of the similarity measure.

The Gaussian kernel function filter is applied to the pixel values of each landmark points and its neighbouring pixels in a 7×7 image patch. Similar to the Gabor filter, each image patch is convoluted with the Gaussian filter, and the responses are used to describe the information in the region of the landmark points. To compare a probe image with the gallery images (original and synthesized), the Gaussian kernel function responses at the landmark points for the probe images are correlated with the gallery images to find a match. The gallery image with the highest correlation coefficient indicates the label for the probe image.

5 Experiments and Results

5.1 Data and Preprocessing

An experiment were conducted on a subset of CMU-PIE [22] face database to test the usefulness of the proposed method. CMU-PIE contains images of 68 subjects, where each subject has 13 different poses, 43 different illuminations and 4 different expressions. For this paper, a subset of cropped images from this database was used [23], where the images are aligned, cropped and resized to 64×64. We used 37 subjects from this database, where all subjects are without glasses.

5.2 Experiment Setup

For each experiment, the 37 subjects were randomly partitioned into 10 subjects for training and 27 subjects for recognition. The training subjects provide the illumination and pose parameters to synthesise novel images. These parameters were obtained from two illumination conditions and three poses (C05, C27 and C22).

For recognition, the gallery set comprises a frontal pose and frontal illuminated image from each of the remaining 27 subjects. Therefore, there were 27 original images in the gallery set. Then it was extended to have 216 synthesised images (8 per subject). The probe set was matched against the extended gallery set and the accuracy was measured by the percentage of probe images that were correctly identified. The probe set contains three images of three poses with random illumination condition for each subject (81 images in total).

5.3 Results and Analysis

Table 1 shows the average recognition result of the experiment with the multi-view face recognition approach (MMMV) to evaluate the usefulness of the synthesised images.

We carried out the experiment independently for five times to eliminate anomalous result. To illustrate the recognition performance advantage of this method, the results are compared to the Eigenface and the Gaussian kernel direct matching without the synthesised images as a benchmark. As shown in the results table, the synthesised images under the proposed Multi-manifold based Multi-view method significantly improves the recognition accuracy. From the study, it is also observed that MMMV consistently obtain good performance (above 85 %) when combined with Gabor filters although the execution time is much slower compared to GKF.

Table 1. Recognition rates for CMU-PIE database.

Method	Recognition results	Variances
MMMV with Gabor filters	91.6 %	3.7
MMMV with GKF	79.8 %	3.9
Eigenfaces	24.2 %	2.2
Gaussian kernel function	43.6 %	3.2

6   Conclusions

Variations in illumination and pose in face images are great challenges for robust and reliable face recognition. In this paper a simple and yet practical approach is proposed to handle the problems through multi-view manifold learning and image synthesis. The proposed method also provides a useful scheme to the problem of limited samples in face database. The novel images recreated with novel variations minimises the difference between gallery and probe images and improves the matching and recognition. We have demonstrated the suitability of this method in improved the recognition performance with solely 2-dimensional images. The next step for this research is to test the performance on larger pose angles and more arbitrary illumination conditions. We are also looking into studying the feasibility of this approach to combine further variations such as facial expression.

References

1. Taigman, Y., Yang, M., Ranzato, M., Wolf, L.: DeepFace: closing the gap to human-level performance in face verification. In: Conference on Computer Vision Pattern Recognition, p. 8 (2014)
2. Lu, C., Tang, X.: Surpassing human-level face verification performance on LFW with GaussianFace. In: Proceedings of the 29th AAAI Conference on Artificial Intelligence (AAAI), pp. 3811–3819 (2014)
3. Turk, M., Pentland, A.: Eigenfaces for recognition. J. Cogn. Neurosci. **3**(1), 71–86 (1991)
4. Hespanha, P., Kriegman, D.J., Belhumeur, P.N.: Eigenfaces vs. fisherfaces: recognition using class specific linear projection. IEEE Trans. Pattern Anal. Mach. Intell. **19**(7), 711–720 (1997)

5. Yang, J., Zhang, D., Frangi, A.F., Yang, J.: Two-dimensional PCA: a new approach to appearance-based face representation and recognition. IEEE Trans. Pattern Anal. Mach. Intell. **26**(1), 131–137 (2004)
6. Tenenbaum, J.B., de Silva, V., Langford, J.C.: A global geometric framework for nonlinear dimensionality reduction. Science **290**, 2319–2323 (2000)
7. Roweis, S.T., Saul, L.K.: Nonlinear dimensionality reduction by locally linear embedding. Science **290**, 2323–2326 (2000)
8. Yin, H.: Advances in adaptive nonlinear manifolds and dimensionality reduction. Front. Electr. Electron. Eng. China **6**(1), 72–85 (2011)
9. Kohonen, T.: The self-organizing map. Proc. IEEE **78**(9), 1464–1480 (1990)
10. Yin, H., Huang, W.: Adaptive nonlinear manifolds and their applications to pattern recognition. Inf. Sci. (Ny) **180**(14), 2649–2662 (2010)
11. Yin, H.: ViSOM—a novel method for multivariate data projection and structure visualization. IEEE Trans. Neural Networks **13**(1), 237–243 (2002)
12. Liu, X., Lu, H., Li, W.: Multi-manifold modeling for head pose estimation. In: 2010 IEEE International Conference on Image Processing, pp. 3277–3280 (2010)
13. Wang, Y., Jiang, Y., Wu, Y., Zhou, Z.-H.: Multi-manifold clustering. In: Zhang, B.-T., Orgun, M.A. (eds.) PRICAI 2010. LNCS, vol. 6230, pp. 280–291. Springer, Heidelberg (2010)
14. Wang, Y., Jiang, Y., Wu, Y., Zhou, Z.-H.: Spectral clustering on multiple manifolds. IEEE Trans. Neural Networks **22**(7), 1149–1161 (2011)
15. Lu, J., Tan, Y., Wang, G.: Discriminative multi-manifold analysis for face recognition from a single training sample per person. In: 2011 International Conference on Computer Vision (ICCV), pp. 1943–1950 (2011)
16. Sun, S.: A survey of multi-view machine learning. Neural Comput. Appl. **23**(7–8), 2031–2038 (2013)
17. Memisevic, R.: On multi-view feature learning. In: Proceedings of the 29th International Conference on Machine Learning (ICML 2012) (2012)
18. Xu, C., Tao, D., Xu, C.: Large-margin multi-view information bottleneck. IEEE Trans. Pattern Anal. Mach. Intell. **36**(8), 1559–1572 (2014)
19. Cootes, T.F., Taylor, C.J., Cooper, D.H., Graham, J.: Active shape models-their training and application. Comput. Vis. Image Underst. **61**(1), 38–59 (1995)
20. Teijeiro-Mosquera, L., Alba-Castro, J.-L., Gonzalez-Jimenez, D.: Face recognition across pose with automatic estimation of pose parameters through AAM-based landmarking. In: 2010 20th International Conference on Pattern Recognition, no. 2, pp. 1339–1342, August 2010
21. Wiskott, L., Kr, N., Von Der Malsburg, C.: Face recognition by elastic bunch graph matching. IEEE Trans. Pattern Anal. Mach. Intell. **19**(7), 1–23 (1999)
22. Sim, T., Baker, S., Bsat, M.: The CMU pose, illumination, and expression (PIE) database 2 capture apparatus and procedure, no. 1, pp. 1–6 (2002)
23. He, X., Cai, D., Niyogi, P.: Laplacian score for feature selection. In: Advances Neural Information Processing, pp. 507–514 (2005)

Neural Network-Based User-Independent Physical Activity Recognition for Mobile Devices

Bojan Kolosnjaji[(✉)] and Claudia Eckert

Faculty of Informatics, Technische Universität München,
Boltzmannstraße 3, 85748 Garching, Germany
kolosnjaji@sec.in.tum.de, claudia.eckert@in.tum.de

Abstract. Activity recognition using sensors of mobile devices is a topic of interest of many research efforts. It has been established that user-specific training gives good accuracy in accelerometer-based activity recognition. In this paper we test a different approach: offline user-independent activity recognition based on pretrained neural networks with Dropout. Apart from satisfactory recognition accuracy that we prove in our tests, we foresee possible advantages in removing the need for users to provide labeled data and also in the security of the system. These advantages can be the reason for applying this approach in practice, not only in mobile phones but also in other embedded devices.

Keywords: Activity recognition · Mobile sensors · Machine learning · Neural networks · Deep learning

1   Introduction

Smartphones are becoming very important tools of our everyday life, because multiple functions of increasing sophistication are present in these mobile devices. One of the crucial drivers that influences the popularity of smartphones is the abundance of sensors, such as the camera, GPS, accelerometer, touch sensor etc. Sensors enable the smartphone to collect the data from and about its environment and use it to create context for its functions. Many smartphone apps are based on the possibility of creating such functionality for the users. One example of this functionality is activity recognition.

Activity recognition is already employed in various information systems for purposes of authentication, surveillance, gesture-based control, industrial robotics etc. Most of activity recognition systems use visual information acquired with different types of camera. However, accelerometers could also provide this functionality. Recent studies suggest the use of accelerometers instead of cameras for activity recognition, because of privacy concerns [3]. Accelerometer data can also be easier to process, as it only measures the signal originating from the subject.

There are multiple papers exploring the idea of using smartphone accelerometers for activity recognition. Most of them use online user-dependent methods for training a machine learning algorithm for recognition, where the recognition

© Springer International Publishing Switzerland 2015
K. Jackowski et al. (Eds.): IDEAL 2015, LNCS 9375, pp. 378–386, 2015.
DOI: 10.1007/978-3-319-24834-9_44

performance improves as the data arrives. However, this method also requires a user to train the algorithm and provide labeled data. Also, the training data can be easily corrupted by a malicious application, which makes the whole system unusable. This is known in adversarial machine learning as a causative attack [8], which effectively disables the recognition system by providing false training samples.

In activity recognition methods for mobile devices there is already a significant amount of research work in recent years. The papers that focus on user-specific activity recognition already achieve high accuracy [1,9]. In the paper of Weiss *et al.* [15] a description can be found about advantages of user-specific training. The work of Prudêncio *et al.* [11] contains a study of performance variation when an activity recognition system trained on one subject is used by multiple other subjects. This variation was proven to be very high, and the success rate is between 35.6 % for walking and 89.9 % for walking upstairs. In further work [10,16], convolutional and deep belief nets are proven to be promising for feature extraction in the activity recognition problems. User-independent training is the focus in only two papers we found [12,13]. The results of these papers are promising, with the recognition performance of over 90 %. However, the tested activities are mutually very different: sitting/standing, walking, cycling, driving, running, on table. In the user-independent recognition of similar activities usually only low accuracy can be achieved with standard machine learning methods, as there are many user-specific features that make up the accelerometer signal. Human gait measurements also include these subject-specific features, which represent noise for the pretrained recognition engines.

In this work we explore the possibility of improving offline and user-independent training for activity recognition using advanced machine learning methods. The offline, user-independent method provides the advantage that the model is ready to use prior to the delivery of the product to a user. It is not necessary for the user to provide labeled data, however it can be done to further increase the recognition accuracy. In this case, the pretrained model is updated by acquiring new patterns. Furthermore, if the classification scheme is only used as pretrained, causative attacks cannot be executed. However, exploratory attacks, where the test set is manipulated, are still possible. We attempt to leverage recent advances in machine learning methods, specifically deep neural networks with Dropout, to provide these advantages. As the user-independent model can be trained offline, there are no resource constraints that apply to embedded systems. Therefore the offline training methods are allowed to be more computationally demanding.

2 Method

To achieve offline recognition, a method needs to be found that is able to extract user-independent activity features and average out features specific to particular users. User-specific features are here considered as noise and extraction of activity recognition features is considered as noise canceling. Machine learning methods exist that are used for feature extraction from a noisy signal.

2.1 Data Preprocessing

We preprocess our data using the Fast Fourier Transform (FFT), since it is a standard tool of extracting properties from time series signals and can be implemented to execute fast and efficient on Graphical Processing Units (GPU) and embedded systems, such as mobile devices.

2.2 Feature Extraction

Autoencoder [6] is a special kind of neural network used for feature extraction, compression and denoising of data. It consists of a standard input and output layer and optionally multiple hidden layers. A linear autoencoder with only one hidden layer is equivalent to linear feature extractors, such as PCA transformation. More hidden layers with nonlinearity can increase the performance of the dimension reduction process and enable more complex features to be extracted from the input dataset. We initially use an autoencoder with three hidden layers to reduce the data dimensionality.

2.3 Activity Recognition

After the use of an autoencoder, we continue to use neural network configurations for activity recognition as well. Standard configuration used by default in most proof-of-concept tests is a perceptron neural network with one hidden layer, in addition to an input and an output layer. This configuration has been proven to be a universal approximator, which means that it can theoretically be trained to approximate any function. However, in recent years it has been shown that increasing number of layers can give a high boost to classification performance of neural networks. This trend of using a high number of layers is known as deep learning [4].

Neural Networks. Motivated by the trend of deep learning, we used a 4-layer neural network configuration. Many research efforts were successful in using an even higher number of layers. However, a high number of layers also means much more training time and more need for computational resources. We have used a 4-layer neural network, since we have noticed a significant increase in accuracy with respect to a standard 3-layer perceptron. Furthermore, no increase in performance has been noticed in case of increasing the number of layers even further. Another problem with complex learning constructs is the possibility of overfitting. Machine learning algorithms can give a very good performance on training data, but can be so specialized with training that it does not generalize well. To cope with this problem we used Dropout [7,14]. Dropout is a modification to deep learning constructs that gives performance enhancements similar to the ones that can be achieved using ensemble learning. The difference is that it does not explicitly use multiple neural networks to achieve this. In Dropout, the ensemble learning effect is achieved by probabilistically turning on and off units of neural network, so that in each mini-batch session a different random subset of neurons is active. We have experimented with multiple values for the probability of "dropping out" single neurons. Although Srivastava *et al.* [14] claim that probability of 50 % is optimal for a wide range of tasks, we have determined that slightly increased probability produces better recognition accuracy on average. For our purpose, best results are achieved by giving each neuron an probability of activation of 70 %, i.e. each neuron gets dropped out with a probability of 30 %.

Random Forest. For comparison, we decided to test the activity recognition performance of another state-of-the-art algorithm. Random forests [5] are used in many machine learning applications and described as a robust universally useful classifier. This method combines the use of multiple decision trees, where every decision tree contains a random subset of features present in the data. As it is built using an ensemble learning approach, this method represents an appropriate choice to compare its performance with the Dropout. To our knowledge, there are no prior publications dedicated to comparison of performance of Dropout and Random Forest on activity recognition datasets.

We performed Leave-One-Out Crossvalidation, where we would use one subject for testing and the rest of the subjects for training the algorithm. The tests are repeated as many times as there were subjects, to get a good overview of the performance distribution.

3 Results

This section is dedicated to the display of results obtained with testing the performance of neural networks and random forest classifiers.

3.1 Datasets

Two public datasets for testing the recognition systems are obtained and used for testing our method. First one is the WISDM dataset [9]. It contains labeled measurements of 6 distinct activities: walking, walking upstairs, walking downstairs, sitting, standing and laying. Measurements were executed on 36 subjects and the results contain 5422 samples in total. The second dataset, UCI HAR dataset [2] is newer and also contains data for 6 activities. 30 subjects were included in their measurement and the dataset contains 10299 samples. The measured activities were: jogging, walking upstairs, walking downstairs, sitting, standing and walking.

We compare the recognition accuracy of these two methods for two datasets and a variety of subjects.

3.2 Accuracy Graphs

First two graphs show the classification accuracy of the neural network and random forest for the WISDM dataset, while the second two graphs show the

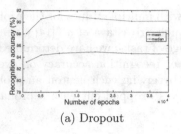
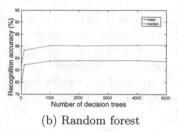

(a) Dropout   (b) Random forest

Fig. 1. Classification accuracy of the neural network with Dropout and random forest methods for the WISDM dataset

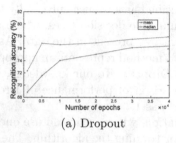
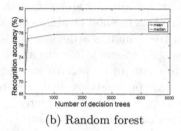

(a) Dropout   (b) Random forest

Fig. 2. Classification accuracy of the neural network with Dropout and random forest methods for the HAR dataset

analogue information for the UCI HAR dataset Fig. 1. The accuracy results are obtained by averaging out (mean and median) the test results for all present subjects in the crossvalidation Fig. 2.

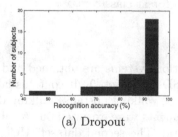
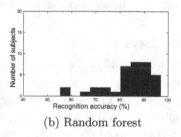

(a) Dropout   (b) Random forest

Fig. 3. Accuracy histogram of the neural network with Dropout and random forest methods for the WISDM dataset

3.3   Distribution of Accuracy

The next group of graphs contains histograms that show the distribution of accuracy in the crossvalidation test Fig. 3. It is noticeable from this, and also from the previous group of graphs that the accuracy of the recognition process varies highly between subjects. Median accuracy is for both datasets significantly higher than the mean accuracy. This means that for the most subjects the accuracy is higher than average, but also that there are a few outliers. Presence

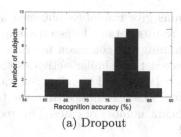

(a) Dropout

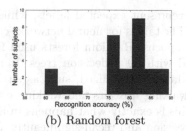

(b) Random forest

Fig. 4. Accuracy histogram of the neural network with Dropout and random forest methods for the HAR dataset

Table 1. Confusion matrices for the HAR dataset

Dropout	Walking	Upstairs	Downstairs	Sitting	Standing	Laying
Walking	1569	49	48	0	0	1
Upstairs	95	1358	106	1	1	1
Downstairs	57	137	1252	0	0	0
Sitting	0	0	0	746	407	270
Standing	1	0	0	732	1447	189
Laying	0	0	0	298	51	1483
Precision	0.9412	0.8694	0.8658	0.5242	0.6108	0.8095
Recall	0.9111	0.8795	0.8905	0.4198	0.7592	0.7629
Random forest	Walking	Upstairs	Downstairs	Sitting	Standing	Laying
Walking	1532	50	14	0	0	0
Upstairs	112	1346	128	1	0	7
Downstairs	78	148	1264	0	0	0
Sitting	0	0	0	833	410	225
Standing	0	0	0	652	1433	106
Laying	0	0	0	291	63	1606
Precision	0.8897	0.8718	0.8990	0.4688	0.7518	0.8261
Recall	0.9599	0.8444	0.8483	0.5674	0.6540	0.8194

of outliers indicates that there are subjects where the subject-specific features contained in the signal are very significant. The histograms for neural networks contain results from using 30000 epochs. For the random forests we display histograms of results obtained with 5000 decision trees Fig. 4.

3.4 Confusion Matrices

The tables below are confusion matrices for WISDM and UCI HAR datasets, for the recognition systems trained with Dropout and random forest Table 1. These tables show specifically which activities get mixed up by the recognition systems.

Rows represent expected labels, while columns give the labels obtained in the tests. The tables for neural networks contain results obtained when training with 30000 epochs. Random forests used for obtaining the confusion matrices were trained with 5000 decision trees. It is expected that similar activities can be easily falsely recognized, such as sitting with laying and standing, or walking with walking upstairs and walking downstairs. One can notice from the tables that this is exactly what happens in a significant number of cases. This reduces the precision and recall significantly Table 2.

Table 2. Confusion matrices for the WISDM dataset

Dropout	Walking	Upstairs	Downstairs	Sitting	Standing	Laying
Walking	1469	142	50	23	6	11
Upstairs	82	341	102	1	1	69
Downstairs	23	70	365	0	0	31
Sitting	2	0	0	266	15	0
Standing	1	1	1	9	226	0
Laying	2	84	54	0	14	1961
Precision	0.9303	0.5345	0.6381	0.8896	0.8626	0.9464
Recall	0.8636	0.5721	0.7464	0.9399	0.9496	0.9272
Random forest	Walking	Upstairs	Downstairs	Sitting	Standing	Laying
Walking	1561	141	59	5	3	6
Upstairs	63	228	77	0	4	28
Downstairs	15	50	221	0	0	18
Sitting	0	0	0	252	25	0
Standing	3	1	0	23	200	0
Laying	59	176	132	3	6	2063
Precision	0.9177	0.3826	0.4519	0.8905	0.8403	0.9754
Recall	0.8794	0.5700	0.7270	0.9097	0.8811	0.8458

4 Conclusion and Future Work

Our work shows that deep learning approaches are a promising alternative for designing user-independent activity recognition systems. Comparison of Dropout with random forest results shows that Dropout can induce improvement in the recognition performance, although the results of using the two algorithms are still comparable. Results show that user-independent training can give accuracy of near 90%, however with some outliers present. This is already close to the results of previous work obtained with online training with labels provided by the user of the device. The success rate is also near the one of Siirtola *et al.* [12], but for much more mutually similar activities in our case. The newer paper by the same authors [13], however, includes a study about recognition accuracy for different phone placement. We plan to do this in the future to further validate our approach. In case of optional additional training by the user, the accuracy can be further improved. We plan to explore this direction and find an accurate method that contains online training as an additional improvement for the optimal pre-trained model. This also requires optimizing the tradeoff between accuracy and resource conservation, especially battery consumption. Furthermore, achieving the recognition abilities for a higher number of diverse activities would give a possibility for a wider range of applications. Another possible improvement is the discovery of new activities. In this paper we recognize a small preassigned set of activities, but the actual number of human activities is much higher and more diverse.

Acknowledgment. The research leading to these results was supported by the Bavarian State Ministry of Education, Science and the Arts as part of the FORSEC research association.

References

1. Anguita, D., Ghio, A., Oneto, L., Parra, X., Reyes-Ortiz, J.L.: Human activity recognition on smartphones using a multiclass hardware-friendly support vector machine. In: Bravo, J., Hervás, R., Rodríguez, M. (eds.) IWAAL 2012. LNCS, vol. 7657, pp. 216–223. Springer, Heidelberg (2012)
2. Anguita, D., Ghio, A., Oneto, L., Parra, X., Reyes-Ortiz, J.L.: A public domain dataset for human activity recognition using smartphones. In: European Symposium on Artificial Neural Networks, Computational Intelligence and Machine Learning, ESANN (2013)
3. Avci, A., Bosch, S., Marin-Perianu, M., Marin-Perianu, R., Havinga, P.: Activity recognition using inertial sensing for healthcare, wellbeing and sports applications: a survey. In: 2010 23rd international conference on Architecture of computing systems (ARCS), pp. 1–10. VDE (2010)
4. Bengio, Y.: Learning deep architectures for AI. Found. Trends Mach. Learn. **2**(1), 1–127 (2009)
5. Breiman, L.: Random forests. Mach. Learn. **45**(1), 5–32 (2001)
6. Hinton, G.E., Salakhutdinov, R.R.: Reducing the dimensionality of data with neural networks. Science **313**(5786), 504–507 (2006)
7. Hinton, G.E., Srivastava, N., Krizhevsky, A., Sutskever, I., Salakhutdinov, R.R.: Improving neural networks by preventing co-adaptation of feature detectors. arXiv preprint arXiv:1207.0580 (2012)
8. Huang, L., Joseph, A.D., Nelson, B., Rubinstein, B.I., Tygar, J.: Adversarial machine learning. In: Proceedings of the 4th ACM Workshop on Security and Artificial Intelligence, pp. 43–58. ACM (2011)
9. Kwapisz, J.R., Weiss, G.M., Moore, S.A.: Activity recognition using cell phone accelerometers. ACM SigKDD Explor. Newslett. **12**(2), 74–82 (2011)
10. Plötz, T., Hammerla, N.Y., Olivier, P.: Feature learning for activity recognition in ubiquitous computing. In: IJCAI Proceedings-International Joint Conference on Artificial Intelligence, vol. 22, p. 1729 (2011)

11. Prudêncio, J., Aguiar, A., Lucani, D.: Physical activity recognition from smartphone embedded sensors. In: Sanches, J.M., Micó, L., Cardoso, J.S. (eds.) IbPRIA 2013. LNCS, vol. 7887, pp. 863–872. Springer, Heidelberg (2013)
12. Siirtola, P., Röning, J.: Recognizing human activities user-independently on smartphones based on accelerometer data. Int. J. Interact. Multimed. Artif. Intell. 1(5), 38–45 (2012)
13. Siirtola, P., Roning, J.: Ready-to-use activity recognition for smartphones. In: 2013 IEEE Symposium on Computational Intelligence and Data Mining (CIDM), pp. 59–64. IEEE (2013)
14. Srivastava, N., Hinton, G., Krizhevsky, A., Sutskever, I., Salakhutdinov, R.: Dropout: a simple way to prevent neural networks from overfitting. J. Mach. Learn. Res. 15(1), 1929–1958 (2014)
15. Weiss, G.M., Lockhart, J.W.: The impact of personalization on smartphone-based activity recognition. In: AAAI Workshop on Activity Context Representation: Techniques and Languages (2012)
16. Zeng, M., Nguyen, L.T., Yu, B., Mengshoel, O.J., Zhu, J., Wu, P., Zhang, J.: Convolutional neural networks for human activity recognition using mobile sensors

Reduction of Signal Strength Data
for Fingerprinting-Based Indoor Positioning

Maciej Grzenda[✉]

Faculty of Mathematics and Information Science,
Warsaw University of Technology,
ul. Koszykowa 75, 00-662 Warszawa, Poland
M.Grzenda@mini.pw.edu.pl
http://www.mini.pw.edu.pl/~grzendam

Abstract. Indoor Positioning Services (IPS) estimate the location of
devices, frequently being mobile terminals, in a building. Many IPS
algorithms rely on the Received Signal Strength (RSS) of radio signals
observed at a terminal. However, these signals are noisy due to both the
impact of the surrounding environment such as presence of other persons, and limited accuracy of strength measurements. Hence, the question arises whether raw RSS data can undergo transformation preserving its information content, but reducing the data and increasing the
resilience of the algorithms to inherent noisiness of raw RSS data.

This work evaluates the use of binning applied to RSS data and proposes the way the reduction of RSS data can be attained. Results show
that the proposed way of using RSS binning causes little or no positioning accuracy degradation. Still, it yields significant RSS data reduction.
Interestingly, in some cases data reduction results even in accuracy gains.

Keywords: Received signal strength · Binning · Fingerprinting

1 Introduction

Indoor positioning systems (IPS) frequently rely on *fingerprinting* approach
[5,7,8]. In the latter case, first measurements of Received Signal Strength (RSS)
are made in a number of points in a building. These measurements bound to
known indoor locations comprise on a *radio map*. Next, in the on-line phase, the
RSS measurements are made in the building. The estimation of the location of a
Mobile Station (MS) is made based on the RSS vector observed in this location.
In terminal-centric approach, the RSS vector may be the RSS vector acquired
by the MS itself. In network-centric approach, the RSS vector may be the vector
of signal strengths emitted by the MS and observed in various points of network
infrastructure. Most frequently RSS of WiFi or GSM signals are used. A mobile
station is usually a mobile phone. In particular, based on the estimated indoor
location of the phones, Location-Based Services (LBS) such as indoor navigation
in shopping centers can be built.

Especially when terminal-centric approach is used, the RSS measurements
are usually made with devices of limited measurement accuracy, such as mobile

© Springer International Publishing Switzerland 2015
K. Jackowski et al. (Eds.): IDEAL 2015, LNCS 9375, pp. 387–394, 2015.
DOI: 10.1007/978-3-319-24834-9_45

phones. In the latter case, the RSS data collected at a terminal side is frequently transmitted to server-side IPS to develop location estimates. This raises the need to reduce the volume of transmitted data. The use of smartphones in IPS results also in limited accuracy of individual RSS values comprising on RSS vectors. First of all, the values are discrete and come from a limited set of cardinality lower than 100. This is even though the true RSS values are continuous. Secondly, some signals are too weak to be reported by a MS. Moreover, changes in the environment such as opening the doors in the nearby area may significantly change the signal strength observed at a terminal [1]. Last, but not least the precision of a MS is limited. In particular, studies show that different MS may report significantly different RSS values in the same point of the same propagation environment [5].

At the same time, distance-based machine learning methods [3] are used to implement positioning algorithms. In the simplest case, the location of a MS is determined based on an RSS vector observed in it and is assumed to be the location of the most similar RSS vector contained in the radio map. Such similarity-based approaches, usually referred to as *fingerprinting*, follow the idea of a well known Nearest Neighbour (NN) technique. Some authors [6] use *database correlation* as an alternative name to *fingerprinting*.

However, distance-based techniques suggests that several minor RSS differences, possibly caused by limited measurement resolution of a MS, may be treated to be equally important as a major difference of a signal strength originating from one source. Hence, the question arises whether further discretisation of the RSS data may contribute to positioning accuracy. In fact, some studies, such as these made in one the best solutions submitted to ICDM 2007 contest [8], suggest that the fact that a signal has been detected may be even more important than its actual value. In line with these works, this study proposes and evaluates the use of RSS binning to attain the discretisation of RSS data, while maintaining the positioning accuracy.

It is important to note that other machine learning techniques can be used in place of NN, including k Nearest Neighbors (kNN) [7] and more sophisticated approaches such as neural networks and random forest. The latter technique has been used by H. Buyruk et al. in their recent study on GSM-based fingerprinting [2]. From a reference point of view, the advantage of Nearest Neighbor technique is that it does not rely on any other parameters except for the selection of a metric function. Some simulation studies [6] show that Euclidean distance can yield virtually optimal results in fingerprinting applications.

This work relies on the extensive data sets collected in the buildings of different architecture, dating from early XXth and XXIst century. All the RSS data have been collected on the terminal side. Both GSM and WiFi Received Signal Strength vectors were collected in a number of reference points. More precisely, the signal strengths of the signals emitted from Base Transceiver Stations (BTS) and WiFi Access Points (AP) have been collected. Every vector contains up to 7 GSM RSS values i.e. the signal strengths from the serving cell and up to 6 other cells and WiFi RSS values [4]. Furthermore, an assumption was made that the orientation of the MS in the testing phase, when its unknown position is being estimated, is not known to the positioning algorithm. Hence, in this study the location estimation is based solely on a single RSS vector containing GSM RSS captured in an unknown location within a building or close to it.

2   The Reference Data

In accordance with the idea of fingerprinting approach, two data sets collected in two different buildings were used in this study. In both cases, a grid of training points, approximately equally spaced was planned. The RSS data collected in these points comprises on a radio map. Hence, the training points will be referred to as Reference Points (RP) in the remainder of this work. In order to ensure objective evaluation of individual data processing techniques, a separate grid of Testing Points (TP) was planned. The location of these points was selected in the same area of a building as the location of the radio map points. These points were also equally spaced. Minor differences in the location of individual points and the number of points in the training and testing set can be observed. Among other reasons, this is due to the presence of the obstacles such as printers on the corridors of the building. Nevertheless, a default grid of both training and testing points had a resolution of 1.5 m. Both training and testing points were placed on different floors of a building. Hence, each point is described by (x, y, f) coordinates standing for (x, y, floor).

In every point, measurements were repeated 10 times in one direction i.e. 10 series of measurements were made in each direction and point. Four directions have been used. Hence, 40 RSS vectors were acquired in each training point. Similarly, 40 RSS vectors were acquired in each testing point. Each RSS vector of the data sets used in this study contained the RSS strength of a signal emitted by a Base Transceiver Station (BTS), uniquely identified by GSM cell identifier. Let $s_{\mathrm{RP}}^{r,d,t}$ denote a signal RSS vector collected in one RP, for one terminal orientation, where r stands for the index of an RP, d denotes direction, and t measurement series. Similarly, $s_{\mathrm{TP}}^{r,d,t}$ denotes a single RSS vector collected in r-th TP. Based on the $s_{\mathrm{RP}}^{r,d,t}$ and $s_{\mathrm{TP}}^{r,d,t}$ vectors, the training set S and the testing set T was created for each building, respectively. Importantly, different signal categories can be recorded in each $s_{\mathrm{RP}}^{r,d,t}$ and $s_{\mathrm{TP}}^{r,d,t}$. Thus, both $s_{\mathrm{RP}}^{r,d,t}$ and $s_{\mathrm{TP}}^{r,d,t}$ take the form of $\mathbf{v} = [\mathbf{v}_{C_1}, \ldots, \mathbf{v}_{C_n}]$ i.e. are composed of potentially many parts, such as WiFi part, GSM part or UMTS part. Let $\mathbf{v}_{\mathrm{GSM}}$ denote GSM RSS vector. While the original data set contains different signal categories, the data used in this study was GSM data only. For this reason, $S_{\mathrm{GSM}} = \{\mathbf{v}_{\mathrm{GSM}} : \mathbf{v} \in S\}$ and $T_{\mathrm{GSM}} = \{\mathbf{v}_{\mathrm{GSM}} : \mathbf{v} \in T\}$ will be used in the remainder of this study.

The data sets from two different buildings were used in this study. The first data set comes from recently constructed, modern premises of the Faculty of Mathematics and Information Science of the Warsaw University of Technology (WUT/FMIS). The other building used as a test site was the building of the Faculty of Physics (WUT/FP), constructed in 1901. The summary of the data collected in both buildings is contained in Table 1. What should be emphasised

Table 1. The building data sets

Data set	Floor count	The number of RP	The number of TP	Signal count C
WUT/FMIS	6	1401	1461	36
WUT/FP	3	728	855	31

is that multifloor sites of different architecture influencing signal propagation were considered. In different parts of each of these sites, different signal sources were observed. Hence, the dimensionality of $\mathbf{v}_{\mathrm{GSM}}$ was 36 in WUT/FMIS data and 31 in WUT/FP data.

3   The Binning Process

The binning process converts raw values of input features into binned values of input features. The objective is to convert observed RSS values into an index of the range, the RSS value is mapped to. Hence, $c()$ function is used, which is defined as follows:

$$c : \mathbb{Z}_{\mathrm{s}}^{S} \longrightarrow \mathbb{Z}_{\mathrm{r}}^{S}, \ x \longrightarrow \tilde{x} = c(x) \tag{1}$$

Both $\mathbb{Z}_{\mathrm{s}} \subset \mathbb{Z}$, and $\mathbb{Z}_{\mathrm{r}} \subset \mathbb{Z}$. However, the key feature of the binning function is that $card(\mathbb{Z}_{\mathrm{r}}) < card(\mathbb{Z}_{\mathrm{s}})$ i.e. that the number of unique values present in the data is reduced. More precisely, assuming $\tilde{x} = [x_1, \cdots, x_S]$, the binned value $c(x_i)$ is defined as: $c(x_i) = min_{j=1,\cdots,k} B_j < x_i$, where $B = [B_1, \cdots, B_k], B_j < B_{j+1}$ is a vector of cut points used in the binning process i.e. a vector defining the ranges $(B_j, B_{j+1}]$ of individual bins. It is worth noting here that the reverse function $d()$ defined as $d : \mathbb{Z}_{\mathrm{r}}^{S} \longrightarrow \mathbb{Z}_{\mathrm{s}}^{S}, \ \tilde{x} \longrightarrow x = d(\tilde{x})$ does not exist.

Two categories of binning methods are considered in this study: interval binning and frequency binning. In both cases, a proposal is made to reserve the first of the bins i.e. ultimate values present in $\mathbb{Z}_{\mathrm{r}}$ for the signals too weak to be detected. Moreover, let us propose to set the upper range of the last bin $B_k, k = card(\mathbb{Z}_{\mathrm{r}})$ to zero. RSS values are defined in dBm and take negative values. Hence, there are no values $x_i = 0$. However, it is possible that the strongest signal observed in the online phase will be stronger than the strongest signal present in the radio map.

Thus, interval binning $c_{\mathrm{I},i}()$ relies on the mapping of RSS values into $C = 1 + \left\lceil \frac{card(\mathbb{Z}_{\mathrm{s}})}{i} \right\rceil$ bin indexes, while assuming equal distance $B_{k+1} - B_k = B_{j+1} - B_j, 2 \leq k, j < C$. Frequency binning $c_{\mathrm{F},i}()$ aims at ensuring similar number of raw detectable RSS values mapped to each bin. Hence, for non-uniform distribution of raw values $\exists k, j : 2 \leq k, j < C \wedge B_{k+1} - B_k \neq B_{j+1} - B_j$. Similarly to interval binning, frequency binning $c_{\mathrm{F},i}()$ relies on the mapping of RSS values into $1 + \left\lceil \frac{card(\mathbb{Z}_{\mathrm{s}})}{i} \right\rceil$ bin indexes.

4   The Impact of Binning on Positioning Accuracy

4.1   Single Floor Positioning

Two different machine learning techniques, namely nearest neighbour technique and random forest composed of 50 trees were used to develop indoor positioning models discussed below. First, each floor in each of the buildings was treated as a separate testbed. The positioning errors were calculated for a positioning technique using the test data for each of the one floor testbeds. Next, Mean Absolute Errors (MAE) were calculated for each floor and averaged over entire building. The details of this process are provided in Algorithm 1. The results of this analysis are summarised in Table 2. It can be observed that the use of interval binning may yield no accuracy degradation compared to the use of raw data. Depending on a building, the best accuracy is attained with $c_{I,2}(D)$ (WUT/FP), or raw data (WUT/FMIS). Importantly, for both buildings the reduction of the cardinality of RSS leaving approximately 50 % ($c_{I,2}(D)$) or even 10 % of unique values ($c_{I,10}(D)$) in the data results in only very limited accuracy degradation. This means that raw RSS data can be replaced with binned data, reducing the volume of data to process and increasing the resilience of IPS methods to fluctuations in RSS readings.

Algorithm 1. The calculation of positioning errors

Input: R - radio map data, T - test data, $N_{\max}$ - the number of floors present
in the data set, R_i and T_i - test and train data collected on i-th floor,
$\phi() = \phi_{\mathrm{GSM}}$ - the function selecting only GSM signals out of available
signals, M - model development technique i.e. kNN or random forest
algorithm, $c()$ - binning function or identity function, in case raw RSS
data is used

Data: $M_{\mathrm{x}}, M_{\mathrm{y}}$ - regression models built using data collected in RP and
predicting x and y coordinates, respectively.

Result: **E** - the vector of accuracy indicators for the testbed

begin

for $i = 1, \ldots, N_{\max}$ **do**

for $s_{\mathrm{TP}}^{r,d,t} \in T_i$ **do**

$\tilde{x} = M_{\mathrm{x}}(c(\phi(R_i)), c(\phi(s_{\mathrm{TP}}^{r,d,t})));$

$\tilde{y} = M_{\mathrm{y}}(c(\phi(R_i)), c(\phi(s_{\mathrm{TP}}^{r,d,t})));$

$e = \mathrm{CalculateIndicators}(x_r, y_r, \tilde{x}, \tilde{y});$

end

$\mathbf{E}_i = \mathrm{CalculateAggregatedIndicators}(\mathbf{e});$

end

$\mathbf{E} = \mathrm{CalculateAveragedIndicators}(\mathbf{E});$

end

Table 2. The average horizontal positioning errors [m] for random forest technique and various interval binning techniques - single floor scenario

Data set	Raw data D	$c_{I,2}(D)$	$c_{I,3}(D)$	$c_{I,5}(D)$	$c_{I,10}(D)$	$c_{I,20}(D)$	$c_{I,50}(D)$
WUT/FP	11.25	11.24	11.32	11.36	11.71	11.90	13.09
WUT/FMIS	11.29	11.35	11.38	11.59	11.77	12.36	12.83

4.2   Multiple Floor Positioning

The question arises whether the tendency observed for single floor testbeds, is also true for multifloor scenario. In the latter case, the positioning within all the building floors covered with the measurements is performed. This scenario is more difficult, since similar combinations of RSS values can be observed on different floors of a building. The MAE values for various interval binning techniques and multifloor testbeds are reported in Table 3. In addition, in Table 4 results for frequency binning method are reported.

Table 3. The horizontal positioning errors [m] for random forest technique and various interval binning techniques - multiple floor scenario

Data set	Raw data D	$c_{I,2}(D)$	$c_{I,3}(D)$	$c_{I,5}(D)$	$c_{I,10}(D)$	$c_{I,20}(D)$	$c_{I,50}(D)$
WUT/FP	11.26	11.27	11.32	11.43	11.65	12.06	13.46
WUT/FMIS	11.75	11.78	11.82	11.90	12.19	12.56	12.82

In this scenario, the possibility to significantly reduce the volume of RSS data is observed as well. In the case of WUT/FP, the binning of 37 unique RSS values observed in raw RSS data to only 5 bins, which is performed by $c_{I,10}(D)$, causes the increase of average positioning error from 11.26 m to 11.65 m i.e. by 39 cm only. This is in spite of the fact that the random forest is built with $c_{I,10}(\phi(R)) = \{\mathbf{x} : \mathbf{x} = [x_1, \cdots, x_S], x_i \in \{1, 2, 3, 4, 5\}\}$ i.e. with vectors having only 5 unique values being indexes of the bins the raw RSS values were mapped to. The scale of this reduction is shown in Fig. 1. In the upper part of the figure the histogram of raw RSS values is shown. The number of RSS values placed in individual bins by $c_{I,10}(D)$ is shown in the lower histogram of the figure. The bin representing signals too weak to be reported has been skipped for clarity reasons, as majority of RSS vector values represent such undetectable values. In the case of frequency binning, reduction performed by $c_{F,10}(D)$ yields the same scale of reduction i.e. a reduction to 5 bins only. Similarly, instead of 11.22 m of average error on raw WUT/FP RSS data, 11.60 m of average error is observed on $c_{F,10}(D)$.

Finally, in Table 5 the summary of MAE errors for NN method is made. It can be observed that error levels are higher in this case. Interestingly, minimum errors for this technique are attained not only when raw data is used, but also when $c_{F,2}(D)$ is used as an input for NN technique. Moreover, the binning process

Table 4. The average horizontal positioning errors [m] for random forest technique and various frequency binning techniques - multiple floor scenario

Data set	Raw data D	$c_{F,2}(D)$	$c_{F,3}(D)$	$c_{F,5}(D)$	$c_{F,10}(D)$	$c_{F,20}(D)$	$c_{F,50}(D)$
WUT/FP	11.22	11.30	11.32	11.45	11.60	12.09	13.43
WUT/FMIS	11.78	11.79	11.80	11.87	12.19	12.42	12.59

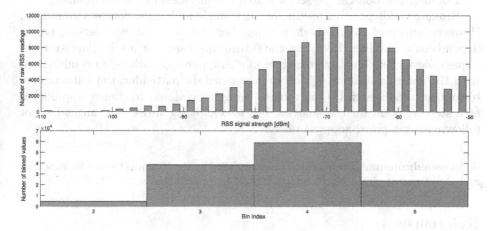

Fig. 1. Histogram of raw (a) and binned with $c_{I,10}(D)$ RSS values (b) (WUT/FP)

Table 5. The horizontal positioning errors [m] for NN technique and various binning techniques, multiple floor scenario

Data set	Raw data D	$c_{F,2}(D)$	$c_{F,3}(D)$	$c_{F,5}(D)$	$c_{F,10}(D)$	$c_{F,20}(D)$	$c_{F,50}(D)$
WUT/FP	14.49	14.34	14.36	14.56	14.90	15.92	17.44
WUT/FMIS	15.03	15.03	14.91	15.31	15.61	16.01	16.71

reduces the risk of paying by a positioning model too much attention to RSS values. Even though these values are noisy and depend on other factors than the location, a random forest performs recursive partitioning using the apparently precise raw RSS values. The precision of these divisions may be too high in view of noisiness of the data or simply not needed. Similarly, distance based techniques such as NN technique can consider several minor RSS strength differences to be equally important to a large strength difference for one signal. In both cases, the risk of overfitting the noisy raw RSS data is present and confirmed by calculation results.

5   Conclusions

This work shows the results of the use of binning methods as a way of reduction of the RSS data. The results of extensive calculations show that binning can be used to reduce the volume and precision of input data, while causing little or no accuracy degradation. This phenomenon is observed both for single floor and multifloor scenarios. By RSS binning, the negative impact of both temporal signal fluctuations and the use of diversity of devices used to perform RSS measurements can be mitigated. Moreover, the volume of the data to be transmitted to server-side backend of IPS solution can be substantially reduced. Finally, the level of discretisation can be set to match the noisiness of the environment.

Binning yields positive results for both single and multi-floor environments. However, which of the methods and number of bins is the best setting partly depends on a building. Hence, in the future the search for a self-adaptive technique selecting the binning procedure in a wrapper approach i.e. in combination with the main positioning algorithm is planned. In particular, the self-adaptive technique could search for bin ranges, using interval and frequency approaches as a starting point for the analysis, but also allowing varied bin ranges i.e. not following any a priori adopted rules.

Acknowledgments. This research was supported by the National Centre for Research and Development, grant No PBS2/B3/24/2014, application No 208921.

References

1. Bento, C., Soares, T., Veloso, M., Baptista, B.: A study on the suitability of GSM signatures for indoor location. In: Schiele, B., Dey, A.K., Gellersen, H., de Ruyter, B., Tscheligi, M., Wichert, R., Aarts, E., Buchmann, A.P. (eds.) AmI 2007. LNCS, vol. 4794, pp. 108–123. Springer, Heidelberg (2007)
2. Buyruk, H., Keskin, A., Sendil, S., Celebi, H., Partal, H., Ileri, O., Zeydan, E., Ergut, S.: RF fingerprinting based GSM indoor localization. In: 2013 21st Signal Processing and Communications Applications Conference (SIU), pp. 1–4, April 2013
3. Flach, P.: Machine Learning: The Art and Science of Algorithms That Make Sense of Data. Cambridge University Press, Cambridge (2012)
4. Grzenda, M.: On the prediction of floor identification credibility in RSS-based positioning techniques. In: Ali, M., Bosse, T., Hindriks, K.V., Hoogendoorn, M., Jonker, C.M., Treur, J. (eds.) IEA/AIE 2013. LNCS, vol. 7906, pp. 610–619. Springer, Heidelberg (2013)
5. Kjargaard, M.B.: Indoor location fingerprinting with heterogeneous clients. Pervasive Mob. Comput. **7**, 31–43 (2011)
6. Machaj, J., Brida, P.: Performance comparison of similarity measurements for database correlation localization method. In: Nguyen, N.T., Kim, C.-G., Janiak, A. (eds.) ACIIDS 2011, Part II. LNCS, vol. 6592, pp. 452–461. Springer, Heidelberg (2011)
7. Varshavsky, A., de Lara, E., Hightower, J., LaMarca, A., Otsason, V.: GSM indoor localization. Pervasive Mob. Comput. **3**(6), 698–720 (2007). perCom 2007
8. Yang, Q., Pan, S.J., Zheng, V.W.: Estimating location using Wi-Fi. IEEE Intell. Syst. **23**(1), 8–13 (2008)

Pattern Password Authentication Based on Touching Location

Orcan Alpar and Ondrej Krejcar[✉]

Faculty of Informatics and Management, Center for Basic and Applied Research,
University of Hradec Kralove, Rokitanskeho 62, 500 03 Hradec Kralove, Czech Republic
orcanalpar@hotmail.com, ondrej@krejcar.org

Abstract. Pattern passwords are one of the embedded authentication method of touchscreen devices, however it has some major drawbacks which briefly are identifiability and imitability. The password of the user is noticeable when entering the pattern due to shining circles. Therefore, what we put forward in this paper is a novel biometric implementation of a hidden system to pattern password authentication for increasing password security. As opposed to general research concept which extracts touch or keystroke durations, we focused on the touching coordinates calculated the distance of the line between the constant pattern node and the touched place as well as the angle. Using these inputs, we trained the neural network by Gauss-Newton and Levenberg-Marquardt algorithms and conducted the experiments with these trained classifiers.

Keywords: Touchscreen · Biometric authentication · Gauss-Newton · Levenberg-Marquardt · Neural network

1 Introduction

Pattern authentication is one of the commonly used method in touchscreen devices where the users connect the nodes to draw the pattern that they had previously determined as the password in enrollment step. Despite the ease of use, it is so easy to catch the password while a user is drawing the pattern due to shining effect of the interfaces. Given these facts, we firstly focus on enhancing the security of the pattern passwords since logging on a touchscreen device like a tablet or a smartphone enables to authenticate the social networks, e-banking and similar systems. Although the majority of biometric keystroke systems deal with the inter-key or inter-node durations as the inputs, we propose a system extracting the analytical coordinates of the touches. Therefore it will be possible for users to create a ghost password since the coordinates on each node could be deliberately designated.

Considering the drawbacks of pattern node authentication, the problem of the system is security when the shining effect is on. Without changing the main dynamics that is briefly touching the nodes, what we dealt with enhancing the system by hidden interface that extracts the touch locations and analyzes the future logins after training. Therefore we wrote an interface in Matlab to emulate 48 × 800 touchscreen Samsung Duo 8262 with a common 9-node format which looks and operates like regular pattern nodes

© Springer International Publishing Switzerland 2015
K. Jackowski et al. (Eds.): IDEAL 2015, LNCS 9375, pp. 395–403, 2015.
DOI: 10.1007/978-3-319-24834-9_46

however it actually collects the touch points stealthily. Moreover we changed the procedure a bit that the user doesn't need to slip from node to node anymore, touching is enough, however the nodes still glow if touched.

Initially, we simulate the touch point by circling the imaginary point and extract the coordinates on a 480 × 800 plane. Afterwards, a line is invisibly drawn between the origin of corresponding node and the touch circle to calculate the angle and the distance. These data are stored to train the neural networks for defining the confidence region. The brief workflow of the proposed system is below in Fig. 1.

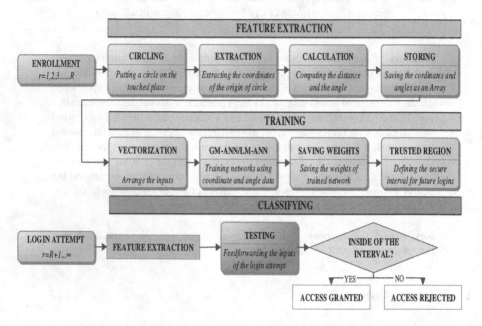

Fig. 1. Workflow of enhanced pattern authentication system

Given the workflow, the system could be divided into three subsystems: feature extraction, training and classifying. In feature extraction, the coordinates of the simulated touch points are extracted to calculate the distance and the angles. After the required number of registration is stored the neural network is trained by optimizing the weights by Gauss-Newton and Levenberg-Marquardt methods, separately to find trusted regions. Once the system is ready for classification, the attempt could be discriminated as "Fraud" or "Real".

In the literature, there are numerous papers related to keystroke and touchscreen authentication and similar such as [1–21] in the past decade. Above all, when comparing this paper with the recent ones, the major performance criteria of the biometric authentication systems are EER; the equal error rate, FAR; the false accept rate and FRR; the false reject rate which are totally different than traditional performance evaluating methods. FAR and FRR correspond the percentages of fraud attempt acceptance and real attempt rejection, respectively.

However these rates are inconstant due to flexible thresholds, for instance if the confidence interval is so large than most of the fraud attacks could be inside. When calculating these rates by a threshold or a confidence interval, there will be two curves usually sketched and the intersection point will reveal EER. These criteria will be thoroughly explained in further sections.

Regarding the similar studies and very briefly; Chang et al. [3] dealt with touchscreen mobile devices and developed a graphical keystroke authentication system. They extracted the pressure feature to enhance the security and reached 6.9 % – 14.6 % EER interval depending on the experiment.

De Luca et al. [5] introduced a touchscreen authentication system to strengthen the regular method by a hidden security layer. They also analyzed the data by a novel method borrowed from speech recognition. Sae-Bae et al. [4] introduced a gesture-based authentication method with five-finger touch gestures. They collected biometric data and implemented a classifier for identifying characteristics of finger gestures and for validating future trials. Shahzad et al.'s [7] paper dealt with a novel authentication method for touchscreen smartphones, disregarding the pattern nodes. They extracted finger velocity, device acceleration and stroke durations as the main features and attained 0.5 % – 6.8 % EER. Angulo and Wästlund [6] dealt with lock pattern dynamics and introduced a hidden interface for the Android mobile platform to collect data. They achieved 10.39 % EER extracting finger-in-dot and finger-in-between nodes features.

Maiorana et al. [10] proposed a statistical classifier with Manhattan and Euclidean distances and achieved 13.59 % – 24.15 % EER. Tasia et al. [17] developed a virtual keyboard and extracted 6 traits and resulted in 8.4 % EER with statistical classifiers. Kang and Cho [22] emulated on-screen keyboard and introduced two interfaces for touchscreens and with several statistical methods they reached 5.64 % EER. Furthermore; Kambourakis et al. [16] implemented a keystroke system for touchscreens and collected four traits: speed, distance, hold-time and inter-time. They achieved 13.6 % – 26 % EER by KNN and Random forest algorithms.

As the final paper we will discuss is our paper [12] where we focused on several intelligent classifiers: back-propagation neural networks (BP-ANN), Levenberg-Marquardt based neural networks (LM-ANN) and adaptive neuro-fuzzy classifiers (ANFIS) and compared them. We simply extracted inter-touch times as inputs and finally achieved EER of 8.75 % by BP-ANN, 2.5 % by ANFIS, 7.5 % by LM-ANN.

In response to these enhancements, what we put forward in this paper as the novelty is disregarding any analyses in time domain and concentrating on the touch locations. Therefore this paper starts with feature extraction with the mathematical calculations of the touch locations. Furthermore, training session using the vectors of distance and angles to each corresponding node is presented. Subsequent to the presentation of the experiments separately conducted for LM and GN algorithms, the results and discussion sections are placed.

2 Feature Extraction

Initially, we wrote a 9-node interface shown in Fig. 1 left, which has a regular pattern password look. To extract the features of the potential ghost password, the touch on the screen with an index finger is simulated by a circle, ordinarily invisible to the user, yet colorized for this research as in Fig. 2 left.

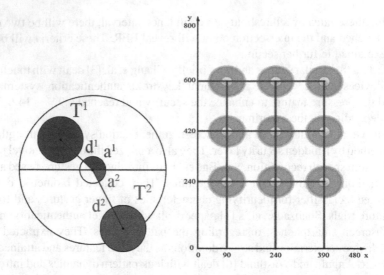

Fig. 2. Touches on a pattern node (on the left) and Node coordinates (on the right)

We already know the coordinates of the origins of the node r, namely

$$N_r = [X_r, Y_r] \tag{1}$$

where N is the node touched, X represents the horizontal axis, Y represents the vertical axis, r is integer number of the node and $r \in [19]$, $X_r, Y_r \in \mathbb{R}^+$, $\forall X_r \in [0480]$, $\forall Y_r \in [0800]$. The positions of the nodes and the coordinates of each node are stated in Fig. 2 right. Therefore for any i^{th} touch T^i on the node r, the coordinates could be extracted as

$$T_r^i = [x_r^i, y_r^i] \tag{2}$$

Given these equations, the distance is analytically calculated by:

$$d_r^i = \sqrt{(x_r^i - X_r)^2 + (y_r^i - Y_r)^2} \tag{3}$$

and the angle by:

$$a_r^i = \tan^{-1} \frac{(y_r^i - Y_r)}{(x_r^i - X_r)} \tag{4}$$

Since the angle itself cannot be perfectly determined by arctangent function, the sine of the angle is also considered to find the region of the angle. The computed couples [d_r^i a_r^i] are concatenated to form an array as inputs of the training system based on the password size.

3 Gauss-Newton and Levenberg-Marquardt Algorithms

As the primer classifiers, we have chosen artificial neural networks to train by several optimization algorithms. The optimizers are selected as Gauss-Newton (GN) and one of the enhancement on this method which is Levenberg-Marquardt (LM). The GN weight training procedure fundamentally is an optimization algorithm, yet it is so practical when the inputs are arrays. Consider that we have a function $V(w)$ which is to be minimized, the GN method adjusts the weights by:

$$\Delta w = - \left[\nabla^2 V(w) \right]^{-1} \nabla V(w) \tag{5}$$

where $\nabla^2 V(w)$ is the Hessian matrix and $\nabla V(w)$ is the gradient. As we purpose to optimize the weights by minimizing the sum of error square function of

$$V(w) = \sum_{i=1}^{N} e_i^2(w) \tag{6}$$

it could be rewritten as;

$$\nabla V(w) = J^T(w) e(w) \tag{7}$$

$$\nabla^2 V(w) = J^T(w) J(w) + S(w) \tag{8}$$

where

$$S(w) = \sum_{i=1}^{N} e_i(w) \nabla^2 e_i(w) \tag{9}$$

and J is the Jacobian matrix;

$$J(w) = \begin{bmatrix} \frac{\partial e_1(w)}{\partial w_1} & \frac{\partial e_1(w)}{\partial w_2} & \cdots & \frac{\partial e_1(w)}{\partial w_N} \\ \frac{\partial e_2(w)}{\partial w_1} & \frac{\partial e_2(w)}{\partial w_2} & \cdots & \frac{\partial e_2(w)}{\partial w_N} \\ \cdots & \cdots & \ddots & \cdots \\ \frac{\partial e_N(w)}{\partial w_1} & \frac{\partial e_N(w)}{\partial w_2} & \cdots & \frac{\partial e_N(w)}{\partial w_N} \end{bmatrix} \tag{10}$$

In GN method, $S(w)$ is assumed to be negligible thus zero, therefore the Eq. (5) could be rewritten as;

$$\Delta w = \left[J^T(w) J(w) \right]^{-1} J^T e(w) \tag{11}$$

The LM modification of GN method can be stated as;

$$\Delta w = \left[J^T(w) J(w) + \lambda I \right]^{-1} J^T e(w) \tag{12}$$

where I is the identity matrix and λ is an adaptive multiplier that forces the Hessian to be positive-definite. If λ = 0, then the method gives GN method however if λ is too low, it will be increased by the adaptive algorithm and vice versa.

4 Training Session

As the main password for the experiments, the four digit pattern is selected likewise in Fig. 1 right. Since the location touched is important as well as the node order in the password, we defined the sequence of 1st node: South-East, 4th node: West, 8th node: South and 6th node: the origin. The pattern nodes are as large to intentionally touch the desired side or the origins in every attempt. We initially extracted the coordinates of the touches, however changed the angles into radians and the distances into kilopixels. We trained the network for 100 epochs with 10 input arrays, one hidden layer with 10 nodes and optimized the weights using GN and LM separately. To establish a confidence interval, the inputs of the training set are individually feedforwarded in the trained network. The achieved results of the training set which also represents the correspondence between the output value and the whole training set, are shown in Fig. 3.

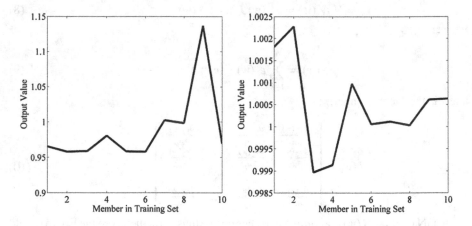

Fig. 3. Feedforward values of the training set by GN (on the left) and LM (on the right)

It is obvious from the Fig. 3 that LM algorithm narrows the interval more than GM. In biometric authentication systems, narrow intervals cause higher False Accept Rate (FAR) however lower False Reject Rate (FRR) and vice versa. The weights optimized by the algorithms are stored individually to test future logins based on the computed lower and upper limits.

5   Results

Initially, 40 real and 40 fraud trials are analyzed to test the performance of the neural network based classifiers. Although it is not mathematically provable, n > 30 for each set will be large enough as sampling size in statistics, therefore deemed sufficient. In these experiments the fraud team is informed of the exact password and the importance of the sequence and locations. In parallel with expectations, outputs the LM based classifier with the narrower interval resulted in 0 % FAR and 7.5 % FRR while the results of GN are a bit unpromising with 22.5 % FAR and 0 % FRR. Although these indicators show the system performance it is always necessary to calculate the Equal Error Rates (EER) for better understanding.

6   Conclusions and Discussions

As a brief summary, we extracted the touch locations instead of touch durations as the main novelty. Two different neural network based classifiers are used to discriminate real attempts from fraud attacks. By Gauss-Newton and Levenberg-Marquardt optimization methods, the weight of the networks are separately tuned to determine confidence intervals.

It is primarily found that the LM algorithm optimizes the weights faster than GN since we can see it is resulted in narrower interval. Although the EERs of each classifier are not adequately low, the results of real attempts are so consistent. The EER mainly depends on the sample space and size as well as the classification algorithm regardless of the threshold. The one and only reason of higher EER is fourth node that is touched in the middle which is also the first weakness observed in the whole system. When touched on the origin, the system calculates the distance perfectly however not the angle, since it is so hard to fix right in the analytical origin. Given this problem, the angle could vary form 0–360° which impedes the training process of the corresponding nodes. Therefore, we recommend to create a ghost password with consistent locations and angles on the nodes which should defy touching in the middle of them. The other weakness is regarding the number of epochs to train the networks. If we had increased the number of iterations, the intervals would have been narrower, which will cause an increase in FRR and decrease in FAR. On the contrary, if the number of epochs is sufficiently low then the networks would not adequately be trained.

Acknowledgment. This work and the contribution were supported by project "SP/2014/05 - Smart Solutions for Ubiquitous Computing Environments" from University of Hradec Kralove, Faculty of Informatics and Management.

References

1. Zheng, N., Bai, K., Huang, H., Wang, H.: You are how you touch: User verification on smartphones via tapping behaviors, Technical report, College of William and Mary (2012)

2. Kwapisz, J., Weiss, G., Moore, S.: Cell phone-based biometric identification. In: Proceedings IEEE International Conference on Biometrics: Theory Applications and Systems (2010)
3. Chang, T.Y., Tsai, C.J., Lin, J.H.: A graphical-based password keystroke dynamic authentication system for touch screen handheld mobile devices. J. Syst. Softw. **85**(5), 1157–1165 (2012)
4. Sae-Bae, N., Ahmed, K., Isbister, K., Memon, N.: Biometric-rich gestures: a novel approach to authentication on multi-touch devices. In: CHI 2012 Proceedings of the 2012 ACM Annual Conference on Human Factors in Computing Systems, New York (2012)
5. De Luca, A., Hang, A., Brudy, F., Lindner, C., Hussmann, H.: Touch me once and i know it's you!: implicit authentication based on touch screen patterns. In: CHI 2012 Proceedings of the 2012 ACM Annual Conference on Human Factors in Computing Systems, New York (2012)
6. Angulo, J., Wästlund, E.: Exploring touch-screen biometrics for user identification on smart phones. In: Camenisch, J., Crispo, B., Fischer-Hübner, S., Leenes, R., Russello, G. (eds.) Privacy and Identity Management for Life. IFIP AICT, vol. 375, pp. 130–143. Springer, Heidelberg (2012)
7. Shahzad, M., Liu, A.X., Samuel, A.: Secure unlocking of mobile touch screen devices by simple gestures: you can see it but you can not do it. In: Proceedings of the 19th Annual International Conference on Mobile Computing & Networking. ACM (2013)
8. Schaub, F., Deyhle, R., Weber, M.: Password entry usability and shoulder surfing susceptibility on different smartphone platforms. In: Proceedings of Mobile and Ubiquitous Multimedia, 2012
9. Shahzad, M., Zahid, S., Farooq, M.: A hybrid GA-PSO fuzzy system for user identification on smart phones. In: ACM, Proceedings of the 11th Annual Conference on Genetic and Evolutionary Computation, pp. 1617–1624 (2009)
10. Maiorana, E., Campisi, P., González-Carballo, N., Neri, A.: Keystroke dynamics authentication for mobile phones. In: Proceedings of the 2011 ACM Symposium on Applied Computing, pp. 21–26. ACM (2011)
11. Rao, M.K., Aparna, P., Akash, G.A., Mounica, K.: A graphical password authentication system for touch screen based devices. Int. J. Appl. Eng. Res. **9**(18), 4917–4924 (2014)
12. Alpar, O.: Intelligent biometric pattern password authentication systems for touchscreens. Expert Syst. Appl. **42**(17), 6286–6294 (2015)
13. Alpar, O.: Keystroke recognition in user authentication using ANN based RGB histogram technique. Eng. Appl. Artif. Intell. **32**, 213–217 (2014)
14. Trojahn, M., Arndt, F., Ortmeier, F.: Authentication with time features for keystroke dynamics on touchscreens. In: De Decker, B., Dittmann, J., Kraetzer, C., Vielhauer, C. (eds.) CMS 2013. LNCS, vol. 8099, pp. 197–199. Springer, Heidelberg (2013)
15. Jeanjaitrong, N., Bhattarakosol, P.: Feasibility study on authentication based keystroke dynamic over touch-screen devices. In: 2013 13th International Symposium on Communications and Information Technologies (ISCIT), pp. 238–242. IEEE (2013)
16. Kambourakis, G., Damopoulos, D., Papamartzivanos, D., Pavlidakis, E.: Introducing touchstroke: keystroke-based authentication system for smartphones. Secur. Commun. Netw. (2014). doi:10.1002/sec.1061
17. Tasia, C.J., Chang, T.Y., Cheng, P.C., Lin, J.H.: Two novel biometric features in keystroke dynamics authentication systems for touch screen devices. Secur. Commun. Netw. **7**(4), 750–758 (2014)
18. Frank, M., Biedert, R., Ma, E., Martinovic, I., Song, D.: Touchalytics: On the applicability of touchscreen input as a behavioral biometric for continuous authentication. IEEE Trans. Inf. Forensics Secur. **8**(1), 136–148 (2013)

19. Sae-Bae, N., Memon, N., Isbister, K., Ahmed, K.: Multitouch gesture-based authentication. IEEE Trans. Inf. Forensics Secur. 9(4), 568–582 (2014)
20. Zhao, X., Feng, T., Shi, W., Kakadiaris, I.: Mobile user authentication using statistical touch dynamics images. IEEE Trans. Inf. Forensics Secur. 9(11), 1780–1789 (2014)
21. Rogowski, M., Saeed, K., Rybnik, M., Tabedzki, M., Adamski, M.: User authentication for mobile devices. In: Saeed, K., Chaki, R., Cortesi, A., Wierzchoń, S. (eds.) CISIM 2013. LNCS, vol. 8104, pp. 47–58. Springer, Heidelberg (2013)
22. Kang, P., Cho, S.: Keystroke dynamics-based user authentication using long and free text strings from various input devices. Inf. Sci. (2014). http://dx.doi.org/10.1016/j.ins.2014.08.070

A New Approach to Link Prediction in Gene Regulatory Networks

Turki Turki[1,2](✉) and Jason T.L. Wang[2](✉)

[1] Computer Science Department, King Abdulaziz University, P.O. Box 80221,
Jeddah 21589, Saudi Arabia
tturki@kau.edu.sa
[2] Department of Computer Science, New Jersey Institute of Technology,
University Heights, Newark, NJ 07102, USA
{ttt2,wangj}@njit.edu

Abstract. Link prediction is an important data mining problem that
has many applications in different domains such as social network analysis and computational biology. For example, biologists model gene regulatory networks (GRNs) as directed graphs where nodes are genes and links
show regulatory relationships between the genes. By predicting links in
GRNs, biologists can gain a better understanding of the cell regulatory
circuits and functional elements. Existing supervised methods for GRN
inference work by building a feature-based classifier from gene expression
data and using the classifier to predict links in the GRNs. In this paper
we present a new supervised approach for link prediction in GRNs. Our
approach employs both gene expression data and topological features
extracted from the GRNs, in combination with three machine learning algorithms including random forests, support vector machines and
neural networks. Experimental results on different datasets demonstrate
the good performance of the proposed approach and its superiority over
the existing methods.

Keywords: Machine learning · Data mining · Feature selection ·
Bioinformatics · Systems biology

1 Introduction

Link prediction is an important data mining problem that finds many applications in social network analysis. One of the methods for solving the link prediction problem is to extract features from a given partially observed network and
incorporate these features into a classifier. The links (i.e., edges) between entities
(i.e., nodes or vertices) in the given partially observed network are labeled [2, 18].
One then uses the classifier built from the given partially observed network to
predict the presence of links for unobserved pairs of entities in the network.
Liben-Nowell and Kleinberg [17] showed that topological features can be used to
increase the accuracy of link prediction in social network analysis. In this paper

© Springer International Publishing Switzerland 2015
K. Jackowski et al. (Eds.): IDEAL 2015, LNCS 9375, pp. 404–415, 2015.
DOI: 10.1007/978-3-319-24834-9_47

we extend their techniques to solve an important bioinformatics problem in systems biology. Specifically, we present a new supervised method to infer gene regulatory networks (GRNs) through link prediction with topological features.

Several authors have developed supervised methods for GRN inference [3,6,27]. For example, Gillani et al. [9] presented CompareSVM, which uses support vector machines (SVMs) to predict the regulatory relationship between a transcription factor (TF) and a target gene where the regulatory relationship is represented by a directed edge (link), and both the TF and target gene are nodes in a gene network. SIRENE [6,19] is another supervised method, which splits the network inference problem into many binary classification problems using one SVM for each TF. The trained SVM classifiers are then used to predict which genes are regulated. The final step is to combine all SVM classifiers to produce a ranked list of TF-gene interactions in decreasing order, and to construct a network based on the ranked list. Cerulo et al. [3] developed a SVM-based method for GRN inference, which uses positive and unlabeled data for training the SVM classifier. Ernst et al. [7] developed a similar semi-supervised approach for GRN inference. In contrast to the above methods, all of which use gene expression data to predict TF-gene interactions [6,12,19], our approach considers features related to the topological structure of a network, which, to the best of our knowledge, is the first of its kind in GRN inference.

Feature extraction is crucial in building efficient classifiers for link prediction [1,8,17]. We adapt topological features employed in social network analysis for network inference in systems biology. We propose to use these topological features alone or combine them with gene expression data. As our experimental results show later, this new approach outperforms the previously developed supervised methods that use only gene expression data for network inference [3,6,9,19].

The rest of this paper is organized as follows. Section 2 presents our approach, explaining the techniques employed for feature extraction and feature vector construction. These features are used in combination with three machine learning algorithms including random forests, support vector machines and neural networks. Section 3 presents experimental results, showing the relative performance of the machine learning algorithms and demonstrating the superiority of our approach over the existing methods. Section 4 concludes the paper and points out some directions for future research.

2   Proposed Approach

2.1   Feature Extraction

Let $G = (V, E)$ be the directed graph that represents the topological structure of a gene regulatory network (GRN) where E is the set of edges or links and V is the set of nodes or vertices in G. Our goal is to build a classifier that includes topological features alone or combined with gene expression data. There are totally sixteen topological features, which are described in detail below.

Node Degree. In considering node degrees, each directed edge $e = (u, v) \in E$ has four topological features, $indeg(u)$, $outdeg(u)$, $indeg(v)$, and $outdeg(v)$, which are defined as the number of edges entering u, leaving u, entering v, and leaving v, respectively.

Normalized Closeness Centrality. Normalized closeness centrality measures the closeness between a node and all other nodes in the graph G. For each node or vertex $v \in V$, the normalized closeness centrality $C(v^{in})$ is defined as

$$C(v^{in}) = \frac{|V| - 1}{\sum_{i \neq v} d(i, v)} \tag{1}$$

where $d(i, v)$, $i \neq v$, is the distance from $i \in V$ to $v \in V$, and $|V|$ is the number of vertices in the graph G [15]. The distance from i to v is the number of edges on the shortest path from i to v. If no such path exists, then the distance is set equal to ∞. Since G is a directed graph, the distance from i to v is not necessarily the same as the distance from v to i. We define the normalized closeness centrality $C(v^{out})$ as

$$C(v^{out}) = \frac{|V| - 1}{\sum_{v \neq i} d(v, i)} \tag{2}$$

where $d(v, i)$, $v \neq i$, is the distance from $v \in V$ to $i \in V$. In considering normalized closeness centrality, each directed edge $e = (u, v) \in E$ has four topological features, $C(u^{in})$, $C(u^{out})$, $C(v^{in})$, and $C(v^{out})$.

Eccentricity. The eccentricity of a vertex $v \in V$ is the maximum distance between v and any other vertex $i \in V$ [24]. For each vertex $v \in V$, the eccentricity $\epsilon(v^{in})$ is defined as

$$\epsilon(v^{in}) = \max_{i \in V} \ d(i, v) \tag{3}$$

The eccentricity $\epsilon(v^{out})$ is defined as

$$\epsilon(v^{out}) = \max_{i \in V} \ d(v, i) \tag{4}$$

In considering eccentricity, each directed edge $e = (u, v) \in E$ has four topological features, $\epsilon(u^{in})$, $\epsilon(u^{out})$, $\epsilon(v^{in})$, and $\epsilon(v^{out})$.

Betweenness Centrality. Betweenness centrality measures the centrality of a vertex in the graph G [28]. For each vertex $v \in V$, the betweenness centrality of v, denoted $Between(v)$, is defined as

$$Between(v) = \sum_{i \neq v \neq j} \frac{\sigma_{i,j}(v)}{\sigma_{i,j}} \tag{5}$$

where $\sigma_{i,j}$ is the total number of shortest paths from vertex i to vertex j and $\sigma_{i,j}(v)$ is the total number of shortest paths from vertex i to vertex j that pass through v [15, 28]. In considering betweenness centrality, each directed edge $e = (u, v) \in E$ has two topological features, $Between(u)$ and $Between(v)$.

Eigenvector Centrality. Eigenvector centrality is another centrality measure where vertices in the graph have different importance. A vertex connected to a very important vertex is different from a vertex that is connected to a less important one. This concept is incorporated into eigenvector centrality [20]. For each vertex $v \in V$, the eigenvector centrality of v, denoted $Eigen(v)$, is defined as [21]

$$Eigen(v) = \frac{1}{\lambda} \sum_{i \in V} a_{v,i} Eigen(i) \qquad (6)$$

where λ is a constant and $\mathbf{A} = (a_{v,i})$ is the adjacency matrix, i.e., $a_{v,i} = 1$ if vertex v is linked to vertex i, and $a_{v,i} = 0$ otherwise. The above eigenvector centrality formula can be rewritten in the matrix form as

$$\mathbf{Ax} = \lambda \mathbf{x} \qquad (7)$$

where $\mathbf{x}$ is the eigenvector of the adjacency matrix $\mathbf{A}$ with the eigenvalue λ. In considering eigenvector centrality, each directed edge $e = (u, v) \in E$ has two topological features, $Eigen(u)$ and $Eigen(v)$.

2.2 Feature Vector Construction

Suppose we are given n genes where each gene has p expression values. The gene expression profile of these genes is denoted by $G \subseteq R^{n \times p}$, which contains n rows, each row corresponding to a gene, and p columns, each column corresponding to an expression value [19]. To train a classifier, we need to know the regulatory relationships among some genes. Suppose these regulatory relationships are stored in a matrix $H \subseteq R^{m \times 3}$. H contains m rows, where each row shows a known regulatory relationship between two genes, and three columns. The first column shows a transcription factor (TF). The second column shows a target gene. The third column shows the label, which is +1 if the TF is known to regulate the target gene or −1 if the TF is known not to regulate the target gene. The matrix H represents a partially observed or known gene regulatory network for the n genes. If the label of a row in H is +1, then the TF in that row regulates the target gene in that row, and hence that row represents a link or edge of the network. If the label of a row in H is −1, then there is no link between the corresponding TF and target gene in that row.

Given a pair of genes g_1 and g_2 where the regulatory relationship between g_1 and g_2 is unknown, our goal is to use the trained classifier to predict the label of the gene pair. The predicted label is either +1 (i.e., a link is predicted to be present between g_1 and g_2) or −1 (i.e., a link is predicted to be missing between g_1 and g_2). Using biological terms, the present link means g_1 (transcription factor) regulates g_2 (target gene) whereas the missing link means g_1 does not regulate g_2.

To perform training and predictions, we construct a feature matrix $D \subseteq R^{k \times 2p}$ with k feature vectors based on the gene expression profile G. For a pair of genes g_1 and g_2, we create their feature vector d, which is stored in the feature matrix D, denoted by D_d and defined as

$$D_d = [g_1^1, g_1^2, \ldots, g_1^p, g_2^1, g_2^2, \ldots, g_2^p] \tag{8}$$

where $g_1^1, g_1^2, \ldots, g_1^p$ are the gene expression values of g_1, and $g_2^1, g_2^2, \ldots, g_2^p$ are the gene expression values of g_2. The above feature vector definition has been used by the existing supervised network inference methods [3,6,9,19]. In the rest of this paper we will refer to the above technique for constructing feature vectors as Ge, indicating that it is based on gene expression data only.

In addition, we propose to construct another feature matrix $D' \subseteq R^{k \times 16}$. Each feature vector d in the feature matrix D', denoted by D'_d, is defined as

$$D'_d = [t_1, t_2] \tag{9}$$

$$t_1 = indeg(g_1), C(g_1^{in}), \epsilon(g_1^{in}), outdeg(g_1), \tag{10}$$
$$C(g_1^{out}), \epsilon(g_1^{out}), Between(g_1), Eigen(g_1)$$

$$t_2 = indeg(g_2), C(g_2^{in}), \epsilon(g_2^{in}), outdeg(g_2), \tag{11}$$
$$C(g_2^{out}), \epsilon(g_2^{out}), Between(g_2), Eigen(g_2)$$

We will refer to this feature vector construction technique as To, indicating that it is based on the sixteen topological features proposed in the paper.

Finally, we construct the third feature matrix $D'' \subseteq R^{k \times (2p+16)}$. Each feature vector d in the feature matrix D'', denoted by D''_d, contains both gene expression data and topological features, and is defined as

$$D''_d = [g_1^1, g_1^2, ..., g_1^p, g_2^1, g_2^2, ..., g_2^p, t_1, t_2] \tag{12}$$

We will refer to this feature vector construction technique as All, indicating that it is based on all the features described in the paper.

3   Experiments and Results

We conduct a series of experiments to evaluate the performance of our approach and compare it with the existing methods for gene regulatory network (GRN) inference [13]. Below, we describe the datasets used in our study, our experimental methodology, and the experimental results.

3.1   Datasets

We used GeneNetWeaver [23] to generate the datasets related to yeast and E. coli. We first built five different networks, taken from yeast, where the networks

Table 1. Yeast networks used in the experiments

Network	Directed	#Nodes	#Edges	#Pos examples	#Neg examples
Yeast 50	Yes	50	63	63	63
Yeast 100	Yes	100	281	281	281
Yeast 150	Yes	150	333	333	333
Yeast 200	Yes	200	517	517	517
Yeast 250	Yes	250	613	613	613

Table 2. E. coli networks used in the experiments

Network	Directed	#Nodes	#Edges	#Pos examples	#Neg examples
E. coli 50	Yes	50	68	68	68
E. coli 100	Yes	100	177	177	177
E. coli 150	Yes	150	270	270	270
E. coli 200	Yes	200	415	415	415
E. coli 250	Yes	250	552	552	552

contained 50, 100, 150, 200, 250 genes (or nodes) respectively. For each network, we generated three files of gene expression data. These files were labeled as knockouts, knockdowns and multifactorial, respectively. A knockout is a technique to deactivate the expression of a gene, which is simulated by setting the transcription rate of this gene to zero [9,23]. A knockdown is a technique to reduce the expression of a gene, which is simulated by reducing the transcription rate of this gene by half [9,23]. Multifactorial perturbations are simulated by randomly increasing or decreasing the activation of the genes in a network simultaneously [23].

Table 1 presents details of the yeast networks, showing the number of nodes (edges, respectively) in each network. The edges or links in a network form positive examples. In addition, we randomly picked the same number of negative examples where each negative example corresponds to a missing link in the network. The networks and gene expression profiles for E. coli were generated similarly. Table 2 presents details of the networks generated from E. coli.

3.2   Experimental Methodology

We considered nine classification algorithms, denoted by RF + All, NN + All, SVM + All, RF + Ge, NN + Ge, SVM + Ge, RF + To, NN + To, SVM + To, respectively. Table 3 lists these algorithms and their abbreviations. RF + All (RF + Ge, RF + To, respectively) represents the random forest algorithm combined with all features including both gene expression data and topological features (RF combined with only gene expression data, RF combined with only topological features, respectively). NN + All (NN + Ge, NN + To, respectively)

Table 3. Nine classification algorithms and their abbreviations

Abbreviation	Classification algorithm and features
RF + All	Random forests with all features
NN + All	Neural networks with all features
SVM + All	Support vector machines with all features
RF + Ge	Random forests with gene expression features
NN + Ge	Neural networks with gene expression features
SVM + Ge	Support vector machines with gene expression features
RF + To	Random forests with topological features
NN + To	Neural networks with topological features
SVM + To	Support vector machines with topological features

represents the neural network algorithm combined with all features (NN combined with only gene expression data, NN combined with only topological features, respectively). SVM + All (SVM + Ge, SVM + To, respectively) represents the support vector machine algorithm combined with all features (SVM combined with only gene expression data, SVM combined with only topological features, respectively). SVM + Ge is adopted by the existing supervised network inference methods [3,6,9,19].

Software used in this work included: the random forest package in R [16], the neuralnet package in R [10], and the SVM with linear kernel in the LIBSVM package [4]. We used R to write some utility tools for performing the experiments, and employed the package, igraph, to extract topological features from a network [5].

The performance of each classification algorithm was evaluated through 10-fold cross validation. The size of each fold was approximately the same, and each fold contained the same number of positive and negative examples. On each fold, the *balanced error rate* (BER) [11] of a classification algorithm was calculated where the BER is defined as

$$BER = \frac{1}{2} \times \left(\frac{FN}{TP + FN} + \frac{FP}{FP + TN} \right) \tag{13}$$

FN is the number of false negatives (i.e., present links that were mistakenly predicted as missing links). TP is the number of true positives (i.e., present links that were correctly predicted as present links). FP is the number of false positives (i.e., missing links that were mistakenly predicted as present links). TN is the number of true negatives (i.e., missing links that were correctly predicted as missing links). For each algorithm, the mean BER, denoted MBER, over 10 folds was computed and recorded. The lower MBER an algorithm has, the better performance that algorithm achieves. Statistically significant performance differences between classification algorithms were calculated using Wilcoxon signed rank tests [13,14]. As in [22,25], we consider p-values below 0.05 to be statistically significant.

3.3 Experimental Results

Table 4 shows the MBERs of the nine classifications on the fifteen yeast datasets used in the experiments. For each dataset, the algorithm having the best performance (i.e., with the lowest MBER) is in boldface. Table 5 shows, for each yeast dataset, the p-values of Wilcoxon signed rank tests between the best algorithm, represented by '-', and the other algorithms. A p-value in boldface ($p \leq 0.05$) indicates that the corresponding result is significant. It can be seen from Table 4 that random forests performed better than support vector machines and neural networks. In particular, random forests combined with all features (i.e., RF + All) performed the best on 10 out of 15 yeast datasets. For the other five yeast datasets, RF + All was not statistically different from the best algorithms according to Wilcoxon signed rank tests ($p > 0.05$); cf. Table 5.

Table 6 shows the MBERs of the nine classification algorithms on the fifteen E. coli datasets used in the experiments. Table 7 shows, for each E. coli dataset, the p-values of Wilcoxon signed rank tests between the best algorithm, represented by '-', and the other algorithms. It can be seen from Table 6 that random forests combined with topological features (i.e., RF + To) performed the best on 6 out of 15 E. coli datasets. For the other nine E. coli datasets, RF + To was not statistically different from the best algorithms according to Wilcoxon signed rank tests ($p > 0.05$); cf. Table 7.

Table 4. MBERs of nine classification algorithms on fifteen yeast datasets

Dataset	RF + All	NN + All	SVM + All	RF + Ge	NN + Ge	SVM + Ge	RF + To	NN + To	SVM + To
Yeast 50 knockouts	16.6	**15.0**	20.0	18.3	15.8	20.0	17.7	16.3	19.4
Yeast 50 knockdowns	**16.6**	17.5	20.0	19.1	22.7	16.9	17.7	18.8	19.4
Yeast 50 multifactorial	**16.1**	17.5	20.8	15.5	24.7	17.2	17.7	16.6	19.4
Yeast 100 knockouts	12.8	13.7	17.3	14.4	20.0	16.7	**11.6**	22.2	14.5
Yeast 100 knockdowns	14.2	14.9	15.2	14.1	29.4	14.1	**11.8**	17.5	14.5
Yeast 100 multifactorial	12.0	17.8	18.0	12.3	21.3	18.4	**11.4**	20.0	14.5
Yeast 150 knockouts	**5.10**	10.7	14.1	5.40	10.4	13.6	6.10	15.4	11.0
Yeast 150 knockdowns	**5.00**	10.5	10.4	5.80	14.7	10.9	5.80	16.0	11.0
Yeast 150 multifactorial	**4.10**	12.4	13.9	**4.10**	15.1	16.1	5.80	10.8	11.0
Yeast 200 knockouts	**1.90**	4.00	5.30	**1.90**	4.90	5.60	2.50	13.7	3.70
Yeast 200 knockdowns	**1.90**	5.10	5.80	**1.90**	6.30	10.5	2.70	9.80	3.70
Yeast 200 multifactorial	**1.90**	6.80	10.1	**1.90**	4.70	14.3	2.50	5.50	3.70
Yeast 250 knockouts	**3.80**	8.40	7.60	4.10	5.50	7.20	7.40	10.6	10.4
Yeast 250 knockdowns	**4.00**	9.70	7.40	**4.00**	6.20	7.30	8.10	7.70	10.4
Yeast 250 multifactorial	4.00	8.50	8.50	**3.90**	6.00	9.00	7.50	11.3	10.4

Table 5. P-values of Wilcoxon signed rank tests between the best algorithm, represented by '-', and the other algorithms for each yeast dataset

Dataset	RF + All	NN + All	SVM + All	RF + Ge	NN + Ge	SVM + Ge	RF + To	NN + To	SVM + To
Yeast 50 knockouts	0.75	-	0.28	0.44	0.67	0.07	0.46	0.78	0.27
Yeast 50 knockdowns	-	0.79	0.46	0.17	0.13	0.52	0.58	0.86	0.33
Yeast 50 multifactorial	-	0.70	0.28	0.89	0.07	0.68	0.58	0.68	0.33
Yeast 100 knockouts	0.34	0.44	**0.05**	0.14	**0.01**	**0.02**	-	**0.04**	0.15
Yeast 100 knockdowns	0.22	0.24	0.16	0.22	**0.00**	**0.03**	-	0.08	0.12
Yeast 100 multifactorial	0.46	0.13	**0.02**	0.27	**0.01**	**0.04**	-	**0.00**	0.12
Yeast 150 knockouts	-	**0.01**	**0.02**	1.00	**0.01**	**0.01**	0.10	**0.01**	**0.02**
Yeast 150 knockdowns	-	**0.02**	**0.02**	0.17	**0.03**	**0.01**	0.17	**0.00**	**0.02**
Yeast 150 multifactorial	-	**0.01**	**0.01**	-	**0.01**	**0.01**	0.17	**0.01**	**0.02**
Yeast 200 knockouts	-	**0.01**	**0.01**	-	**0.01**	**0.02**	0.50	**0.01**	0.17
Yeast 200 knockdowns	-	**0.04**	**0.01**	-	**0.01**	**0.02**	0.10	**0.01**	0.17
Yeast 200 multifactorial	-	**0.04**	**0.01**	-	**0.06**	**0.00**	0.50	**0.01**	0.17
Yeast 250 knockouts	-	**0.03**	0.20	0.17	0.06	0.10	0.17	0.09	**0.02**
Yeast 250 knockdowns	-	**0.01**	**0.10**	-	0.06	**0.05**	0.07	0.20	**0.05**
Yeast 250 multifactorial	1.00	**0.05**	0.11	-	**0.01**	**0.03**	0.18	**0.05**	**0.03**

Table 6. MBERs of nine classification algorithms on fifteen E. coli datasets

Dataset	RF + All	NN + All	SVM + All	RF + Ge	NN + Ge	SVM + Ge	RF + To	NN + To	SVM + To
E. coli 50 knockouts	14.5	5.70	9.80	18.8	22.0	21.5	**5.00**	11.6	18.4
E. coli 50 knockdowns	14.5	9.50	10.7	19.1	19.0	16.7	**5.00**	10.8	18.4
E. coli 50 multifactorial	15.3	10.3	8.20	15.7	22.9	18.0	**5.00**	10.3	18.4
E. coli 100 knockouts	10.3	9.50	14.2	12.0	14.4	17.7	**5.60**	6.00	13.6
E. coli 100 knockdowns	10.6	11.6	14.2	11.4	14.1	18.0	**5.40**	9.80	13.6
E. coli 100 multifactorial	9.80	10.4	11.4	9.80	12.4	13.6	**7.10**	9.90	13.6
E. coli 150 knockouts	**2.40**	4.40	2.90	**2.40**	8.80	5.90	2.70	5.90	3.30
E. coli 150 knockdowns	**2.20**	4.40	2.70	**2.20**	8.10	3.70	2.70	3.80	3.30
E. coli 150 multifactorial	**2.20**	3.80	2.40	**2.20**	8.70	2.40	2.70	3.30	3.30
E. coli 200 knockouts	5.50	5.50	5.10	5.50	**4.60**	**4.60**	6.40	11.0	6.00
E. coli 200 knockdowns	5.30	5.00	5.80	6.10	**4.40**	5.50	6.40	8.50	6.00
E. coli 200 multifactorial	5.00	4.20	3.90	4.90	**2.80**	3.40	6.40	8.00	6.00
E. coli 250 knockouts	**6.60**	10.3	7.00	7.70	8.80	8.00	6.30	12.7	10.0
E. coli 250 knockdowns	5.30	9.30	5.90	**5.10**	6.70	7.50	6.20	11.9	10.0
E. coli 250 multifactorial	5.40	11.1	8.00	**5.20**	11.0	9.70	6.30	10.6	10.0

Table 7. P-values of Wilcoxon signed rank tests between the best algorithm, represented by '-', and the other algorithms for each E. coli dataset

Dataset	RF + All	NN + All	SVM + All	RF + Ge	NN + Ge	SVM + Ge	RF + To	NN + To	SVM + To
E. coli 50 knockouts	0.06	1.00	0.46	**0.03**	**0.00**	**0.01**	-	0.10	**0.04**
E. coli 50 knockdowns	0.06	0.07	0.46	**0.01**	**0.01**	**0.01**	-	0.46	**0.04**
E. coli 50 multifactorial	0.06	0.49	0.46	0.08	**0.01**	**0.04**	-	**0.04**	**0.04**
E. coli 100 knockouts	0.08	**0.04**	**0.01**	**0.01**	**0.00**	**0.01**	-	0.68	**0.04**
E. coli 100 knockdowns	**0.02**	0.16	**0.01**	**0.01**	**0.01**	**0.01**	-	0.27	**0.04**
E. coli 100 multifactorial	0.67	0.22	0.17	0.67	0.11	0.06	-	0.79	0.17
E. coli 150 knockouts	-	0.10	0.25	-	**0.02**	0.06	0.65	**0.04**	0.65
E. coli 150 knockdowns	-	**0.04**	0.17	-	0.06	0.06	1.00	0.10	1.00
E. coli 150 multifactorial	-	0.06	1.00	-	0.10	1.00	1.00	0.10	1.00
E. coli 200 knockouts	0.91	0.50	0.46	0.91	0.68	-	0.41	**0.03**	0.46
E. coli 200 knockdowns	0.89	0.50	0.27	0.75	-	0.46	0.46	0.11	0.68
E. coli 200 multifactorial	0.67	0.09	0.14	0.67	-	0.28	0.13	0.13	0.20
E. coli 250 knockouts	-	**0.02**	0.23	0.65	**0.00**	0.12	0.71	0.12	**0.05**
E. coli 250 knockdowns	0.17	**0.02**	**0.02**	-	**0.02**	**0.02**	0.10	**0.02**	**0.01**
E. coli 250 multifactorial	1.00	**0.02**	**0.01**	-	**0.01**	**0.01**	0.72	**0.01**	**0.02**

These results show that using random forests with the proposed topological features alone or combined with gene expression data performed well. In particular, the RF + All algorithm achieved the best performance on 14 out of all 30 datasets. This is far better than the SVM + Ge algorithm used by the existing supervised network inference methods [3,6,9,19], which achieved the best performance on one dataset only (i.e., the E. coli 200 knockouts dataset in Table 6).

It is worth pointing out that, for a fixed dataset size (e.g., 200), the SVM+To algorithm always yielded the same mean balanced error rate (MBER) regardless of which technique (knockout, knockdown or multifactorial) was used to generate the gene expression profiles. This happens because these different gene expression profiles correspond to the same network, and SVM+To uses only the topological features extracted from the network without considering the gene expression data. On the other hand, due to the randomness introduced in random forests and neural networks, RF+To and NN+To yielded different MBERs even for the same network.

4   Conclusion

We present a new approach to network inference through link prediction with topological features. Our experimental results showed that using the topological features alone or combined with gene expression data performs better than the existing network inference methods that use only gene expression data. Our work assumes that there are exactly the same number of positive examples (i.e., links that are present) and negative examples (i.e., links that are missing) in the datasets. In many biological networks, however, negative datasets (majority class) are usually much larger than positive datasets (minority class). In future work we plan to extend our previously developed imbalanced classification algorithms and boosting algorithms [26,29,30] to tackle the imbalanced link prediction problem for gene network construction. We have adopted default parameter settings for the three machine learning algorithms studied in the paper. We also plan to explore other parameter settings (e.g., different kernels with different parameter values in SVMs) in the future.

References

1. Al Hasan, M., Chaoji, V., Salem, S., Zaki, M.: Link prediction using supervised learning. In: SDM06: Workshop on Link Analysis, Counter-terrorism and Security (2006)
2. Al Hasan, M., Zaki, M.J.: A survey of link prediction in social networks. In: Aggarwal, C.C. (ed.) Social Network Data Analytics, pp. 243–275. Springer, New York (2011)
3. Cerulo, L., Elkan, C., Ceccarelli, M.: Learning gene regulatory networks from only positive and unlabeled data. BMC Bioinformatics 11(1), 228 (2010)
4. Chang, C.C., Lin, C.J.: Libsvm: a library for support vector machines. ACM Trans. Intell. Syst. Technol. (TIST) 2(3), 27 (2011)
5. Csardi, G., Nepusz, T.: The igraph software package for complex network research. Int. J. Complex Syst. 1695(5), 1–9 (2006)
6. De Smet, R., Marchal, K.: Advantages and limitations of current network inference methods. Nat. Rev. Microbiol. 8(10), 717–729 (2010)
7. Ernst, J., Beg, Q.K., Kay, K.A., Balázsi, G., Oltvai, Z.N., Bar-Joseph, Z.: A semi-supervised method for predicting transcription factor-gene interactions in escherichia coli. PLoS Comput. Biol. 4(3), e1000044 (2008)
8. Fire, M., Tenenboim, L., Lesser, O., Puzis, R., Rokach, L., Elovici, Y.: Link prediction in social networks using computationally efficient topological features. In: 2011 IEEE Third International Conference on Privacy, Security, Risk and Trust (PASSAT) and 2011 IEEE Third Inernational Conference on Social Computing (SocialCom), pp. 73–80. IEEE (2011)
9. Gillani, Z., Akash, M.S., Rahaman, M., Chen, M.: CompareSVM: supervised, support vector machine (SVM) inference of gene regularity networks. BMC Bioinformatics 15(1), 395 (2014)
10. Günther, F., Fritsch, S.: neuralnet: training of neural networks. R Journal 2(1), 30–38 (2010)
11. Guyon, I., Alamdari, A.R.S.A., Dror, G., Buhmann, J.M.: Performance prediction challenge. In: International Joint Conference on Neural Networks, IJCNN 2006, pp. 1649–1656. IEEE (2006)
12. Hecker, M., Lambeck, S., Toepfer, S., Van Someren, E., Guthke, R.: Gene regulatory network inference: data integration in dynamic models: a review. Biosystems 96(1), 86–103 (2009)
13. Japkowicz, N., Shah, M.: Evaluating Learning Algorithms. Cambridge University Press, Cambridge (2011)

14. Kanji, G.K.: 100 Statistical Tests. Sage, London (2006)
15. Kolaczyk, E.D.: Statistical Analysis of Network Data: Methods and Models. Springer, New York (2009)
16. Liaw, A., Wiener, M.: Classification and regression by randomforest. R News **2**(3), 18–22 (2002). http://CRAN.R-project.org/doc/Rnews/
17. Liben-Nowell, D., Kleinberg, J.: The link-prediction problem for social networks. J. Am. Soc. Inform. Sci. Technol. **58**(7), 1019–1031 (2007)
18. Menon, A.K., Elkan, C.: Link prediction via matrix factorization. In: Gunopulos, D., Hofmann, T., Malerba, D., Vazirgiannis, M. (eds.) ECML PKDD 2011, Part II. LNCS, vol. 6912, pp. 437–452. Springer, Heidelberg (2011)
19. Mordelet, F., Vert, J.P.: Sirene: supervised inference of regulatory networks. Bioinformatics **24**(16), i76–i82 (2008)
20. Newman, M.: Networks: An Introduction. Oxford University Press, Oxford (2010)
21. Newman, M.E.: The mathematics of networks. In: The New Palgrave Encyclopedia of Economics, vol. 2, pp. 1–12 (2008)
22. Nuin, P.A.S., Wang, Z., Tillier, E.R.M.: The accuracy of several multiple sequence alignment programs for proteins. BMC Bioinformatics **7**, 471 (2006). http://dx.doi.org/10.1186/1471-2105-7-471
23. Schaffter, T., Marbach, D., Floreano, D.: Genenetweaver: in silico benchmark generation and performance profiling of network inference methods. Bioinformatics **27**(16), 2263–2270 (2011)
24. Takes, F.W., Kosters, W.A.: Computing the eccentricity distribution of large graphs. Algorithms **6**(1), 100–118 (2013)
25. Thompson, J.D., Plewniak, F., Poch, O.: A comprehensive comparison of multiple sequence alignment programs. Nucleic Acids Res. **27**(13), 2682–2690 (1999). http://dx.doi.org/10.1093/nar/27.13.2682
26. Turki, T., Wei, Z.: IPRed: Instance reduction algorithm based on the percentile of the partitions. In: Proceedings of the 26th Modern AI and Cognitive Science Conference MAICS, pp. 181–185 (2015)
27. Wang, J.T.L.: Inferring gene regulatory networks: challenges and opportunities. J. Data Min. Genomics Proteomics **06**(01), e118 (2015). http://dx.doi.org/10.4172/2153-0602.1000e118
28. Ye, J., Cheng, H., Zhu, Z., Chen, M.: Predicting positive and negative links in signed social networks by transfer learning. In: Proceedings of the 22nd International Conference on World Wide Web, pp. 1477–1488. International World Wide Web Conferences Steering Committee (2013)
29. Zhong, L., Wang, J.T.L., Wen, D., Aris, V., Soteropoulos, P., Shapiro, B.A.: Effective classification of microRNA precursors using feature mining and adaboost algorithms. OMICS **17**(9), 486–493 (2013)
30. Zhong, L., Wang, J.T.L., Wen, D., Shapiro, B.A.: Pre-mirna classification via combinatorial feature mining and boosting. In: 2012 IEEE International Conference on Bioinformatics and Biomedicine, BIBM 2012, Philadelphia, PA, USA, 4–7 October 2012, pp. 1–4 (2012). http://doi.ieeecomputersociety.org/10.1109/BIBM.2012.6392700

On Stability of Ensemble Gene Selection

Nicoletta Dessì, Barbara Pes[✉], and Marta Angioni

Dipartimento di Matematica e Informatica, Università degli Studi di Cagliari,
Via Ospedale 72, 09124 Cagliari, Italy
{dessi,pes}@unica.it, martaangioni@gmail.com

Abstract. When the feature selection process aims at discovering useful knowledge from data, not just producing an accurate classifier, the degree of stability of selected features is a very crucial issue. In the last years, the ensemble paradigm has been proposed as a primary avenue for enhancing the stability of feature selection, especially in high-dimensional/small sample size domains, such as biomedicine. However, the potential and the implications of the ensemble approach have been investigated only partially, and the indications provided by recent literature are not exhaustive yet. To give a contribution in this direction, we present an empirical analysis that evaluates the effects of an ensemble strategy in the context of gene selection from high-dimensional micro-array data. Our results show that the ensemble paradigm is not always and necessarily beneficial in itself, while it can be very useful when using selection algorithms that are intrinsically less stable.

Keywords: Feature selection stability · Ensemble paradigm · Gene selection

1 Introduction

Biomedical data are often characterized by a very high dimensionality, i.e. a huge number of features, coupled with a relatively small sample size. In this context, the identification of a small set of highly discriminative biomarkers, e.g. a group of disease associated genes, requires the application of suitable feature selection methods [1]. So far, high classification accuracy has been the main (and often the only) criterion used to assess the quality of selected biomarkers, leading to a generally low level of reproducibility of the reported results (that turn out highly dependent on the specific set of samples used in the selection stage).

In the last years, the need of making the detected biomarkers more reproducible, and hence more useful for clinical applications, has led to an increasing attention to stable feature selection protocols. Specifically, the term *stability* is used to describe the extent to which a feature selection method is sensitive to changes in the training data, e.g. adding or removing a given percentage of samples [2]. Stable feature selection is of paramount importance in many biological studies since biologists may have more confidence on selection results that do not change significantly when data are perturbed in some way.

Recent literature [3, 4] has discussed a number of approaches that may be potentially useful for improving the stability of feature selection methods. Among them, ensemble

© Springer International Publishing Switzerland 2015
K. Jackowski et al. (Eds.): IDEAL 2015, LNCS 9375, pp. 416–423, 2015.
DOI: 10.1007/978-3-319-24834-9_48

feature selection has been recommended [5] as a very promising paradigm that does not require prior knowledge nor complex transformations of the original feature space [3]. Basically, an ensemble selection strategy involves two fundamentals steps: (i) creating a set of different feature selectors (ensemble components) and (ii) combining the outputs of the different selectors into a single final decision. Different approaches can be applied both for creating the feature selectors as well as for their combination [4].

Indeed, a number of studies [5, 6] have shown the potential of the ensemble paradigm in the context of biomarker discovery, opening the way to a very important and challenging research area. However, as observed in [7], the effectiveness of ensemble approaches has not been evaluated extensively and, to date, there is not a consensus on the superiority of ensemble methods over simple ones [8, 9].

To give a contribution to this field, our paper presents an empirical study aiming at providing indications on benefits and limitations of the ensemble paradigm across different selection algorithms. Specifically, we considered eight selection methods, representatives of different selection approaches (both univariate and multivariate), and investigated if, and to which extent, they get more stable when used in an ensemble fashion. The analysis was conducted on a high dimensional micro-array dataset [10] that provides an interesting benchmark for the selection of genes relevant to cancer diagnosis.

The rest of this paper is structured as follows. Section 2 provides some useful background on the ensemble paradigm. Section 3 describes all materials and methods involved in our empirical study, while Sect. 4 presents and discusses the experimental results. Finally, Sect. 5 provides concluding remarks.

2   Ensemble Feature Selection

The ensemble paradigm has been extensively applied in the context of classification problems. Basically, it involves the construction of a set of classifiers whose individual decisions are combined in some way (e.g., by weighted voting) to classify new data points. As discussed in [11], an ensemble learning strategy may expand the hypothesis space, reduce the risk of choosing a wrong hypothesis and give a better approximation to the true unknown function.

Recent research efforts [3–6] have explored the extension of this paradigm to the feature selection process, the rationale being to exploit the output of a set of different selectors, instead of using a single one. There are two main approaches for creating the different selectors to be included in the ensemble, i.e. (i) by applying a single feature selection method to different perturbed versions of a dataset (*data-diversity* approach) or (ii) by applying different selection methods to the same dataset (*functional-diversity* approach). Hybrid ensemble strategies are also possible where diversity is injected both at the data level and at the function level (i.e. different selection methods are applied to different versions of a dataset).

Specifically, we focus here on ensemble techniques that use, as ensemble components, a set of *rankers*, each one providing in output a *ranked list* where the original features appear in descending order of relevance. Indeed, in the context of biomarker discovery (e.g. gene selection from high dimensional micro-array data), ranking techniques are widely used for identifying a given number of potentially relevant features to be further validated by domain experts [12]. In an ensemble selection perspective, different ranked lists can be aggregated into a single *ensemble list* that would hopefully provide a better representation of the underlying domain.

A number of aggregation strategies have been experimented in the context of ensemble feature ranking [13], but there is no evidence of which strategy may be more appropriate for a given task. Usually, simple functions are adopted that assign each feature an overall score based on the feature's rank across all the lists to be combined. In our study, we consider the following approaches:

– *Mean/median aggregation.* The overall score of each feature is obtained as the mean/median value of the feature's rank across all the original lists. Since the feature's rank in a given list corresponds to its ranking position (the most relevant feature has rank 1), the smallest the overall score the higher the overall importance of the feature.
– *Exponential aggregation.* This strategy consists in deriving, for each feature, a local score (within each of the original lists) that is expressed as an exponentially decreasing function of the rank, namely *exp(-rank/thr)*, where *thr* is a suitable threshold [8, 13]. Then, the overall score for each feature is obtained by summing up the corresponding local scores: in this case the higher the overall score, the higher the overall importance of the feature.

Hence, the final ensemble list is obtained by ordering the features from the most important to the least important (according to their overall scores).

3   The Case Study

The aim of our study is to evaluate when, and to which extent, the use of an ensemble selection method is to be preferred to a simple one. Specifically, we concentrate here on comparing the stability of simple ranking techniques with the stability of ensemble ranking techniques in the context of micro-array data analysis. As ensemble strategy, we adopt a *data-diversity* approach, i.e. we apply a given ranking method to different perturbed versions of a dataset so as to obtain a number of ranked lists that are potentially different from each other, though produced by the same algorithm. Based on the aggregation strategies discussed in Sect. 2, these lists are then combined to obtain an overall ensemble ranking from which to derive the final feature subset. While suggested as a primary avenue for improving the stability of the selection process [3, 5], this approach has been so far experimented for a limited number of selection methods, with no evidence of a generalized superiority of ensemble techniques over simple ones. This motivates further experimental investigations, as the one here presented. In what follows, more details are given about the methodology we adopted for stability evaluation (Sect. 3.1) as well as about the selection methods and the benchmark used in the empirical study (Sect. 3.2).

3.1   Methodology for Stability Evaluation

The stability of a selection method can be defined as its sensitivity to changes in the dataset composition [2]. In other words, a method is stable when produces the same (or almost the same) output despite some modifications in the set of records used for training. Hence, given a dataset D, we adopted a sub-sampling strategy to create a number K of reduced datasets D^i ($i = 1, 2, \ldots, K$), each containing a percentage X of instances randomly drawn from D. A simple ranking method, denoted here as R, was then applied to each D^i so as to obtain a ranked list L^i as well as a feature subset S^i with the n highest ranked features (i.e., the most relevant ones). The stability analysis was then performed by comparing the resulting K feature subsets: indeed the more similar they are, the more stable the ranking method R.

Specifically, we applied a proper consistency index [14] to derive a similarity value for each pair of subsets: this value basically reflects the degree of overlapping between the subsets (i.e., $|S^i \cap S^j|/n$), with a correction term that takes into account the probability that a feature is included in both subsets simply by chance. By averaging the resulting similarity values over all pair-wise comparisons, we obtained a global evaluation of the degree of stability of the ranking method R.

The same stability analysis was then conducted on the ensemble version of the ranker R. Specifically, each of the reduced datasets D^i ($i = 1, 2, \ldots, K$) was further sampled (but with replacement) to get a number B of *bootstrap* datasets from which to obtain B ranked lists (each produced by R) to be aggregated into an ensemble list $L^i_{ensemble}$. Each ensemble list, in turn, originated a feature subset $S^i_{ensemble}$ containing the n highest ranked features. The resulting K subsets were finally compared, in a pair-wise fashion, according to the above cited consistency index [14] and the average similarity (over all pairs of subsets) was used as a measure of the ensemble stability.

3.2 Ranking Methods and Micro-Array Benchmark

As ranking methods, we considered four univariate techniques, which assess the relevance of each single feature independently from the others, namely:

- *Information Gain* (IG), *Symmetrical Uncertainty* (SU) and *Gain Ratio* (GR), that are grounded on the concept of entropy.
- *OneR* (OR), that relies on a simple rule-based classifier.

We also experimented with four multivariate methods, which take into account the inter-dependencies among features, namely:

- *ReliefF-Weighted* (RFW), that evaluates the relevance of features based on their ability to distinguish between instances that are near to each other.
- *SVM_ONE*, that relies on a linear SVM classifier to derive a weight for each feature.
- *SVM_RFE*, that weights the features using a linear SVM classifier, as SVM_ONE, but adopts a backward elimination strategy that iteratively removes the features with the lowest weights and repeats the overall weighting process on the remaining features. The percentage of features removed at each iteration was set in our experiments as 10 % (SVM_RFE10) and 50 % (SVM_RFE50).

More details about the above methods and their pattern of agreement can be found in [15]. For all of them, we exploited the implementation provided by WEKA [16].

As benchmark for the empirical study we used the DLBCL tumor dataset [10], which contains 77 biological samples, each of them described by the expression level (measured by micro-array technology) of 7129 genes. The task, in terms of biomarker discovery, is to identify the genes useful in discriminating two types of lymphoma, i.e. follicular lymphoma (FL) and diffuse large b-cell lymphoma (DLBCL).

4   Experimental Results and Discussion

According to the methodological approach previously described, we compared the stability of each ranking method with the stability of the ensemble version of the method itself. Specifically, for each ranker, we implemented three ensembles that differ for the inner aggregation strategy, i.e. *mean*, *median* and *exponential* (see Sect. 2). Note that, after some tuning experiments, the threshold value of the exponential function was set as 5 % of the original number of features (7129). The other parameters of our methodology (see Sect. 3.1) were set as follows: $X = 90$ % (sample size), $K = 50$ (number of samples), $B = 50$ (number of bootstraps). The results of the stability analysis are summarized in Fig. 1 for feature subsets of different sizes (n), i.e. for different percentages of selected features.

As regards the four univariate methods (IG, SU, GR, OR), we can see that (Fig. 1(a)) the higher the stability of the simple method (the red curve), the less it benefits from the ensemble approach. Indeed, the IG ranker, which is the most stable among the considered univariate approaches, does not benefit at all from the adoption of an ensemble strategy (irrespective of the aggregation function). SU, in turn, seems to benefit only slightly from the ensemble approach (here, mean aggregation is somewhat better), but the positive impact of the ensemble paradigm is very strong for the methods that are less stable in their simple version, i.e. GR and OR. In particular, for GR, ensemble ranking is much better than simple ranking, especially for small percentages of selected features (with a clear superiority of mean aggregation), while OR turns out more stable in its ensemble version for all the considered subset sizes (irrespective of the aggregation strategy).

A similar trend emerges from the stability curves of the four multivariate methods (RFW, SVM_ONE, SVM_RFE10, SVM_RFE50). Indeed, as shown in Fig. 1(b), the ensemble approach is not beneficial for the most stable ranker, i.e. RFW, while it gives some improvement for SVM_ONE (a bit less stable than RFW). On the other hand, both SVM_RFE10 and SVM_RFE50, which are the least stable among the multivariate approaches, strongly benefit from the adoption of an ensemble strategy, especially per small percentages of selected features. Here, no aggregation function clearly outperforms the others, though mean aggregation seems to be slightly better.

Overall, the results in Fig. 1 seem to indicate that the ensemble paradigm is not always and necessarily beneficial in itself, but only in dependence on the "intrinsic" stability on the considered method.

Further insight can be derived from the summary results in Fig. 2, where the stability curves of all methods in their simple form (on the left) are compared with the stability curves of the corresponding ensembles (on the right). Specifically, we considered here the ensemble implemented with mean aggregation, for both univariate (Fig. 2(a)) and multivariate approaches (Fig. 2(b)). Quite surprisingly, despite the strong differences

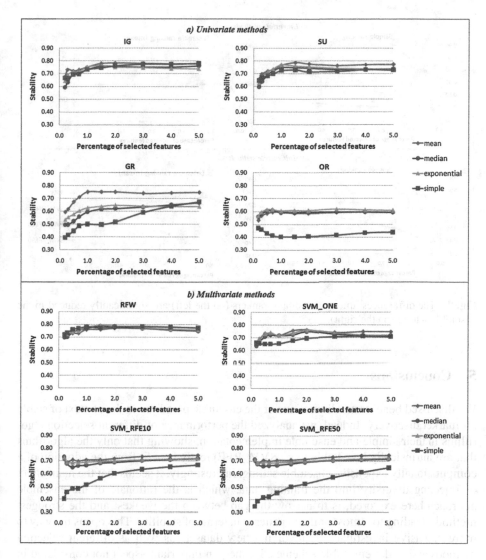

Fig. 1. Stability analysis: simple ranking vs ensemble ranking (mean, median and exponential) (Color figure online)

observed among the ranking methods in the simple form, these methods produce very similar patterns in the ensemble setting. Indeed, as regards the univariate approaches, the stability of SU and GR reaches that of IG (the most stable method) and even OR (the least stable) gets more close to the other univariate rankers. As well, among the multivariate approaches, the weak methods reach the strongest one, i.e. RFW. This is an interesting, so far neglected, implication of the ensemble paradigm: when trained in a sufficiently diversified data space (according to the *data-diversity* approach here adopted), even quite dissimilar methods can produce very similar (sometimes coincident) results in terms of stability.

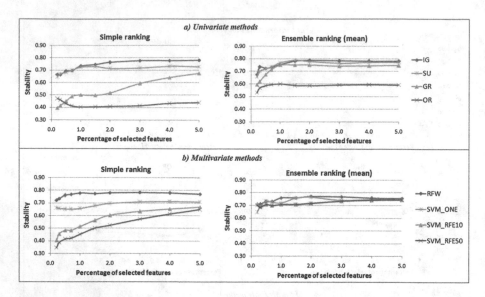

Fig. 2. The differences among the simple rankers (on the left) are significantly reduced in the ensemble setting (on the right).

5   Conclusions

We discussed benefits and limitations of the ensemble paradigm in the context of stable biomarker discovery. Indeed, we analyzed the performance of different selection algorithms in their simple and ensemble implementation, showing that only the algorithms that are intrinsically less stable (e.g., SVM_RFE) really benefit from the adoption of a computationally expensive ensemble setting. Interestingly, it seems that the main effect of injecting diversity into the training data, which is the rationale of the ensemble approach here explored, is to narrow the gap between the weakest and the strongest methods, leading to almost uniform patterns in terms of stability. This opens the way to more extensive investigations, involving new datasets as well as different selection methods and wider ensemble schemes. Further, an important aspect not considered in this study is the extent to which the ensemble paradigm affects the predictive quality of the selected feature subsets, besides their stability. In this regard, new experiments are currently in progress, and our future research will be devoted to obtain a more complete and clear picture on the effects and the implications of ensemble feature selection in terms of both stability and predictive performance (and their trade-off).

Acknowledgments. This research was supported by Sardinia Regional Government (project CRP-17615, *DENIS: Dataspaces Enhancing the Next Internet in Sardinia*).

References

1. Saeys, Y., Inza, I., Larranaga, P.: A review of feature selection techniques in bioinformatics. Bioinformatics **23**(19), 2507–2517 (2007)
2. Kalousis, A., Prados, J., Hilario, M.: Stability of feature selection algorithms: a study on high-dimensional spaces. Knowl. Inf. Syst. **12**(1), 95–116 (2007)
3. Zengyou, H., Weichuan, Y.: Stable feature selection for biomarker discovery. Comput. Biol. Chem. **34**, 215–225 (2010)
4. Awada, W., Khoshgoftaar, T.M., Dittman, D., Wald, R., Napolitano, A.: A review of the stability of feature selection techniques for bioinformatics data. In: IEEE 13th International Conference on Information Reuse and Integration, pp. 356–363. IEEE (2012)
5. Abeel, T., Helleputte, T., Van de Peer, Y., Dupont, P., Saeys, Y.: Robust biomarker identification for cancer diagnosis with ensemble feature selection methods. Bioinformatics **26**(3), 392–398 (2010)
6. Saeys, Y., Abeel, T., Van de Peer, Y.: Robust feature selection using ensemble feature selection techniques. In: Daelemans, W., Goethals, B., Morik, K. (eds.) ECML PKDD 2008, Part II. LNCS (LNAI), vol. 5212, pp. 313–325. Springer, Heidelberg (2008)
7. Kuncheva, L.I., Smith, C.J., Syed, Y., Phillips, C.O., Lewis, K.E.: Evaluation of feature ranking ensembles for high-dimensional biomedical data: a case study. In: IEEE 12th International Conference on Data Mining Workshops, pp. 49–56. IEEE (2012)
8. Haury, A.C., Gestraud, P., Vert, J.P.: The influence of feature selection methods on accuracy, stability and interpretability of molecular signatures. PLoS ONE **6**(12), e28210 (2011)
9. Dessì, N., Pes, B.: Stability in biomarker discovery: does ensemble feature selection really help? In: Ali, M., Kwon, Y.S., Lee, C.-H., Kim, J., Kim, Y. (eds.) IEA/AIE 2015. LNCS, vol. 9101, pp. 191–200. Springer, Heidelberg (2015)
10. Shipp, M.A., Ross, K.N., Tamayo, P., Weng, A.P., et al.: Diffuse large B-cell lymphoma outcome prediction by gene-expression profiling and supervised machine learning. Nat. Med. **8**(1), 68–74 (2002)
11. Dietterich, T.G.: Ensemble methods in machine learning. In: Kittler, J., Roli, F. (eds.) MCS 2000. LNCS, vol. 1857, pp. 1–15. Springer, Heidelberg (2000)
12. Dessì, N., Pascariello, E., Pes, B.: A comparative analysis of biomarker selection techniques. BioMed Res. Int. 2013, Article ID 387673 (2013)
13. Wald, R., Khoshgoftaar, T.M., Dittman, D., Awada, W., Napolitano, A.: An extensive comparison of feature ranking aggregation techniques in bioinformatics. In: IEEE 13th International Conference on Information Reuse and Integration, pp. 377–384. IEEE (2012)
14. Kuncheva, L.I.: A stability index for feature selection. In: 25th IASTED International Multi-Conference: Artificial Intelligence and Applications, pp. 390–395. ACTA Press Anaheim, CA, USA (2007)
15. Dessì, N., Pes, B.: Similarity of feature selection methods: An empirical study across data intensive classification tasks. Expert Syst. Appl. **42**(10), 4632–4642 (2015)
16. WEKA. http://www.cs.waikato.ac.nz/ml/weka/

Early Alzheimer's Disease Prediction in Machine Learning Setup: Empirical Analysis with Missing Value Computation

Sidra Minhas[1(✉)], Aasia Khanum[2], Farhan Riaz[1],
Atif Alvi[2], and Shoab A. Khan[1]
and Alzheimer's Disease Neuroimaging Initiative

[1] Computer Engineering Department, College of E&ME,
National University of Science and Technology (NUST), Islamabad, Pakistan
sidra.minhas@ceme.nust.edu.pk
[2] Computer Science Department, Forman Christian College, Lahore, Pakistan

Abstract. Alzheimer's Disease (AD) is the most prevalent progressive neurodegenerative disorder of the elderly. Prospective treatments for slowing down or pausing the process of AD require identification of the disease at an early stage. Many patients with mild cognitive impairment (MCI) may eventually develop AD. In this study, we evaluate the significance of using longitudinal data for efficiently predicting MCI-to-AD conversion a few years ahead of clinical diagnosis. The use of longitudinal data is generally restricted due to missing feature readings. We implement five different techniques to compute missing feature values of neuropsychological predictors of AD. We use two different summary measures to represent the artificially completed longitudinal features. In a comparison with other recent techniques, our work presents an improved accuracy of 71.16 % in predicting pre-clinical AD. These results prove feasibility of building AD staging and prognostic systems using longitudinal data despite the presence of missing values.

Keywords: Machine learning · Alzheimer's Disease (AD) · Mild Cognitive Impairment (MCI) · ADNI · Longitudinal data · Missing value · Support vector machine · Classification · AUC

1 Introduction

Alzheimer's Disease (AD) is the most common form of dementia diagnosed in elder persons [1]. AD results from neuro-degeneration which leads to cognitive failure and other behavioral deficiencies in daily activities of patients. From 2000–2010, number of deaths due to AD increased by 68 %, whereas deaths caused by other diseases decreased (Alzheimer's Association, 2014). Hence, there is a dire need of developing systems for early diagnosis of AD. For effective treatment, it is necessary to identify the disease process at an early stage so that its progress can be halted or slowed down. Mild Cognitive Impairment (MCI) is an early stage of deteriorating memory and intellectual functionality. Though patients with MCI do not necessarily worsen, they are at high risk of progressing to AD. To observe and monitor the people effected by or at risk of

© Springer International Publishing Switzerland 2015
K. Jackowski et al. (Eds.): IDEAL 2015, LNCS 9375, pp. 424–432, 2015.
DOI: 10.1007/978-3-319-24834-9_49

being effected by AD, a longitudinal study named Alzheimer's Disease Neuroimaging Initiative (ADNI) [2] was launched in USA by National Institute of Aging (NIA), in association with the National Institute of Biomedical Imaging and Bioengineering (NIBIB), the Food and Drug Association (FDA) and several other pharmaceutical companies in 2003.

MCI is a heterogeneous group consisting of instances which regain to normal cognition, stay stable as MCI (MCIs) or progress to AD (MCIp). ADNI data, which consists of serial Neuropsychological measure (NM), images (MRI, PET scans), biochemical findings and genetic factors, has been used in numerous machine learning setups to identify MCIp from within MCI subjects. Most of the studies have used multimodal baseline data as input for learning disease outcome prediction [3, 4]. Moradi et al. [5] used a semi-supervised machine learning framework using baseline imaging and NM data. Mattila et al. [4] developed a statistical Disease State Index method to estimate progression from healthy to disease on bases of various data sources. A good summary of multimodal performance for AD prediction can be found in [5]. However, it is now realized that for effective staging of the disease, it is useful to employ information contained in time-sampled longitudinal data [6]. A recent implementation by Runtti et al. [7] used longitudinal heterogeneous multimodal predictors to differentiate between MCIp and MCIs. In view of such research findings, [8, 9] have published to use both cognitive tests and biomarkers for AD staging.

Longitudinal studies, such as ADNI, are faced with the problem of missing data which may be significant for slowly progressing diseases such as AD. Missing data in these initiatives arise due to patients missing one or more follow up visits or completely dropping out of the study. Consequently, a reduction in sample size is observed which may bias the classifier performance. Lo and Jagust [10] suggests that missing data in ADNI is not Missing Completely at Random; hence, may contribute to further understanding of AD progression. Generally, machine learning packages resort to immediately dropping the instances with missing values. Our hypothesis is that by imputing missing data values instead of instantaneous dropping it, and effectively utilizing longitudinal features- important information differentiating MCIp from MCIs will be retained and superior predictive performance will be delivered.

A number of studies exist utilizing multimodal (mostly baseline) biomarkers for AD conversion prediction in machine learning framework [3–7]. The present paper however employs unimodal longitudinal data for disease prediction. We select only NM data modality due to its cheap and non-invasive nature as opposed to more expensive and complicated methods of imaging and bio specimen sampling. We incorporate longitudinal data in our machine learning setup, and attempt to enhance their prognostic power by effectively handling missing values. Using our proposed approach, we foretell which of the MCI patients will progress to AD within 36 months of the baseline visit and which of them will retain a stable diagnosis. Through extensive experimentation, we demonstrate the efficacy of longitudinal data while accounting for missing values. The paper is organized in four sections. Section 2 describes the materials and proposed method. Experimental results are discussed in Sect. 3, followed by the conclusion in Sect. 4.

2   Materials and Methods

Figure 1 shows the major steps of this study. As a first step, missing data, within the dataset, were identified and substituted using various techniques. The resulting artificially completed longitudinal features were represented using two different summary measures. Support Vector Machine (SVM) classifier is employed for quantitative performance evaluation. SVM is a commonly used technique in the domain under study, preferred due to its high accuracy. Summary measures are later used to identify each feature's relevance in MCIp vs. MCIs segregation (*). The impact of sequentially reducing the number of features on grouping performance is also studied.

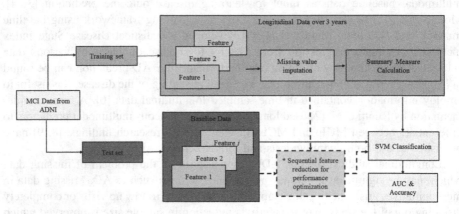

Fig. 1. Proposed method

2.1   Data

ADNI1 focused on MCI subjects, for which follow ups were conducted biannually after baseline visit for the first two years. Afterwards, follow ups were conducted annually. Over a period of 3 years, 200 of the MCI patients were reported to have progressed to AD (MCIp) whereas 100 of them remained stable at MCI (MCIs). The remaining 100 patients had unstable diagnosis. Group wise patient IDs used in this study were provided by [5]. From the ADNI website, neuropsychological test results of MCI patients were downloaded. The sixteen neuropsychological biomarkers used in this study are the responses of the tests listed in Table 1. For each feature, approximately 30 % of the patients had missed at least one follow up visit. It is worth noticing that the features Geriatric Depression Scale (GDS) and Immediate Recall Total Score (LIMM) are considered 100 % missing as they were recorded annually only; 6th and 18th month readings were missing.

2.2   Missing Data Computation

Each of the j features were arranged in $i \times v$ matrix depicting i training instances, each with v follow up visits. Following techniques were used for data enhancement through completion:

Table 1. NM used and percentage of missing data

Feature number	Test performed	Abbreviation	Missing data/%
1	AD Assessment Scale – 11 items	ADAS_11	30.71
2	AD Assessment Scale – 13 items	ADAS_13	35.58
3	Rey Auditory Verbal Test –Trail 1	AVTOT1	30.34
4	Rey Auditory Verbal Test –Trail 2	AVTOT2	30.34
5	Rey Auditory Verbal Test –Trail 3	AVTOT3	30.34
6	Rey Auditory Verbal Test –Trail 4	AVTOT4	30.71
7	Rey Auditory Verbal Test –Trail 5	AVTOT5	30.71
8	Clock Drawing Test	CLOCKSCOR	30.34
9	Clock Copy Test	COPYSCOR	31.09
10	Functional Assessment Questionnaire	FAQ	31.09
11	Geriatric Depression Scale	GDS	100
12	Immediate Recall Total Score	LIMM	100
13	Mini Mental State Examination	MMSE	29.59
14	Neuropsychiatric Inventory Questionnaire	NPIQ	29.21
15	Trail Making Test A	TRAA	29.59
16	Trail Making Test B	TRAB	37.45

- *Case Deletion (CD):*
  Using this technique, the instances which have one or more missing value are removed from the training set. As a result only those cases for which all follow up readings for all features were recorded were retained for training.
- *Case Ignorance (CI):*
  In this case, the missing values were ignored and all semi-known data was included in the training set.
- *Group Mean Replacement (GMR):*
  The unknown feature values were replaced with the mean of known values at that time point.
- *Last Observation Carried Forward (LOCF):*
  As ADNI reports the natural progression of MCI patients without any intervention, it can be assumed that the patients retain previous scores for missed follow up visits. One possible scheme for filling up unavailable values is by carrying forward the last known value. For the cases, where baseline reading was missing, the next known value was assigned to it.
- *K Nearest Neighbor (KNN):*
  For each instance in the training set, every missing value is substituted by the value in its 'k' nearest neighbor column. The nearest neighbor is defined as the instance having most similar value on the previous visit. Similarity, in this case, is measured using absolute difference between feature readings. Parameter 'k' is set to 1 in this scenario.

2.3   Experimental Setup

10-fold cross validation method is employed for performance investigation. To cumulatively represent longitudinal data, two summary measures are devised in this study. Summary measures demonstrate the projection of longitudinal readings at baseline (t = 0) which are calculated according to (1) and (2):

$$SM1_{ji} = \frac{\sum_{n=0}^{v} i_n}{v} \tag{1}$$

$$SM2_{ji} = \frac{1}{2}\left(i_{\frac{v}{2}} + i_{\frac{v}{2}+1}\right) \tag{2}$$

$SM1_{ji}$ and $SM2_{ji}$ respectively correspond to arithmetic mean and median value of jth feature of ith instance over v follow up visits. The final data matrix is of size ixj. The temporal factor of conversion is out of scope of this study.

Standard SVM with Radial Basis Function Kernel and soft margin set to 1 is used for training and prediction. In order to study the impact of complete and incomplete data on prediction performance, a total of eleven experiments are designed. First experiment is restricted to using only the baseline feature readings for training. Whereas in rest of the experiments, partially known feature sets are completed using the computation techniques and are coupled with the two summary measures.

Finally, for performance optimization and removal of redundant features, feature selection is performed. Each summarized feature is ranked according to class separability criteria. Classifier independent feature evaluation is performed using Information Gain measure which quantifies the reduction in entropy caused by partitioning the data using that attribute. Higher the information gain, higher the rank. This selection pertains to both training and test set. Lowest ranked features are deleted one by one. Final feature set consists of features resulting in maximum Area under Curve (AUC) of Receiver Operating Characteristic (ROC) Curve.

3   Results and Discussion

From a total of 400 MCI patients initially recruited in ADNI, there were only 167 instances of MCIp class and 100 members of MCIs class that had at least one reading for all the j features. However, when CD was performed, the sample size reduced by approximately 50 % as only 78 samples of MCIp and 67 cases of MCIs had complete follow up readings for all biomarkers. Due to imbalanced dataset, performance quantification pivots on AUC instead of accuracy. Table 2 provides the results of prediction without performing feature selection. It can be noticed that using only baseline readings for conversion prediction produces low scores for both AUC and accuracy as compared to longitudinal data. It is also visible that CD underperformed all other techniques in terms of both accuracy and AUC and presented a high variance across validation folds. Overall best performance (AUC: 74.43 %, Accuracy: 68.16 %) is clearly displayed by GMR scheme using SM1 as a summary measure.

Table 2. AUC, Accuracy (Acc) and their variance (var.) over CV folds, without feature selection

Technique	AUC (%)	AUC var. (%)	Acc (%)	Acc var. (%)
BL	70.44	0.63	66.29	0.12
CD-SM1	72.53	2.16	60.69	0.24
CD-SM2	68.13	2.48	58.62	0.11
CI-SM1	73.23	0.84	67.79	0.01
CI-SM2	72.63	0.87	67.42	0.04
GMR-SM1	**74.43**	**0.90**	**68.16**	**0.04**
GMR-SM2	73.90	0.86	67.04	0.02
LOCF-SM1	72.69	0.86	67.42	0.04
LOCF-SM2	73.90	1.00	67.04	0.04
KNN-SM1	69.89	1.03	67.42	0.03
KNN-SM2	72.05	0.91	67.04	0.07

Table 3 demonstrates the prediction results after feature set purification. Table 3 (Row: 3–18, col: 2–12) reveal the feature ranks under each experiment. The top ranked features required to provide best AUC are highlighted and corresponding count is mentioned in Row 19. N/A for LIMM and GDS exhibit absence of complete cases in CD, GMR and KNN setups. Rey's Auditory Verbal Test scores are pointed out as most significant feature in all experiments. AUC reading achieved using the reduced feature set are stated in Row 20 and its variance over 10 cross validation folds is mentioned in Row 21. Last row of Table 3 exhibits the accuracy obtained by using the reduced feature set. It can be seen that maximum AUC of 76.42 % was displayed by LOCF-SM1 with a corresponding Accuracy of 71.16 %, by using top 13 features only. LOCF-SM2, CD-SM1 and CI-SM1 also deliver competitive AUC results but display high variance in AUC values over validation folds and poor accuracy, hence indicating their instability for AD prediction. However, overall best accuracy of 72.28 % is displayed by employing baseline data only. This signifies that artificially generated data does introduce minor impurities in longitudinal data. Even though, maximum accuracy is achieved by using baseline data only, it cannot be concluded as the most appropriate one due to a considerably lower and highly variable AUC reading. Resultantly, it can be established that LOCF using SM1 as summary measure is a promising technique for MCIp vs. MCIs identification.

Figure 2 shows the ROC curves for some selected experiments. Figure 2(a) shows the ROC curves for competing techniques in terms of AUC whereas Fig. 2(b) shows the resulting curves of feature selection under the best concluded technique (LOCF-SM1).

A brief accuracy comparison between the proposed and previously published techniques applied on unimodal ADNI data is given in Table 4. Varying data modalities, validation methods and follow up times make direct comparison difficult. Most relevant comparison of our work is with Ewers et al. [14] and Casanova et al. [15] in terms of all variables. Significant amount of improvement can be detected when complete, longitudinal data are used. It can also be observed that by effective handling of missing longitudinal data, NM alone are competitive with studies using more advanced and expensive modality data like imaging (MRI).

Table 3. Performance results of feature selection

Feature number	BL	CD		CI		GMR		LOCF		KNN	
	–	SM1	SM2	SM1	SM2	SM1	SM2	SM1	SM2	SM1	SM2
1	7	**10**	10	10	**10**	10	10	**10**	**10**	10	10
2	6	6	6	6	6	6	7	6	7	7	7
3	5	7	7	7	7	7	6	7	6	6	6
4	1	2	2	2	2	2	2	2	2	2	2
5	15	**11**	1	5	1	5	5	5	1	5	1
6	4	1	11	1	5	11	1	1	5	1	5
7	14	5	5	13	12	1	11	13	13	11	11
8	3	4	4	12	13	4	4	12	12	4	4
9	13	14	14	4	4	3	8	4	4	8	8
10	12	8	8	8	16	8	3	8	8	9	14
11	8	N/A	N/A	16	8	N/A	N/A	16	16	N/A	N/A
12	11	N/A	N/A	9	15	N/A	N/A	9	15	N/A	N/A
13	9	3	3	3	3	14	14	3	3	14	13
14	2	13	12	15	9	13	13	15	**14**	13	3
15	10	12	13	14	14	12	12	14	9	3	9
16	16	9	9	11	**11**	9	9	**11**	11	12	12
# of feat.	9	11	8	8	13	14	4	13	14	2	8
AUC	74.44	76.09	73.73	75.67	73.81	74.43	73.95	**76.42**	76.09	70.16	73.98
AUC Var.	1.73	1.69	1.01	1.63	1.38	0.94	1.15	**0.60**	0.79	0.71	0.50
Accuracy	72.28	66.21	67.59	68.91	66.67	68.16	65.54	**71.16**	70.79	62.17	64.79

C 1: (R 2–17) Feature number corresponding to Table 1. C 2–12: (R 3–18) Feature ranks under each experiment, (R 19): Number of top ranked features (feat.) required for max AUC, (R 20): AUC (%) achieved using selected features, (R 21): Variance (Var) of AUC (%) over 10 cross validation folds, (R 22): Accuracy (%) corresponding to max AUC

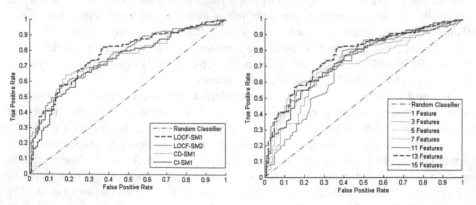

Fig. 2. ROC Curves, (a) competing techniques, (b) feature selection under LOCF-SM1 scheme

Table 4. Accuracy (Acc) Comparison

Author	Data	Validation method	Follow up (months)	Acc (%)
Moradi et al. [5]	MRI	k fold cv	0–36	69.15
Cuingnet et al. [11]	MRI	Ind. Test set	0–18	67
Wolz et al. [12]	MRI	k fold cv	0–48	68
Ye et al. [13]	MRI	k fold cv	0–15	56.10
Ewers et al. [14]	NM	k fold cv	0–36	64.60
Casanova et al. [15]	NM/MRI	k fold cv	0–36	65/62
Present work (LOCF-SM1)	NM	k fold cv	0–36	71.16

4 Conclusions and Future Work

In this paper, we presented a machine learning framework for prediction of AD in MCI patients using longitudinal, neuropsychological scores from ADNI. Proper selection of relevant neuropsychological biomarkers, effective substitution of missing feature readings and a strong summary measure for the longitudinal data helps retain significant differences between MCIp and MCIs. We aim to repeat this scheme with multimodal data while considering longer follow up times to further enhance predictive power of this system.

Acknowledgements. We would like to thank all investigators of ADNI listed at: https://adni. loni.usc.edu/wp-content/uploads/how_to_apply/ADNI_Acknowedge-ment_List.pdf, for developing and making their data publically available.

References

1. Duthey, B: Background paper 6.11: Alzheimer disease and other dementias. A Public Health Approach to Innovation, Update on 2004 Background Paper, pp. 1–74 (2013)
2. Alzheimer's Disease Neuroimaging Initiative. http://adni.loni.ucs.edu. Accessed April 2015
3. Asrami, F.F.: AD Classification using K-OPLS and MRI. Masters' Thesis, Department of Biomedical Engineering, Linkoping University (2012)
4. Mattila, J., Koikkalainen, J., Virkki, A., Simonsen, A., van Gils, M., Waldemar, G., Soininen, H., Lötjönen, J.: ADNI: a disease state fingerprint for evaluation of AD. J. Alzheimer's Dis. **27**, 163–176 (2011)
5. Moradi, E., Pepe, A., Gaser, C., Huttunen, H., Tohk, J.: Machine learning framework for early MRI-based Alzheimer's conversion prediction in MCI subjects. NeuroImage **104**, 398–412 (2015)
6. Zhang, D., Shen, D.: Predicting future clinical changes of MCI patients using longitudinal and multimodal biomarkers. PLoS ONE **7**(3), e33182 (2012)
7. Runtti, H., Mattila, J., van Gils, M., Koikkalainen, J., Soininen, H., Lötjönen, J.: Quantitative evaluation of disease progression in a longitudinal mild cognitive impairment cohort. J. Alzheimer's Dis. **39**(1), 49–61 (2014)

8. Sperling, R.A., Aisen, P.S., Beckett, L.A., Bennett, D.A., Craft, S., Fagan, A.M., Iwatsubo, T., Jack Jr., C.R., Kaye, J., Montine, T.J., Park, D.C., Reiman, E.M., Rowe, C. C., Siemers, E., Stern, Y., Yaffe, K., Carrillo, M.C., Thies, B., Morrison-Bogorad, M., Wagster, M.V., Phelps, C.H.: Toward defining the preclinical stages of AD: recommendations from the National Institute on Aging-Alzheimer's Association workgroups on diagnostic guidelines for Alzheimer's disease. Alzheimers Dement **7**, 280–292 (2011)

9. Albert, M.S., DeKosky, S.T., Dickson, D., Dubois, B., Feldman, H.H., Fox, N.C., Gamst, A., Holtzman, D.M., Jagust, W.J., Petersen, R.C., Snyder, P.J., Carrillo, M.C., Thies, B., Phelps, C.H.: The diagnosis of mild cognitive impairment due to AD: recommendations from the National Institute on Aging-Alzheimer's Association workgroups on diagnostic guidelines for Alzheimer's disease. Alzheimers Dement **7**, 270–279 (2011)

10. Lo, R.Y., Jagust, W.J.: Predicting missing biomarker data in a longitudinal study of AD. Neurology **78**(18), 1376–1382 (2012)

11. Cuingnet, R., Gerardin, E., Tessieras, J., Auzias, G., Lehéricy, S., Habert, M.O., Chupin, M.: Automatic classification of patients with AD from structural MRI: a comparison of ten methods using the ADNI database. Neuroimage **56**(2), 766–781 (2011)

12. Wolz, R., Julkunen, V., Koikkalainen, J., Niskanen, E., Zhang, D.P., Rueckert, D., Soininen, H., Lötjönen, J.: Multi-method analysis of MRI images in early diagnosis of AD. PLoS ONE **6**(10), 25446 (2011)

13. Ye, D.H., Pohl, K.M., Davatzikos, C.: Semi-supervised pattern classification: application to structural MRI of AD. In: 2011 International Workshop on Pattern Recognition in NeuroImaging (PRNI), pp. 1–4. IEEE (2011)

14. Ewers, M., Walsh, C., Trojanowskid, J.Q., Shawd, L.M., Petersene, R.C., Jack Jr., C.R., Feldmang, H.H., Bokdeh, A.L.W., Alexanderi, G.E., Scheltens, P., Vellas, B., Dubois, B., Weinera, M., Hampe, H.: Prediction of conversion from mild cognitive impairment to AD dementia based upon biomarkers and neuropsychological test performance. Neurobiol. Ageing **33**(7), 1203–1214 (2012)

15. Casanova, R., Hsu, F.C., Sink, K.M., Rapp, S.R., Williamson, J.D., Resnick, S.M., Espeland, M.A.: AD risk assessment using large-scale machine learning methods. PLoS ONE **8**(11), e77949 (2013)

Description of Visual Content in Dermoscopy Images Using Joint Histogram of Multiresolution Local Binary Patterns and Local Contrast

Sidra Naeem, Farhan Riaz[✉], Ali Hassan, and Rida Nisar

Department of Computer Engineering, College of Electrical
and Mechanical Engineering, National University of Sciences and Technology
(NUST), Islamabad, Pakistan
farhan.riaz@gmail.com

Abstract. Melanoma is a deadly form of skin lesion for which the mortality rate
can be significantly reduced, if detected at an early stage. Clinical findings have
shown that an early detection of melanoma can be done by an inspection of visual
characteristics of some specific regions (lesions) of the skin. This paper proposes
a pattern recognition system that includes three vital stages to conform the
analysis of skin lesions by the clinicians: segmentation, feature extraction, and
classification. Segmentation is performed using active contours with creasness
features. The feature extraction phase consists of a variant of local binary pattern
(LBP) in which joint histogram of LBP pattern along with the contrast of the
patterns are used to extract scale adaptive patterns at each pixel. Classification
was performed using support vector machines. Experimental results demonstrate
the superiority of the proposed feature set over several other state-of-the-art
texture feature extraction methods for melanomas detection. The results indicate
the significance of contrast of the pattern along with LBP patterns.

Keywords: Texture · Local binary patterns · Dermoscopy

1 Introduction

Melanoma is the deadliest cancerous skin lesion and its occurrence has been constantly
increasing in the recent years [1]. The detection of melanoma at its early stage can
increase the probability of patient's survival significantly [2]. Dermoscopy is a clinical
procedure, widely used by the dermatologists to carry out an in vivo observation of skin
lesions. The physicians apply gel to the skin of the patients and then examine the skin
with a dermoscope allowing the inspection of surface and sub-surface structures that are
invisible to the naked eye. Skin cancer diagnosis is carried out using an assessment of
these structures for which several medical rules have been defined, the most common are
the ABCD rule [3], Menzies' method [4] and seven- point check list [5]. The ABCD rule
assigns a score to a lesion. This score is a combination of four different features, namely,
(A)symmetry, (B)order, (C)olors, and (D)ifferential structures. The Menizes' method
identifies two different types of features in dermoscopy images: negative feature as

© Springer International Publishing Switzerland 2015
K. Jackowski et al. (Eds.): IDEAL 2015, LNCS 9375, pp. 433–440, 2015.
DOI: 10.1007/978-3-319-24834-9_50

symmetrical pattern indicates the absence of melanoma, while positive feature is present in the form of blue-white veil, atypical dots and network, etc., and they specify the presence of melanoma. Finally, the seven point check-list method assigns a score for the presence of a lesion. This check-list identifies the presence of differential structures in a lesion. Various computer aided diagnosis (CAD) systems have been proposed that facilitates the early recognition of melanomas, in which some systems try to mimic the performance of dermatologists by detecting and extracting several dermoscopic structures, for example pigment network, irregular streaks, regression structures, granularities and blotches [6]. CAD systems adopt various pattern recognition approaches i.e. by the extraction of various features (color, texture and shape) from each dermoscopy image followed by the training of a classifier (Fig. 1).

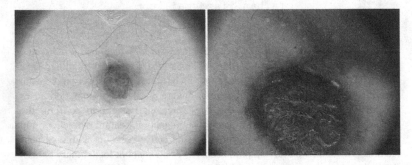

Fig. 1. Dermoscopy images: the left figure shows a normal skin lesion whereas the right one indicates a melanoma

In this article, we propose a technique for automated system for the detection of melanomas that gives better classification results than various state of the art methods. The automated system involves three different steps: (1) segmentation using level sets method, (2) then feature extraction using joint histogram of LBP pattern and contrast of the pattern (3) classification using support vector machine (SVM). Rest of the paper is organized as follows: Sect. 2 contains methodology that is adopted in the paper followed by dataset description (Sect. 3). Afterwards, we discuss our experimental results (Sect. 4) followed by conclusions (Sect. 5).

2 Methodology

A complete pattern recognition system as illustrated in Fig. 2 comprises three main parts; image segmentation, feature extraction and classification. For automatic image segmentation we have used level set creasness functions and manual annoatation is provided by the physicians. For feature extraction we have used joint histogram of multiresolution LBP and local contrast. Finally classification is done using Support Vector Machine.

2.1 Segmentation

Segmentation is an essential step in the automatic classification of skin lesions because accuracy of this block effects the successive steps. But, segmentation is a vital step because of the extensive variety of lesion shapes, colors, sizes and various skin tones. Additionally, the smooth transition of between the skin and the lesion can make correct segmentation of the lesion a challenging task to handle. To address these challenges, various segmentation techniques have been presented, which can be categorized as threshold-based, edge-based and region-based methods. When the contrast between lesion and the skin is high, thresholding based methods illustrate very good results. Edge based methods fail when the skin and the lesion are separated by smooth boundaries, which happens commonly in dermoscopic images. In this paper, we implement level sets curve evolution for the segmentation of dermoscopy images given its significance in the literature [7].

2.1.1 Automatic Segmentation Using Level Sets Curve Evolution Function

Active contours are dynamic fronts which move within the images towards object boundaries. In the level set formulation, dynamic fronts C are usually illustrated using zero level set $C(t) = \{(x, y)|\varnothing(t, x, y) = 0\}$ of an underlying function $\varnothing(t, x, y)$. The generic form of a level set function $\varnothing$ is shown in Eq. 1

$$\frac{\partial \varnothing}{\partial t} = F|\nabla \varnothing| \tag{1}$$

The function F is known as the speed function and $\nabla \varnothing$ is the curve evolution. Traditional level sets have a limitation that is the active contour can have sharp or/and flat shapes, shocks, etc., where the derivative of the curve may be undefined. This problem is resolved by initializing $\varnothing$ as a signed distance function and during evolution periodically reinitialize it as a signed distance. Effectively, the curve evolution can be achieved by:

$$\in (\varnothing) = \mu P(\varnothing) + \in_e (\varnothing) \tag{2}$$

$P(\varnothing)$ is defined as internal energy, which is a signed distance function and $\in_e (\varnothing)$ term is known as external energy function. This term usually depends on the image data. For the calculation of external energy terms in the level sets framework, we have used creasness features which are obtained using the multilocal level set extrinsic curvature with improvement by structure tensor features (MLSEC-ST) that was originally presented by Lopez et al. [8]. This operator has the ability to enhance ridges and valleys in an image. We can find the magnitude of creasness from the images using Eq. 3:

$$\mathcal{F}(m, n) = 1 - e^{-(\lambda_1(m,n) - \lambda_2(m,n))^2} \tag{3}$$

Where, $\lambda_1 > \lambda_2 > 0$ indicates the eigen values of the structure tensor function, while n are the rows and m are columns that indicate the pixel position in the gray scale image at which creasness is calculated. Similar magnitudes of the eigen values at a particular pixel indicate a flat region having no transitions in the gray level values whereas a high difference in magnitudes of the eigen values indicates the presence of ridges and valleys. Empirical findings indicate that creaseness features improve lesion boundaries while concealing the local image texture.

2.2   Feature Extraction

Segmentation is followed by next step that is feature extraction. Based on the clinical studies, we have devised a variant of local binary pattern for differential structure detection.

2.2.1   Standard Local Binary Pattern (LBP)

The differential structures in dermoscopy images can be analyzed using the local binary pattern (LBP) texture descriptor. It unifies the structural and statistical information of the texture using a histogram of the LBP codes. Originally proposed by Ojala et al. [9] LBP is gray scale and also invariant texture descriptor that creates LBP codes at every pixel in the image by thresholding its neighborhood pixel with central pixel value and then concatenate the results in the form of a pattern. The thresholding function can be obtained using Eq. 4:

$$\text{LBP}_{P,R} = \sum_{p=0}^{P-1} s(g_p - g_c) 2^P, \quad s(x) = \begin{cases} 1 & x \le 0 \\ 0 & x > 0 \end{cases} \tag{4}$$

Where g_c and g_p show gray level values of the central pixel and also its neighbor respectively, where p is the neighbor's index. P represents the total number of neighbors in a circular set surrounding a pixel at a radius R from g_c. The LBP pattern used for detection in this research is a uniform pattern that is basically an extension of the original operator. An LBP is uniform if its uniformity measure is at most 2. To quantify the uniformity of the LBP $U(\text{LBP}_{P,R})$ is a parameter defined in Eq. 5:

$$U(\text{LBP}_{P,R}) = \sum_{p=1}^{P-1} |sign(g_p - g_c) - s(g_{p-1} - g_c)| \tag{5}$$

The motivation for using uniform LBPs is their ability to detect the significant intrinsic characteristics of textures like spots, line edges, edges and corners.

2.2.2   Joint Histogram of Multiresolution LBP Patterns and Contrast

LBP can be regarded as a two-dimensional phenomenon characterized by two orthogonal properties: spatial structure (patterns) and contrast (the strength of the patterns). Contrast is a property of texture usually regarded as a very important cue for human vision, but the standard LBP operator totally ignores the contrast of gray level patterns. In a more general view, texture is distinguished not only by LBP patterns but also the strength of the patterns. These two measures supplement each other in a very useful way. An enhancement in the standard LBP texture can be obtained if the contrast of the LBP patterns is also taken into account. This is mainly because; image texture is described by the local contrast and underlying pattern at a particular spatial location. More specifically, the observers typically examine the images with different levels of visual attention (detail) depending on the richness of texture content (contrast). Thus, an enhancement in the description of texture content can be obtained if the strength of LBPs is also incorporated in the texture descriptor. To take the strength of LBPs into account, we have used various scales of LBPs for feature extraction. The multi scale LBPs can be obtained by varying the values of R to obtain the patterns at various radii from the center pixel. Let $r_1, r_2 \ldots, r_k$ be various radius at which the LBPs are calculated. At each pixel position, the absolute difference between the center pixel and its neighbors can be given as:

$$C^k(g_c) = \sum_{i=0}^{P-1} |g_p^k - g_c| \tag{6}$$

Where $C^k(g_c)$ represents the contrast of an underlying LBP patterns at the resolution r^k. The LBP having the highest strength consists of the most relevant pattern and can be obtained as follows:

$$ULBP(g_c) = arg_{C^k(g_c)} \max\{ULBP(P, r_k)\} \tag{7}$$

Therefore, the LBP at g_c is represented by the pattern, which exhibits the maximum strength i.e., $c^k(g_c)$ when analyzed at various resolutions at the pixel g_c. The LBP codes of an image along with the contrast of the patterns are used to create a joint histogram. Joint histogram of LBP patterns and contrast is created visiting each (x; y) location once. At each x, y location, increment histogram bin (b1, b2), where b1 is the 1D histogram bin for the value at x; y in LBP pattern and b2 is the 1D histogram bin for the value at (x; y) in contrast. Result of joint histogram is (b1 * b2) two dimensional array. This generated joint histogram is further used as a feature.

2.3   Classification

For classification, we have used a support vector machine (SVM) [10]. An SVM classifies the data by finding the hyper plane which maximizes the margin of separation between two distinct classes. Given a training set of samples $X_{1...N}$ having N different training samples and the coefficients $\alpha_{1...N}$ learned in the training stage, SVM decision function is as follows:

$$y(x) = \sum_i \alpha_i K(X_I, x) + b \tag{8}$$

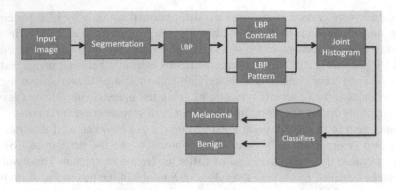

Fig. 2. Overview of the proposed system

$K(:)$ denotes the kernel function while x is the input vector. We have used the SVM linear kernel in our implementation. For classification we have used Weka (http://www.cs.waikato.ac.nz/ml/weka/) data mining tool in our experimental results.

3   Materials

The dataset that we have used is composed of 200 dermoscopy images with the following composition: 80 % (160) nevus and 20 % (40) melanoma. The images were acquired at the Hospital Pedro Hispano, Matosinhos [11]. All images have been acquired through clinical exams using a dermoscope at magnification of 20× with a resolution of 765 × 573 pixels. Each image was manually segmented (to identify the lesions) and classified by an experienced dermatologist as being normal, atypical nevus (benign) or melanoma (malignant).

4   Experimental Results

For our classification experiments, we have compared the performance obtained by the manual annotations with the lesion obtained after segmentation. The objective of the classification task is to identify the presence or absence of melanoma in the images. For feature extraction, we have used joint histogram of LBP patterns with the local contrast of the pattern. LBP with 8 neighbors at multiple scales having radii of 1, 1.5 and 2.0 are used in this research. Feature extraction is followed by classification using SVMs with linear kernel and 10-fold cross validation. We have imbalanced data so we used average overall accuracy (Acc), sensitivity (SE) and specificity (SP) as appropriate performance measures to assess the performance of the classifier. We attribute the superior performance of joint histogram of LBP patterns with the contrast of the pattern in order to capture the micro-structures in the images which are effectively representative of the differential structures in the images. Local contrast in multi scales of LBP approximates the human observers (clinicians) who try to visualize the image texture at various levels of attention. For performing this comparison, we select state of the art techniques i.e. LBP and Homogeneous Texture (HT) along with our proposed method i.e. joint histogram of multiresolution LBP and local contrast (JHLBP_C). The test bed used for all these methods is the same as that used for our proposed method. Our experiments demonstrate that the homogenous texture shows worst performance amongst state-of-the-art methods. Standard LBP shows relatively low classification rates. Our novel descriptor shows better performance as compared to the other descriptors considered in this paper (Table 1).

Table 1. Comparison of SVM classification results obtained using proposed features and state-of-the-art feature extraction methods.

Methods	Manual segmentation			Automatic segmentation		
	Acc %	SE %	SP %	Acc %	SE %	SP %
HT	81.01	64.1	86.2	79	79.1	63.0
LBP	86	88.3	71.4	82.5	86.8	64.7
JHLBP_C	90.5	91.2	86.2	89.5	91.6	78.7

5   Discussion and Conclusion

In this paper, an automatic system for the detection of melanomas in dermoscopy images using LBP patterns and contrast has been proposed. The full pipeline of a pattern recognition system including segmentation, feature extraction and classification has been used. For the segmentation of dermoscopic images we have used active contours with creasness function followed by feature extraction using LBP patterns and contrast at multi scale to create a joint histogram. This selection is based on the clinical significance of differential structures in the detection of melanomas. The performance of the proposed automated system is tested on manual and automatic segmented images using overall accuracy, sensitivity and specificity. The best performance of the system, SE = 91.2 and SP = 86.2 was achieved by manual annotated masks. While the performance of the system SE = 91.6 and SP = 78.7 was achieved by automatic segmented masks. In view of the classification results, minor degradation was observed when automatic segmentation methods are used, instead of manual ones. According to these results, we conclude that joint histogram of LBP pattern and contrast features have an important role in the detection of melanomas. In addition, our results show the superiority of the proposed descriptor as compared to other state-of-the-art feature extraction methods.

Although good results are obtained, there is still scope for improvement in each step of the pattern recognition pipeline for melanoma detection. In dermoscopy images, there are some situations where there is a smooth transition between the skin and the lesion. Active contours fail to detect true boundaries of the lesions in such cases. The need to devise more adequate image features for use by active contours may improve the segmentation results. Since the dermoscopy images suffer from reduced color spaces, therefore more adequate color descriptors can be constructed by adapting the color spaces to the specific scenario of dermoscopy potentially giving better results for the identification of melanoma.

Acknowledgements. This paper was financially supported by National University of Sciences and Technology (NUST) Islamabad, Pakistan. We would like to thank NUST for providing us vital financial assistance for carrying out this research.

References

1. Alcón, J.F., Ciuhu, C., Kate, W.T., Heinrich, A., Uzunbajakava, N., Krekels, G., Siem, D., Haan, G.D.: Automatic imaging system with decision support for inspection of pigmented skin lesions and melanoma. IEEE J. Sel. Top. Sig. Process. **3**, 14–25 (2009)
2. Maglogiannis, I., Doukas, C.N.: Overview of advanced compute revision systems for skin lesions characterization. IEEE Trans. Inf. Technol. Biomed. **13**, 721–733 (2009)
3. Stolz, A.R., Cognetta, A.B.: Abcd rule of dermatoscopy:a new practical method for early recognition of malignant melanoma. Eur. J. Dermatol. **4**, 521–527 (1994)
4. Menzies, M.S., Ingvar, C.: Frequency and morphologic characteristics of invasive melanomas lacking specific surface microscopic. Arch. Dermatol. **132**, 1178–1182 (1996)
5. Delfino, M., Argenziano, G., Fabbrocini, G., Carli, P., Giorgi, V.D., Sammarco, E.: Epiluminescence microscopy for the diagnosis of doubtful melanocytic skin lesions: comparison of the ABCD rule of dermatoscopy and a new 7-point checklist based on pattern analysis. Arch. Dermatol. **134**, 1563–1570 (1998)
6. Barata, C., Marques, J., Rozeira, J.: Detecting the pigment network in dermoscopy images: a directional approach. In: Proceedings of 33rd IEEE EMBS Annual International Conference, pp. 5120–5123 (2011)
7. Silveira, M., Nascimento, J.C., Marques, J.S., Marcal, A.R.S., Mendonca, T., Yamauchi, S., Maeda, J., Rozeira, J.: Comparison of segmentation methods for melanoma diagnosis in dermoscopy images. IEEE J. Sel. Top. Sig. Process. **3**, 35–45 (2009)
8. Lopez, J.S., Lumbreras, F., Villanueva, J.J.: Evaluation of methods for ridge and valley detection. IEEE Trans. Pattern Anal. Mach. Intell. **21**, 327–335 (1999)
9. Ojala, T., Pietikainen, M., Maenpaa, T.: Multiresolution gray-scale and rotation invariant texture classification with local binary patterns. IEEE Trans. Pattern Anal. Mach. Intell. **24**, 971–987 (2002)
10. Guyon, I., Weston, J., Vapnik, S.B.V.: Gene selection for cancer classification using support vector machines. Mach. Learn. **46**, 389–422 (2002)
11. Mendoncÿa, T., Ferreira, P.M., Marques, J., Marcÿal, A.R.S., Rozeira, J.: A dermoscopic image database for research and benchmarking. Accepted for Presentation in Proceedings of PH^2 IEEE EMBC (2013)

Modeling the Behavior of Unskilled Users in a Multi-UAV Simulation Environment

Víctor Rodríguez-Fernández[1](✉), Antonio Gonzalez-Pardo[2,3],
and David Camacho[1]

[1] Universidad Autónoma de Madrid (UAM), 28049 Madrid, Spain
victor.rodriguez@inv.uam.es, david.camacho@uam.es
[2] Basque Center for Applied Mathematics (BCAM), Bilbao, Spain
[3] TECNALIA, OPTIMA Unit, 48160 Derio, Spain
agonzalezp@bcamath.org

Abstract. The use of Unmanned Aerial Vehicles (UAVs) has been growing over the last few years. The accelerated evolution of these systems
is generating a high demand of qualified operators, which requires to
redesign the training process and focus on a wider range of candidates,
including inexperienced users in the field, in order to detect skilled-
potential operators. This paper uses data from the interactions of multiple unskilled users in a simple multi-UAV simulator to create a behavioral
model through the use of Hidden Markov Models (HMMs). An optimal
HMM is validated and analyzed to extract common behavioral patterns
among these users, so that it is proven that the model represents correctly the novelty of the users and may be used to detect and predict
behaviors in multi-UAV systems.

Keywords: Unmanned Aerial Vehicles (UAVs) · Human-Robot
Interaction (HRI) · Hidden Markov Model (HMM) · Behavioral model

1 Introduction

The study of Unmanned Aerial Vehicles (UAVs) is currently a growing area in the
research community. These new technologies offer many potential applications in
multiple fields such as infrastructure inspection, monitoring coastal zones, traffic
and disaster management, agriculture and forestry among others [9].

The work of UAV operators is extremely critical due to the high costs involving any UAV mission, both financial and human. Thus, lot of research in the field
of human factors, and more specifically, in Human Supervisory Control (HSC)
and Human-Robot Interaction (HRI) systems, have been carried out, in order
to understand and improve the performance of these operators [7].

In recent years, two topics are emerging in relation to the study of Unmanned
Aircraft System (UAS). The first is the effort to design systems where the current many-to-one ratio of operators to vehicles can be inverted, so that a single

V. Rodríguez-Fernández and D. Camacho—AIDA Group: http://aida.ii.uam.es.

K. Jackowski et al. (Eds.): IDEAL 2015, LNCS 9375, pp. 441–448, 2015.
DOI: 10.1007/978-3-319-24834-9_51

operator can control multiple UAVs. The other one is related to the fact that accelerated UAS evolution has now outpaced current operator training regimens, leading to a shortage of qualified UAS pilots. Due to this, it is necessary to re-design the current intensive training process to meet that demand, making the UAV operations available for a less limited pool of individuals, which may include, for example, high-skilled video-game players [8].

This work is focused on modeling and analyzing the behavior of users, unexperienced in the field of UAS-based simulations, when monitoring and interacting with a multi-UAV mission running in a simple and gamified simulation environment. This model will help us to extract common behavioral patterns which identify low-level operators, as well as to detect, based on the weaknesses of these users, elements that should be added in a simulation environment so that it is suitable for novice UAS training.

While many techniques have been used to model several aspects of human behavior in HRI systems [6], Hidden Markov Models (HMMs) have proven to obtain small and comprehensive behavioral models, as well as to achieve good results discovering unobservable states in many fields including speech recognition [10], psychology [11], and, what is more relevant for this work, aviation and pilot performance analysis [5].

The rest of the paper is structured as follows: Sect. 2 gives a theoretical overview of HMMs. Then, in Sect. 3 we detail how to process the data from the simulations to create and validate HMMs. Section 4 shows the best model resulted from this process and analyze it. Finally, we present several conclusions and future work lines.

2   Basics on Hidden Markov Models (HMMs)

HMMs are stochastic models mainly used for the modeling and prediction of sequences of symbols. They are characterized by a set of N discrete (*hidden*) states $S = \{S_1, \ldots, S_N\}$, which can be interpreted as phases in a cognitive process which each produce typical behaviors. The term *Markov* in a HMM pertains to the time-dependence between the consecutive states S_t, which follows a Markov process. This means that the current state S_t only depends on the previous state S_{t-1} and not on earlier states, i.e.: $P(S_t|S_1 \ldots, S_{t-1}) = P(S_t|S_{t-1})$. The transition probabilities between states are denoted by the matrix A, with entries: $a_{ij}(t) := P(S_{t+1} = j|S_t = i), \quad i, j = 1, \ldots, N$. Another set of parameters governing the probability distribution over the states of the HMM is the set of initial state probabilities, π, defined as: $\pi_i := P(S_1 = i), i = 1, \ldots, N$.

On the other hand, the term *hidden* in a HMM indicates that the underlying states S_t cannot be observed directly. This means that, let $V = \{V_1, \ldots, V_M\}$ be the set of all the M possible *observation symbols* in our data domain (also called *dictionary*), the distribution function $f(V_t|S_t = i)$ is not a deterministic function but rather a probability density function, also known as *emission function*. $B = \{b_1(c), \ldots, b_N(c)\}, c \in V$ represents the emission functions of the HMM for each of the N states. In sum, a HMM is defined as the tuple:

$$H := \{S, V, A, B, \pi\} \tag{1}$$

Three main computational issues need to be addressed with HMMs: The first one relates to how to compute the probability (likelihood) that a given observation sequence O^s is produced by a model H. This value is useful because it allows to compare and select which model fits better to a given dataset. This (log-) likelihood is usually computed by the so-called *forward-backward* algorithm [10]. The second issue consists in determining, given a sequence of observation symbols O^s, which corresponding sequence of hidden states is most likely to produce it. This problem is addressed by the use of the popular *Viterbi* algorithm [4].

Finally, the last problem is how to learn the model, which refers to how can we calculate a set of model parameters (See Eq. 1) that maximizes the likelihood to produce a given set of observation sequences. Although there are supervised learning techniques for HMMs, in the domain of multi-UAV systems there is still a high shortage of experts able to label data from this type of missions objectively, in order to develop a supervised analysis. Thus, we can only work in this field by using unsupervised learning techniques [2]. The most commonly used unsupervised algorithm for estimating HMMs parameters is a form of expectation maximization (EM) called the *Baum Welch algorithm* [1].

One important aspect to consider when fitting a HMM to a given dataset is that the number of hidden states, N, must be known in advanced, which is often unrealistic. To choose an optimal number of states without prior knowledge, several statistic metrics are used to compare and select models, of which Bayesian Information Criterion (BIC) is the best known [3]. These metrics penalize the likelihood of a model by a complexity factor proportional to number of parameters in the model and the number of training observations.

3   Modeling User Behavior in a Multi-UAV Simulation Environment Through HMMs

The experimentation carried out in this paper consist in applying HMMs to model how a set of unskilled users behave when interacting with a multi-UAV simulation environment.

The simulator used as the basis for this work has been designed following the criteria of availability and usability. It is known as Drone Watch and Rescue (DWR), and its complete description can be found in [12]. DWR might be considered a serious game, since it gamifies the concept of a multi-UAV mission, challenging the player to capture all mission targets while avoiding at the same time the possible incidents that may occur during a mission. To avoid these incidents, a user in DWR can perform multiple interactions to alter both the status of the UAVs in the mission and the waypoints composing their mission plan. Below are described the set of interactions that can be performed in DWR, along with its identifier (O_i), used for the analysis carried out in this paper:

- *Select UAV (O_0)*: Allows a use r to monitor and control a specific UAV.
- *Set UAV Speed (O_1)*: Change the speed of a selected UAV.

- *Set simulation speed (O_2)*: Users can "accelerate" time during a simulation.
- *Change UAV Path (O_3)*: Add/change/remove waypoints of any UAV.
- *Select waypoint table (O_4)*: Waypoints can be edited in a waypoints table.
- *Set control mode (O_5)*: Control modes in DWR manage how a user can change a UAV path (*Monitor mode, Add waypoints, Manual mode*).

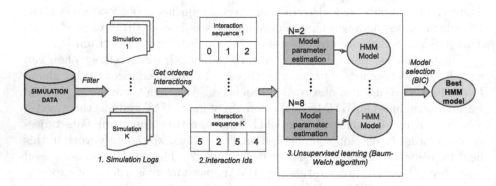

Fig. 1. Data processing steps to fit the simulation data with a HMM

Figure 1 shows the steps taken in this experiment to create and validate a model from the interactions of users in the simulator DWR. First, raw simulation data is filtered as explained above to obtain a number of K useful simulation logs, each of them representing one execution of the simulator. Since we want to model the behavior of users through the interactions with the system, we extract, from each simulation log, the sequence of interactions performed chronologically by the users. Each interaction in a sequence is identified by a integer, from 0 to $M - 1$, where M is the number of possible interactions with the simulator, in this case $M = 6$ (See list of interactions above). The set $V = \{0, 1, \ldots, M - 1\}$ is the *dictionary* of the HMMs we will build.

The next step is to use the K interaction sequences to build some HMMs using the *Baum-Welch* algorithm, so that the likelihood for the model to produce our dataset is maximized. Since we do not have knowledge of the number of cognitive states N governing the behavior of the users, we train the models with different values of N, ranging from 2 to 8. Finally, we must select which of the HMMs created fits better to our data, analyzing the values of log-likelihood and BIC for each model. Once we have selected the best HMM, we can analyze it and extract behavioral information about the users. To build and fit the HMMs, the *depmixS4* R package has been used [13].

4   Experimentation and Model Analysis

The dataset used in this experiment is composed of **80 distinct simulations**, played in the simulator DWR by a total of **25 users**, all of them inexperienced

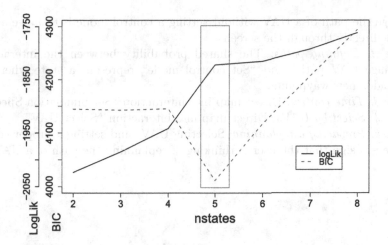

Fig. 2. HMM Model selection using log-likelihood and BIC values

Table 1. Emission Probabilities for the chosen 5-state HMM.

	O_0	O_1	O_2	O_3	O_4	O_5
State 1	0.000	0.002	0.000	**0.950**	0.047	0.000
State 2	0.008	0.004	0.028	**0.687**	0.000	**0.273**
State 3	0.000	0.075	**0.925**	0.000	0.000	0.000
State 4	**0.900**	0.017	0.017	0.065	0.000	0.000
State 5	**0.260**	0.029	0.000	0.000	0.000	**0.710**

in HSC systems. Taking into account the level of difficulty of the test missions, we have considered as useless those simulations with less than 10 interactions, obtaining a total of **48 valid simulations** ($K = 48$).

The results for the model comparison between different number of states are shown in Fig. 2. As can be seen, models with larger number states obtain better values of log-likelihood (the higher the better), although the increase is significantly lower from $N = 5$. Besides, the BIC value achieves a global minimum in the 5-state model, which indicates a good and feasible fit for the data (the lower the better). For these reasons, we conclude that **the 5-state HMM is the best model for our dataset**.

Once we have chosen an optimal number of states, we must give sense to them in the context of the experiment. This process is called *state-labeling* and it is based on analyzing the emission functions of each states, i.e., how the probability of producing an interaction is distributed among each state. Based on the plots shown in Fig. 3 and Table 1, we can label the states of the 5-state HMM as follows:

– *State 1: Moving waypoints*: The emission function is almost completely focused on the interaction "Change UAV path". In the simulator, the only way to change the path of a UAV without setting a control mode is to move a way-point directly through the screen.

- *State 2: Defining paths*: The shared probability between the interactions "Change UAV path" and "Set control mode" represent a state where the user adds new waypoints.
- *State 3: Time control*: Direct map into interaction "Set Simulation Speed".
- *State 4: Selecting UAVs*: Direct map into interaction "Select UAV".
- *State 5: Preparing a replanning*: Selecting UAVs and setting a proper control mode is a sign that the user is thinking of replanning the path of a UAV.

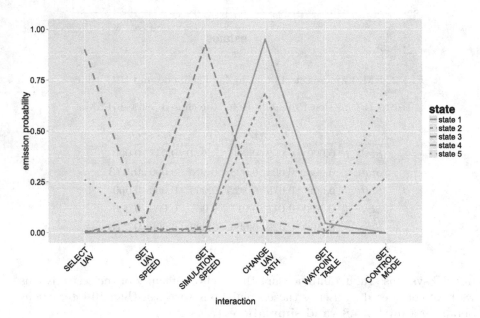

Fig. 3. Emission functions for the chosen 5-state HMM model

To complete the model analysis, the Markov chain of the 5-state HMM must be analyzed to extract information about the underlying process. Figure 4 shows this chain, labeling each state with the results of the state-labeling made above. Below are discussed the main behavioral patterns derived from this model:

- The user is high likely to start a simulation by selecting UAVs. Then, we find a high-probability cycle between the states "Selecting UAVs" and "Moving way-points", which fits with the behavior of a novice user, since these interactions are direct, simple and do not suppose a big change to the simulation.
- The low self-transition probability of the state "Preparing a replanning" is a sign that the time spent by these users thinking of how to act is short, i.e., they usually feature an incautious replanning behavior.

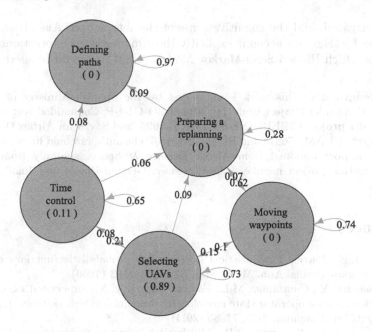

Fig. 4. Markov chain for the chosen 5-state HMM model. Transition probabilities below 0.05 are omitted for legibility purposes.

- The presence of a value of 0.11 as initial probability for the state "Time Control" indicates that some users start a simulation by accelerating the time speed, without examining the mission scenario first. This clearly represents an impatient behavior appropriate of unaware users.
- The state "Defining paths" has no likely chance of going outside it. Such a state is called an *absorbing state*. This is a sign that these users, when defining new and complex paths, fall into long and unsuccessful replanning which leads to aborting and finishing the simulation.

5   Conclusions and Future Work

This paper has presented a way for modeling the behavior of unskilled users interacting with a multi-UAV simulator, through the use of Hidden Markov Models. Several HMMs, with different number of states, have been created using unsupervised learning strategies, and the best of them has been validated and chosen for analysis. Using that model, we have deducted the sense of the cognitive states involving the process, and extracted some behavioral patterns which characterize well the novelty of the users.

As future work, we intend to expand the possible dictionaries feeding the HMMs, creating complex grammars of observation symbols that not only identify an interaction with a number, but with a list of parameters depending on the UAV treated, and the cognitive sense of the interaction. Also, it would be interesting to take into account explicitly the time between interactions in the model, for which Hidden Semi-Markov Models (HSMMs) could be used.

Acknowledgments. This work is supported by the Spanish Ministry of Science and Education under Project Code TIN2014-56494-C4-4-P, Comunidad Autonoma de Madrid under project CIBERDINE S2013/ICE-3095, and Savier an Airbus Defense & Space project (FUAM-076914 and FUAM-076915). The authors would like to acknowledge the support obtained from Airbus Defence & Space, specially from Savier Open Innovation project members: José Insenser, Gemma Blasco and Juan Antonio Henríquez.

References

1. Baum, L.E., Petrie, T.: Statistical inference for probabilistic functions of finite state Markov chains. Ann. Math. Stat. **37**, 1554–1563 (1966)
2. Boussemart, Y., Cummings, M.L., Fargeas, J.L., Roy, N.: Supervised vs. unsupervised learning for operator state modeling in unmanned vehicle settings. J. Aerosp. Comput. Inf. Commun. **8**(3), 71–85 (2011)
3. Burnham, K.P., Anderson, D.R.: Multimodel inference understanding AIC and BIC in model selection. Sociol. Methods Res. **33**(2), 261–304 (2004)
4. Forney Jr, G.D.: The viterbi algorithm. Proc. IEEE **61**(3), 268–278 (1973)
5. Hayashi, M.: Hidden Markov models to identify pilot instrument scanning and attention patterns. In: 2003 IEEE International Conference on Systems, Man and Cybernetics, vol. 3, pp. 2889–2896. IEEE (2003)
6. Leiden, K., Laughery, K.R., Keller, J., French, J., Warwick, W., Wood, S.D.: A review of human performance models for the prediction of human error. Ann Arbor **1001**, 48105 (2001)
7. McCarley, J.S., Wickens, C.D.: Human factors concerns in UAV flight. University of Illinois at Urbana-Champaign Institute of Aviation, Aviation Human Factors Division (2004)
8. McKinley, R.A., McIntire, L.K., Funke, M.A.: Operator selection for unmanned aerial systems: comparing video game players and pilots. Aviat. Space Env. Med. **82**(6), 635–642 (2011)
9. Pereira, E., Bencatel, R., Correia, J., Félix, L., Gonçalves, G., Morgado, J., Sousa, J.: Unmanned air vehicles for coastal and environmental research. J. Coast. Res. **2009**(56), 1557–1561 (2009)
10. Rabiner, L.: A tutorial on hidden Markov models and selected applications in speech recognition. Proc. IEEE **77**(2), 257–286 (1989)
11. Schmittmann, V.D., Visser, I., Raijmakers, M.E.: Multiple learning modes in the development of performance on a rule-based category-learning task. Neuropsychologia **44**(11), 2079–2091 (2006)
12. Rodríguez-Fernández, V., Héctor, D., Menéndez, D.C.: Design and development of a lightweight multi-uav simulator. In: 2015 IEEE International Conference on Cybernetics (CYBCONF). IEEE (Paper accepted, 2015)
13. Visser, I., Speekenbrink, M.: depmixs4: an R-package for hidden Markov models. J. Stat. Softw. **36**(7), 1–21 (2010)

Multistep Forecast of FX Rates Using an Extended Self-organizing Regressive Neural Network

Yicun Ouyang$^{(\boxtimes)}$ and Hujun Yin

School of Electrical and Electronic Engineering, The University of Manchester,
Manchester M13 9PL, UK
yicun.ouyang@postgrad.manchester.ac.uk, hujun.yin@manchester.ac.uk

Abstract. In this paper, an extended self-organizing regressive neural network is proposed for multistep forecasting of time series. The main features of this method are building input segments with various lengths for training and the network capable of learning multiple regressive models for forecasting various horizons. The inter dependencies among future points are preserved and this results in all forecasting tasks naturally. Experiments on foreign exchange rates show that the new method significantly improves the performance in multistep forecasting compared to the existing methods.

Keywords: Self-organizing neural network · Time series · Mixture autoregressive models · Multistep forecasting

1 Introduction

Foreign exchange (FX) market is a global market for currency trading and it is by far the largest market in the world with volume surpassing that of commodities, financial futures and stocks. According to a survey of Bank for International Settlements (BIS) in 2013 [1], trading in FX markets averaged $5.3 trillion per day. The participants of this market include international banks (major participants % of overall volume according to Euromoney FX survey 2014 [2]: Citi 16.04 % Deutsche Bank 15.67 % and Barclays Investment Bank 10.91 %), hedge funds, commercial companies as well as individual traders.

In FX trading, there is a conception called leverage, which makes one be able to control a much larger amount of currency than that he/she has on deposit. That means the profits or losses in the FX trading account are magnified through leverage. As a high-risk, high-returned market, there seems to be so much money that can be made in FX trading. However, the volatility in currency markets is so high that it may cause an extremely high loss within a short period. To analyze and control the risks of trading, FX rates modelling and forecasting play an important role.

Many statistical models and methods exist for linear autoregressive analysis [3] such as autoregressive (AR), autoregressive moving average (ARMA), autoregressive integrated moving average (ARIMA), as well as heteroskedastic models

© Springer International Publishing Switzerland 2015
K. Jackowski et al. (Eds.): IDEAL 2015, LNCS 9375, pp. 449–456, 2015.
DOI: 10.1007/978-3-319-24834-9_52

autoregressive conditional heteroskedastic (ARCH) and generalized autoregressive conditional heteroskedastic (GARCH). Although these methods have been dominating force in time series analysis, they are mainly for one-step prediction. Multistep or long term forecasting is much more difficult than one-step prediction, but accurate or reliable multistep models are highly desirable in many applications. Few models or methods have been devised for multistep forecast due to the difficulties.

There are mainly two categories of the existing approaches to multistep forecasting: iterative and independent [4]. In an iterative method, the previous one-step forecasting result is used as the input for the next-step by the same model; while individual models are built for different horizons in an independent method. Errors are accumulated in the former while the latter generally has high computational complexity. There are also some other methods for multistep forecasting, including the multiple support vector regression (MSVR) model [5] combining the iterative and independent methods, the multiple output approach [6] preserving the stochastic dependencies, and the Kohonen-maps-based double quantization method modelling both the original space and a deformation space [7].

In addition to statistical models, recently more forecasting methods are based on neural networks [8], including the widely used multilayer perceptron (MLP) [9], recurrent neural network (RNN) [10] and support vector machine (SVM) [11]. Besides, the neural networks such as self-organizing map (SOM) [7], self-organizing mixture autoregressive (SOMAR) [12] and the neural gas mixture autoregressive (NGMAR) [13] have also attracted much interest in the last decades [14]. Both SOMAR and NGMAR are based on mixture AR models and employ a new similarity measure, termed the sum of autocorrelation (SAC), for selecting the best matching unit (BMU). Their better performance for one-step prediction was demonstrated. This work considers extending them for long term prediction.

The extended networks are based on heterogeneous mixture AR models and nodes of various orders of AR models are used for multistep forecasting. The inputs are segmented with various lengths for forecast horizons and the reference vectors update weight components according to the length of input segments. The experimental results show that the proposed neural network outperforms other competitors.

The remaining of the paper is organized as follows. In Sect. 2, methodology of the extended self-organizing neural network is described in detail. Section 3 shows the experimental results and conclusions are drawn in Sect. 4.

2 Methodology

In the basic SOM-based approaches, consecutive time points are grouped into segments of fixed length and then used as the input vectors for training SOM, which is termed as vector SOM (VSOM) [12]. In an early attempt to combine SOM and AR models, the self-organizing autoregressive (SOAR) [15] model was proposed with each neuron representing an AR model and its parameters as the reference vector. However, SOAR has difficulty to converge to the correct AR models. To solve this problem, SOMAR was introduced to learn a mixture of AR processes by a self-organizing procedure. Based on the principle that a stochastic process is characterized by white noise residuals, SOMAR employs the autocorrelation of errors to measure the similarity between input and reference vectors. It was demonstrated that SOMAR obtained significantly better performance in modelling and predicting on a variety of time series than regressive and other SOM-based models.

2.1 NGMAR

To free the arrangement of neurons in the SOMAR, the NGMAR [13] model was proposed by replacing SOM with neural gas (NG). Given the current input x_t and the previous segment $\mathbf{x}(t-1)$, the modelling error of each node is defined as,

$$e_i(t) = x_t - \mathbf{w}_i^T \mathbf{x}(t-1) \tag{1}$$

Similar to SOMAR, the autocorrelation coefficient of the error sequences is also used in NGMAR and the BMU is chosen with the minimum sum of autocorrelation of coefficients (SAC) as,

$$i^* = arg\,min_i(\textstyle\sum_{v=-s}^{s} \|R_i(v)\|), i = 0, 1, ..., M - 1 \tag{2}$$

$$R_i(v) = \tfrac{1}{s\sigma^2} \textstyle\sum_{j=0}^{s-v-1} (e_i(t-j) - \mu)(e_i(t+v-j) - \mu) \tag{3}$$

where M is the total number of neurons, v is the lag, s is the batch size, μ and σ^2 are the mean and variance of the error sequences respectively. The reference vectors are then updated by,

$$\Delta\mathbf{w}_{i_k} = \gamma(t)exp(-k/\lambda(t))e_{i_k}(t)\mathbf{x}(t-1) \tag{4}$$

$$\gamma(t) = \gamma_0(\gamma_T/\gamma_0)^{t/T} \tag{5}$$

$$\lambda(t) = \lambda_0(\lambda_T/\lambda_0)^{t/T} \tag{6}$$

where k is the ranking of neuron i_k, the learning rate $\gamma(t)$ and the neighbourhood range $\lambda(t)$ decay to zero with time, and T is the total number of iterations. It has been demonstrated that NGMAR further improves the performance of time series predition [13].

2.2 Segmenting Input Vectors

For building multistep forecasting models, usually there are two ways to build input vectors, corresponding to two multistep forecasting methods mentioned before. In the way used in the iterative method, each current time point value and its past $l-1$ values are grouped into a segment for training the same model. However, in the independent method, individual models are trained for different horizons and the current value and its some-steps-ago values are segmented as input vectors. That means the current value and the past values are not adjacent in time.

In the proposed method, a new way of building input vectors is adopted by segmenting the consecutive future points with different lengths as

$$\mathbf{x}^{(h)}(t) = [x_{t-l+1}, \cdots, x_t, \cdots, x_{t+h-2}, x_{t+h-1}] \tag{7}$$
$$(h = 1, \cdots, H)$$

where H is the maximum horizon to be predicted. When $H = 1$, the method becomes the one step method SOMAR or NGMAR.

2.3   Extended Self-organizing Regressive Neural Networks

Since the input vectors are built with different lengths $l, l+1, \cdots, l+H-1$, the reference vectors are of the maximum length, $(l + H - 1)$,

$$\mathbf{w}_i = [w_{i,0}, \cdots, w_{i,l-2}, w_{i,l-1}, \cdots, w_{i,l+H-2}] \tag{8}$$

Different components of the reference vectors are updated by the corresponding elements of the input vectors. For the input vector $\mathbf{x}^{(1)}(t)$, the components $[w_{i,0}, \cdots, w_{i,l-1}]$ are updated; For $\mathbf{x}^{(2)}(t)$, $[w_{i,0}, \cdots, w_{i,l}]$ are updated; and likewise, for the input vector $\mathbf{x}^{(H)}(t)$, all components $[w_{i,0}, \cdots, w_{i,l+H-2}]$ are updated. The updating rule is similar to that of the NGMAR,

$$\Delta\mathbf{w}_{i_k} = \gamma(t)exp(-k/\lambda(t))e_{i_k}(t)\mathbf{x}^{(h)}(t) \tag{9}$$

This new proposed method is termed as extended NGMAR, and the pseudocode code is described in Algorithm 1. If it is applied with SOMAR when topological grid structure is required, it results in extended SOMAR.

Algorithm 1. The pseudocode of extended NGMAR

1: Initialise the weight matrix $\mathbf{w}_i$, and specify training iterations N;
2: **for** $t = 1$ to N **do**
3:    Build individual input vectors $\mathbf{x}^{(h)}(t)$ for each forecasting horizon h;
4: **end for**
5: **for** $t = 1$ to N **do**
6:    Calculate SAC value of each batch of input vectors and all reference vectors;
7:    Build RANK based on SAC values;
8:    Update the weights $\mathbf{w}_i$ of neuron i;
9: **end for**
10: **for** $h = 1$ to H **do**
11:    Predict $\hat{x}_{t+h}$ by using BMU i^* and its corresponding components of $\mathbf{w}_{i*}$;
12: **end for**

---

3   Experimental Results and Comparison

In order to evaluate its prediction performance, the proposed method has been applied to three foreign exchange (FX) rates downloaded from the PACIFIC exchange rate services.

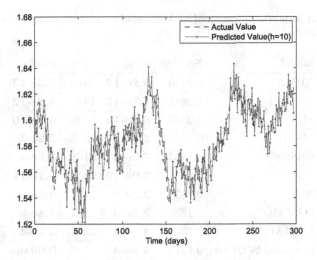

Fig. 1. Actual and predicted values (of 10 step ahead) by the extended NGMAR on GBP/USD.

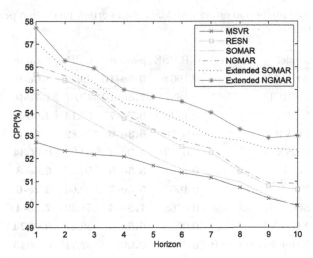

Fig. 2. Performance comparisons (CPP) of all the horizons ($h = 1, 2, \cdots , 10$) by various models on GBP/USD.

3.1 Foreign Exchange Rates

In the experiments, the daily closing prices of GBP/USD, GBP/EUR and GBP/JPY over 12 years (2002–2013) were used. Each FX rate series was divided into a training set (first 90 % points) and a test set (last 10 % points), and 10 % of the training data was used as the validation set to terminate the training and to decide model orders.

Table 1. Averaged prediction errors (NRMSE) and correct prediction percentages (%) on GBP/USD ($h = 1, 2, \cdots, 10$).

Model	NRMSE	p-value	CPP	p-value
ARMA	0.2955	**5.8e-12**	49.68	**7.3e-12**
GARCH	0.2837	**7.7e-12**	49.92	**5.6e-12**
MLP	0.2367	**6.3e-10**	50.49	**4.1e-10**
MSVR	0.1672	**3.9e-7**	51.43	**3.2e-7**
RESN	0.1353	**7.2e-4**	53.10	**3.8e-4**
SOM	0.2396	**7.0e-10**	50.22	**5.1e-11**
NG	0.1835	**2.1e-8**	50.63	**2.9e-9**
SOMAR	0.1533	**3.3e-6**	52.18	**7.8e-5**
NGMAR	0.1364	**4.1e-4**	53.25	**3.1e-4**
Extended SOMAR	0.1375	**4.2e-4**	54.15	**0.0086**
Extended NGMAR	**0.1159**	N/A	**54.68**	N/A

Table 2. Averaged prediction errors (NRMSE) and correct prediction percentages (%) on GBP/EUR ($h = 1, 2, \cdots, 10$).

Model	NRMSE	p-value	CPP	p-value
ARMA	0.1964	**9.6e-10**	49.29	**3.0e-9**
GARCH	0.1882	**7.5e-10**	49.67	**1.7e-9**
MLP	0.1343	**5.7e-7**	50.14	**4.9e-8**
MSVR	0.0878	**6.3e-6**	50.82	**5.5e-6**
RESN	0.0749	**4.9e-4**	52.22	**0.0034**
SOM	0.1402	**3.5e-8**	50.17	**5.5e-8**
NG	0.1085	**5.5e-7**	50.66	**3.2e-6**
SOMAR	0.0768	**1.3e-4**	51.39	**7.5e-4**
NGMAR	0.0740	**5.8e-4**	52.37	**0.0024**
Extended SOMAR	0.0678	**0.0038**	52.71	**0.0043**
Extended NGMAR	**0.0573**	N/A	**53.26**	N/A

Figure 1 shows the GBP/USD rates of the test set and the estimated values by extended NGMAR over ten steps. A chart to show the performance comparisons of all the horizons is given in Fig. 2.

There are several ways to measure prediction performance [16], of which two were applied to the experiments: the normalized root mean square error (NRMSE) and the correct prediction percentage (CPP)

$$NRMSE = \sqrt{\frac{(1/N_{test}) \sum_{t=1}^{N_{test}} (x_t - \hat{x}_t)^2}{(1/N_{test}) \sum_{t=1}^{N_{test}} (x_t - \bar{x})^2}} \tag{10}$$

$$CPP = \frac{Number\ of\ Correct\ Direction\ Predictions}{Total\ Number\ of\ Predictions} \tag{11}$$

where N_{test} is the number of test examples, $\hat{x}_t$ and x_t are the predicted and actual values respectively, and $\bar{x}$ denotes the sample mean.

The extended SOMAR and extended NGMAR have been compared with the ARMA, GARCH, MLP, MSVR, RESN [17], SOM, NG, SOMAR and NGMAR for predicting these FX rates. Tables 1, 2 and 3 show the averaged performances (NRMSE and CPP) over $h = 1, 2, \cdots, 10$ horizons and their t-test values. The proposed methods, extended SOMAR and extended NGMAR, outperform other neural networks as shown in the tables in terms of both NRMSE and CPP. The significant tests (p-value) show that the improvements by the proposed methods over other methods are significant, about 1.7–9.0 %. SOMAR and NGMAR are the closest models to the proposed methods, but the difference between them are still 1.7–2.8 % in CPP. The better performance obtained by extended NGMAR than extended SOMAR can be explained by adopting the dynamic neuron structure and neighbourhood ranking, which makes weights updated more accurately.

Table 3. Averaged prediction errors (NRMSE) and correct prediction percentages (%) on GBP/JPY ($h = 1, 2, \cdots, 10$).

Model	NRMSE	p-value	CPP	p-value
ARMA	0.2131	**8.5e-11**	49.37	**3.0e-8**
GARCH	0.1992	**3.4e-10**	49.80	**5.2e-7**
MLP	0.1565	**6.5e-8**	50.13	**5.5e-6**
MSVR	0.1067	**3.7e-7**	50.96	**6.2e-4**
RESN	0.0926	**3.5e-5**	51.54	**0.0026**
SOM	0.1509	**6.1e-8**	50.15	**4.4e-6**
NG	0.1326	**2.5e-7**	50.51	**8.7e-5**
SOMAR	0.1015	**4.7e-6**	51.29	**2.6e-4**
NGMAR	0.0912	**3.8e-5**	51.87	**0.0051**
Extended SOMAR	0.0835	**4.5e-4**	52.31	**0.0075**
Extended NGMAR	**0.0670**	N/A	**52.85**	N/A

4   Conclusions

In this paper, a new extended self-organizing neural network has been introduced for multistep prediction. It considers dependencies between consecutive points by including all these points for training. Moreover, different components of reference vectors are updated according to horizons. In the experiments, the extended self-organizing neural network is compared with other neural networks for FX rates time series prediction. The results confirm that the proposed method significantly outperforms the existing methods.

References

1. Triennial Central Bank Survey of foreign exchange and derivatives market activity in 2013 (2013). http://www.bis.org/publ/rpfx13.htm
2. Euromoney FX survey 2014 (2014). http://www.euromoney.com/
3. Tsay, R.S.: Analysis of Financial Time Series. Wiley, New York (2005)
4. Sorjamaa, A., Hao, J., Reyhani, N., Ji, Y., Lendasse, A.: Methodology for long-term prediction of time series. Neurocomputing **70**, 2861–2869 (2007)
5. Zhang, L., Zhou, W.-D., Chang, P.-C., Yang, J.-W., Li, F.-Z.: Iterated time series prediction with multiple support vector regression models. Neurocomputing **99**, 411–422 (2013)
6. Taieb, S.B., Sorjamaa, A., Bontempi, G.: Multiple-output modeling for multistep-ahead time series forecasting. Neurocomputing **73**, 1950–1957 (2010)
7. Simon, G., Lendasse, A., Cottrell, M., Fort, J.-C., Verleysen, M.: Double quantization of the regressor space for long-term time series prediction: method and proof of stability. Neural Netw. **17**, 1169–1181 (2004)
8. Ouyang, Y., Yin, H.: Time series prediction with a non-causal neural network. In: Proceedings of the IEEE Conference on Computational Intelligence for Financial Engineering & Economics, London, UK, pp. 25–31 (2014)
9. Hornik, K., Stinchcombe, M., White, H.: Multilayer feedforward networks are universal approximators. Neural Netw. **2**, 359–366 (1989)
10. Giles, C.L., Lawrence, S., Tsoi, A.C.: Noisy time series prediction using recurrent neural networks and grammatical inference. Mach. Learn. **44**, 161–183 (2001)
11. Cortes, C., Vapnik, V.: Support-vector networks. Mach. Learn. **20**, 273–297 (1995)
12. Ni, H., Yin, H.: A self-organising mixture autoregressive network for FX time series modelling and prediction. Neurocomputing **72**, 3529–3537 (2009)
13. Ouyang, Y., Yin, H.: A neural gas mixture autoregressive network for modelling and forecasting FX time series. Neurocomputing **135**, 171–179 (2014)
14. Yin, H.: Self-organising: maps background, theories, extensions and applications. In: Fulcher, J. (ed.) Computational Intelligence: A Compendium, pp. 715–762. Springer, Heidelberg (2008)
15. Lampinen, J., Oja, E.: Self-organizing maps for spatial and temporal AR models. In: Proceedings of the 6th Scandinavian Conference on Image Analysis, Helsinki, Finland, pp. 120–127 (1989)
16. De Gooijer, J.G., Hyndman, R.J.: 25 years of time series forecasting. Int. J. Forecast. **22**, 443–473 (2006)
17. Li, D., Han, M., Wang, J.: Chaotic time series prediction based on a novel robust echo state network. IEEE Trans. Neural Netw. Learn. Syst. **23**, 787–799 (2012)

Natural Gesture Based Interaction with Virtual Heart in Augmented Reality

Rawia Frikha[✉], Ridha Ejbali, Mourad Zaied, and Chokri Ben Amar

REGIM-Lab: REsearch Groups in Intelligent Machines,
University of Sfax, ENIS, BP 1173, 3038 Sfax, Tunisia
{frikha.rawia.tn,ridha_ejbali,mourad.zaied,ckokri.benamar}@ieee.org

Abstract. Augmented reality AR is a relatively new technology that blends digital content such as information, sound, video, graphics, or GPS data into our real world.The natural interaction of users with virtual content is one of the big challenges of the AR application. In this paper, we present a new approach to natural interaction with virtual objects employing users' hands and fingers. We use a real-time image processing system to track gestures from users as a first step and to convert the gestures' shapes on object commands as a second step. The paper describes how these techniques were applied in an interactive AR heart visualization interface. Our aim is to provide an interactive learning tool that will help students to learn about the components of the heart. Experimental results showed the effectiveness of the proposed method.

Keywords: Augmented reality · Virtual objects · Hand tracking · Hand gestures recognition · Human computer interaction

1 Introduction

AR has become, in a few years, a new mass medium and an emerging technology field. It allows users to view digital content clearly superimposed on the real environment in real time [1,2]. Moreover, there are many possible areas that could benefit from the use of AR technology such as engineering, entertainment and education.

Intuitive interaction techniques are a major aspect of AR applications that allow users to manipulate virtual objects without the aware utilization of prior knowledge. For that purpose, different interaction techniques and concepts have emerged in the field of AR [3]. Vision-based hand gesture recognition is a major driver for intuitive interaction [4,5]. Computer vision algorithms process a video stream, detect the user's hands, and determine a gesture. The gesture launches an interaction function of the AR application that manipulates a virtual object. However, hand gestures recognition is still a challenging research field. In the field of AR, the techniques have not left the search laboratories until today. One reason is the need for technical devices that are attached to the user's hand in order to track it. The user also still acts as an operator of a machine. So,

© Springer International Publishing Switzerland 2015
K. Jackowski et al. (Eds.): IDEAL 2015, LNCS 9375, pp. 457–465, 2015.
DOI: 10.1007/978-3-319-24834-9_53

the interaction is not intuitive. Hand gestures and computer vision-based hand gesture recognition are used to ensure the natural interaction [7]. A user can interact with a virtual object as s/he interacts with a physical object using his/her hands. One advantage is, that a user does not need to wear or carry any technical device in his/her hand.

In this paper, we have developed a set of interaction techniques specifying how the user can manipulate the AR virtual content with free-hand gestures, such as rotation, scaling, etc. By using computer vision algorithms to capture the hand and identify the fingers gestures, we can map the motion of the hand to commands. No additional devices need to be attached to the hands of the user. Our technique aims at improving intuitiveness by supporting natural gesture interaction.

We organize this paper into four parts: first, we place our work in perspective of related work, discussing how hand interaction has previously been implemented in AR environments. Next, we explain the implementation of our approach. Then, we describe the results of our informal usability tests. Finally, we conclude with a summary and an outlook for future works.

2   Related Work

There were a number of previous research projects that have explored hand- and finger-based interaction in Augmented Reality interfaces. The most traditional approach for interacting with AR content was the use of vision tracked fiducial markers as interaction devices [6–9]. Piekarski et al. [6] developed a wearable outdoor AR system. They used special gloves for interaction with the system and the environment. Fiducial markers on the user's thumbs are visually tracked by a camera mounted on a head mounted display (HMD). Buchmann et al. [7] developed a technique for natural, fingertip-based interaction with virtual objects in AR environments. Similarly to Piekarski's system, fiducial markers were attached on each finger to track gestures from the user. These capabilities are demonstrated in an application of AR urban planning. However, these two works used markers and the fingers looked awkward with the markers attachments and they didn't have to occlude each other in order to be detected. Other works, such as [8,9] partially addressed encumbrance issues due to multi-marker use by attaching color stickers on fingertips. Hurst et al. [8] presented various interaction techniques on mobile phones by using the system's sensor data, and color markers on fingertips. They used single fingers and couples of fingers for interactions. Thus, the user can apply some operations such as translation, rotation, scaling on the virtual object based on finger gestures. Mistry et al. [9] developed a wearable gestural interface that tracks the location of coloured markers or visual tracking fiducial on user's fingers and projects visual information on a surface. Reifinger et al. presented also a similar system [10] by replacing fiducial markers at the fingertips with optical sensors. They used an infrared hand tracking system and gesture recognition. The tracking system detects the markers and a computer-internal hand model is built using this data. Thus, the user becomes able to manipulate virtual objects as real objects.

The accuracies in gesture detection by marker-based or optical-based inputs are high, however, privacy devices, for example, optical sensors and data gloves are required and their attachments to the fingers limit the freedom of hand movements. In order to keep the AR set-up costs low and to avoid the hand movement limitations, computer-vision based hand gesture recognition was suggested as an alternative.

3   Approach

In this paper, we propose a new approach to natural interaction with virtual objects using hand gestures in AR environments. Our proposed approach uses a real-time image processing system to interpret gestures. The flowchart of our prototype system consists of three main modules (Fig. 1).

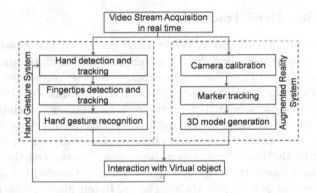

Fig. 1. The overall scheme of the 3D virtual objects' command.

- AR system: This module is responsible for connecting and capturing images from the camera and then processing this output to track the marker. The output of this module is a 3D model superimposed on the real world.
- Hand gesture system: This module is responsible for the detection and the tracking of hands and fingers. The output of this module is the hand gesture.
- Interaction module: This module is responsible for mapping hand gestures into functional input and interacting with the virtual heart.

3.1   AR System

In this work, we used a monitor-based AR system. The hardware required is a webcam for image acquisition, computing equipment to process the video and a monitor for display. In AR research, marker-based camera pose estimation approaches [11] have shown effective recording of virtual objects with the aid of robust detection of fiducial markers. For this reason, we use ARToolKit, an open source marker based on a tracking software. The operating principle of this software is to calculate the basic geometric transformations to position 3D objects in a real-time image. First, it operates by searching black and white planar markers for each frame of the image. Second, it calculates the position and orientation of the marker relative to the camera. Finally, the computer generates 3D objects and superimposes it on the marker. The first and the important setup of an AR application is the camera calibration that is used for alignment of the marker and the object to obtain correctly the combination of the real and the virtual worlds. So, a camera calibration application included with the ARToolKit library is used.

For rendering, we used OpenVRML (www.openvrml.org), an open-source parser and renderer for VRML97 and X3D files, including support for texturing, animation and networked content. The output of this system is an augmented 3D objects that the user can observe on the screen of the monitor.

3.2   Real-Time Hand Tracking

In order to command a virtual object by hand gesture, it is necessary to detect the hand in a stream video first. So, to accomplish this task, we applied the skin color detection algorithm on each input image. Several researches have shown that the main variance is in intensity rather than chrominance. Generally, HSV and YCbCr color spaces help to retrieve from the intensity variations [12]. In our work, we used HSV color space. A big advantage of this color space is that it is less sensitive to shadow and uneven lighting. Generally, the acquired images are presented in the RGB (red, green, blue) color space. For that reason, we convert the input image to HSV color space. Then, a threshold is applied to all the pixels of the image to detect the skin areas. In our implementation, we used the value suggested by Y. Wang [13] for the discrimination of skin and non-skin pixel. It's defined by this equation:

$$
\begin{aligned}
&((H >= 0) \ and \ (H <= 50)) \\
&and \ ((S >= 0.20) \ and \ (S <= 0.68) and \ ((V >= 0.35) \ and \ (S <= 1.0))
\end{aligned}
\tag{1}
$$

Next, filter median and smoothing are applied on the segmented image to eliminate the small noise regions. The output of this step is a binary image whose skin areas are represented by the white color and the other non-skin areas are represented by the black color.

After skin regions are detected, we must determine which regions correspond to the hand. For this purpose, we first extract the contours of all skin regions detected in the binary image using contour detection operations. For each region R_i, we obtain a set of coordinates $C_{i(j)} = (x_j, y_j)$ of the perimeter which outlines each region. N_i represents the total number of perimeter coordinates in the contour C_i. We then select the most significant outline to represent the outline of the hand, using Ni as a measure of the contour size. N_i must exceed a threshold S to ensure that the contour C_i is considered (S = 55 in our case).

Once the hand is detected, it must be tracking. The idea is to use a tracking method based on the detection of points of interests. So, the idea is to look for areas of the image where there is a strong contrast change, and this is particularly the case with the points lying on the edges. We will concern ourselves with the track corners, areas where there is a strong gradient in two orthogonal directions. We use the cvGoodFeaturesToTrack() method, implemented in the OpenCV library.

In the tracking phase, simple images that do not have multiple hands do not cause problems. However, the tracking problems worsen with the increase of the hands detected in the image and especially in the case of overlapping faces and hand tracking, they can give false results.

3.3 Real-Time Finger Tracking

In order to detect the fingertips, we use the k-curvature algorithm [14] which is used to find pixels that represent peaks along the contour perimeters of hands detected. First, at each pixel i in a hand contour, we compute the angle α between the two vectors of three points $[C(i), C(i-k)] and [C(i), C(i+k)]$ using the following equation where k is a constant that was fixed after some tests:

$$\alpha_{C_i} = \arccos \frac{a^2 + b^2 - c^2}{2ab} \qquad (2)$$

where a is the distance between $[C(i), C(i-k)]$, b is the distance between $[C(i), C(i+k)]$, c is the distance between $[C(i-k), C(i+k)]$ and α_{C_i} is the angle between $[C(i), C(i-k)]$ and $[C(i), C(i+k)]$. The idea here is to find points that represent potential peaks or valleys along the perimeter as shown in Fig. 2.

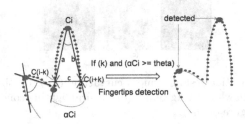

Fig. 2. Illustration of the k-curvature

To evaluate the stage of fingertips detection, a base of 100 test images is created. For this test we have chosen images containing: one hand, two hands and more than two hands in several forms, under different lighting conditions and from different skin colors. The evaluation protocol at this phase consists in testing the 100 test images by changing the value of k in the range [10..30] and by varying α in the range [30 %–90 %]. This stage has as goal to determine the rate of correct detections of the fingertips based on the two variables α and k. The good detection rates are calculated as follows:

$$Good\ detection\ rate\ = \frac{number\ of\ detected\ fingers}{total\ number\ of\ fingers\ in\ images} \quad (3)$$

The results of applying the evaluation protocol used in the detection will be displayed in the Fig. 3 below. According to the figure above, we notice that the rate of good detection depend on the two variables α and k. So, by varying α and k, we obtained a good detection rate (89 %) with k set to 30 and α between 70 % and 90 %. Figure 4 shows the results of fingertips detection, the fingertip is marked with a blue point. Our method can detect the fingertips under different lighting conditions (Fig. 4(a) and 4(b)) and background clutter (Fig. 4(c)).

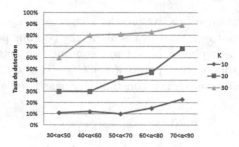

Fig. 3. Variation in the rate of good detections based on the values of K and α

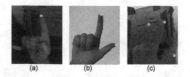

Fig. 4. Fingertips detection results using k-curvature. (a) Dark situation. (b) Natural lighting condition with white background. (c) Natural lighting condition with cluttered background (Color figure online).

3.4   Interaction Module

This section describes a set of interaction techniques specifying how the user can manipulate the AR virtual heart. All interactions are performed by free-hand gestures without any devices attached to the user's interaction. The proposed system is tested using an Intel $CORE^{TM}$ i3 CPU with a 4-GO of RAM and the windows 7 operating system. The test image sequences were grabbed from a Web camera (labtec webcam) with a resolution of 640 by 480. The camera stands opposite to the user. It flows the user's hand in two dimensions and according to the shape of the hands the interaction is made. We used five gestures in our system. Figure 5 shows the command of a 3D virtual heart via hand gesture recognition. When the three, two, one and zero fingers are recognized, the model is rotated, enlarged, attenuated and stopped interaction, respectively. In addition, we displayed heart's components when the four finger are presented. We notice that the result of the command of the AR virtual heart is excellently related to the fingertips recognition performances. The average processing time per frame, which includes preprocessing, rendering of the AR virtual heart, fingertips detection and tracking, and recognition of gestures, is 18.35 ms. This processing time (less than 20 ms) is fast enough for real-time application.

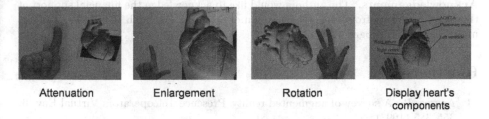

| Attenuation | Enlargement | Rotation | Display heart's components |

Fig. 5. Command of a virtual heart via hand gestures

Our system has been tested by a group of users. The aim of the test was to explore whether the interaction techniques facilitate the manipulation of virtual heart or not. We also want to know if our system complies to user expectations. A group of 15 users participated in the study. We noted that no user has a previous experience with hand gesture based interaction. Participants are between 20 and 55 years old. Before starting the test, the various interaction techniques were presented. Each user has taken few minutes to practise the technique of interaction. During the test, the user may choose which interaction they want to apply in the heart. The user can choose among five types of interaction. He/She must present their hands in front of the camera and use one of five gestures proposed by our system. So, according to the introduced gesture, interaction is made with the heart.

In general most users had no problem using gestures to manipulate virtual object even when they had no a previous experience with AR interaction. In addition, most users were able to interact in an intuitive way with the virtual heart and told us that the interaction was easy and we could manipulate virtual object in the same way as real objects. The results of user test indicated that the use of AR were helpful and facilitate learning components of the heart. Furthermore, our approach not only provides more natural and intuitive interaction but also offers an economical and convenient way of interaction.

4   Conclusion

In this work, we presented a learning tool to controlling the virtual heart via natural hand interaction without grasble devices. For this purpose, we use skin color segmentation to detect the hand region. Then, the k-curvature algorithm is implemented to find fingertips and to use it to define the hand gesture. So, a user can interact with the virtual heart by using these techniques. The future work has two objectives. First, we will create a computerized model of a heart and orient it with a hand gesture. Until now, a simple 3D model of heart has been only used. The purpose of this model is to provide a means of visualization and guidance to surgeons during surgery and to facilitate the learning and the work of surgeons. In the next step, the surgeons should be able to simulate the different hand-made gestures during surgery.

Acknowledgements. The authors would like to acknowledge the financial support of this work by grants from General Direction of Scientific Research (DGRST), Tunisia, under the ARUB program.

References

1. Azuma, R.: A survey of augmented reality. Presence Teleoperators Virtual Env. **6**, 355–385 (1997)
2. Azuma, R., Baillot, Y., Behringer, R., Feiner, S., Julier, S., MacIntyre, B.: Recent advances in augmented reality. IEEE Comput. Graph. Appl. **21**(6), 34–47 (2001)
3. Zhou, F., Been-Lirn Duh, H., Billinghurst, M.: Trends in augmented reality tracking, interaction and display: a review of ten years of ISMAR. In: Proceedings of the 7th IEEE/ACM ISMAR, pp. 193–202 (2008)
4. Radkowski, R., Stritzke, C.: Interactive hand gesture-based assembly for augmented reality applications. In: Proceedings of the 2012 International Conference on Advances in Computer-Human Interactions, pp. 303–308 (2012)
5. Ejbali, R., Zaied, M., Ben Amar, C.H.: A computer control system using a virtual keyboard. In: ICMV (2015)
6. Piekarski, W., Thomas, B.H.: Using ARToolKit for 3D hand position tracking in mobile outdoor environments. In: Proceedings of 1st International Augmented Reality Toolkit Workshop (2003)
7. Buchmann, V., Violich, S., Billinghurst, M., Cockburn, A.: FingARtips: gesture based direct manipulation in augmented reality. In: Proceedings of the 2nd International Conference on Computer Graphics and Interactive Techniques in Australasia and South East Asia, GRAPHITE 2004, pp. 212–221. ACM, New York (2004)
8. Hurst, W., van Wezel, C.: Gesture-based interaction via finger tracking for mobile augmented reality. Multimedia Tools Appl. **62**, 1–26 (2012)
9. Mistry, P., Maes, P., Chang, L.: WUW - Wear ur World: a wearable gestural interface. In: CHI 2009 Extended Abstracts on Human Factors in Computing Systems, pp. 4111–4116. ACM (2009)
10. Reifinger, S., Wallhoff, F., Ablassmeier, M., Poitschke, T., Rigoll, G.: Static and dynamic hand-gesture recognition for augmented reality applications. In: Jacko, J.A. (ed.) HCI 2007. LNCS, vol. 4552, pp. 728–737. Springer, Heidelberg (2007)

11. Kato, H., Billinghurst, M.: Marker tracking and HMD calibration for a video-based augmented reality conferencing system. In: Proceedings of the 2nd International Workshop on Augmented Reality, IWAR 1999 (1999)
12. Chitra, S., Balakrishnan, G.: Comparative study for two color spaces HSCbCr and YCbCr in skin color detection. Appl. Math. Sci. **6**, 4229–4238 (2012)
13. Wang, Y., Yuan, B.: A novel approach for human face detection from color images under complex background. Pattern Recogn. **34**(10), 1983–1992 (2001)
14. Segen, J., Kumar, S.: Fast and accurate 3D gesture recognition interface. IEEE International Conference on Pattern Recognition (1998)

Qualitative and Quantitative Sentiment Proxies: Interaction Between Markets

Zeyan Zhao[1] and Khurshid Ahmad[2](✉)

[1] Trinity College Dublin, Dublin, Ireland
zhaoz@tcd.ie
[2] Copenhagen Business School, Copenhagen, Denmark
kahmad@scss.tcd.ie

Abstract. Sentiment analysis is a content-analytic investigative framework for researchers, traders and the general public involved in financial markets. This analysis is based on carefully sourced and elaborately constructed proxies for market sentiment and has emerged as a basis for analysing movements in stock prices and the associated traded volume. This approach is particularly helpful just before and after the onset of market volatility. We use an autoregressive framework for predicting the overall changes in stock prices by using investor sentiment together with lagged variables of prices and trading volumes. The case study we use is a small market index (Danish Stock Exchange Index, OMXC 20, together with prevailing sentiment in Denmark, to evaluate the impact of sentiment on OMXC 20. Furthermore, we introduce a rather novel and quantitative sentiment proxy, that is the use of the index of a larger market (US S&P 500), to see how the smaller market reacts to changes in the larger market. The use of larger market index is justified on economic/financial grounds in that globalisation has introduced a degree of interdependence, and allow us to explore global influences as a proxy for sentiment. We look at the robustness of our prediction. (Local) Negative sentiment (as articulated in Danish newspapers over a 7 year period (2007–2013), does have an impact on the local markets, but the global market (S&P 500) has an even greater impact.

Keywords: Sentiment analysis · Equity index markets · Time series analysis · Vector Autoregressive · Granger causality

1 Preamble

The study of equities and commodities has benefited from the use of regression models pioneered by statisticians like Udny Yule (1922) [1] and Maurice Kendall (1976) [2]. Auto-regression, the dependence of the current value of a variable on its past (few) values (lagged values), has been used to explore fluctuations in a variety of time series, especially in economics and finance; usually this dependence is linear in nature. Any deviation from the model can be attributed to a number of causes: for example, there is a steady growth in the value of a

© Springer International Publishing Switzerland 2015
K. Jackowski et al. (Eds.): IDEAL 2015, LNCS 9375, pp. 466–474, 2015.
DOI: 10.1007/978-3-319-24834-9_54

stock prices or there maybe events that are external to the system that may have a disruptive influence on the values of stock prices - this disruption may be due to changing fashions, market sentiment, extreme weather, disasters or discoveries of many kinds.

Charles Sims (1980) [3] extended the scope of regression analysis considerably by specifying a framework where a regressand (say, today's stock price return) maybe regressed against its past values and against any number of other regressors - including traded volume and indeed the sentiment. Sims's framework [3] allows us to look at the simultaneous, interdependent analysis of how a sentiment variable is influenced by the changes in the equity time series. This kind of analysis helps to answer basic questions of causality. There are statistical tests of the directionality of causality, especially due to Clive Granger (see Granger and Newbold, 1977 [4]).

A number of previous studies have mentioned that investor sentiment leads the market price changes. Investor sentiment is not quantifiable like stock prices and volume changes. In recent years, researchers intend to include a proxy for the investor sentiment. There are three noteworthy studies here: (i) Antweiler and Frank (2004) [5] downloaded messages from two large message boards and used the number of messages as sentiment proxy. They generated bullishness measures from the messages using computational linguistics methods and tested the effect of investor sentiment of 45 companies in Dow Jones Industrial Average and Dow Jones Internet Index. They concluded that sentiment impacts trading volume of Dow Jones averages and to the volatility of change in prices. (ii) Tetlock (2007) [6] generated his sentiment proxy from the word count of negative and weak affect words in a comment column in the *Wall Street Journal*. Then through principal components factor analysis generated a media pessimism factor. He tested the media pessimism against market price and found that high pessimism measure leads a price decreasing. He also concluded that unusual high or low pessimism measures result in high trading volume. (iii) Garcia (2013) [7] counted the affect words of two comment columns in *New York Times*, over a period of 100 years (1905–2005) and extracted negative polarity words from the columns and used the frequency as the proxy for investor sentiment. Garcia showed that negative sentiment has a greater impact during recessionary periods (12 basis points) when compared to expansionary periods (3.5 basis points).

In each of these studies the so-called investor sentiment is computed by looking at the frequency count of pre-selected words in newspaper texts or just the number of messages in a message board. Tetlock and Garcia analysed texts using a bag-of-words model and matched the individual words in the news with negative affect words in a dictionary or affect words categorised according to affect categories. The researchers deliberately selected opinion columns (or in the case of Antweilwer and Frank the choice was the texts extracted from online message boards) that comment on the news about the markets which can be factual or anticipatory - such comments usually comprise affect words. The market movement is proxied using an aggregate index which is a measure of the value of a section of a stock market; the assumption here is that the aggregate index will reflect a consensus about the state of the market. The choice of the newspaper texts, that of the affect categories, and the choice of the aggregate index are the important biases in these studies.

We have chosen to study equity market in a medium-sized, diverse economy, that is representative of economies within the EU, and has close connections with the USA. Denmark is amongst the top 40 world economies, with close trading relationships with other Scandinavian countries, and has close economic and political ties with the USA. We look at the changes in Danish Stock Exchange index - OMX Copenhagen 20[1], together with news about Denmark in Danish news sources for computing sentiment proxies. The impact of one of the major economies (USA) is felt throughout the world and nowhere more so in the EU: we look at the price fluctuations of major US indices, especially S&P 500. S&P 500 is used here as a potential source of affect within the Danish market due to the country's relationship with the USA.

The interaction between return and investor sentiments is tested using Vector Autoregressive models (VAR). We find that OMXC 20 return does not have significant autocorrelation in our data sample period after using robust standard errors. The S&P 500 return influences OMXC 20 return in a reversing way. In addition, the Danish negative sentiment has significant impact on OMXC 20 return.

In this paper we follow the tradition of financial sentiment analysis but with two important distinctions. First, in order to calculate the impact of affect, we look at any available quantitative proxies, including movements in the world markets as these movements may induce affect changes in investors, and compare and contrast the quantitative and affect (or qualitative) proxies. Second, we look almost exclusively on newspaper reports as this is the most widely circulated type of text and is easily available to traders; perhaps, this type of text is not the main receptacle of affect when compared to, say, opinion columns, but reduces our bias in examining the impact of sentiment if any; we motivate the choice of exogenous variables like sentiment within a Vector Autoregressive formalism by looking at the stylised facts related to OMXC 20 and S&P 500. While it is possible that affect impacts market movements to a larger or smaller degree; here we use statistical methods like Chi-square interdependence test to look at the cause and effect relationship between affect and market movement.

2   Text Sampling: Our 'Danish' Corpus

The Danish economy, with a population of 5.6 million, is ranked 34th in the world in 2014 by the International Monetary Fund(IMF) and 10th in the Organisation for Economic Co-operation and Development(OECD) in 2012. The stock exchange turnover is 172.5 billion Euro from Q4 2006 to Q3 2007, the key index is OMXC 20. Economic and financial news is published in 36 newspapers (in Danish and available in translation) on a daily basis.

---

[1] OMX Copenhagen 20 (OMXC 20) is an equity market index for the Copenhagen Stock Exchange, a part of the NASDAQ OMX Group traded on NASDAQ stock exchange.

Table 1. Summary of our Danish corpus from 01/01/2007 to 31/12/2013

	Danish corpus	(St. dev)
Coverage (days)	1,985	
Number of articles	11,770	
Total number of tokens	1,593,400	
Number of tokens per year	227,629	($\pm$84,699)
Number of tokens per day	803	($\pm$570)

2.1 Text Collection

The **"Representiative" Corpus of Danish in English** are downloaded from the business news provider *"LexisNexis News and Business"*. We search for maximum availability of news articles ending on 31 Dec 2013. We use *"Denmark"* or *"Danish"* as keywords and *"Banking & Finance"* or *"Economy & Economic Indicators"* as industry and subject in the search criteria. We collect all the news in English and the resources are *"Danish Business Digest"*[2] and *"Esmerk Denmark News"*[3]. The Danish corpus is available back to Jan 2007. It has nearly 1,700 articles every year. After combining news articles from different sources, the Danish economic news corpus comprise over 11,000 news articles which include over one and half millions tokens[4] (Table 1).

2.2 Affect Dictionary and a Proxy of Investor Sentiment

Harvard GI is a historically important lexicon for content analysis of textual data originally developed in the 1960s by Philip Stone and his colleagues [8] at the Harvard Laboratory of Social Relations. The collection of Danish news text corpus is processed by *Rocksteady*[5] to generate the statistical data of investor sentiment. We use *Harvard GI* dictionary as the reference lexicon to extract the *"negative"* sentiment term frequencies using the *"bag-of-words"* method.

3 Our Price Return Series

We use OMXC 20 index daily prices. The index is a weighted index comprising 20 most-traded stocks in Denmark. The index price time series contain information over the period 01/01/2007 through 30/12/2013. The daily return series r_t for

---

[2] Danish Business Digest is a daily abstracting service in English language, which covers Denmark and provides corporate, industry and economic news.

[3] Esmerk Denmark News provides English-language summaries on key business issues abstracted from local language sources (including 96 different news agencies: Børsen, Jyllands-Posten and Politiken etc.).

[4] A **token** is an individual occurrence of a linguistic unit in speech or writing.

[5] A sentiment analysis system developed at Trinity College Dublin.

Table 2. Summary statistics for times series of returns for OMXC 20 and S&P 500. N = 1656: The observations start in Jan 2007 and end in Dec 2013. G %, A % and A* % denote constant annual return, arithmetic annual return and consecutive annual return respectively. The last columns has the z-score for each series.

Series	$10^4 \bar{r}$	$10^2 sd$	Skewness	Kurtosis	G %	A %	A* %	z
OMXC 20	1.24	1.49	−0.21	6.16	2.99	5.97	5.73	0.34
S&P 500	1.02	1.50	−0.31	8.50	2.44	5.66	5.19	0.28

the price is the logarithm of the ratio of price today divided by price yesterday. Ideally, the return is expected to follow a random walk - for every one tick followed by a tick-down. The mean of the return series for OMXC 20 is close to zero (1.24×10^{-4}), however the distribution of OMXC 20 return is not normal, as its excess kurtosis is greater than zero and the series is skewed negatively. (Table 2). When compared with similar stylised facts of the S&P 500, we find that the return (1.02×10^{-4}) is smaller than that of OMXC 20 but has similar stylised facts like OMXC 20.

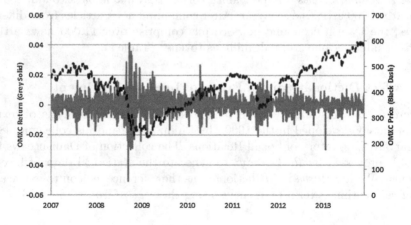

Fig. 1. OMXC 20 price and return

The return time series for OMXC 20 (Fig. 1) shows a number of periods of clustering, especially around well known market downturns. Clustering suggest that conventional wisdom, that for every up-tick there is a down-tick, does not quite work, especially when the market starts showing instability and within the unstable zone, and one must look for other causal explanations. There are a number of formalisms for compensating for the loss of the so-called random walk in return time series. We have chosen the formalism due to Sims [3], the Vector Autoregressive, to include other variables that may be responsible.

3.1   Clustering in OMXC 20

Essentially, what we are looking is the time domain development of OMXC 20 and S&P 500 in terms of their past historical values on a daily basis over a 7 year period (Jan 2007–Dec 2013) together with the evolution of sentiment variable. Following the practice in econometrics and in finance, we posit a linear relationship between these different variables and their historical values. This model is a simplification of what happens in dynamical systems like financial markets, so there will be a residual error. The expectation here is that the error is independently and identically distributed ($i.i.d$) (Heteroskedasticity and autocorrelation in the residuals are dealt using HC robust standard errors. (Newey and West, 1987) [9]). The endogenous variables we use including five lags of indices returns and negative sentiments, to control their autocorrelation. Exogenous variables include five lags of the detrended log turnover of OMXC 20 and the index volatility proxies[6] and dummy variables for day-of-the-week and month-of-the-year effects.

Table 3. Regressand(Response/explained)(Market return)

Step	Regressor (predictor/controlled)	Equation
I. Internally consistent model	Lagged values of Regressand ($L_5\beta r_t^{DK}$)	$r_t^{DK} = \alpha + L_5\beta r_t^{DK} + Exog_t + \epsilon_t$
II. Check impact of key external quantitative constraint on the regressand	$L_5\beta r_t^{DK}$ + lagged values of external constraint ($L_5\delta r_t^{US}$)	$r_t^{DK} = \alpha + L_5\beta r_t^{DK} + L_5\delta r_t^{US} + Exog_t + \epsilon_t$
III. Check impact of key qualitative constraint on the regressand	$L_5\beta r_t^{DK}$ + lagged values qualitative variable ($L_5\gamma Neg_t^{DK}$)	$r_t^{DK} = \alpha + L_5\beta r_t^{DK} + L_5\gamma Neg_t^{DK} + Exog_t + \epsilon_t$
IV. Aggregate quantitative and qualitative variables	$L_5\beta r_t^{DK} + L_5\delta r_t^{US} + L_5\gamma Neg_t^{DK}$	$r_t^{DK} = \alpha + L_5\beta r_t^{DK} + L_5\delta r_t^{US} + L_5\gamma Neg_t^{DK} + Exog_t + \epsilon_t$

Note: $L_n\beta(x_t) = \beta_1 x_{t-1} + \beta_2 x_{t-2} + \ldots + \beta_n x_{t-n}$.

What we have is a multivariate time-series model. We treat all the variables as endogenous and regress each of the variables on past lags of these variables. The models are fitted using ordinary least squares estimation techniques. Following methodological conventions, we test hypotheses by assessing the joint statistical significance of the coefficients on single variables and use Chi-square tests.

We define a 5th order Vector Autoregressive model with the error term ϵ_i. Four tests are designed to evaluate the quantitative and qualitative constraint impacts step by step (Table 3).

The coefficients (α, β, δ, and γ) help in quantifying the impact of historic values of the dependent variable and that of the control variables like sentiment

---

[6] 1st difference of conditional variances of GARCH(1,1) models.

and stock market movements elsewhere. The regression equations help in determining the values of these coefficients. The values of these coefficients, together
with the values of their statistical significance, are an indication of the impact
of each of the variables on right hand side of equations.

Table 4. Impact of S&P 500 and negative sentiment on OMXC returns: *,
** and *** denote values of coefficients' (α, β, γ, and δ) statistical significance at 0.1,
0.05 and 0.01 levels respectively. All coefficients are in **basis points**.

| | Tests | Dependent variable: r_t^{DK} | | | |
		Step I	Step II	Step III	Step IV
βr_t^{DK}	L_1	541	−1540 ***	530	−1545 ***
	L_2	−353	−500	−344	−507
	L_3	−505	−411	−517	−432
	L_4	454	301	457	300
	L_5	−669	−614	−657	−612
δr_t^{US}	L_1		3824 ***		3810 ***
	L_2		805		811
	L_3		323		342
	L_4		−182		−170
	L_5		525		535
γNeg_t^{DK}	L_1			−6.6 *	−5.4
	L_2			3.6	1.3
	L_3			0.0	−0.8
	L_4			1.2	1.9
	L_5			5.3	4.7
r_t^{US}	$\chi^2(5)$[joint]		184.81 ***		182.78 ***
Neg_t^{DK}	$\chi^2(5)$[joint]			4.63	3.31

3.2   Impact of Quantitative Sentiment

We regressed the values of OMXC 20 returns (2007–2013, N = 1650) against its
past 5 days values in the first instance to see the impact of the historic values
of OMXC 20 on its present value. The first day lag has a positive contribution
followed by negative contributions from the 2nd, 3rd and 5th day lags and a
positive contribution from the 4th day lag. The overall contribution is negative.
However, none of the day lag has a statistically significant contribution level
($p < 0.1$) confirming the property of stock returns that there is no correlation
between returns for different days. The results are shown in Table 4 - "Step I".

We use five lags of S&P 500 returns as the control variable to test the impact
of world market. The 1st day lag of the S&P 500 returns positively contributes

the regressand. It also influence the 1st day lag of the OMXC 20 returns - becoming negatively autocorrelated. Chi-square test shows that S&P 500 returns Grange cause OMXC 20 at 0.01 level (Table 4 - "Step II").

3.3 Sentiment Feedback

When we add negative sentiment to the first OMXC model, the impact of the past values of OMXC 20 returns on its present value is slightly lower. However, the negative impact of 1st day lag negative sentiment is 6.6 basis points and statistically significant at 0.1 level ("Step III" in Table 4); and it is fully recovered within the 2nd to 5th days lag.

If we add both S&P 500 returns and negative sentiment to the OMXC model, the influence of US market causes the statistical significance of 1st day lag of local negative sentiment effects to vanish ("Step IV" in Table 4). The causality of S&P 500 returns remains however the robustness drops 1 % due to the interdependence between S&P 500 returns and Danish negative sentiment.

4   Conclusions

This paper used asset measures from OMXC 20 and S&P 500 indices and measures of investor sentiments based on Danish economic news corpus from Lexis Nexis. We tested the previous week's impacts of investor sentiment proxies on stock returns using Vector Autoregressive models. We found the absence of autocorrelation in OMXC 20 returns confirming the property of return series, however it is not the case in the S&P 500 returns. As a consequence, 1st day lag of S&P 500 return positively impacts OMXC 20 returns and forces the 1st day lag of OMXC 20 returns becoming negative autocorrelated. On the other hand, OMXC 20 returns are negatively affected by 1st day lag of negative sentiments in Danish newspapers. In addition, the presence of systematical influence from the world major index to a comparatively smaller index controls the sentiment impacts. After we took into account the S&P 500 returns influence on OMXC 20 returns, the impact of local news sentiment becomes weak. We also studied the Granger causality relationship between these measures using VAR models. Our main finding is that 5 day lags of S&P 500 returns, as proxies of world market sentiments, cause OMXC 20 returns at 0.01 significance level.

We are continuing our studies at firm-level sentiment that is we are looking at the impact of sentiment on individual firms that typically comprise OMXC 20.

References

1. Yule, U.G.: An Introduction to the Theory of Statistics. Charles Griffin, London (1922)
2. Kendall, M.: Time Series. Charles Griffin, London (1976)
3. Sims, C.A.: Macroeconomics and reality. Econometrica **48**, 1–48 (1980)

4. Granger, C.W.J., Newbold, P.: Forecasting Economic Time Series. Academic Press, New York (1977)
5. Antweiler, W., Frank, M.Z.: Is all that talk just noise? The information content of internet stock message boards. J. Financ. **59**(3), 1259–1294 (2004)
6. Tetlock, P.C.: Giving content to investor sentiment: the role of media in the stock-market. J. Financ. **62**(3), 1139–1168 (2007)
7. Garcia, D.: Sentiment during recessions. J. Financ. **LXVIII**(3), 1267–1300 (2013). doi:10.1111/jofi.12027. http://dx.doi.org/10.1111/jofi.12027
8. Stone, P.J., Dunphy, D.C., Smith, M.S., Olgilvie, D.M., with associates: A Computer The General Inquirer: Approach to Content Analysis, The MIT Press, Cambridge (1966)
9. Newey, W.K., West, K.D., Simple, A.: Positive semi-definite, heteroscedastic and autocorrelation consistent covariance matrix. Econometrica **55**, 703–709 (1987)

Propagating Disaster Warnings on Social and Digital Media

Stephen Kelly[(⊠)] and Khurshid Ahmad

Trinity College Dublin, Dublin, Ireland
{kellys25,kahmad}@scss.tcd.ie

Abstract. A nexus of techniques including information extraction techniques, including a bag of words model, web and social media search and time series analysis, are discussed that may reveal the potential of social media and social networks. Social aspects of data privacy are discussed to ensuring that the data collected, filtered, and then used. This work is the effort of Trinity College Dublin and other universities.

Keywords: Information extraction · Sentiment analysis · Web and social media search · Security and privacy · Time series analysis

1 Introduction

Disaster management invariably involves few disaster operatives, with limited information, having to help many victims who need vast quantities of information. For example, after a major disaster event, civil protection agencies are expected to estimate the extent of damage including fatalities, victims needing medical attention, food, water and shelter, but this information is either not available or comprises much noise. The victims equally need more information: information about recovery schedules, places of refuges, food and water sources; again this information is scarce. The advent of social media has increased the flow of data. However this data has to be collated systematically, processed reliably and in real time, and the resultant information broadcast in a manner that not only increases information but engenders trust. Furthermore, the social media data has to be reconciled with information available on traditional sources such as government websites.

Much of the data available on microblogging sites, social networks and formal media, comprises names of people, place, things and events.Processing this information allows disaster managers to target vulnerable members of a disaster-struck community and deploy scarce and valuable resources more effectively. Herein lies the inherent tension in the whole exercise: disaster managers wish to have sufficient data about people, places and things that can, in principle, intrude into the privacy of the individual. We describe techniques of information retrieval and extraction to process the high-volume throughput data, data which is mainly textual in nature. We extract named entities from social media data and create a time series of these categories of terms that are used in outlining

© Springer International Publishing Switzerland 2015
K. Jackowski et al. (Eds.): IDEAL 2015, LNCS 9375, pp. 475–484, 2015.
DOI: 10.1007/978-3-319-24834-9_55

an ontology of disaster. From the time variation in the term count in different categories, we can examine the velocity of a disater reporting on social media. Currently our system developed logs the quantity of personal data it has and forms what we call intrusion index.

Leveraging social media in emergency response and disaster management is not a new concept. The fast dissemination of information on social media has meant a new medium of communication is being used by the public, providing potentially unused information that can benefit emergency management and disaster planning [3,4].

Social media text and data inherently contain information that would be considered private. However, technology and users can also benefit from sharing this data. Disaster management and aid efforts can benefit greatly from this increased level of publicly sourced information. For example, location information is used in emergency management by detecting and identifying geographical location from social media. For the wider scope of the projects that we are working on, we integrate this geolocation meta data and specific Twitter sources together with text analytics and time-series visualisation to improve the efficiency of emergency management.

2 Methods and Data

The following section outlines the data and methods used on the text collections, their method of aggregation and construction of the corpus, lexicons and glossaries used in the text analysis method. Additional methods presented in the following section pertain to creating an index of intrusion which attempts to identify and monitor the frequency of entities and information that would potentially be considered sensitive. Corresponding information associated with these entities in text would thereby be useful and susceptible to being collected and mined. Event detection in a corpus of twitter messages is also presented while considering how robust the method is towards the potential noise evident in social media data.

2.1 Twitter Messages as a Text Corpus

Three different collections of text of Twitter messages were collected. The first corpus, described as "Public Media" was obtained from Twitter messages collected by Newstex[1] archived by Lexis Nexis database. The majority of sources in this corpus are considered to be authoritative corporate and independent publishers. The second corpus is a concise selection of authoritative sources on Twitter concerned with Irish related media and services. These sources include messages from weather, emergency services, local authorities, and Irish police forces giving total 19 sources. These sources were chosen as they are the most active authorities on social media disseminating official information to the public

---

[1] http://newstex.com/about/what-is-authoritative-content/.

regarding general safety concerns, weather warnings, and major announcements. The last corpus was collected by specifying a location with corresponding geo coordinates when running the Twitter API streamer. The location set was a bounding box around the region of Ireland, tweets geotagged as being from Ireland were then returned. The sample text for the sources and location collections were collected using Twitter's public API. This restricts users in the volume of messages that are allowed to be collected. The criteria of data collection and search, for instance location-based tweets, further restricts the sample of text. However a more specific corpus of text can thus be collected.

2.2 Text Analysis and Entity Recognition

The method of text analysis used is a dictionary based approach to analysing text content. The bag of words model focuses on counting the frequency of terms that have been categorised into a subset of dimensions. These categories are then defined in a lexicon. The overall frequency of these categories can then be interpreted as estimating the tone of a text and thus acts as a foundation for creating news sentiment and following topic discussion. Such an approach has previously been successful and the basis of the General Inquirer system using the Harvard-IV psychosocial dictionary [7]. The *Rocksteady* content analysis system, developed at Trinity College Dublin, builds on this principle. In this study we use a glossary of weather terminology with the *Rocksteady* system to analyse term occurrence. The weather terminology was collected from a glossary[2] made pubic by the National Weather Service(NWS) under the U.S Department of Commerce[3].

Named entity recognition has been implemented in the Cicui system developed in Trinity College Dublin and utilises the Stanford Named Entity Recognizer(NER) [9]. This implementation uses the 7-class model, the types listed as: Time, Location, Organisation, Person, Money, Percent, Date, trained for the Message Understanding Conference(MUC)[4]. Previous work on identifying entities for the development of an index of intrusion involved hand-tagging of proper nouns and led to the use of *CiCui* with NER, and static lists containing prior information [5]. An additional static list of places and person names was used as a contrast. The lists of places and names are contained in the gazetteer lists and names corpus are both contained in the Natural Languages Toolkit(NLTK) [1]. For the source and location text collection a list of Irish localities was added as prior information is known about these corpora.

2.3 Event and Anomaly Detection

Outlier detection involves identifying observations that are inconsistent with other observations in an observed data set. This observation may have a low

---

[2] http://www.erh.noaa.gov/er/box/glossary.htm.
[3] http://www.weather.gov/erh/.
[4] http://nlp.stanford.edu/software/CRF-NER.shtml.

probability of occurrence but ultimately the discordant observation may contain potentially useful information. Outliers may come from a shift or change in the scale of a parameter or sub-sample of a dataset, resulting from innovations, new information, or simply measurement error. Shifts in the parameters of a time series occur due to these outliers also. Detecting these changes, although difficult, can prove beneficial with the aid of robust statistical methods.

Anomaly detection is being performed by social media companies to detect spikes in the volume of media being posted on their services. Seasonal events are often seen to have anomalies as the number of images posted increases due to these events. Local anomalies can be seen in smaller frequencies over a larger time series. The Extreme Studentised Deviate is a test for identifying observations as outliers in a given dataset [6]. The Seasonal Hybrid ESD (S-H-ESD) algorithm is an extension to the original algorithm that decomposes a series into components. The particular decompositions produce a cyclical or seasonal component that can be used to define smaller periods. Combined with robust methods, such as median based estimation which are less sensitive to outliers, cyclical effects in a time series or sequence can be detected to determine local and global anomalies. An open source implementation has been released by Twitter Inc[5].

Noise and outliers in the series can have a large impact on the mean of the series and subsequent models that use mean estimations. Median estimation of a series and models, which use such parameters, are then less susceptible to these outliers and give a more robust estimation of statistical changes. Any changes in the series can thus be determined with more confidence; these change points can be highlighted in a series and used to determine potential events or anomalies. Using the E-Divisive with medians algorithm changes in the mean are detected using statistically robust methods [2]. The statistical distance between probability distributions is measured and the discrepancy between the parameters, mean or median, is measure to check for equality in distributions [8]. An R package for the implementation of the EDM algorithm has also been released as open source by Twitter Inc[6].

3   Results

We begin by looking at the overall volume of messages in each of the three corpora defined. From there we can approximate potential events using robust statistical methods. In each case anomalies are detected in the time series that relate to the volume of the tweets. By analysing the content, we also identify potential entities. These entities become the contributing unit of information when constructing the intrusion index. The simple frequency of tweets is then examined where conclusions are drawn between messages volume and terminology. The frequency of words during the events is then examined and the frequency of terms and contrasting differences examined. Finally conclusions

---

[5] https://github.com/twitter/AnomalyDetection.
[6] https://github.com/twitter/BreakoutDetection.

are drawn regarding the potential events, volume, and content of the Twitter messages.

3.1   Text Corpora

Three collections of Twitter messages are used and collected according to different criteria from *public media*, public body *sources* and *location*. Each shows the time periods for each of the collections, where messages and terms have been aggregated on a daily frequency (Table 1).

Table 1. Summary of daily message volume for each of the collections of Twitter messages aggregated.

	Public media	Sources	Location
Terms	526350	47930	225841
Volume	45263	2290	16207
Start	17/11/2014	01/01/2015	06/01/2015
End	14/01/2015	18/01/2015	18/01/2015

3.2   Intrusion Index - Entities in Social Media Text

We compare the use of NER to a keyword list of location and places names for the three Twitter message collections. We examine the percentage occurrence of entity terms, the list of place and person names and the trained NER implemented in *CiCui*, on each of the potential terminologies extracted from each of the three collections. After text processing the terms extracted using either method shows the pre-trained NER retrieves a higher number of potential place and person names (See Table 2).

Table 2. Occurrence of person names and location names on each of the three Twitter message collections.

% Frequency	Public media	Sources	Location
	Word List		
Names	7.9 %	4.7 %	4.5 %
Places	2.1 %	1.2 %	0.7 %
	CiCui NER		
Names	17.8 %	10.9 %	7.6 %
Places	21.4 %	7.2 %	12.8 %

A particular difference of using the trained NER and keyword list is the potential of the former to detect twitter handles. Popular, highly frequent names in general language are more readily detected with the NER. From previous work done on the occurrence of entities in social media messages [5], it was found that half of all highly frequent potential entities were Twitter handles or sources. Overall the use of NER in *CiCui* finds more up to date references for places and names, in particular finding Twitter sources.

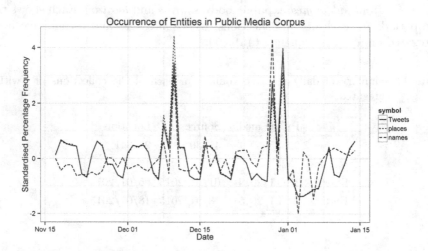

Fig. 1. The percentage frequency of automatically extracted entities, all location and person names, and volume of Twitter messages occurring in the corpus

We plot the occurrence of the extracted entity list from the NER in the public media message collection in time. The frequency occurrence of the entities coincides with the large event spikes, suggesting the volume consists of references to such entities (Fig. 1).

3.3   Event Detection in Social Media

Anomaly detection was performed by using the Seasonal Hybrid ESD (S-H-ESD) algorithm with the aim of identifying global anomalies across the entire sample, and potential local anomalies in cyclical or seasonal periods. It can be applied to a data sample of arbitrary order, it is applied in this study to the public media collection of Twitter messages.

From the plot of the series and potential anomalies, it is determined that two periods are shown as being positive anomalies. We report that less periods show two additional positive anomalies. However, using robust anomaly detection two distinct observations are shown. The seasonal effect of the series can be viewed by observation of the plot, of which it is clear through the use of statistical methods some occurrence or increase in tweets is observed (Fig. 2).

Potential state transitions are detected using the E-Divisive with Medians(EDM) algorithm, where potential mean shifts and trends are detected. This

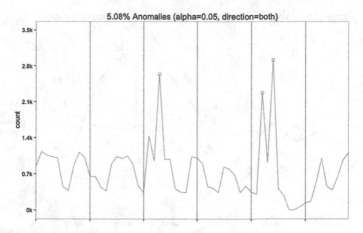

Fig. 2. Application of the S-H-ESD algorithm to detect possible global and local anomalies in the volume of Tweets in the Public Media corpus. Potential anomalies are indicated as point estimates on the series.

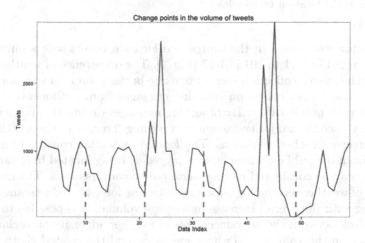

Fig. 3. Application of the EDM algorithm to detect possible change points in the volume of Tweets in the Public Media corpus.

particular assessment of the series is to demonstrate any particular discrepancy between anomaly detection and any potential "breakouts". Identifying any difference is beneficial as the objective is to find a numerical way of finding potential events of interest in the time series of Twitter messages.

The application of the algorithm on the series demonstrates four distinct change points in the series, considered to be breakout points (Fig. 3). The EDM algorithm attempts to account for anomalies when it determines breakouts and therefore is not just a repetition of the previous results demonstrated for anomaly detection.

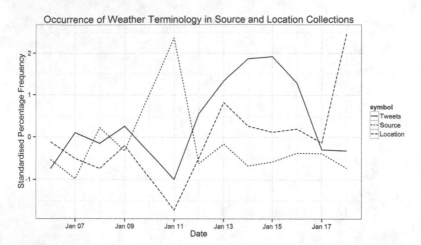

Fig. 4. Shows the volume of Tweets from the source collection, the occurrence of weather terminology in the source collection as compared to the weather terminology occurrence in the location terminology

The volume of tweets in the source and location tweets sees a jump in the period from 2015-01-11 to 2015-01-17 (Fig. 4). The occurrence of weather terminology in the *source* collection sees a large rise in the volume of weather related messages. The *source* collection contains messages from authoritative sources concerned with public safety. By observing messages during this period in this collection we see the majority are about a coming forecast in the weather, predicting adverse weather conditions. The *location* collection consists of messages sampled from the public Twitter stream tagged as being located in Ireland. This collection may be considered more general public conversations. The number of mentions of weather terminology also rises after the forecast and announcements, correlating with the overall increase in message volume. It is possible to make a tentative link between the announcements of weather adversity on social media, the increase in the volume of Twitter messages, and the related chatter of the public.

4   Conclusions and Future Work

This paper has presented the construction of an intrusion index using entities extracted from a corpus of Twitter messages and the determination of events in the corpus over a period of time. The construction of three individual collections of Twitter messages using different criteria has provided the basis if the case studies. The construction of the intrusion index, composed of entities extracted from the corpus of text, and annotated in time has been a central contribution in this study. The detection of events and anomalies has also shown promising results, particularly in the application to emergency management. Combining an automatic method of event detection and the intrusion index can help locate potential areas of information and represent important entities in those topics. In the case of a disaster related event, places badly effected would be more readily detected, with vital information being extracted.

The occurrence of entities during event periods shows that the intrusion index is a valuable contribution and avenue of exploration for a framework that monitors emergency events unfolding on social media. The structure of Twitter specifically, and social media in general, means sharing information with people is integral to its function. Future work involving NER being trained on Twitter data has shown promising results in other studies and may help increase the retrieval of entities for the instrusion index.

In our research we have partnered with a technology company specialising in emergency management software. Their system, *SIGE*, uses visualisation and GIS information in order to predict vulnerable areas and allow emergency responders to react quickly with real-time information. The construction of data layers for use with the system and WebGis is essential to its functionality. By analysing and processing relevant social media data a new data layer can be incorporated adding potentially vital information. Social media must be processed and analysed accordingly, using methods as outlined in this paper, filtering messages and text, supplying the output to the system and aiding the decision-making processes. The system then leverages this information to create a clear, trustworthy message that can be disseminated to disaster management personnel and citizens. The integration of processing geolocated Twitter messages, text analytics, (*CiCui* system) and time-series visualisation (*Rocksteady* system) into a traditional disaster management alarm system would leverage a new form of real time data that hasn't been fully utilised before.

Finally, these automatic tasks augmented with traditional data in emergency management platforms would benefit greatly from an integrated decision making process, where the integration of the systems would benefit more than their use in isolation. An application of this is described in with integration into a disaster management system.

Acknowledgments. The authors would like to thank the EU sponsored Slaindail Project (FP7 Security sponsored project #6076921) and Xiubo Zhang for use of the *CiCui* system.

References

1. Bird, S.: Nltk: the natural language toolkit. In: Proceedings of the COLING/ACL on Interactive presentation sessions, pp. 69–72. Association for Computational Linguistics (2006)
2. Huber, P.J.: Robust Statistics. Springer, Heidelberg (2011)
3. Hughes, A.L., Palen, L.: Twitter adoption and use in mass convergence and emergency events. Int. J. Emerg. Manage. **6**(3), 248–260 (2009)

4. Hui, C., Tyshchuk, Y., Wallace, W.A., Magdon-Ismail, M., Goldberg, M.: Information cascades in social media in response to a crisis: a preliminary model and a case study. In: Proceedings of the 21st International Conference Companion on World Wide Web, pp. 653–656. ACM (2012)
5. Kelly, S., Ahmad, K.: Determining levels of urgency and anxiety during a natural disaster: noise, affect, and news in social media. In: DIMPLE: DIsaster Management and Principled Large-Scale information Extraction Workshop Programme, p. 70
6. Rosner, B.: Percentage points for a generalized esd many-outlier procedure. Technometrics **25**(2), 165–172 (1983)
7. Stone, P.J., Dunphy, D.C., Smith, M.S.: The General Inquirer: A Computer Approach to Content Analysis. MIT Press, Cambridge (1966)
8. Székely, G.J., Rizzo, M.L.: Energy statistics: a class of statistics based on distances. J. Stat. Plann. Infer. **143**(8), 1249–1272 (2013)
9. Toutanova, K., Klein, D., Manning, C.D., Singer, Y.: Feature-rich part-of-speech tagging with a cyclic dependency network. In: Proceedings of the 2003 Conference of the North American Chapter of the Association for Computational Linguistics on Human Language Technology, vol. 1, pp. 173–180. Association for Computational Linguistics (2003)

A Distributed Approach to Flood Prediction Using a WSN and ML: A Comparative Study of ML Techniques in a WSN Deployed in Brazil

Gustavo Furquim[1,5], Gustavo Pessin[2,5]($\boxtimes$), Pedro H. Gomes[3,5], Eduardo M. Mendiondo[4,5], and Jó Ueyama[1,5]

[1] Institute of Mathematics and Computer Science (ICMC), University of São Paulo (USP), São Carlos, São Paulo, Brazil
gafurquim@usp.br, joueyama@icmc.usp.br
[2] Vale Institute of Technology, Belém, Pará, Brazil
gustavo.pessin@itv.org
[3] Autonomous Network Research Group (ANRG), University of Southern California, Los Angeles, CA, USA
pdasilva@usc.edu
[4] São Carlos School of Engineering (EESC), University of São Paulo (USP), São Carlos, São Paulo, Brazil
[5] Center of Monitoring and Early Warning of Disasters, Ministry of Science, Technology & Innovation, São Carlos, Brazil
emm@sc.usp.br, emm@cemaden.gov.br

Abstract. Natural disasters (e.g. floods, landslides and tsunamis) are phenomena that occur in several countries and cause a great deal of damage, as well as a serious loss of life and materials. Although very often these events cannot be avoided, their environments can be monitored and thus predictions can be made about their likely occurrence so that their effects can be mitigated. One feasible way of carrying out this monitoring is through the use of wireless sensor networks (WSNs) since these disasters usually occur in hostile environments where there is a lack of adequate infrastructure. This article examines the most recent results obtained from the use of machine learning techniques (ML) and adopts a distributed approach to predict floods using a WSN deployed in Brazil to monitor urban rivers. It also conducts a comparative analysis of ML techniques (e.g. Artificial Neural Networks and Support Vector Machines) for the task of flood prediction and discusses the results obtained from each type of technique explored so far. Finally, in the discussion of the results, a suggestion is made about how to improve accuracy in forecasting floods by adopting a distributed approach, which is based on allying computing intelligence with WSNs.

Keywords: Wireless sensor networks · Machine learning · Distributed systems · Flash flood forecast

1 Introduction

In recent years, there has been an increase in the frequency of natural disasters throughout the world. This rise is mainly caused by the increase in the number

© Springer International Publishing Switzerland 2015
K. Jackowski et al. (Eds.): IDEAL 2015, LNCS 9375, pp. 485–492, 2015.
DOI: 10.1007/978-3-319-24834-9_56

of occurrences of climate calamities, such as severe storms and floods. These disasters lead to a large number of victims, as well as incurring financial losses that, directly or indirectly, affect the lives of millions of people [3]. During 2013, for example, 330 disasters were recorded that were triggered by natural events throughout the world. These disasters affected 96.5 million people and caused the deaths of 21.610 victims, together with damage estimated at US\$ 118.6 billion [9]. This situation has been aggravated by changes in the climatic conditions of the planet and is largely found in urban districts such as those that prevail in the State of São Paulo, Brazil. In these regions the destruction of the ecosystem caused by pollution and a lack of planning is more acute and affects the environment by altering the local climate. In this context, WSNs are an attractive way of carrying out the monitoring of urban rivers and other natural environments because of the following reasons: low costs particularly with regard to infrastructure; accessibility to inhospitable environments and the fact that they offer the prospect of using high-precision sensors; and their adaptability to environmental changes [8]. However, the collected data must be studied and their behavior understood before suitable flood predictive models can be created with a high degree of accuracy. ML techniques are very useful for this since they can produce predictive models without the need for a precise knowledge of the hydrological processes involved or the physical variables of the hydrographic basin of the region [6]. In São Carlos, Brazil, a WSN has been installed which is called REDE (*REde de sensores sem fio para Detectar Enchentes* – WSN for Flood Detection), developed by the Institute of Mathematics and Computer Sciences (ICMC) - University of São Paulo (USP) [8], which seeks to undertake the monitoring of urban rivers. One of the key features of this article is to explore this area by using and analysing data collected by the REDE system. A comparative study will also be carried out between the ML techniques and a distributive prediction which makes it possible to forecast the occurrence of floods and create an alert system for taking preventive measures.

The initial analysis of the data was conducted by employing the concepts of Time Series or more precisely, the Immersion Theorem of Takens [7]. This theorem creates a time series with overlapping data in a simplified and multidimensional representation and following this procedure, employs modelling techniques that can allow observations to be related over a period of time and for predictions to be made with a greater degree of accuracy. Furthermore, three ML techniques were employed for this modelling: Support Vector Machine (SVM), the Gaussian Process (GP) and Multilayer Perceptron (MLP). In this way, this article examines these ML techniques when used for flood prediction and show the comparative results of this study. The local forecasts carried out by models present in some of the sensor nodes, can help in the forecast carried out by the other nodes. In this way, an analysis was conducted that sought to determine when the predictions carried out by some nodes, could be used to increase the accuracy of the prediction of the other nodes, leading to a distributed approach for flood prediction. The adoption of this distributed predictive control approach, together with the comparative analysis of the ML techniques, made it possible to increase the accuracy of the predictions as well as to partition the processing needed to implement the various models that had been created. The rest of this article is divided in the following manner: Sect. 2 outlines the studies related to flood prediction. Section 3 discusses the methods employed and describes the following: (a) how the data collected by the WSN was handled, (b) the Immersion Theorem of Takens and (c) the ML techniques that were employed and compared in the studies. The article concludes with an analysis of the obtained results (Sect. 4); the final considerations and suggestions for further studies are made in Sect. 5.

2   Related Works

Hossain et al. [5] make use of a system based on adaptive neuro-fuzzy inference (ANFIS) to make predictions of floods, as well as the likely persistence of this flooding in the River Meghna, Bangladesh, by using data collected from a WSN. Although good results were attained in this article, only a system based on a WSN was used and no wireless sensor network was in fact installed in the region. In addition, there is also a description in this article of the features that the sensors used in the experiment, which leads to their application being restricted.

Still on the question of flood prediction, Damle and Yalcin [2] describe an approach based on a combination of chaos theory and data mining that is adopted to obtain features of the time series containing the daily output of the Mississipi River in the US. This methodology follows a case flow which is initiated by using the Takens Theorem to reconstruct the time series. The reading of the data was carried out through a seasonal collection and not by means of WSN. Despite achieving good results, the method only predicts the use of data from a point on the river and thus ends up by not analyzing the existing relationship between the different collection points of the whole river. Elizabeth et al. [1] set out a system for prediction and flood alerts by using statistical models and a WSN made up of 4 types of different nodes. The predictive system was tested with historical data collected over a period of 7 years from the Blue River, in Oklahoma, although the article does not state how the collection was carried out. In addition, a reduced version of the proposed WSN, which has few nodes and a predictive model for centralized processing, was installed in the River Charles, in Massachussets, and River Agúan, located in the north of Honduras. The simplified installation of the WSN served to test the communications system and data collection.

3   Methods of Flood Forecasting

Takens [7] observed that a given time series $x_0, x + 1, ..., x_{n-1}$, can be reconstructed in multidimensional space $x_n(m, \tau) = (x_n, x_{n+\tau}, ..., x_{n+(m-1)\tau})$, called time-delay coordinate space, where m represents the embedded dimension and τ represents the dimension of separation. In other words, the time series is reconstructed in vectors with m values, in which each value corresponds to an observation spaced with an equal time delay to τ. These vectors represent the interdependence relationship that exists between the observations and simplifies the study of the behaviour of the time series. After this reconstruction, ML can be employed to produce prediction models of the time series with a greater degree of accuracy. It is difficult to determine the values of m and τ with precision in some kinds of time series, in particular those extracted from real-world environments and subject to noise. However, they can be estimated by means of methods like Auto-Mutual Information (for τ) and False Nearest Neighbors (for m).

Currently, the WSN used for this study (the REDE system) consists of 7 sensors; six of them are level sensors which are used to measure the water pressure and change the height in centimeters and 1 sensor measures the pluviosity and shows the volume of rainfall in the region (Fig. 1).

Fig. 1. A REDE WSN node deployed in São Carlos - SP, Brazil.

The data considered in this article were obtained by three sensors from the REDE system (called here $s1$, $s2$ and $s3$) during the whole month of May 2014. The readings of the water level were carried out at 5 min intervals and the sensors were arranged in the following way: Sensors $s1$ and $s2$ are in different rivers which converge to form the water flow that passes by Sensor $s3$. If the proposed approach is adopted, the first stage is to measure the overlapping of the time series formed by the readings of the river levels given by the three sensors. Auto-Mutual Information techniques were employed for this, and these were designed to estimate the separation dimension and the False Nearest Neighbors, to estimate the embedded dimension. In this way, the separation and embedded dimensions that were used to estimate the overlapping of the time series that was read are as follows: $\tau = 1$ and $m = 2$ for sensors s1, s2 and s3. For sensor s2 we also evaluate $\tau = 5$ with $m = 2$. After carrying out the time series with overlapping data, three ML techniques were employed and evaluated before being implemented to produce prediction models, namely: SVM (with epsilon-SVR and nu-SVR kernels), GPs and the MLP. These techniques were employed and compared to produce local forecast models which only use the data from the sensors themselves. This analysis made it possible to determine the best prediction model and the overlapping parameters for each sensor, as well as to allow comparisons to be made with the distributed approach described in the following paragraph. If the current time is considered to be t, the predictions were carried out for the $t + 1$ instant, when $\tau = 1$ (5 min in the future), or $t + 5$ (25 min in the future), when $\tau = 5$. In the second stage of the experiments, a distributed approach for predictions was employed, which explores the readings and the processing capacity of multiple sensors (instead of only one). In this way, the processing capacity of the WSN nodes can be explored in a more satisfactory way as well as giving the sensors a greater degree of freedom for the task of forecasting floods, particularly in situations where communication with the central database has been lost. In this approach, the $s1$ an $s2$ carry out a part of the processing and send the results to sensor $s3$. This processing involves making forecasts of the river level for instant $t + 1$ (5 min in the future), with regard to its respective locations and only using the readings themselves. After being sent to sensor $s3$, the local predictions $(t + 1)$ of sensors $s1$ and $s2$ were combined with the local readings of sensor $s3$, where the rest of the processing was executed. In this study, the ML techniques were applied again to create forecast models for sensor $s3$, this time by making use of the combined data and predictions of the 3 sensors. The distributed approach to flood prediction set out here, also seeks to analyze whether the predictions made by the groups of sensors ($s1$ and $s2$, in this case), can improve the more long-term forecasting carried out by the other sensors ($s3$, in this case). As well as creating forecast models for the $t + 1$ instant (5 min in the future), the same combination of data and data techniques was used for this in sensor $s3$, to create predictive models for instant $t + 2$ (10 min in the future). It should be noted that this article only makes use of the data collected by the real-world WSN to simulate and analyze the approach put forward, since the predictive models and distributed approach are not implemented in the nodes of the REDE system. Nonetheless, the nodes of the REDE system have the capacity to execute the forecast models and carry out the necessary communication. In addition, the results obtained from these experiments (which are shown in the next section), have encouraged us to make a real-world implementation of the proposed approach.

4   Results and Discussion

After the extent of the overlap of the time series was read by sensors $s1$, $s2$ and $s3$, the SVM implementations (epsilon-SVR and nu-SVR kernels), GP and MLP were used to create flood prediction models. For overlapping values equal to $\tau = 1$ and $m = 3$, two readings of the river level were used as input. The output is the forecast of a point in the future ($m = 3$), where all these values (from the input to the prediction) are consecutive readings above the river levels ($\tau = 1$). The Waikato Environment for Knowledge Analysis (WEKA) [4] was employed to model and evaluate both techniques. The results are obtained by employing a 10-fold cross validation procedure.

As can be seen in Table 1, the best results for all the sensors were obtained with values of $\tau = 1$ and $m = 2$. Thus in the experiments that follow, where more data for distributed prediction sensors were used, these parameters ($\tau = 1$ and $m = 2$) were kept to carry out the overlapping of the time series of sensors $s1$ and $s2$ and the technique was used to ensure a greater degree of accuracy for both sensors (in the case of the GP). In the distributed predictive approach, the best forecasting models for instant $(t + 1)$ were executed in sensors $s1$ and $s2$ that consider the current instant as being t. The forecasts made by these sensors

Table 1. Performance of the ML techniques that consider the estimated values for overlapping data.

Sensor	τ	m	SVM (epsilon-SVR)		SVM (nu-SVR)		GP		MLP	
			R^2	RMSE	R^2	RMSE	R^2	RMSE	R^2	RMSE
$s1$	1	2	0.5456	0.2132	0.5439	0.1956	**0.6069**	**0.1804**	0.5935	0.1836
$s2$	1	2	0.3484	1.4142	0.3481	1.4257	**0.6051**	**1.0934**	0.5679	1.1547
	5	2	0.2829	1.4699	0.2825	1.4807	0.3583	1.3815	0.2948	1.4563
$s3$	1	2	0.6059	1.7499	0.6040	1.7486	0.8731	0.9796	**0.8794**	**0.9334**

Table 2. Results of the ML techniques for the distributed prediction in $t + 1$.

	SVM (epsilon-SVR)		SVM (nu-SVR)		GP		MLP	
	R^2	RMSE	R^2	RMSE	R^2	RMSE	R^2	RMSE
$s3$	0.6059	1.7499	0.6040	1.7486	0.8731	0.9796	0.8794	0.9334
Distributed	0.5864	1.7979	0.5844	1.7986	**0.8858**	0.9534	0.8805	**0.9270**

were then sent to sensor $s3$, where new forecast models were created for instants $t + 1$ and $t + 2$ using the data itself and the forecasts made by sensors $s1$ and $s2$. Table 2 shows the performance of the ML techniques to create new forecast models of a distributed approach for instant $t + 1$. The parameters of the ML techniques were left in the Weka standard deviation, since the MLP used the learning rate $= 0.3$ and a single layer with 6 hidden neurons. The first line of the table shows the results obtained without the use of distribution for purposes of comparison and the better performance is highlighted in bold.

As can be seen in Table 2, the performance indices diverge with regard to what is the best model created by the ML techniques. This is because the GP achieves a better performance when account is taken of R^2 and MLP has a better performance when the RMSE is considered. Another point that is worth highlighting is that both of the forecast models for $t + 1$, produced by MLP or the GP, improved their performance when a distributed approach was adopted. However, both the SVM kernels had a worse performance with the proposed approach. Table 3 makes use of the same distributed approach by using the forecasts for instant $t + 1$ carried out by sensors $s1$ and $s2$. However, in these experiments, the forecast models created by the ML techniques for sensor $s3$, make forecasts for instant $t + 2$. In this way, the forecasts were made 10 min in

Table 3. Results of the ML techniques of distributed prediction in $t + 2$.

	SVM (epsilon-SVR)		SVM (nu-SVR)		GP		MLP	
	R^2	RMSE	R^2	RMSE	R^2	RMSE	R^2	RMSE
$s3$	0.5487	1.8463	0.5482	1.8453	0.7661	1.3167	**0.7784**	**1.2613**
Distributed	0.5872	1.7943	0.5856	1.7937	0.7961	1.2302	**0.8065**	**1.1787**

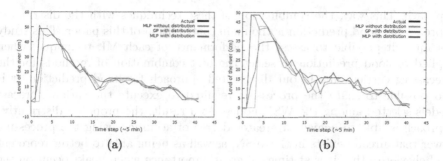

Fig. 2. Graphic comparison between the real-world readings and the forecasts made by the (a) $t+1$ and (b) $t+2$ instants.

advance, which allows warnings to be given with more time for decision-making and thus go further in saving lives and reducing financial losses. The parameters of the ML techniques were kept the same as in the previous experiment. For comparative purposes, the first line of the Table 3 shows the results obtained without the use of the distributed approach and the better performance was highlighted in bold. In a different way from the previous experiment (and as can be seen in Table 3), all the models formed by the ML techniques achieved a better performance when a distributed approach was adopted to make forecasts for instant $t+2$. Since the best results were obtained from the MLP, this time it improved its performance in both the performance indexes followed by the GP. The SVM, which had both the kernels, managed to improve its performance by adopting a distributive approach despite creating models that had a performance that was inferior to that of the other techniques. The performance of the forecast models created by the ML techniques can be illustrate graphically as shown in Fig. 2(a) (for the predictions in $t+1$) and Fig. 2(b) (for the predictions in $t+2$). As can be seen, in both cases there is a discrepancy between the values estimated by the MLP and the real-world readings of the river level when the distributed approach is not employed. This discrepancy is corrected when the distributed approach is adopted and as well as this, the peak values are represented in a better way. However, even when the distributed approach is employed, the GP ends up by underestimating the peak values, and shows a better performance when the readings are more stable. This analysis is of great value because in examining flood prediction, the peak values are very important since they are at the points where the flooding occurs.

5   Conclusion and Future Works

A distributed approach has been set out and examined to increase the accuracy of flood prediction in urban rivers by using data collected by means of WSNs and ML. In addition, three ML techniques were analyzed and compared (SVM with two different kernels, GP and MLP). These techniques were employed and evaluated both to create local forecast models and to create models for a distributive approach. This way, the combination of these techniques with the distributed approach for flood prediction is the main contribution of this paper. This study has made it possible to assess the performance of each ML technique, when applied to flood prediction to select the best combination of methods for the forecasting carried out. In our distributed approach for flood prediction, it is also possible to share the processing required to execute the various forecast models selected among the WSN nodes. As a result, the proposed distributive approach is able to achieve a greater degree of accuracy using the processing power that already exists in the WSN, as well as being able to better represent the behavior of the river at times of great importance when peaks occur on the water level. It thus offers a promising means of tackling the problem of flood prediction. In future work we intend to analyze in greater depth the relationship between the readings and predictions made by the sensors at different points of the hydrographic basin. Our aim is also to embed the distributed approach proposal in the nodes of the REDE system.

References

1. Basha, E.A., Ravela, S., Rus, D.: Model-based monitoring for early warning flood detection. In: Proceedings of 6th ACM SenSys (2008)
2. Damle, C., Yalcin, A.: Flood prediction using time series data mining. J. Hydrol. **333**(2–4), 305–316 (2007)
3. Guha-Sapir, D., Hoyois, P., Below, R.: Annual disaster statistical review 2013: the numbers and trends. Technical report, Centre for Research on the Epidemiology of Disasters (2014)
4. Hall, M., Frank, E., Holmes, G., Pfahringer, B., Reutemann, P., Witten, I.H.: The weka data mining software: an update. SIGKDD Explor. Newsl. **11**(1), 10–18 (2009)
5. Hossain, M., Turna, T., Soheli, S., Kaiser, M.: Neuro-fuzzy(nf)-based adaptive flood warning system for bangladesh. In: ICIEV International Conference (2014)
6. Kar, A., Winn, L., Lohani, A., Goel, N.: Soft computing-based workable flood forecasting model for ayeyarwady river basin of myanmar. J. Hydrol. Eng. **17**(7), 807–822 (2012)
7. Takens, F.: Detecting strange attractors in turbulence. Dyn. Syst. Turbul. **898**, 366–381 (1981)
8. Ueyama, J., Hughes, D., Man, K.L., Guan, S., Matthys, N., Horre, W., Michiels, S., Huygens, C., Joosen, W.: Applying a multi-paradigm approach to implementing wireless sensor network based river monitoring. In: 2010 CDEE International Symposium (2010)
9. Wallemacq, P., Herden, C., House, R.: The human cost of natural disasters 2015: a global perspective. Technical report, Centre for Research on the Epidemiology of Disasters (2015)

A Learning Web Platform Based on a Fuzzy Linguistic Recommender System to Help Students to Learn Recommendation Techniques

Carlos Porcel[1]($\boxtimes$), Maria Jesús Lizarte[2], Juan Bernabé-Moreno[2],
and Enrique Herrera-Viedma[2]

[1] Departament of Computer Science, University of Jaén, Jaén, Spain
cporcel@ujaen.es
[2] Department of Computer Science and Artificial Intelligence, University of Granada,
Granada, Spain
{mjlizarte,jbernabemoreno}@gmail.com, viedma@decsai.ugr.es

Abstract. The rapid advances in Web technologies are promoting the development of new pedagogic models based on virtual teaching. To achieve this personalized services are necessary to provide the users with relevant information, according to their preferences and needs. Recommender systems can be used in an academic environment to improve and assist users in their teaching-learning processes. In this paper we propose a fuzzy linguistic recommender system to facilitate learners the access to e-learning resources interesting for them. By suggesting didactic resources according to the learner's specific needs, a relevance-guided learning is encouraged, influencing directly the teaching-learning process. We propose the combination of the relevance degree of a resource for a user with its quality in order to generate more profitable and accurate recommendations. In addition to that, we present a computer-supported learning system to teach students the principles and concepts of recommender systems.

Keywords: Recommender system · Teaching and learning · Fuzzy linguistic modeling

1 Introduction

The great advances in Web technologies are promoting the development of new pedagogic models, known as e-learning, that complement the existing classical education [7,10] and involve an active role of the students. The new technologies improve the teaching-learning processes, aiding the information broadcasting in an efficient and easy manner, and providing tools for the personal and global communications that allow for the collaborative learning [3,8]. In this academic scope it is essential to provide the students with relevant information, according with their preferences or needs [8].

Recommender systems seek to discover information items that are valuable to the users. In some way, we can see them as personalized services because they

© Springer International Publishing Switzerland 2015
K. Jackowski et al. (Eds.): IDEAL 2015, LNCS 9375, pp. 493–500, 2015.
DOI: 10.1007/978-3-319-24834-9_57

have an independent profile for each user [2,4,12]. As a prerequisite for that, the system needs some information about every user, such as the ratings provided by the users about the viewed or purchased items. One of the most extended methods to generate recommendations is the *collaborative approach* in which the recommendations to a particular user are based upon the ratings provided by those users with similar profiles. In this sense, we can use the connections inherent to an educational community to support collaborative approach, where students rate resources and these ratings are shared with a large community [3,4].

However, this approach tend to fail when the system has few ratings: *cold start problem*. Due to this reason, we suggest to combine collaborative approach with a content-based one to obtain a hybrid recommendation scheme (see Sect. 2.1). Moreover, we also consider that taking into account the quality of the items leads to more sounded recommendations. In everyday life we usually buy well known products or popular products from well-known brands. These products are popular because the common opinion considered that they are high quality and this satisfies consumers. The idea is to combine the estimated relevance through the hybrid approach with the quality of items.

On the other hand, recommender systems are becoming popular tools for reducing information overload and to improve the conversion rate in e-commerce web sites [2,12]. However these are quite complex techniques not necessarily accessible to those student lacking computer skills. For this reason, we also propose to incorporate a library of functions to assist the students in the learning of these systems.

Then, in this work we present a learning Web platform that incorporates:

- A fuzzy linguistic recommender system based on the quality of items to facilitate learners the access to e-learning resources interesting for them. The system measures resources' quality and takes it into account as new factor in the process of generating recommendations.
- A tool to help students to learn recommendation techniques, which consists of a library of functions implementing several recommendation generation approaches. So, the usage of our library allows learners to forget the implementation details and focus only on their use and operation.

The paper is structured as follows. In Sect. 2, the preliminaries are presented. Next in Sect. 3, we describe our proposal. Finally, some concluding remarks and future works are pointed out in Sect. 4.

2   Preliminaries

2.1   Basis of Recommender Systems

Recommender systems try to guide the user in a personalized way towards suitable tasks among a wide range of possible options [2,12]. Personalized recommendations rely on knowing users' characteristics, which might be tastes, preferences as well as the ratings of previously explored items. The way of acquiring this information may vary from implicit information, that is, analyzing users behavior, or explicit information, where users directly provide their preferences.

Other aspect to take into account is the way of generating recommendations. In the literature we can find them mainly pooled in two categories [2]. In the first one authors consider two different approaches: On one side, the *content-based approaches* generate the recommendations taking into account the characteristics used to represent the items and the ratings that a user has given to them. On the other side, the *collaborative approaches* generate recommendations using explicit or implicit preferences from many users, ignoring the items representation. The second one extends the categorization with another three approaches: *Demographic systems*, *Knowledge-based systems* and *Utility-based systems* [2].

Since each approach has certain advantages and disadvantages, depending on the scope settings. In order to combine different approaches to reduce the disadvantages of each one and to exploit their benefits, a widespread solution is the combination of approaches, in what is known as *hybrid approach* [2].

2.2 Fuzzy Linguistic Approach

The fuzzy linguistic approach is a tool based on the concept of linguistic variable proposed by Zadeh [13]. This theory has given very good results to model qualitative information and it has been proven to be useful in many problems. We describe the approaches used in our proposal.

The 2-tuple Fuzzy Linguistic Approach. In order to reduce the loss of information of other methods such as classical or ordinal, in [5] is proposed a continuous model of information representation based on 2-tuple fuzzy linguistic modelling. To define it both the 2-tuple representation model and the 2-tuple computational model to represent and aggregate the linguistic information have to be established.

Let $S = \{s_0, ..., s_g\}$ be a linguistic term set with odd cardinality. We assume that the semantics of labels is given by means of triangular membership functions and consider all terms distributed on a scale on which a total order is defined. In this fuzzy linguistic context, if a symbolic method aggregating linguistic information obtains a value $\beta \in [0, g]$, and $\beta \notin \{0, ..., g\}$, we can represent β as a 2-tuple (s_i, α_i), where s_i represents the linguistic label, and α_i is a numerical value expressing the value of the translation between numerical values and 2-tuple: $\Delta(\beta) = (s_i, \alpha)$ and $\Delta^{-1}(s_i, \alpha) = \beta \in [0, g]$ [5].

In order to establish the computational model negation, comparison and aggregation operators are defined. Using functions Δ and Δ^{-1}, any of the existing aggregation operators can be easily be extended for dealing with linguistic 2-tuples without loss of information [5]. For instance arithmetic mean, weighted average operator or linguistic weighted average operator could be used.

Multi-granular Linguistic Information Approach. A problem modelling the information arises when different experts have different uncertainty degrees on the phenomenon or when an expert has to evaluate different concepts. Then, several linguistic term sets with a different granularity of uncertainty are necessary. In such situations, we need tools to manage multi-granular linguistic information [6]. In [6] a multi-granular 2-tuple fuzzy linguistic modelling based on the concept of linguistic hierarchy is proposed. A *Linguistic Hierarchy LH*, is a set of levels $l(t, n(t))$, where each level t is a linguistic term set with different granularity $n(t)$. In [6] a family of transformation functions between labels from different levels was introduced. To establish the computational model we select a level that we use to make the information uniform and thereby we can use the defined operator in the 2-tuple model. This result guarantees that the transformations between levels of a linguistic hierarchy are carried out without loss of information.

3   Web Platform Description

3.1   Fuzzy Linguistic Recommender System Based on the Quality

In this work we face the recommendation process of pedagogical resources as a task with two elements: the resource estimated relevance along with its quality. In order to represent the different concepts to be assesses, we will use different label sets $(S_1, S_2, ...)$ selected from a LH [6]. In our system the concepts represented are the following:

- **Importance degree** of a discipline regarding to a resource scope, which is assessed in S_1.
- **Similarity degree** among resources or students, which is assessed in S_2.
- **Relevance degree** estimated of a resource for a student, which is assessed in S_3.
- **Satisfaction degree** of a learner regarding to a recommended resource, which is assessed in S_4.
- **Preference degree** of a resource regarding to other, which is assessed in S_5.

Level 2 (5 labels) is used to represent importance and preference degrees ($S_1 = S^5$ and $S_5 = S^5$) and level 3 (9 labels) is used to represent similarity, relevance and satisfaction degrees ($S_2 = S^9$, $S_3 = S^9$ and $S_4 = S^9$).

Resources Representation. When a new resource is added, the system obtains an internal representation mainly based in its scope. We use a *vector model* to represent the resource scope and a classification composed by 25 disciplines (see Fig. 1), i.e., a resource i is represented as $VR_i = (VR_{i1}, VR_{i2}, ..., VR_{i25})$, where each component $VR_{ij} \in S_1$ is a linguistic assessment that represents the importance degree of the discipline j with regard to the scope of i. These importance degrees are assigned by the teachers when they add new resources.

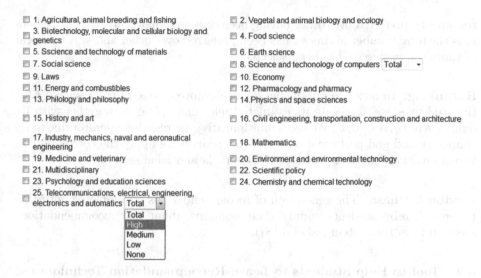

Fig. 1. Interface to define the disciplines of the resource scope.

Students' Profiles. To acquire students' preferences, we use the proposed method in [11]. It consists of requesting students to provide their preferences over 5 resources (label of S_5), using a incomplete fuzzy linguistic preference relation. Furthermore, in accordance with results presented in [1], it is enough for students to provide a row of the relation. Then, we use the method presented in [1] to complete the relation and then we can obtain a vector that represent students' preferences. That is, the preferences of student x are represented as a vector $VU_x = (VU_{x1}, VU_{x2}, ..., VU_{x25})$ (the same 25 disciplines used to represent the resources). Moreover, in this way we manage to reduce the cold start problem, since thanks to the supplied information by students when they register into the system, they can already start receiving recommendations.

Relevance Estimation. In order to estimate the relevance of the resource i for the student e, i.e. $i(e) \in S_3$, we implement a hybrid approach which switches between content-based and collaborative approaches [2]. The former is applied when a new item is inserted and the later when a new student is registered. We rely on a matching process by similarity measures among vectors (label of S_2). Particularly, we use the standard cousin measure, but defined it in a linguistic context.

Quality Estimation. Following the method described to acquire the students' preferences, we can count the number of times a resource is chosen to be shown among the outstanding resources, as well as the number of times it has been chosen over the rest. The displayed resources will vary over time, so that the system records each time a resource is chosen and the number of times it is preferred to other. Hence, we compute the quality of a resource i as the probability of this resource be preferred over other having been selected, that is $q(i) = p_i/s_i$, where p_i is the total number of times i has been preferred over other and s_i is the total of times the resource i has been selected.

Reranking. In order to obtain the final relevance degree of the resource i for the student e, we aggregate its estimated relevance $i(e) \in S_3$ together with its quality score $q(i) \in [0,1]$. We use a multiplicative aggregation operator due to its simplicity and god performance. In all this process we apply the corresponding transformation functions to adapt the intervals and label sets.

Feedback Phase. The generation of recommendations is completed with this phase, whereby students supply their opinions about the recommendations received from the system (label of S_4).

3.2 Tool to Help Students to Learn Recommendation Techniques

We also propose a web-hosted tool to support the learning, testing and improvement around recommender systems. With this tool, students can work with different recommendation schemas without having to deal with implementation details. So, our tool speeds up the learning of recommendation approaches by abstracting away the technicalities related to their implementation. At present, we have implemented the following schemas: content-based, collaborative and trust-based. The programming language was R and for the integration of the schemas in the web platform we used Shiny[1]. At present, we have a trial version. When the user logs in, following options are displayed:

- Recommendations estimation. The user can select the recommendation approach of his choice as well as the data set for the training and testing (at present we offer Epinions[2] [9] and MovieLens[3] data sets). After that, the compatibility of the selected data set and the recommendation schema is checked: the MovieLens data set is just compatible with collaborative and content-base schemas, while the Epinions data set is just suitable for the trust-based one. At last, the user is offered the choice of showing a list of recommendations or estimating the suitability of a resource for a particular student.
- Approach assessment. The user selects a recommendation approach, a data set (given the restrictions we have mentioned above) and proceeds with the parameterization of the selected schema. Under this step, the performance of the recommender is assessed executing different experiments and providing the MAE (Mean Absolute Error).
- Recommender systems related documentation. Under this menu point, we gather all relevant information about recommender systems, including presentations, definitions, working examples, etc.

---

[1] http://shiny.rstudio.com/.
[2] http://www.trustlet.org/wiki/Epinions.
[3] http://grouplens.org/datasets/movielens/.

3.3 System Evaluation

The Web tool has been validated with the pupils of the Economic Faculty in University of Jaén.

Web Platform Evaluation. The goal is to perform a set of tests, to try to get obtaining a system without errors, ensuring software quality. To achieve this we conducted a series of tests known as *black box*. Through these tests we studied the behavior of the trial version from the point of view of the inputs and outputs without getting into details about how it internal workings. Therefore, black box testing is especially suitable for those modules that will be user interface (as in our case), to check that the functionality is desired. Each of the expected responses of the system are called checkpoints.

Once all test cases have been designed, we perform every single tests to check each of the existing checkpoints and so we were able to verify that all the responses were in line with expected responses.

Recommendation Approaches Evaluation. We have developed online experiment to test the accuracy of the system predicting the rating that student would give to a recommended resource. To do that, we used a data set with 200 pedagogical resources of different areas and 30 students. Later, 100 new resources were added and recommended. The rating given by students over these recommendations of the new resources were registered in the system. This rating was compared with the estimated rating by the system, it allowed us to estimate the Mean Absolute Error (MAE), which measures the mean absolute deviation between a estimated value and the real value assigned by the student. In our case, we obtained an average MAE of *0.69*. To compare results, we repeated the experiments but without taking into account the quality of resources, obtaining an average MAE of *0.74*. The result implies that through the application of the new approach, we get an improvement of 7 %.

4 Conclusions

The advances in Web technologies are promoting the development of new pedagogic models which improve the teaching-learning processes, aiding the information broadcasting in an efficient and easy manner, and providing tools that implement the collaborative learning. In this paper we have proposed a fuzzy linguistic recommender system with a dual perspective: not only finding relevant resources, but also imposing that the resource shall be valid from the standpoint of the quality of items. This facilitates learners the access to e-learning resources interesting for them. Suggesting didactic resources according to the learner's specific needs, a relevance-guided learning is encouraged, influencing directly the teaching-learning process. Moreover, a computer-supported learning system is presented to teach students the principles and concepts of recommender systems. We developed online tests and obtained satisfying experimental results.

As future work, we consider to studying automatic techniques to establish the internal representation of resources or users' profiles, such as educational data mining techniques.

Acknowledgments. This paper has been developed with the financing of Projects UJA2013/08/41, TIN2013-40658-P, TIC5299, TIC-5991, TIN2012-36951 co-financed by FEDER and TIC6109.

References

1. Alonso, S., Chiclana, F., Herrera, F., Herrera-Viedma, E., Alcalá-Fdez, J., Porcel, C.: A consistency-based procedure to estimating missing pairwise preference values. Int. J. Intell. Syst. **23**, 155–175 (2008)
2. Burke, R., Felfernig, A., Göker, M.: Recommender systems: an overview. AI Mag. **32**, 13–18 (2011)
3. Dascalu, M., Bodea, C., Moldoveanu, A., Mohora, A., Lytras, M., Ordoñez de Pablos, P.: A recommender agent based on learning styles for better virtual collaborative learning experiences. Comput. Hum. Behav. **45**, 243–253 (2015)
4. Goga, M., Kuyoro, S., Goga, N.: A recommender for improving the student academic performance. Procedia - Soc. Behav. Sci. **180**, 1481–1488 (2015)
5. Herrera, F., Martínez, L.: A 2-tuple fuzzy linguistic representation model for computing with words. IEEE Trans. Fuzzy Syst. **8**(6), 746–752 (2000)
6. Herrera, F., Martínez, L.: A model based on linguistic 2-tuples for dealing with multigranularity hierarchical linguistic contexts in multiexpert decision-making. IEEE Trans. Syst. Man Cybern. Part B: Cybern. **31**(2), 227–234 (2001)
7. Kearsley, G.: Online Education: Learning and Teaching in Cyberspace. Wadsworth, Belmont (2000)
8. Mamat, N., Yusof, N.: Learning style in a personalized collaborative learning framework. Procedia - Soc. Behav. Sci. **103**, 586–594 (2013)
9. Massa, P., Avesani, P.: Trust metrics in recommender systems. In: Golbeck, J. (ed.) Computing with Social Trust, pp. 259–285. Springer, London (2009)
10. Popovici, A., Mironov, C.: Students' perception on using elearning technologies. Procedia - Soc. Behav. Sci. **180**, 1514–1519 (2015)
11. Porcel, C., Herrera-Viedma, E.: Dealing with incomplete information in a fuzzy linguistic recommender system to disseminate information in university digital libraries. Knowl.-Based Syst. **23**, 32–39 (2010)
12. Tejeda-Lorente, A., Porcel, C., Peis, E., Sanz, R., Herrera-Viedma, E.: A quality based recommender system to disseminate information in a university digital library. Inf. Sci. **261**, 52–69 (2014)
13. Zadeh, L.: The concept of a linguistic variable and its applications to approximate reasoning. Part I, Inf. Sci. **8**, 199–249 (1975). Part II, Inf. Sci. **8**, 301–357 (1975). Part III, Inf. Sci. **9**, 43–80 (1975)

Comparison of Clustering Methods
in Cotton Textile Industry

Dragan Simić[1]([⊠]), Konrad Jackowski[2], Dariusz Jankowski[2],
and Svetlana Simić[3]

[1] Faculty of Technical Sciences, University of Novi Sad,
Trg Dositeja Obradovića 6, 21000 Novi Sad, Serbia
dsimic@eunet.rs
[2] Department of Systems and Computer Networks,
Wroclaw University of Technology,
Wybrzeże Wyspiańskiego 27, 50-370 Wrocław, Poland
{konrad.jackowski,dariusz.jankowski}@pwr.edu.pl
[3] Faculty of Medicine, University of Novi Sad,
Hajduk Veljkova 1-9, 21000 Novi Sad, Serbia
drdragansimic@gmail.com

Abstract. Clustering is the task of partitioning data objects into groups, so that
the objects within a cluster are similar to one another and dissimilar to the
objects in other clusters. The efficiency random algorithm for good k is used to
estimate the optimal number of clusters. In this research two important clustering algorithms, namely centroid based k-means, and representative object
based fuzzy c-means clustering algorithms are compared in the original
real-world U.S. cotton textile and apparel imports data set. This data set is not
analyzed very often, it is dictated by business, economics and politics environments and its behaviour is not well known. The analysis of several different
real-world economies and industrial data sets of one country is possible to
predict it's economic development.

Keywords: Data clustering · Number of clusters · k-means algorithm · Fuzzy
c-means · Random algorithm

1 Introduction

Clustering is deemed to be one of the most difficult and challenging problems in
machine learning, particularly due to its unsupervised nature. The unsupervised nature
of the problem implies that its structural characteristics are not known, except if there is
some sort of domain knowledge available in advance. The purpose of clustering is to
identify natural groupings of data from a large data set to produce a concise representation of a system's behavior. One of the main difficulties for cluster analysis is that,
the optimal and correct number of clusters of different types of datasets is seldom
known in practice.

Clustering is the process of assigning data objects into a set of disjoint groups
called clusters so that objects in each cluster are more similar to each other than objects

© Springer International Publishing Switzerland 2015
K. Jackowski et al. (Eds.): IDEAL 2015, LNCS 9375, pp. 501–508, 2015.
DOI: 10.1007/978-3-319-24834-9_58

from different clusters. The goal of clustering is to group similar objects in one cluster and dissimilar objects in different clusters. Data Clustering is a technique of finding similar characteristics among the data set which are always hidden in nature and dividing them into groups, called clusters. Different data clustering algorithms exhibit different results, since they are very sensitive to the characteristics of original data set especially noise and dimension. Originally, clustering was introduced to the data mining research as the unsupervised classification of patterns into groups [1].

Clustering techniques offer several advantages over manual grouping process. First, a clustering algorithm can apply a specified objective criterion consistently to form the groups. Second, a clustering algorithm can form the groups in a fraction of time required by manual grouping, particularly if long list of descriptors or features is associated with each object. The speed, reliability and consistency of clustering algorithm in organizing data represent an overwhelming reason to use it. Obviously, this type of problem is encountered in many applications, such as text mining, gene expressions, customer segmentations, and image processing, to name just a few.

Synthetic or artificial dataset on one side and the industrial data on other are two data types which are used in recent researches. In many industrial cases it may be difficult to obtain large datasets. Moreover, the measurements may contain a high noise level and often is asynchronous. For that reason synthetic data sets are very often used. The synthetic datasets, usually initially partitioned into segments the magnitude of the eigenvalues and their cumulative rate to their sum show that two segments are sufficient to approximate the distribution of the data with 98 % accuracy. Obviously, this analysis can be fully automatized.

But the real-life is not artificial, and this research is focused on industrial data set. The U.S. cotton textile and apparel import, real-world dataset investigated in this research is originally taken from TRADE DATA U.S. Textiles and Apparel Imports [2]. But, in this research the used data sets encompasses period between 1990 and 2012, and includes 200 countries. Some data are eliminated because countries no longer exist or because data are incomplete.

The motivation for this research is to analyse real-world data set, which is produced in real-life, the behaviour is not well known and the data set is dictated by business, economics and politics environments. The future data set behaviour is not well known and it is not easy to predict. The contribution of this research is to analyze one of real-world data set, whose behavior is not well known, and not easily predicted. This data set is not analyzed very often as are, for example, data sets from UCI Machine Learning Repository, which are mostly artificial data sets. The contribution of this research is important for the business and economics. Also, analyzing real-world data set it is possible, partially, to predict economic development.

The rest of the paper is organized in the following way: Sect. 2 provides some approaches of data clustering techniques and related work. Section 3 presents k-means clustering method, fuzzy c-means clustering method, random algorithm for good k and data collection of the original data set. The experimental results of comparison of k-means and fuzzy c-means clustering algorithms on original real-world U.S. cotton textile imports data set of twenty countries are presented in Sect. 4. Finally, Sect. 5 gives concluding remarks.

2 Clustering Method and Related Work

Clustering is widely used in different fields such as biology, psychology, and economics. Several clustering algorithms have been proposed in this paper. Cluster analysis doesn't use category labels that tag objects with prior identifications. The absence of category labels distinguishes cluster analysis from discriminant analysis, pattern recognition and decision analysis. Clustering differs from classification in that there is no target variable for clustering. Instead, clustering algorithms seek to segment the entire data set into relatively homogeneous subgroups or clusters [3]. It doesn't use category labels that tag objects with prior identifications. The absence of category labels distinguishes cluster analysis from discriminant analysis, pattern recognition and decision analysis. The objective of clustering is simply to find a convenient and valid data organization, not to establish rules for separating future data into categories.

General references regarding data clustering are presented in [4]. A very good presentation of contemporary data mining clustering techniques can be found in the textbook [5]. Clustering has always been used in statistics and science. The classic introduction to pattern recognition framework is given in [6]. Machine learning clustering algorithms were applied to image segmentation and computer vision [7]. Statistical approaches to pattern recognition are discussed in [8]. Clustering is also widely used for data compression in image processing, which is also known as vector quantization [9].

Clustering in data mining was brought to life by intense developments in information retrieval and text mining [10], GIS or astronomical data [11], Web applications [12], texture clustering and classification [13], DNA analysis [14] and computational biology [15].

The applicability of clustering is manifold, ranging from market segmentation [16] and image processing [17] through document categorization and web mining [18]. An application field that has been shown to be particularly promising for clustering techniques is bioinformatics [19]. Indeed, the importance of clustering gene-expression data measured with the aid of microarray and other related technologies has grown fast and persistently over the recent years [20]. Clustering methods are widely used in application in finance industry [21] and chemical industrial [22].

3 Data Clustering

Data Clustering is a technique of partitioning a data set without prior known information. It finds its use in most of the applications where unsupervised learning occurs. The different kinds of data clustering algorithms when used for varying data sets produce different kinds of results based on the initial input parameters, environment conditions, and the data set nature. In such a scenario, since there are no predefined classes or groups known in clustering process, finding whether an appropriate metric for measuring in found cluster configuration, number of clusters, cluster shapes is acceptable or not, has always been an issue.

3.1   *K*-means Clustering Method

K-means is a centroid-based partitioning technique which uses the centroid, also called center of a cluster. Conceptually, the centroid of a cluster is its center point. The centroid can be defined in various ways such as by the mean of the objects assigned to the cluster. *K*-means is one of the most widely used and studied clustering formulations [23]. Among them *k*-means method is a simple and fast clustering technique, and probably the most frequently used clustering algorithm.

3.2   Fuzzy C-means Clustering Method

Fuzzy c-means clustering method is introduced in [24], extend from hard c-mean clustering method. Fuzzy c-means is an unsupervised clustering algorithm that is applied to wide range of problems connected with feature analysis, clustering and classifier design.

3.3   Data Collection

The U.S. cotton textile and apparel import, real-world dataset investigated in this research is originally taken from TRADE DATA U.S. Textiles and Apparel Imports [2]. The entire data set presented in U.S. cotton textile and apparel imports, originally contains cotton textile and apparel import data for the period between 1988 and 2012 and includes 238 countries. In this research three data sets, for years: 1990, 2000, 2010 and 200 countries are used. Some data, in defined data sets, are eliminated because countries no longer exist or the data are incomplete. But data set used here is quite sufficient to show changes in world's cotton production and changes in export cotton textile to the USA.

3.4   Random Algorithm for Good *k*

A major challenge in data clustering is the estimation of the optimal number of clusters. There are many methods where some researches were suggested. Some of them are: Rule of thumb for number of clusters; Information Criterion Approach – Akaike's information criterion (AIC) [25], Bayesian inference criterion (BIC) [26]; Choosing *k* using the silhouette [27]; Gap statistics [28].

Random algorithm for good *k* is used in this research. It is done in the following way: a random subset of data points is selected and used as input to the random algorithm for clustering. Then, the cluster centers as mean/median are computed for each cluster group. It is important to mention that the number of clusters is less or equal ($\leq$) to subset data points. Usually, the data points in subset data make less than 5 % of all objects in data set. Finally for each instance that was not selected in the subset, its distance to each of the centroids is simply computed and assigned to the closest one. Considering that data set contains $n = 200$ data points the expected number of clusters is less or equal to 10.

4 Experimental Results

The experimental results can be presented in two steps. First, experimental results from *Random Algorithm for good k*, optimal number of clusters, will be presented. According to the nature of algorithm, it is repeated 300 times, hundred times for every data set (1990, 2000, 2010), and experimental results for optimal number of clusters are summarized and presented in Table 1.

Table 1. Frequency of number of clusters on 'Random algorithm for good k'

Number of clusters	Frequency 1990	Frequency 2000	Frequency 2010
10	33	13	3
9	34	19	10
8	23	23	15
7	6	23	29
6	3	10	24
5	1	9	11
4	0	2	6
3	0	1	2
Experiments	100	100	100
Suggestion clusters	**7.69**	**7.88**	**6.28**

On the other side, *Rule of thumb for number of clusters* is defined as $k \approx \sqrt{n/2}$ where n is number of data points and k is number of clusters. Considering that data set contains $n = 200$ data points, it can be assumed that the optimal number of clusters is $k = 10$. Suggested number of clusters on 'Random algorithm for good k' are (7.69 7.88 6.28) respectable the year 1990, 2000, 2010. According to previous results it is decided to use eight clusters for the rest of this research.

Table 2 shows experimental results of clusters, number of countries in clusters, and cluster value in (thousands of pounds) for k-means and fuzzy c-means clustering method for years 1990, 2000, and 2010.

Table 2. Eight clusters, number of countries in the cluster (Co.), cluster value (millions of pounds) for k-means and fuzzy c-means for 1990, 2000 and 2010 of U.S. cotton textile imports

	1990				2000				2010			
	k-means		Fuzzy c-means		k-means		Fuzzy c-means		k-means		Fuzzy c-means	
Cl.	Co.	Value	Co.	Value	Co.	Value	Co.	Value	Co.	Value	Co.	Value
1	2	343,431	2	343,469	1	1,087,002	1	1,086,993	1	2,147,483	1	2,147,483
2	3	147,247	3	140,117	5	362,971	5	364,656	3	792,821	3	788,281
3	4	82,174	3	90,112	3	213,168	3	211,710	3	518,589	3	524,083
4	7	50,885	6	56,429	5	158,490	5	156,511	3	306,170	3	305,713
5	10	29,678	12	31,292	5	104,215	6	98,526	7	151,995	6	156,016
6	14	12,841	12	13,838	5	66,958	5	55,796	10	82,002	8	91,948
7	17	4,752	19	4,974	19	25,351	22	21,237	13	30,443	13	36,891
8	143	262	143	221	157	1,340	153	891	160	737	163	812

In this research selected and used data set consists of U.S. cotton textile imports for years 1990, 2000, and 2010, 200 countries and 8 clusters. Every country is clustered depending on the value of U.S. cotton textile imports, for every year respectively. Experimental results of clustering and ranking twenty out of two-hundred countries in eight clusters for four measurements of U.S. cotton textile imports are presented in Table 3.

Table 3. Experimental results of clustering and ranking 20 of 200 countries in 8 clusters for 1990, 2000 and 2010 of U.S. cotton textile imports (thousands of pounds) (**first cluster bold**, second cluster underline, *third cluster italic*), Cl. 'k' – position in k-means cluster, Cl. 'k' – position in fuzzy c-means cluster

		1990			2000			2010		
	Country	cotton	Cl.'k'	Cl.'c'	cotton	Cl.'k'	Cl.'c'	cotton	Cl.'k'	Cl.'c'
1	China	**362,211**	1	1	<u>417,299</u>	2	2	**3,310,700**	1	1
2	Hong Kong	**324,650**	1	1	*321,282*	3	3	16,630	7	8
3	China (Taiwan)	<u>180,525</u>	2	2	147,089	5	4	40,881	6	7
4	Pakistan	<u>131,029</u>	2	2	<u>541,966</u>	2	2	925,986	2	2
5	India	<u>130,189</u>	2	2	*367,701*	3	2	<u>788,252</u>	2	2
6	South Korea	*94,847*	3	3	112,650	6	5	85,823	5	7
7	Mexico	*89,710*	3	3	**1,416,516**	1	1	<u>628,563</u>	2	3
8	Philippines	*77,208*	3	3	130,464	6	5	71,522	6	7
9	Indonesia	*66,930*	3	4	172,558	5	4	*380,960*	3	4
10	Dominican Rep.	60,720	4	4	253,419	4	3	70,711	6	7
11	Thailand	58,393	4	4	173,370	5	4	133,693	5	6
12	Bangladesh	53,975	4	4	247,243	4	3	<u>628,189</u>	2	3
13	Brazil	51,565	4	4	60,455	7	6	33,025	7	7
14	Turkey	48,101	4	4	226,583	4	3	69,778	6	7
15	Sri Lanka	42,525	4	5	100,885	6	5	89,768	5	6
16	Malaysia	40,917	4	5	71,831	7	6	36,603	6	7
17	Costa Rica	36,937	5	5	118,369	6	5	14,140	7	8
18	Singapore	33,078	5	5	29,730	7	7	1,018	8	8
19	Canada	32,389	5	5	272,652	4	3	40,065	6	7
20	Guatemala	31,622	5	5	150,651	5	4	117,633	5	6
...	...	...	...	...	...	...	...	...	...	...
28	Honduras	16,429	6	6	<u>454,554</u>	2	2	*429,508*	3	4
30	El Salvador	14,261	6	6	263,241	4	3	*284,013*	3	5
...	...	...	...	...	...	...	...	...	...	...
113	Cambodia	12	8	8	92,702	6	5	*272,825*	3	5

Experimental results show that it is possible to cluster and rank countries in economy and business. Presented results also show trend of cotton production in the world. Cell backgrounds for countries with different clusters are marked in light grey. This research confirms that different data clustering methods, in this case k-means clustering algorithm and c-means means clustering algorithm exhibit different results for the same data set. This is evident for all tested data sets (1990, 2000, 2010.) and it is particularly characteristic of data set for year 2010.

It can be seen that some countries develop much more than the others, such as China. Also, it can be seen that the number of high producers in 2010 is smaller than in 1990. In 1990, in first five clusters there were 26 countries but in 2010, in the same five clusters, there were only 14 countries, which means that big countries produced even more, while the smaller countries produced less.

5 Conclusion and Future Work

This paper offers a comparison between k-means clustering method and c-means clustering method. The U.S. cotton textile and apparel import, real-world data set investigated in this research is, originally, taken from TRADE DATA U.S. Textiles and Apparel Imports.

This research confirms the following: Different clustering algorithms produce different partitions because they impose different structure on the data, and that no single clustering algorithm is optimal.

Experimental results encourage further research. In future work, comparison between these and new clustering algorithms in larger data set of cotton textile industry will be attempted. Some new ensemble algorithms will be added in future researches.

Acknowledgments. The authors acknowledge the support for research project TR 36030, funded by the Ministry of Science and Technological Development of Serbia.

References

1. Jain, A.K., Dubes, R.C.: Algorithms for Clustering Data. Prentice Hall, Upper Saddle River (1988)
2. http://otexa.trade.gov/Msrcat.htm. Accessed 29 April 2015
3. Larose, D.T.: Discovering Knowledge in Data: An Introduction to Data Mining. Wiley, New York (2005)
4. Spath, H.: Cluster Analysis Algorithms. Ellis Horwood, Chichester (1980)
5. Han, J., Kamber, M.: Data Mining. Morgan Kaufmann Publishers, Burlington (2001)
6. Duda, R., Hart, P.: Pattern Classification and Scene Analysis. Wiley, New York (1973)
7. Jain, A.K., Murty, N., Flynn, P.J.: Data clustering: a review. ACM Comput. Surv. **31**(3), 264–323 (1999)
8. Dempster, A., Laird, N., Rubin, D.: Maximum likelihood from incomplete data via the EM algorithm. J. R. Stat. Soc. Ser. B **39**(1), 1–38 (1977)
9. Gersho, A., Gray, R.M.: Vector quantization and Signal Compression. Communications and Information Theory. Kluwer Academic Publishers, Norwell (1992)
10. Steinbach, M., Karypis, G., Kumar, V.: A comparison of document clustering techniques. In: 6th ACM SIGKDD, World Text Mining Conference, Boston (2000)
11. Ester, M., Frommlet, A., Kriegel, H.P., Sander, J.: Spatial data mining: database primitives, algorithms and efficient DBMS support. Data Min. Knowl. Discov. **4**(2–3), 193–216 (2000)

12. Heer, J., Chi, E.: Identification of web user traffic composition using multimodal clustering and information scent. In: 1st SIAM ICDM, Workshop on Web Mining, Chicago, pp. 51–58 (2001)
13. Petrov, N., Georgieva, A., Jordanov, I.: Self-organizing maps for texture classification. Neural Comput. Appl. **22**(7–8), 1499–1508 (2013)
14. Tibshirani, R., Hastie, T., Eisen, M., Ross, D., Botstein, D., Brown, P.: Clustering methods for the analysis of DNA microarray data. Department of Statistics, Stanford University, Stanford, Technical report. http://statweb.stanford.edu/~tibs/ftp/jcgs.ps. Accessed 29 April 2015
15. Piórkowski, A., Gronkowska–Serafin, J.: Towards precise segmentation of corneal endothelial cells. In: Ortuño, F., Rojas, I. (eds.) IWBBIO 2015, Part I. LNCS, vol. 9043, pp. 240–249. Springer, Heidelberg (2015)
16. Bigus, J.P.: Data Mining with Neural Networks. McGraw-Hill, New York (1996)
17. Jain, A.K., Dubes, R.C.: Algorithms for Clustering Data. Prentice Hall, Upper Saddle River (1988)
18. Mecca, G., Raunich, S., Pappalardo, A.: A New algorithm for clustering search results. Data Knowl. Eng. **62**(3), 504–522 (2007)
19. Valafar, F.: Pattern recognition techniques in microarray data analysis: a survey. Ann. N. Y. Acad. Sci. **980**, 41–64 (2002)
20. Jiang, D., Tang, C., Zhang, A.: Cluster analysis for gene expression data: a survey. IEEE Trans. Knowl. Data Eng. **16**(11), 1370–1386 (2004)
21. Das, N.: Hedge fund classification using k-means clustering method. In: 9th International Conference on Computing in Economics and Finance (2003) http://www.ijarcsms.com/docs/paper/volume1/issue6/V1I6-0015.pdf. Accessed 25 June 2015
22. Shi, W., Zeng, W.: Application of k-means clustering to environmental risk zoning of the chemical industrial area. Front. Environ. Sci. Eng. **8**(1), 117–127 (2014)
23. Fukunaga, K.: Introduction to Statistical Pattern Recognition. Academic Press, San Diego (1990)
24. Bezdek, J.C.: Pattern Recognition with Fuzzy Objective Function Algorithms. Springer, New York (1981)
25. Akaike, H.: A new look at statistical model identification. IEEE Trans. Autom. Control **19**(6), 716–723 (1974)
26. Schwarz, G.: Estimating the dimension of a model. Ann. Stat. **6**(2), 461–464 (1978)
27. Milligan, G.W., Cooper, M.C.: An examination of procedures for determining the number of clusters in a data set. Psychometrika **50**(2), 159–179 (1985)
28. Tibshirani, R., Walther, G., Hastie, T.: Estimating the number of clusters in a data set via the gap statistic. J. R. Stat. Soc. **63**(2), 411–423 (2001)

A Belief Function Reasoning Approach to Web User Profiling

Luepol Pipanmaekaporn and Suwatchai Kamonsantiroj[✉]

Department of Computer and Information Science,
King Mongkut's University of Technology North Bangkok, Bangkok 10800, Thailand
{luepolp,suwatchaik}@kmutnb.ac.th

Abstract. This paper presents a novel approach to web user profiling. Our proposed approach consists of two main parts. The first part focuses on discovering user interests in a user feedback collection, usually including relevant and irrelevant documents. Frequent pattern mining widely used in data mining community is applied to extract user feedback information. The second part is to represent user profiles. We introduce a novel user profile model based on belief function reasoning. In this model, the user profile is described by a probability distribution over the user feedback information extracted. Experimental results on an information filtering task show that the proposed approach clearly outperforms several baseline methods.

Keywords: Web user profile · User interests discovery · Belief function reasoning · Relevance feedback

1 Introduction

With the explosive growth of information available on the Web, there is an imminent need for more effective and efficient technologies to support web users to find information relevant to them. As each user has different topics of interest, user profiles play an important role in describing the difference of information needs. A user profile captures user's interests and preferences in providing a user with relevant information. The most commonly used representation of Web user profiles is based on *keyword-based* approach [3] that extracts a finite set of keywords (or terms) from a user feedback collection, usually including *relevant* and *irrelevant* documents, to describe user's interests. A profile vector of terms can be accurately learnt from this data using a variety of machine learning techniques. Despite this, the keyword-based profiling approach does have limitations. First of all, large amounts of terms in the feedback text result in not only high-dimensional profile vectors extracted, but also inappropriate terms easily chosen. Secondly, the keyword-based approach always suffers from failure to capture semantic relationships among terms.

Pattern-based approaches to Information filtering (IF) have given more attention [8,9,11]. The idea is to mine closed sequential patterns widely used in data

© Springer International Publishing Switzerland 2015
K. Jackowski et al. (Eds.): IDEAL 2015, LNCS 9375, pp. 509–516, 2015.
DOI: 10.1007/978-3-319-24834-9_59

mining community to capture semantic information hidden in feedback texts and then use them to construct a user profile model. For example, in [9], a collection of closed sequential patterns discovered in a collection of relevant documents is used to score a document based on their appearance in the document. In [8], a novel method for using the closed patterns in text was proposed to solve the problem of using long patterns, where the discovered patterns in text is deployed into a distribution of frequent terms and then use the term distribution for the purpose of user profiling. A revising method of pattern deploying that incorporates negative feedback information was proposed [11]. Experimental results demonstrated that pattern-based approaches achieve encouraging performance in comparing with keyword-based models.

Nevertheless, there are two major issues related to the pattern-based scheme. The first issue is how to deal with the large number of the discovered closed patterns that may include meaningless (or noisy) patterns in describing user interests. The second issue is how to utilise the discovered patterns to significantly improve the system performance. This paper presents a novel approach to Web user profiling. It basically consists of two main parts. The first part involves pattern discovery in user feedback data. Instead of closed sequential patterns, we mine closed frequent patterns from the feedback collection. We propose a novel pattern pruning method that makes use of negative feedback information. The second part is to use the discovered patterns to represent a user profile. We introduce a novel user profile model based on belief function reasoning. In this model, the user profile is described as a probability distribution over the discovered patterns. Experimental results on an information filtering task show that the proposed approach clearly outperforms several baseline methods.

2 Basic Definitions

2.1 Closed Patterns in Text

Let D be a training set of documents, including a set of positive (relevant) documents, D^+, and a set of negative (irrelevant) ones, D^-. We assume that all documents are split into paragraphs. Thus, a given document d consists of a set of paragraphs $PS(d)$. Let $T = \{t_1, t_2, \ldots, t_m\}$ be a set of terms which are extracted from D^+. Given X be a set of terms (called a *termset*) in document d, $coverset(X)$ denotes the covering set of X for d, which includes all paragraphs $dp \in PS(d)$ where $X \subseteq dp$, i.e., $coverset(X) = \{dp | dp \in PS(d), X \subseteq dp\}$. The *absolute* support of X is the number of occurrences of X in $PS(d) : sup_a(X) = |coverset(X)|$. The *relative* support of X is the fraction of the paragraphs that contain the pattern: $sup_r(X) = \frac{|coverset(X)|}{|PS(d)|}$. A termset X called *frequent pattern* if its sup_a (or sup_r) $\geq min\_sup$, a minimum support.

Given a set of paragraphs $Y \subseteq PS(d)$, we can define its *termset*, which satisfies $termset(Y) = \{t | \forall dp \in Y \Rightarrow t \in dp\}$ By defining the closure of X as: $Cls(X) = termset(coverset(X))$ a pattern (or termset) X is *closed* if and only if $X = Cls(X)$. Let X be a closed pattern. We have $sup_a(X_1) < sup_a(X)$ for all patterns $X_1 \supset X$.

2.2 The Transferable Belief Model

Transferable Belief Model (TBM) [6] is a model for the representation of quantified beliefs held by an agent. In TBM, beliefs can be held at two levels: (1) a *credal* level where beliefs are quantified and represented by belief functions and (2) a *pignistic* level where beliefs can be used to make decisions by maximizing expected utilities. In the credal level, TBM deals with a finite set of exclusive and exhaustive propositions, called the *frame of discernment* (denoted by Ω). All the subsets of Ω belong to the power set of Ω, denoted by 2^Ω. A strength of subset of elements in Ω is given by the definition of a mass distribution function $m : 2^\Omega \rightarrow [0, 1]$, which provides a measure of uncertainty, applied over all the subsets of elements in the frame of discernment. The mass function also satisfies the following properties: (1) $m(\emptyset) = 0$ and (2) $\sum_{A \in 2^\Omega} m(A) = 1$. In the pignistic level, probability masses applied over all the subsets of elements in the frame of discernment can be used to infer the mass for the single elements as the means to make decisions [5]. The masses are represented by probability functions called *Pignistic* probabilities. The pignistic probability is defined as: $BetP(A) = \sum_{B \subseteq \Omega} \frac{|A \cap B|}{|B|} \frac{m(B)}{(1-m(\emptyset))}$ for all subsets $A \subseteq \Omega$.

3 User Interests Discovery in Relevance Feedback

We firstly apply closed pattern mining to the positive feedback data to capture relevant information of the user. The pattern mining allows us to capture syntactic and semantic relationships among terms in sentences or paragraphs without language knowledge. Furthermore, many noisy terms and patterns can be naturally removed with the minimum support threshold and the closure constraint. In this work, we adopt $SPMining$ algorithm [9] to effectively discover frequent closed patterns from both the positive D^+ and negative data D^-. For all positive documents $d_i \in D^+$, the $SPMining$ algorithm discovers all patterns, P_i, based on a given min_sup. Let $P_1, P_2, \ldots, P_n$ be the sets of patterns for all documents $d_i \in D^+ (i = 1, \ldots, n)$, where $n = |D^+|$. The result of this algorithm is a collection of patterns P^+ collected from these sets, i.e., $P^+ = P_1 \cup P_2 \cup \cdots \cup P_n$.

3.1 Negative Feedback Selection

Negative training data contains useful information for clearly estimating the boundary for positive one. However, the unbalanced number of training positive and negative samples could affect to obtain good features. In this work, we propose the idea to balance the classes using offenders in the negative documents, where an *offender* is a negative document that shares common terms with positive ones. Given a set of size-1 patterns in P^+, say P_1^+, we use all the patterns in P_1^+ to identify the offenders in negative documents as the following function: $S(nd) = \sum_{t_i \in nd \cap P_1^+} tf_i$ where tf_i is the term frequency of term t_i in the negative document nd for size-1 patterns in P_1^+. The document nd is *offender* if its score is positive. We select the top-n offenders in the negative documents to balance positive documents, D^-_{off}, where $n = |D^+|$. After class balancing, we discover closed patterns from the offender set D^-_{off} to obtain a single set of patterns P^- as we did in the positive one.

3.2   Pattern Pruning in Relevance Feedback

The large number of discovered patterns often hinder their effective use since among them are redundant and contain noise. A lot of research in data mining has been conducted to shrink the large set of discovered frequent patterns, among which the concept of closed patterns [10] has been widely applied for text mining applications [8,9]. However, the closed set method usually targets on the summarization of frequent patterns discovered in a given category. This may be insufficient to deal with noise in the feedback texts in cases where some discovered patterns in positive documents can be irrelevant or ambiguous. In this work, we propose a novel pattern pruning method for user feedback. The idea is to check which patterns in a target relevance have been used in the other context. We define two types of patterns appearing in a given category R:

Definition 1 (total conflict). *Given a pattern $p \in R$, pattern p is called total conflict with other category $\overline{R}$ if $\exists q \in \overline{R}$ and $termset(p) \subseteq termset(q)$.*

Definition 2 (partial conflict). *Given a pattern $p \in R$, p is called partial conflict with a category $\overline{R}$ if $\exists q \in \overline{R}$ and $termset(p) \cap termset(q) \neq \emptyset$.*

For the positive data D^+, total conflict patterns identified in this data contain termsets that may be used in other categories since they are subsequences of some patterns in the non-relevant data D^-. Partial conflict patterns in the underlying data are ones that overlap with, *but* not any subsequences of those patterns. The other patterns are *non-conflict*, and considered highly relevant to the target category. To obtain informative patterns in the positive data, we remove all total conflict patterns. Some patterns in the negative category provide useful information to perform specialization. In this context, all conflict patterns (i.e., total conflict and partial conflict) identified in this category are used to estimate weights of terms for representing user's interests. We will describe this in the next section. We also remove all non-conflict patterns in this category since they are irrelevant. Based on the above definitions, we define a set of decision rules to classify patterns into one of the following groups:

$$S^+ = \{p|p \in P^+, \forall q \in P^- \Rightarrow p \nsubseteq q\}$$
$$S^- = \{q|q \in P^-, \exists p \in S^+ \Rightarrow q \cap p \neq \emptyset\}$$
$$N = (P^+ \cup P^-) - S^+ - S^-$$

where $S^+ \cap S^- \cap N = \emptyset$. P^+ and P^- are two categories of patterns in the relevant and non-relevant data respectively. Based on the decision rules: a collection S^+ consists of non-conflict patterns and partial conflict patterns from the positive data while a collection S^- contains patterns from the negative data which are identified are one of the two kinds of conflict patterns. Finally, all patterns classified into a collection N are removed.

4 Representation of User Profile

It is important to define a frame of discernment for a domain of interests and its mass functions. In this case, a vector of terms extracted from discovered patterns can be considered as the set of hypotheses for user profiling. Let Ω consists of n terms extracted from all patterns in S^+ and S^-, i.e., $\Omega = \{t_1, t_2, \ldots, t_n\}$. We define a set-valued mapping $\psi :: S^+ \cup S^- \rightarrow 2^\Omega$ to generate the relationship between patterns and the term space Ω for generating mass functions. By applying this mapping ψ, we define a mass function $m^+ : 2^\Omega \rightarrow [0,1]$ on Ω, the set of terms, called *positive mass* function, which satisfies:

$$m^+(A) = \begin{cases} 0 \ if \ A = \emptyset; \\ \frac{S(\{p|p \in S^+, \psi(p)=A\})}{\sum_{B \subseteq \Omega} S(\{q|q \in S^+, \psi(q)=B\})} , otherwise \end{cases} \quad (1)$$

for all $A \subseteq \Omega$, where $S(p)$ is a significant function that returns the relevance of pattern p with respect to its class category. This function could be obtained with significant measures in data mining such as Information Gain or Document Frequency (DF). However, we estimated the pattern's relevance by using the weight function proposed in [9]. We also define mass functions from negative data that are generated by patterns in S^-, defined as *negative mass* functions ($m^-(A)$). After mass functions were generated, pignistic transformation is applied to score the weight of each term in the profile vector Ω. For each term $t_i \in \Omega$, we first transfer positive mass functions into a *pignistic probability* as the following functions: $Pr_{m^+}(t_i) = \sum_{\emptyset \neq A \subseteq \Omega, t_i \in A} \frac{m^+(A)}{|A|}$ where A denotes a subset on Ω space and $|A|$ is the number of elements in A. The pignistic value represents a probability given by mass functions and is assigned to single elements in the frame of discernment for betting. In our case, we use the resulting probability as the means to score terms in the profile vector corresponding to their distribution the term dependency data (i.e., positive and negative data). The high value assigned to a term represents the importance of the term in the underlying data. The pignistic value of the term t_i assigned by negative mass functions is also calculated in the same means of positive mass functions. Finally, we combine two pignistic transforms in order to evaluate the weight of term t as the following equation:

$$w(t_i) = \frac{Pr_{m^+}(t_i) \times (1 - Pr_{m^-}(t_i))}{1 - \min \{Pr_{m^+}(t_i), Pr_{m^-}(t_i)\}} \quad (2)$$

The weight reflects how important the term for representing a user's topic of interest. The higher weight assigned to the term represents that the term is more important. A document evaluation function is built for the use of profile in web filtering. Given a new document d, the weight of the document d is assigned as the total support of profile terms matched in the document.

5 Experimental Evaluation

We conduct experiments on Reuters Corpus Volume 1 (RCV1) data collection and TREC topics used in TREC "filtering track" [7] for evaluating a robust

IF system. It includes 50 topics of Reuters Corpus articles from 1996-08-20 to 1997-08-19. The relevance judgements of the articles have been made by human assessors of the National Institute of Standards and Technology (NIST), called *assessor* topics. For each assessor topic, its data collection is split into two sets: a training set and a test set. All documents are marked in XML and some meta-data information. In order to avoid bias in experiments, we remove the meta-data information. A common basic text processing is used for all documents, which includes stop-words removal according to a given stop-words list and stemming terms.

5.1   Baseline Models and Settings

We grouped baseline models into two categories. The first category includes a number of data mining (DM) based methods, including PTM [9], PDS [8] and IPE [11] while the second category includes two effective keyword-based IF models (i.e. Rocchio [2] and SVMs). In data mining-based models, both PTM and PDS that use only positive features from positive data to formulate user profile models while IPE uses both positive and negative features obtained by discovered closed patterns. The minimum support threshold (min_sup) is an important parameter and is sensitive for a specified data set. We set $min\_sup = 0.2$ (20 % of the number of paragraphs in a document) for all models that use patterns since it was recommended as the best value for this data collection [8, 9, 11].

In keyword-based models, the Rocchio algorithm has been widely adopted in text categorization and filtering [2]. The Rocchio builds a centroid for representing user profiles. There are two important parameters (i.e., α and β) in the Rocchio model. According to [1, 2], there are two recommendations for setting the two parameters: $\alpha = 16$ and $\beta = 4$; and $\alpha = \beta = 1.0$. We have tested both accommodations on assessor topics and found the latter recommendation was the best one. Therefore, we let $\alpha = \beta = 1.0$. SVM is one of state-of-the-art text classifiers [4]. However, the SVM here is used to rank documents rather than to make a binary decision, and it only uses terms based features extracted from training documents. For this purpose, we choose the linear SVM modified for document retrieval in [11]. For each topic, we also selected 150 terms in the positive documents, based on tf × idf values for all keyword-based models.

5.2   Experimental Results

The effectiveness is resumed by five common IR measures: The precision of the top 20 returned documents ($top - 20$), F_1 measure, Mean Average Precision (MAP), the break-even point (b/p), and Interpolated Average Precision (IAP) on 11-points [1]. The belief function reasoning model (BRM) is firstly compared with all DM-based models. BRM is also compared with keyword-based models underpinned by Rocchio and SVM for each measuring variable over all the 50 assessing topics, respectively. The results of overall comparisons between BRM and all DM based models have shown in Table 1(a). The most important findings revealed in this table are that both PDS and IPE models outperforms PTM model over all the standard measures while the slight increase in IPE as compared to PDS. The results support the effective use of closed patterns in text for user profiling. We also compare BRM with IPE. As seen in Table 1(a), BRM significantly increases for all the evaluation measures with +9.14 % (max +11.35 % on $top - 20$ and min +5.90 % on F_1) in percentage change on average over the standard measures. The encouraging improvements of BRM is also consistent and significant on 11-points as shown in Fig. 1. The results illustrate the highlights of the belief function reasoning approach to reduce uncertainties involved in estimating term weights.

Table 1. Comparison performance results on all assessor topics

Methods	top-20	MAP	b/p	$F_{\beta=1}$
BRM	**0.549**	**0.484**	**0.470**	**0.466**
IPE [11]	0.493	0.441	0.429	0.440
PDS [8]	0.496	0.444	0.430	0.439
PTM [9]	0.406	0.364	0.353	0.390
%chg	+11.35	+9.75	+9.55	+5.90

(a) BRM vs DM-based models

Methods	top-20	MAP	b/p	$F_{\beta=1}$
BRM	**0.549**	**0.484**	**0.470**	**0.466**
Rocchio[2]	0.474	0.431	0.420	0.430
SVM [11]	0.453	0.408	0.421	0.409
%chg	+15.82	+12.29	+11.90	+8.37

(b) BRM vs keywords-based models

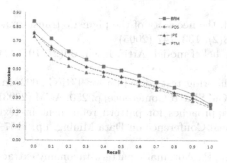

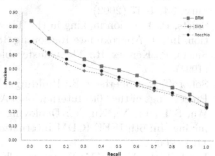

(a) BRM vs DM-based models   (b) BRM vs keywords-based models

Fig. 1. Comparison performance results on IAP 11-points

As shown in Table 1(b), both Rocchio and SVM that are based on keyword-based models perform over PTM model, excepting for PDS and IPE. This illustrates keywords remain the very effective concept for text retrieval. However, the results compared between the keywords-based models and IPE (also PDS) support that patterns are much effective to select useful terms. In comparisons with Rocchio and SVM, BRM performs better than Rocchio with +12.09 % increasing in average (max +15.82 % on $top - 20$ and min +8.37 % on F_1). The excellent performance of BRM is also obtained as compared to SVM.

6   Conclusions

This paper presents a novel approach to Web user profiling based on belief function reasoning. We have presented a unified model for acquiring and representing user's interests in Web text data. We discover frequent closed patterns in text collections and show how to deal with the extracted patterns for effective user profiling. Many experiments are conducted on Reuters Corpus Volume 1 and TREC standard text collections for information filtering task. We compare the proposed approach with the state-of-the-art information filtering models. The experiment results illustrate that our proposed approach achieve encouraging improvements of the filtering performance.

References

1. Buckley, C., Salton, G., Allan, J.: The effect of adding relevance information in a relevance feedback environment. In: ACM SIGIR 17th International Conference, pp. 292–300 (1994)
2. Joachims, T.: A probabilistic analysis of the rocchio algorithm with TFIDF for text categorization. In: Proceedings of the 4th International Conference on Machine Learning, pp. 143–151 (1997)
3. Schiaffino, S., Amandi, A.: Intelligent user profiling. In: Bramer, M. (ed.) Artificial Intelligence. LNCS (LNAI), vol. 5640, pp. 193–216. Springer, Heidelberg (2009)
4. Sebastiani, F.: Machine learning in automated text categorization. ACM Comput. Surv. **34**, 1–47 (2002)
5. Smets, P.: Decision making in the TBM: the necessity of the pignistic transformation. Int. J. Approximate Reasoning **38**(2), 133–147 (2005)
6. Smets, P., Kennes, R.: The transferable belief model. Artif. Intell. **66**(2), 191–234 (1994)
7. Soboroff, I., Robertson, S.: Building a filtering test collection for TREC 2002. In: Proceedings of the 26th International ACM SIGIR Conference, p. 250. ACM (2003)
8. Wu, S.T., Li, Y., Xu, Y.: Deploying approaches for pattern refinement in text mining. In: 6th IEEE ICDM International Conference on Data Mining, pp. 1157–1161 (2006)
9. Wu, S.T., Li, Y., Xu, Y., Pham, B., Chen, P.: Automatic pattern-taxonomy extraction for web mining. In: 3th IEEE/WIC/ACM WI International Conference on Web Intelligence, pp. 242–248 (2004)
10. Zaki, M.J.: Generating non-redundant association rules. In: 6th ACM SIGKDD International Conference on Knowledge Discovery and Data Mining, pp. 34–43 (2000)
11. Zhong, N., Li, Y., Wu, S.-T.: Effective pattern discovery for text mining. IEEE Trans. Knowl. Data Eng. **24**(1), 30–44 (2012)

The Belief Theory for Emotion Recognition

Halima Mhamdi[1]([✉]), Hnia Jarray[2], and Med Salim Bouhlel[1]

[1] UR.SETIT, Sfax University, Road Soukra km 4, 3038 Sfax, Tunisia
mhemdi.halima@gmail.com
[2] Higher Institute of Computer and Multimedia of Gabes, Gabes University,
BP 122, 6033 Gabes, Tunisia

Abstract. This paper presents a facial expression classification system
based on a data fusion process using the theory of belief. Such expressions
correspond to the six universal emotions (happiness, surprise, disgust,
sadness, anger, and fear) as well as the neutral expression. The suggested
algorithm rests on the decision fusion of both approaches: the global
analysis and the local analysis of facial components. The classification
result, throughout these two approaches, will be enhanced by fusion. The
performance and the limitations of the recognition system and its ability
to deal with different databases are identified through the analysis of a
large number of results on the FEEDTUM database.

Keywords: Data-fusion · The belief theory · Emotions classification ·
Local analysis · Global analysis

1 Introduction

Facial expression is physiologically identified as the result of the deformation
of the facial features caused by emotion. The importance of the information
of an expression is contained in the main permanent deformation of the facial
features; namely, the eyes, the nose, the eyebrows, and the mouth. This fact
was corroborated, from a psychological point of view, in different works [1,2].
These works demonstrated, through psychological experiments, that the information held by a facial expression is not contained in a particular trait but in
the combination of the facial features mentioned above. To recognize the facial
expression, indeed, one has to classify the deformations of the facial structures
in abstract classes using visual information. To analyze the facial expressions
automatically, one has to go through two steps. The first step is the extraction of the facial structures (optic flow [3], 3D deformable motion model [4],
Monitoring structure points [5]). The second step is the classification of the
facial structures (Markov hidden model, recurrent neural network, Gabarits
energy movement, Feed-forward neural network, Nearest neighbor with distance
measurement, SVM...) Additionally, some hybrid systems have also come into
existence [6,7]. In the recognition step, imperfect information involves several
concepts. The first concerns imprecise information and it is usually well controlled. The second concept is uncertainty. It aims at distinguishing imprecision;

© Springer International Publishing Switzerland 2015
K. Jackowski et al. (Eds.): IDEAL 2015, LNCS 9375, pp. 517–526, 2015.
DOI: 10.1007/978-3-319-24834-9_60

in other words, it does not refer to the content of information but to its "quality". Finally, incompleteness can occur in many applications. Modeling such heterogeneous knowledge quickly becomes a critical issue. The scientific community also focuses on such issues. The problem is getting more and more complicated when there is a fusion of information in order to make the best decision in the sense of a criterion. In the same mathematical formalism, different theories for data fusion are available for modeling information from one or more sources. The most common and the oldest of these is the theory of probability. However, the latter is deficient especially when knowledge is incomplete, uncertain or imprecise. Thereafter, different theories appeared. They include the possibility theory and the evidence theory. These two theories have recently been the focal point of extensive research. A synthesis of these three approaches has been the subject of a suggested study. Such a study is aimed to choose the most appropriate approach to the data fusion domain in the context of facial-expression analysis [8]. This study paved us the way to choose the theory of evidence as the best fusion tool and as the only tool (among the fusion methods) to model the doubt. The evidence theory [8–11], is well adapted to this type of problem. In fact, this model facilitates the integration of knowledge and can handle the uncertain and imprecise data stemming from the automatic segmentation algorithms. It is well adapted when there is a lack of data. In addition, it can shape the intrinsic doubt that may occur between the facial expressions in the classification process. The classification of a facial expression into a simple category of emotion is not realistic because human expressions vary according to the individual.

2   A Facial Expression Analysis System

As shown in Fig. 1, our emotion recognition system is based on facial expressions. It analyzes the movement of different facial features in a video sequence to determine the emotion of a person.

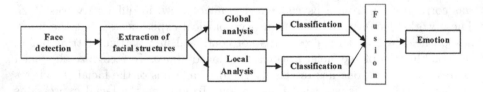

Fig. 1. Overview of our emotion recognition system

2.1   Face Detection

Face detection is the first step in our facial expression recognition system which is intended to demarcate the area of interest by a rectangle. To do this task, we used a fast and robust face detector implemented in the OpenCV library. It was originally developed by P. Viola and M. Jones [12] and, then, improved and implemented by R. Lienhart [13]. Actually, this detector is based on the Haar descriptors and classifiers in cascade.

- *Haar descriptors:* the information on the luminance and color of a given point is generally in the values of a pixel. It is therefore wiser to find sensors based on more global characteristics of the object. This is the case of the Haar descriptors which are functions allowing to know the difference in contrast between a plurality of rectangular areas in an image.They thus encode existing contrasts in the face and spatial relations. Eventually, a cascade classifiers is used to classify an area as a face or not according to face value of its descriptor.
- *cascade of classifiers:* Every step, a simple classifier (also called weak because of their low discriminative power) is built. The combination of all the weak classifiers will form a strong classier that can recognize any kind of object it was trained with. The problem is to search for this particular sized window over the full picture, applying the sequence of weak classifiers on every subwindow of the picture.

2.2 Locating Facial Feature Points

After detecting the face with the Viola-Jones detector, we located the facial feature points in the frame that encompasses the face. There are many different techniques dedicated to the extraction of the facial features [14]. In fact, we chose a quick method. Such a method is adapted to real-time processing which rests on the points of interest. The different regions were selected using the *eye.DetectMultiScale* and the *mouth.DetectMultiScale*. For each one, we can extract the characteristic points by the *FeaturesPointsDetec* function. Algorithms have been developed to solve this problem. As the Canny algorithm calculates the gradient of an image and finds the contours, so the corners can be detected by selecting the best detected contours. The latter correspond to the two corners of each eye, the four corners of the mouth, the characteristic points of the eyebrows, the nose, and the lower part of the face by using the *DetectFacialFeaturesInRegion* function which detects the points in the regions.

2.3 Extraction of the Facial Structures

The use of optical flow, the approach to the global analysis of the facial structures, provides motion information for each pixel of the picture. Thus, it measures the movement vectors according to the intensity of the pixels of two consecutive or temporally close pictures. In a motion detection context, unlike the pixels belonging to dynamic objects, inactive pixels possess zero velocity (Fig. 2).

In fact, the movements of the facial structures are described by a coding system called FACS (Facial coding system) [15]. These movements are divided into action-units (AUs); therefore, the optical flow is calculated so as to know the AUs. The algorithm of Horn and Schunk [16] is used to calculate the optical flow. Furthermore, for a monitoring application, we keep the same points for longer time, but some tracked points are lost because of error in the time line.

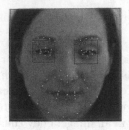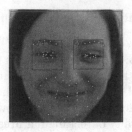

Fig. 2. Extraction of characteristic points of the face

In our case, the conservation of the points is held and the direction of the flow is towards the AUs. These flows are used to recognize and identify the AUs which were previously marked by the characteristic points obtained by the *DetectFA-cialFeatureInRegion* function. Figure 3 shows the calculation of the optical flow of the eyebrow and the mouth. For example, an upward global movement of the pixels of an eyebrow from the beginning to the end of an AU corresponds to AU1. Locally, when using the parametric models of the regions of interest [17–19], the motion models were also used to model the movements of the mouth, the nose, the eyelids and the eyebrows. These methods allow expressing, with few basic parameters, the facial movements. In our case, the facial components are structured by the features (Fig. 4). The distance between the characteristic points, which form the structure of the face, is calculated.

Fig. 3. Structured facial components

Fig. 4. Structured facial components

2.4   Classification

After extracting the characteristic parameters described in the previous section, we used a supervised statistical classifier named SVM (Support Vector Machine) to establish correspondence between the calculated parameters and the corresponding emotions. An apprenticeship program is realized by SVM in order to solve an optimization problem involving a quadratic programming resolution system in an area of significant dimension. The library LibSVM [20] implements the SVM algorithm. Such a program is used so as to select a set of core functions and to regulate the parameters of these functions (e.g. the exponent of the polynomial kernel functions or the standard deviation of the radial basis functions). These choices are usually made by a cross-validation technique.

3   Data Fusion for Emotion Recognition

Information fusion can take different forms according to the time it is performed [21]. It is possible to fuse data just after extracting the signals, to fuse the attributes coming from different modalities or to fuse information during the decision phase. We are interested in the third level of fusion.

3.1   The Belief Theory

Unlike the probability theory, the theory of evidence, also known as the theory of belief or the Dempster Shafer theory, is relatively recent. Indeed, the work of Dempster and Shafer gave birth to this theory [22]. The innovation and the power of such a theory attract a lot of interest.

3.2   Modeling and Estimation of the Mass Function

Let us assume that $D = \{d_1, d_2...d_n\}$ where each d_i denotes the emotions taken by each approach. In this case of research, the decisions are the seven basic emotions and the sources are the two methods of analysis of facial structures: the optical flow and the movement models. A mass function is defined as a function of 2^D. Generally, it requires $m(\phi) = 0$, and for the source S_j of the mass function m_j a standardization of the form $\sum_{A \subset D} m_j(A) = 1$

$M_i^j(x)$ can be estimated by a conditional probability $p(f_j(x)/d_i)$. The probabilistic models are based on the estimation of $p(f_j(x)/d_i)$ noted as $p(S_j/d_i)$.

3.3   Bayesian Combination

The combination by the Bayesian approach consists in determining the probabilities $p(d_i/S_1, ..., S_m)$. These probabilities can directly be estimated by the Bayes rule at the level of modeling:

$$p(d_i/S_1, ..., S_m) = \frac{p(S_1, ..., S_m/d_i)\, p(d_i)}{p(S_1, ..., S_m)} \tag{1}$$

As a matter of fact, the difficulty of estimating the different probabilities in these equations (a large number of apprenticeship data is needed) leads to a statistical independence assumption of the sources. Each source must be conditioned to a decision.

Then, we get:

$$p(di/S1, ..., Sm) = \frac{\prod_{j=1}^{m} p(S_j/d_i)p(d_i)}{\prod_{j=1}^{m} p(S_j)} \tag{2}$$

After combination, the result is a fused information stemming from two approaches.

The fused information must be used to improve decision making. The last step, thus, is about decision. This decision must be in conformity with the quality measurement (decision criteria), which might lead to its rejection. The most common rule for the probabilistic and Bayesian decision is the posterior maximum. Such a rule has been obtained in order to complete the fusion diagram.

4   Results and Discussion

4.1   Description of the Used Data-Base

In order to assess our work, we used The FEEDTUM database.

This is a Munich Technical University database [23] which contains color video sequences of the facial expressions of men and women. The orientation of the camera is frontal. The small head movements are present. The size of the image is 640 by 480 pixels. The facial expressions are generated by using emotion-induction movies.

To calculate the recognition rate, the selection of the pictures is done randomly in order to build the apprenticeship corpus as well as the test corpus. The random selection is made in such a way that both the apprenticeship and test corpora contain samples of all classes and all people. All the results of the recognition-rate tables are obtained by repeating the operation five times.

4.2   Implementation and Results

The implementation of the SVM method is based on the libSVM library [20]. The libSVM integrates the classification of the support vectors SVC (C-SVC, nu-SVC) which will be used later to classify the facial expressions, the regression (epsilon-SVR, nu-SVR) and the estimation of the distribution in order to use the SVM tool easily. Indeed, the libSVM supports the multi-class classification. We focus on the multi-class classification because we have six emotions.

Similarly, the SVM technique is used for the local analysis. The distance between the feature points which form the structure of the face is calculated.

Table 1. Classification rate using the global analysis method

	Anger	Disgust	Surprise	Happiness	Sadness	Fear	Neutral
Anger	**16 %**	15 %	8 %	10 %	19 %	15 %	12 %
Disgust	8 %	**19 %**	8 %	11 %	24 %	15 %	13 %
Surprise	0 %	2 %	**97 %**	0 %	0 %	0 %	0.75 %
Happiness	3 %	8 %	3 %	**95 %**	17 %	10 %	3 %
Sadness	3 %	13 %	3 %	6 %	**73 %**	19 %	6 %
Fear	1 %	2 %	1 %	2 %	3 %	**83 %**	5 %
Neutral	7 %	20 %	5 %	8 %	32 %	18 %	**7 %**

Table 2. Classification rate using the local analysis method

	Anger	Disgust	Surprise	Happiness	Sadness	Fear	Neutral
Anger	**89 %**	0 %	2 %	1 %	41 %	1 %	3 %
Disgust	0.6 %	**95 %**	0.7 %	0.5 %	1 %	0.6 %	0.8 %
Surprise	0 %	0 %	**98 %**	0.8 %	0 %	0 %	0 %
Happiness	0 %	0 %	0.7 %	**98 %**	0 %	0 %	0 %
Sadness	1 %	2 %	1 %	1 %	**86 %**	5 %	3 %
Fear	0.6 %	1 %	1 %	0.5 %	2 %	**91 %**	2 %
Neutral	0 %	0.5 %	0 %	0 %	2 %	1 %	**95 %**

Table 3. Classification results of the different approaches

Emotion \ Approche	Holistic approach	Local approach	Our approach	Hammal
Anger	19%	85%	90%	–
Disgust	24%	91%	95%	43.10%
Surprise	97%	98%	98%	84.44%
Happiness	95%	98%	97%	76.36%
Sadness	73%	86%	88%	–
Fear	83%	80%	93%	–
Neutral	7%	86%	100%	88%

To assign the exact class to every emotion, a comparison with the support vectors of our model is made.

Table 1 shows an example of the classification results, subject 6 of the FEED-TUM database, using the global analysis method. We notice that the emotions surprise, happiness, sadness and fear are well classified. In contrast, the emotions of disgust, anger and neutral have low percentages of recognition. These emotions converge towards other emotions. By applying the second method, the local analysis, the emotions are well categorized (Table 2) Table 3 expresses the classification results of the two approaches. In the case of expressions of disgust,

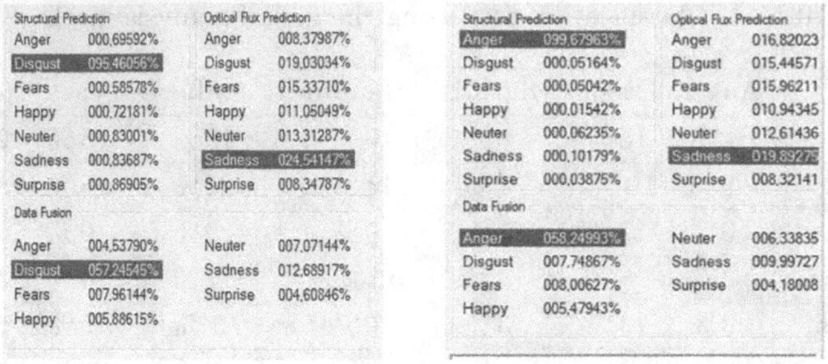

(a) doubt between disgust emotion and sadness

(b) doubt between anger emotion and sadness

Fig. 5. Recognized emotion

anger and neutral, the results are particularly poor with the global analysis approach and, relatively, low compared to the local approach. The local recognition algorithm gives better results in most of the videos of the test database. In order to remedy the inadequacies of the first approach, we resort to decision-fusion. Moreover, the belief theory can model the intrinsic doubt which may occur between the facial expressions in the classification process. By implementing the fusion diagram, through the use of the mass function as a belief context, we keep the encouraging and promising results. Indeed, although it is shown through the results given by Fig. 5 and Table 3 that our approach significantly softens or improves decision analysis. The fusion result, therefore, describes a decision or possibly obtained data with objectivity and moderation trying to reach and to better mimic reality by taking account of the existence of two decisions from the local analysis and the global analysis.

The table above gives also the result of Hammal's approach [24,25]. Four expressions are available: Joy, Surprise, Disgust and Neutral. We note that the expression Disgust led to the lowest recognition rate. But compared to the results of Hammal, it is higher. A great advantage is that Neutral expression recognition rate is 100 % We only test only on Neutral sequences and we ignore the Neutral moments in expression sequences.

5   Conclusion

We have presented our approach to emotion recognition by applying the belief theory. The principle rests on the fusion of decisions. The first step is the detection of the face. The latter is based on the descriptors of Haar. The extraction of the facial features inside the frame which delimits the face is based on the fact that the human faces are constructed in the same geometrical configuration. An anthropometric model is developed for the detection of the interest points combined with the Shi-Thomasi method for better location.

The control of the characteristic points in the pictures of the sequence is provided by the pyramidal algorithm of Lucas and Kanade for the global analysis method and the local analysis motion model. By relying on these two methods, the emotions are recognized by our large margin separating classifier (SVM). For a better decision, the classification results are fused by the belief theory.

References

1. Ekman, P.: Facial expression. In: The Handbook of Cognition and Emotion (1999)
2. Bassili, J.N.: Facial motion in the perception of faces and of emotional expression. Exp. Psychol. Hum. Percept. Perform. **4**, 373–379 (1978)
3. Cohn, J., Zlochower, A., James Lien, J.-J., Kanade, T.: Feature-point tracking by optical flow discriminates subtle differences in facial expression. In: Proceedings of the International Conference on Automatic Face and Gesture Recognition, pp. 396–401 (1998)
4. DeCarlo, D., Metaxas, D., Stone, M.: An anthropometric face model using variational techniques. In: Proceedings of the SIGGRAPH, pp. 67–74 (1998)
5. Wang, M., Iwai, Y., Yachida, M.: Expression recognition from time-sequential facial images by use of expression change model. In: Proceedings of the International Conference on Automatic Face and Gesture Recognition, pp. 324–329 (1998)
6. Fasel, B., Luettin, J.: Automatic facial expression analysis: a survey. Pattern Recogn. **36**, 259–275 (2003)
7. Meulders, M., De Boeck, P., Van Mechelen, I.: Probabilistic feature analysis of facial perception of emotions. Appl. Statist. **54**, 781–793 (2005)
8. Ramasso, E., Panagiotakis, C., Rombaut, M., Pellerin, D.: Human action recognition in videos based on the transferable belief model - application to athletics; jumps. Pattern Anal. Appl. J. (2007)
9. Girondel, V., Caplier, A., Bonnaud, L., Rombaut, M.: Belief theory-based classifiers comparison for static human body postures recognition in video. Int. J. Sig. Process. **2**, 29–33 (2005)
10. Denoeux, T., Smets, Ph.: Classification using belief functions: the relationship between the case-based and model-based approaches. IEEE Trans. Syst. Man Cybern. **36**(6), 1395–1406 (2006)
11. Mercier, D.: Information Fusion for automatic recognition of postal addresses with belief functions theory. University of Technologie of Compiegne, December 2006
12. Viola, P., Jones, M.: Robust real-time object detection. In: 2nd International Workshop on Statistical and Computational Theories of Vision - Modeling, Learning, Computing, and Sampling Vancouver, Canada (2001)
13. Lienhart, R., Maydt, J.: An extended set of haar-like features for rapid object detection. In: IEEE ICIP, vol. 1, pp. 900–903, September 2002
14. Yuille, A.L., Hallinan, P.W., Cohen, D.S.: Feature extraction from faces using deformable templates. Int. J. Comput. Vis. **78**, 99–111 (1992)
15. Ekman, P., Friesen, W.: Facial Action Coding System: A Technique for the Measurement of Facial Movement. Consulting Psychologists Press, Palo Alto (1978)
16. Lobry, S.: Improving Horn and Schunks Optical flow algorithm. Laboratoire de Recherche et Dveloppement et lEpita (2012)
17. Yacoob, Y., Davis, L.S.: Recognizing human facial expression from long image sequences using optical flow. IEEE Trans. Pattern Anal. Mach. Intell. **18**, 636–642 (1996)
18. Black, M., Yacoob, Y.: Recognizing facial expressions in image sequences using local parametrized models of image motion. Int. J. Comput. Vis. **25**, 23–48 (1997)

19. Huang, C., Huang, Y.: Facial expression recognition using model-based feature extraction and action parameters classification. J. Vis. Commun. Image Represent. **8**, 278–290 (1997)
20. Chang, C.-C., Lin, C.-J.: Libsvm: library for support vector machines. Department of Computer Science. National Taiwan University, Taipei (2001)
21. Martin, A.: La fusion dinformations, Polycopie de cours, ENSIETA (2005)
22. Arif, M.: Fusion de Donnees: Ultime Etape de Reconnaissance de Formes, Applications lIdentication et lAuthentication. Universite de Tours (2005)
23. Wallhoff, F.: Feedtum: facial expressions and emotion database. Technische Universitat Munchen, Institute for Human-Machine Interaction (2005)
24. Hammala, Z., Couvreurb, L., Capliera, A., Rombaut, M.: Facial expression classification: an approach based on the fusion of facial deformations using the transferable belief model. Int. J. Approximate Reasoning **46**, 542–567 (2007)
25. Hammal, Z., Couvreur, L., Caplier, A., Rombaut, M.: Facial expression recognition based on the belief theory: comparison with different classifiers. In: Roli, F., Vitulano, S. (eds.) ICIAP 2005. LNCS, vol. 3617, pp. 743–752. Springer, Heidelberg (2005)

Using a Portable Device for Online Single-Trial MRCP Detection and Classification

A. Hassan[1]([⊠]), U. Ghani[1], F. Riaz[1], S. Rehman[1], M. Jochumsen[2],
D. Taylor[3], and IK. Niazi[2,3,4]

[1] Department of Computer Engineering, College of Electrical and Mechanical Engineering,
National University of Sciences and Technology, Islamabad, Pakistan
{alihassan,usman.ghani77,farhan.riaz,
saad.rehman}@ceme.nust.edu.pk
[2] Department of Health Science and Technology, Aalborg University, Aalborg, Denmark
{imrankn,mj}@hst.aau.dk
[3] Health and Rehabilitation Research Institute, Auckland University of Technology,
Auckland, New Zealand
{iniazi,denise.taylor}@aut.ac.nz
[4] Centre for Chiropractic Research, New Zealand College of Chiropractic,
Auckland, New Zealand
imran.niazi@nzchiro.co.nz

Abstract. In the past decade, the use of movement-related cortical potentials (MRCPs) for brain computer interface-based rehabilitation protocols has increased manifolds. Such systems suffer severely from high frequency colored noise making it extremely difficult to recognize these signals with high accuracy on a single-trial basis. All previous work in this domain has mainly focused on offline systems using computing power of lab computers in which the detection of the MRCPs is done independent to the classification of the type of movement. The main focus of this work is to test the detection of the presence of the MRCP signal as well as its classification into different types of movements in a single online system (portable Raspberry Pi II) where the classification system takes over only after the presence of MRCP signal has been detected. To achieve this, the MRCP signal was first spatially (Laplacian) and later band pass filtered to improve the signal to noise ratio, then a matched filter was applied to detect the signal. This was obtained with a detection latency of -458 ± 97 ms before the movement execution. Then six temporal features were extracted from 400 ms data after the point of detection to be classified by a standard linear support vector machine. The overall accuracy of 73 % was achieved for the online detection and classification for four different types of movements which is very close to the base line accuracy of 74 % using the offline system. The whole system was tested on Matlab and verified on a Raspberry Pi II as a portable device. The results show that the online implementation of such a system is feasible and can be adapted for stroke patient rehabilitation.

Keywords: Movement-related cortical potentials · Rehabilitation · Detection of movement onset · Online brain computer interface system

© Springer International Publishing Switzerland 2015
K. Jackowski et al. (Eds.): IDEAL 2015, LNCS 9375, pp. 527–534, 2015.
DOI: 10.1007/978-3-319-24834-9_61

1 Introduction

Use of electroencephalography (EEG) for brain computer interfacing (BCI) application has shown a lot of promise in the gaming industry [1] as well as neurological rehabilitation of patients suffering from disorders such as stroke or spinal cord injury [3]. Movement-related cortical potentials (MRCPs) are essentially EEG signals encoding movement-related information such as kinetic profiles of different movements for different limbs. MRCPs can be recorded using surface electrodes; the greatest signal-to-noise ratio for the MRCP is obtained from electrodes overlying the motor cortical representation of the limb performing the movement [2].

The MRCP is the negative deflection of EEG signal from normal as shown in Fig. 1 where the first part of the signal (between −6 s to −3 s) is the colored noise generated from several sources like background brain activity, eye blinks, impedance of wires etc. Many researchers have tried to remove this colored noise using band pass filters with several cutoff frequencies [4]. It is a general consensus that the MRCP is a low frequency signal with frequencies less than 10 Hz. Usually a high pass filter with cutoff of 0.05 Hz is used to remove the DC drift from the signal while several cutoffs have been suggested for the low pass filters ranging from 1 Hz to 10 Hz [4]. Figure 1 shows the negative deflection of the MRCP signal between −3 s to 0 s (with 0 s being the movement onset) which is the negative potential developing in the brain before the execution of actual movement. This negative potential can be used to detect the intention of the movement. The last phase of the MRCP, between 0 s to 2 s, is related to the inflow of afferent feedback coming back to the brain to monitor the movement that has been executed. This afferent feedback is missing or impaired patients who suffer from neural disorders such as stroke in which the neural connections between the muscles and brain are damaged. In [5], the authors have shown that MRCP signals can be used to modulate the brain's plasticity (the proposed mechanism for relearning movements); this shows that BCI's has the potential to be used for stroke rehabilitation. In this study, it was also shown that better BCI performance led to greater plastic changes. Based on this observation, there is an incitement to investigate techniques to improve the performance of BCI systems to optimize the rehabilitation procedures.

Most of the research in this area is to devise powerful signal processing methods to remove and suppress the unwanted noise as a first pre-processing step. After noise removal, the next step is to detect the presence of the MRCP signal. As the MRCP signal is severely corrupted by noise, many researchers have opted to use average of several MRCP trials for better detection accuracy. Recently, acceptable accuracy for single-trial MRCP detection was shown [6], which paves the way of using these signal for closed-loop rehabilitation paradigms as suggested by [5]. However, most of the work done on classification of MRCPs into different types of movements is independent of the signal detection as done by [6, 7]. The main goal of this research paper is to develop a frame work for online detection of MRCP signals associated with movements with different kinetic profiles. This was extended to a portable devices (Raspberry Pi) to show the potential for real-time decoding of brain signals to be used in rehabilitation clinics.

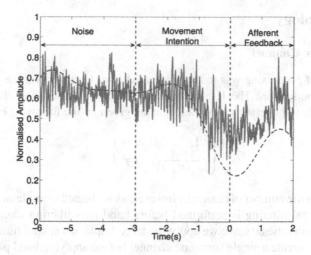

Fig. 1. Single-trial MRCP signal associated with a dorsiflexion of the right ankle joint. The blue signal is the noise corrupted signal while the red is the superimposed EEG signal after band pass filtering with cutoff 0.05 Hz–0.4 Hz from Laplacian Cz channel (surrogate). Time t = 0 is the time when the motor execution is performed by the subject.

2 Dataset

Six healthy subjects (two females, 25 ± ←3 years old) participated in this study. The procedures were approved by the local ethics committee (N-20100067). All subjects gave their informed consent prior to participation.

2.1 Signal Acquisition

All signals were acquired using 10 standardized locations i.e., FP1, F3, F4, Fz, C3, C4, Cz, P3, P4 and Pz. The signals were referenced to the right ear lobe and grounded at nasion. All subjects were seated comfortably with their right foot in a pedal with a force transducer attached. They were directed to undertake isometric dorsiflexion of right ankle. The signals were sampled at 500 Hz and were digitized using 32 bit accuracy. For further details regarding the recording setup, the readers are referred to [8].

Each subject performed 50 repetitions of four different types of cued movements categorized upon the time taken to complete the movement (fast and slow) and the percentage of maximum voluntary contraction (MVC) applied by the subject (20 % MVC and 60 % MVC). Data was collected for following four types of movements: (i) 0.5 s to achieve 20 % MVC (fast20), (ii) 3 s to achieve 20 % MVC (slow20), (iii) 0.5 s to achieve 60 % MVC (fast60) and (iv) 3 s to achieve 60 % MVC (slow60). This makes a total of 1200 trials. Trials corrupted with electrooculography (EOG) were kept in the dataset.

3 Methodology

3.1 Surrogate Channel

The first step of processing was to apply a spatial filter to correct for the blurring from e.g. volume conduction. The large Laplacian filter was on all nine channels of the recorded signals to generate a surrogate channel given by the following equation:

$$x_i = \begin{cases} 1, & i = 1 \\ -\frac{1}{(N_{ch}-1)}, & i \neq 1 \end{cases} \tag{1}$$

where N_{ch} is the number of channel. However, as we intend to implement this system online, the spatial filtering is performed before band pass filtering since they exhibit commutative properties. Hence we apply the large Laplacian spatial filter before band pass filtering to create a single surrogate channel before applying band pass filtering.

3.2 Pre-processing with Band Pass Filtering

As the MRCP signals are very low frequency signals with frequencies less than 1 Hz as shown previously in [0], the acquired signals are pre-processed with a band pass filter from 0.05–0.4 Hz. The high pass filter with cutoff 0.05 Hz is used to remove the DC drift from these signals which corrupts these low frequency signals. Then a 5 tap low pass Butterworth filter with cutoff 0.4 Hz is applied to remove colored noise components from the signal. This gives us a noise free clean signal to be used for signal detection.

3.3 Template Extraction and Signal Detection

The MRCP signals are detected by template matching using matched filters (MF). The templates for the signals are extracted from the initial negative phase of the MRCP signal. A MF gives the maximum peak when the input signal $x[t]$ matches with the template $h[t]$ given by:

$$y[n] = \sum_{-\infty}^{\infty} h[n-t]x[t] \tag{2}$$

A threshold is applied to the output of the MF to make the decision regarding the signal detection. In our experiments, we have used a threshold of 67 % template match for the true detection. Figure 2 shows the output of the matched filter with peaks exceeding the threshold marked as red.

3.4 Feature Extraction and Classification

Six temporal features are extracted from each frame: (i) point of maximum negativity, (ii) mean amplitude, (iii) and (iv) slope and intersection of a linear regression from −2 s to the point of detection or −0.1 s, (v) and (vi) slope and intersection of a linear regression from −0.5 s to point of detection or −0.1 s. These features are passed on to a linear support vector machine with fixed penalty C = 0.1. The decision of SVM is taken by

$$y' = sign\left(\langle \omega^*, x' \rangle + b^*\right)$$
(3)

where y' is the predicted label, ω^* and b^* are the parameters defining the separation boundary learnt from the training data. The training is done offline on the training data and the trained parameters are used to implement the decision boundary on the Raspberry Pi.

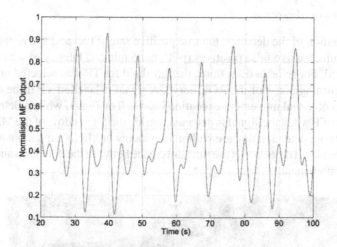

Fig. 2. Normalized output of the matched filter where the signal is marked as red when the output goes beyond 67 %.

4 Evaluation

4.1 Test Bed

The whole system is tested in two different scenarios with the same features and classification methodology. To establish the baseline results, first the detection and classification is applied as two separate disjoint steps. In the second scenario, the classification step is applied only after the signal has been detected. The features are extracted from the signal 400 ms after and 600 ms before the point of detection. Figure 3 shows our actual test bed for online testing of the system. It consists of a Raspberry Pi II, a stepper motor to control the movement of the limb to simulate the electric stimuli being provided to the limb along with some electronics to control the motor.

4.2 Raspberry Pi Setup

The main focus of this research work is to implement the detection and classification of MRCP signals on a portable device for which a Raspberry Pi II is used. The whole algorithm of pre-processing for noise removal, detection using matched filters and classification using support vector machines is implemented on this device. The raw data from 12 subjects performing 4 different types of movements is transmitted to this device in the form of frames each containing 2500 samples using socket programming of TCP/IP protocol. The total transmission delay of sending each frame to the Raspberry Pi via TCP/IP is on average 125 ms. The processing time of each frame on the Raspberry Pi from receiving the buffer to making the final class decision is 52 ms per frame.

4.3 Evaluation Measures

For the evaluation of the detector, the true positive rate (TPR) and active false positive (aFP) per 5 min, passive false positives (pFP) per 5 min and latency of signal detection was calculated. For reliable detections, the threshold for TP was selected on the elbow point of the turning phase of the ROC curve. A signal is labelled TP if the detection is between ± 1 s of actual movement execution i.e., $t = 0$ on Fig. 1, while a detected signal is deemed as pFP when the signal is detected outside the 5 s window of the MRCP signal and aFP are those detection that are within 5 s window of MRCP signal but are not TPs. Latency of the signal detection is the time difference between the detection and the actual onset of the movement.

Fig. 3. Actual test bed showing the current setup of controlling an external limb via Raspberry Pi and a Texas instruments H-bridge IC (TI-L293D).

5   Results

5.1   Baseline Detection Results

Table 1 shows the baseline detection accuracy where the signal is detected using the matched filter output. The templates for the MF are extracted from the first 40 % data while the remaining 60 % data is used for testing the results. The last column of the table shows the latency of the detection, which is the time of detection before the actual motor execution i.e., t = 0. Interestingly, the pFP are equal to zero indicating that the system has not detected any FP when there was no real movement done by the subject which is very encouraging.

Table 1. Baseline results of MRCP detection using matched filters with 67 % threshold.

	TPR	aFP	pFP	Latency
s20	0.95 ± 0.17	12.67 ± 4.6	0.00 ± 0.0	−536.5 ± 247
s60	0.79 ± 0.09	13.50 ± 3.8	0.00 ± 0.00	−345.8 ± 329
f20	0.82 ± 0.08	7.83 ± 5.5	0.00 ± 0.00	−409.2 ± 268
f60	0.83 ± 0.09	12. 50 ± 5.1	0.00 ± 0.00	−541.2 ± 360

5.2 Classification Results

The Table 2 shows the classification results for 2-class problem where the detection and classification of MRCP signals are done in two separate scenarios: offline using 10-fold cross validation (10-FCV), in which the detection step is completely independent of the classification, and an online scenario in which the features are only extracted from the point of detection of signal as indicated by the matched filter. The interesting fact is that the results of online detection system are not too much different from the offline detection system supporting the idea that such online systems can be practical.

Table 2. Classification accuracy for 2-class problem for two separate setups: offline and online

	Baseline classification		Online classification	
	10-FCV	30–70 split	10-FCV	30–70 split
s20s60	0.606	0.603	0.596	0.574
f20s20	0.795	0.801	0.782	0.772
f60s20	0.823	0.829	0.796	0.829
f20s60	0.811	0.81	0.786	0.761
f60s60	0.823	0.828	0.783	0.775
f20f60	0.563	0.556	0.586	0.659
Average	0.74 ± 0.12	0.74 ± 0.12	0.72 ± 0.10	0.73 ± 0.09

6   Discussions

In this paper we have shown the detection of MRCPs in two different scenarios: offline and online. Both scenarios have been tested in Matlab and verified on a Raspberry Pi II as a proof of concept. Results given in Table 2 show that it is possible to run the detection for MRCP signals with different kinetic profiles on a single-trial basis and obtain a similar performance on a portable system as well. These results as well as their implementation on a commercially available portable device like a Raspberry Pi paves the way for developing portable devices that can decode brain activity. This is an important step to promote the technology transfer of BCI systems from the laboratory to the clinic. The next steps are to extend this work to from two class problem to four class where two different types of force and speed can be detected and then classified, which will extend the current binary brain switch (on/off) to have more degrees of freedom. Moreover, it should be tested using the intended user group motor – motor impaired patients and preferably in a clinical setting.

References

1. Wang, Q., Sourina, O., Nguyen, M.: EEG-based "serious" games design for medical applications. In: IEEE International Conference on Cyberworlds (CW), pp. 270–276 (2010)
2. Nascimento, O., Farina, D.: Movement-related cortical potentials allow discrimination of rate of torque development in imaginary isometric plantar flexion. IEEE Trans. Biomed. Eng. 55(11), 2675–2678 (2008)
3. Mohseni, H., Hamid, R., Maghsoudi, A., Shamsollahi, M.: Seizure detection in EEG signals: a comparison of different approaches. In: Proceedings of the 28th Annual International Conference of the IEEE Engineering in Medicine and Biology Society, vol. supplement, pp. 6724–6727 (2006)
4. Garipelli, G., Chavarriaga, R., Millán, J.: Single trial analysis of slow cortical potentials: a study on anticipation related potentials. J. Neural Eng. 10(3), 036014 (2013)
5. Niazi, I., Mrachacz-Kersting, K., et al.: Peripheral electrical stimulation triggered by self-paced detection of motor intention enhances motor evoked potentials. IEEE Trans. Neural Syst. Rehabil. Eng. 20(4), 595–604 (2012)
6. Niazi, I., Jiang, N., Tiberghien, O., Nielsen, J., Dremstrup, K., Farina, D.: Detection of movement intention from single-trial movement-related cortical potentials. J. Neural Eng. 8(6), 066009 (2011)
7. Hassan, A., Riaz, F., Rehman, S., Jochumsen, M., Niazi, I., Dremstrup, K.: An empirical study to remove noise from single trial MRCP for movement intention. In: 28th IEEE Canadian Conference on Electrical and Computer Engineering, Halifax, Nova Scotia, Canada, May 2015
8. Jochumsen, M., Niazi, I., Mrachacz-Kersting, N., Farina, D., Dremstrup, K.: Detection and classification of movement-related cortical potentials associated with task force and speed. J. Neural Eng. 10(5), 056015 (2013)

The Impact of News Media and Affect in Financial Markets

Stephen Kelly[✉] and Khurshid Ahmad

Trinity College Dublin, Dublin, Ireland
{kellys25,kahmad}@scss.tcd.ie
http://www.springer.com/lncs

Abstract. Literature on financial analysis is increasingly focused on what market participants do with information on price changes and whether all participants get the same price influencing information. The efficient market hypothesis first posed by Eugene Fama says all influential information is incorporated into price. Doubts and limitations of this theory have been expressed, leading Fama to accept a weak form of efficiency in financial markets. A variety of data such as key announcements, trading volume, consumer surveys and qualitative information like sentiment and affect may contain information not accounted for in price. For such qualitative information and news media, the challenge is in collating, processing, and aggregating this information with traditional financial time series. The influence of news media, quantified by computing methods, is not definitively described by economic theory. As such, we rely on inferences made from the modelling of the data to evaluate any potential explanatory power. To assess whether the information from news media is fully incorporated into price, we use six different statistical models and evaluate a proxy for news information against that of returns for the Dow Jones Industrial Average and New York Stock Exchange trading volume.

Keywords: Sentiment analysis · Text processing · OLS · Statistical proxy

1 Introduction

The search for sentiment about financial markets in general, and about individual equities in particular, has been in full swing for the last 20 years. Various proxies for such sentiment have been explored like searching for affect words in a time-ordered collection of texts. The affect words are essentially a cluster of words pre-assigned to an affect category like evaluation, including the oft-cited negative and positive *sentiment* words [1]. Other more transparent proxies have included news flow - number of texts per unit of time [7], timings of key macro-economic announcements [2], or stock market performance indicators [3]. The choice of texts can be quite revealing about the design of the sentiment search systems. Sentiment is expected to be found in opinion forming texts published either in comment columns in financial newspapers [4,6] or as online messages [7].

© Springer International Publishing Switzerland 2015
K. Jackowski et al. (Eds.): IDEAL 2015, LNCS 9375, pp. 535–540, 2015.
DOI: 10.1007/978-3-319-24834-9_62

The question we have been pondering is this: Economic theories about changes in financial markets now allow room for qualitative aspects of economic and financial transactions. The argument is if information extracted from text has explanatory power for returns of a financial asset. We wish to investigate if the explanatory power of a quantified metric, negative sentiment from text, has confounding or cross sectional effects with a more traditional measure of market sentiment such as asset or exchange trading volume.

2   Review

The research and adoption of content analysis of news in financial literature has gained momentum in recent times. One present challenge is in combining econometric methods with that of content extraction and sentiment analysis. While more recent studies have focused on the automated extraction of information and affect from text to represent investor sentiment [4,6–8]. Previous methods of creating a proxy for investor sentiment have made use of consumer surveys, trading volume, different investor trades, dividend premiums, and implied volatility of options amongst others [3]. The influence of announcements on assets and markets is well known, with a previous study by the authors of this paper showing the predictive power of the oil inventory announcement on a crude oil benchmark as compared to a sentiment proxy for news and a more traditional model [9]. While more recent studies have focused on the automated extraction of information and affect from text to represent investor sentiment. In many of these studies, distinctions and similarities in the explanatory information of the different proxies are examined.

The tasks involved in the work presented in this paper involve the retrieval of multiple streams of data, extracting quantitative measure of information from qualitative data, and using this in traditional statistical models. The data in this case is a quantified measure of news stories organised as a time series of sentiment.

3   Research Design

The basic strategy is to use sentiment proxies as conditioning variables, and see whether the manner in which characteristics spread future returns depends on the conditioning variable. This will help us to answer whether sentiment, broadly defined, has cross-sectional effects.

3.1   Data

Previous studies have highlighted that a simple count of negative terms in the column *Abreast of the market* published daily in the Wall Street Journal has predictive power for returns of the Dow Jones Industrial Average (DJIA) index [4]. It is from this example of using content analysis and econometrics that we draw our study.

The *Abreast of the Market* column (AOTM) is concerned with news regarding market activity, specifically large-capitalisation stocks quoted on the DJIA. We construct a corpus from 03/01/1989 to 16-11-1999 of daily AOTM articles, consisting of over 2717 news pieces. From this text collection we use the General Inquirer dictionary [1] with the *Rocksteady Affect Analysis System* (developed at Trinity College Dublin) to construct a time series of negative sentiment from the AOTM corpus. We investigate the potential impact that this metric has on predicting DJIA returns.

Market activity has often been linked to trading volume. Previous studies have suggested that news media and online messages are related to asset and exchange volume either directly or by influencing sentiment and the behaviour of market participants [4,7,10,11]. We use the New York Stock Exchange (NYSE) daily trading volume to assess the predictive power of volume on DJIA returns. We use a transformed measure of volume to obtain a stationary series, we use the detrended log volume of NYSE exchange as outlined in previous studies [4,11].

We calculate returns of the DJIA index as the log difference of price. All series used in this study are stationary according to an Augmented DickeyFuller test for several lags.

3.2 Constructing a Proxy for News

To assess the impact of our sentiment metric and whether it is representative of any information, we extract a quantified measure of information from our corpus of text. The approach used here counts the frequency of negative terms occurring in the corpus of text. The negative terms are categorised according to the General Inquirer dictionary [1]. As most text documents used are annotated or have some form of time stamping, it is possible to create a time series of the frequency of affect. Using the bag-of-words model of content analysis, the tokens for all news in a day are counted and the total terms of a category from a predefined dictionary are also counted, the category chosen is negative sentiment. By comparing these two measures, the relative frequency of negative sentiment for a day can be observed and subsequently a time series is formed.

Mapping data from raw and unstructured format; as a collection of text, to a convenient form; a time series of observations, is necessary to perform the statistical modelling. The time series of affect, or negative sentiment, is aggregated with that of financial data, in this case the returns of the DJIA index. Much of the financial data is freely available on a daily frequency, providers such as Quandl[1] provide an API for major formats and languages which aggregate financial data from sources and exchanges ranging from the CBOE, SEC to Yahoo and Google finance. Financial data and the sentiment time series are aligned according to available data, in the case of weekends and holidays where no trading data is available, news data is omitted from the aggregated data table. This is done so as to simplify any assumptions and any causal effects between the variables.

---

[1] http://www.quandl.com/help/packages.

3.3 Model Specification

To determine the impact of our exogenous variables on the DJIA returns with rely on regression models. We use several parsimonious regression models with some consideration given to robustness of results when detecting statistically significant effects using Newey-West standard errors [5]. Several models are run in an incremental manner to examine potential corr-sectional effects between variables:

$$r_t = \alpha_0 + \sum_{i=1}^{5} \alpha_i r_{t-i} + \varepsilon_t \tag{1}$$

$$r_t = \alpha_0 + \sum_{i=1}^{5} \beta_i s_{t-i} + \varepsilon_t \tag{2}$$

$$r_t = \alpha_0 + \sum_{i=1}^{5} {}_i Vol_{t-i} + \varepsilon_t \tag{3}$$

$$r_t = \alpha_0 + \sum_{i=1}^{5} \alpha_i r_{t-i} + \sum_{i=1}^{5} \beta_i s_{t-i} + \varepsilon_t \tag{4}$$

$$r_t = \alpha_0 + \sum_{i=1}^{5} \alpha_i r_{t-i} + \sum_{i=1}^{5} {}_i Vol_{t-i} + \varepsilon_t \tag{5}$$

$$r_t = \alpha_0 + \sum_{i=1}^{5} \alpha_i r_{t-i} + \sum_{i=1}^{5} \beta_i s_{t-i} + \sum_{i=1}^{5} {}_i Vol_{t-i} + \varepsilon_t \tag{6}$$

where r_t represents the returns Dow Jones Industrial Average Index, and s_t is the negative sentiment proxy extracted from the corpus of text, Vol_t is the trading volume for New York Stock Exchange. In each case we include five lags of the variable where $i = 5$ to account for a week of trading. By approaching the modelling of each of the series in this way, it is possible to observe possible confounding effects between variables and note the robustness of the sentiment and other independent variables on returns.

4 Result

We present the results of various combinations of the two proxies in question, that of negative sentiment and the NYSE exchange volume. We examine the impact of these variables from a historical perspective using an explanatory model. The coefficients of the model are presented in basis points, where one basis point equals a change of 0.01 % percentage points. This normalises the variables and resulting coefficients so as to have zero mean and unit variance making it possible to interpret the independent coefficients as one standard deviation changes to returns.

In Table 1 we report the resulting coefficient values for all variables in the defined equations in Sect. 3.3. We report the basis point impact and statistical significance of each of the lagged coefficients of the variable of the DJIA returns. We also perform a hypothesis test for the inclusion of the proxy variables in each model, this is the chi-squared test statistic ($\chi^2[NegSent]$, $\chi^2[Volume]$) reported for both sentiment and exchange volume. The hypothesis test acts as a Granger-causality test to assess the explanatory power of the proxy variable in each model.

Table 1. Coefficients for models defined by Eqs. 1 to 6 where α_{t-i} are the coefficients for DJIA returns, β_{t-i} represents negative sentiment, γ_{t-i} the NYSE trading volume, the mode was constructed using data for the period of 1989-01-03 to 1999-11-16 ($n = 2717$). Statistical significance at 90 %(*), 95 %(**), and 99 %(***) are shown accordingly. Newey and West [5] standard errors are used to account for heteroskedasticity and auto-correlation that may occur in estimation.

Coefficient	1	2	3	4	5	6
α_0	7.57***	7.29	6.17	7.03	6.16	7.00
α_1	−0.77			0.93	1.94	0.83
α_2	2.08			−2.12	−1.20	−1.94
α_3	4.90**			−3.47*	−3.70*	−3.74
α_4	4.15**			−0.29	−1.96	−0.64
α_5	−0.28			−0.18	−1.68	0.10
β_1		−2.85		−2.66		−3.11
β_2		−0.61		−1.59		−1.86
β_3		2.21		1.04		0.65
β_4		5.08**		5.09**		5.28**
β_5		4.54**		4.67**		5.39***
γ_1			−0.02		−0.28	−0.65
γ_2			−2.41		−2.35	−2.48
γ_3			1.56		1.85	2.28
γ_4			3.80*		3.79	4.12*
γ_5			−1.81		−1.86	−2.81
$Adjusted - \bar{R}^2$	0.3 %	0.4 %	0.1 %	0.5 %	0.2 %	0.6 %
AIC	−17955	−17932	−17948	−17920	−17914	−17920
$\chi^2[NegSent]$	na	17.06***	na	13.85**	na	16.04***
$\chi^2[Volume]$	na	na	6.96	na	7.34	9.54*

From the results, we see that negative sentiment has a greater impact on returns than NYSE exchange trading volume. The statistical confidence is higher for the inclusion of negative sentiment as an explanatory variable for returns as evident from the higher and statistically significant chi-squared test. Negative sentiment is also seen to be robust to the inclusion of NYSE volume also, with coefficients remaining significant and little change in magnitude. This suggest no confounding effects with exchange volume and supports the case that news media as an independent variable has a predictive impact for DJIA returns that is different and unique to trading volume. The impact of negative sentiment on DJIA returns is statistically significant and dispersed throughout the week over the five lags of sentiment used.

5   Conclusion

It is possible that the information being quantified from text is an indirect representation for other contemporaneous information that can be explained using other indirect or direct variables. However the method of using text analysis to generate a variable that adds explanatory power still aids statistical analysis and econometrics. With the ever growing interest in automated methods of information analysis and proxy estimation, these methods will benefit greatly from more elaboration and improvements in efficiency.

References

1. Stone, P.J., Dunphy, D.C., Smith, M.S.: The General Inquirer: A Computer Approach to Content Analysis. MIT Press, Cambridge (1966)
2. Andersen, T.G., Bollerslev, T.: Deutsche mark-dollar volatility: intraday activity patterns, macroeconomic announcements, and longer run dependencies. J. Finan. 53(1), 219–265 (1998)
3. Baker, M., Wurgler, J.: Investor sentiment and the cross-section of stock returns. J. Finan. 61(4), 1645–1680 (2006)
4. Tetlock, P.C.: Giving content to investor sentiment: the role of media in the stock market. J. Finan. 62(3), 1139–1168 (2007)
5. Newey, W.K., West, K.D.: Hypothesis testing with efficient method of moments estimation. Int. Econ. Rev. 28, 777–787 (1987)
6. Garcia, D.: Sentiment during recessions. J. Finan. 68(3), 1267–1300 (2013)
7. Antweiler, W., Frank, M.Z.: Is all that talk just noise? The information content of internet stock message boards. J. Finan. 59(3), 1259–1294 (2004)
8. Lechthaler, F., Leinert, L.: Moody Oil-What is Driving the Crude Oil Price? Eidgenössische Technische Hochschule Zürich, CER-ETH-Center of Economic Research at ETH Zurich (2012)
9. Kelly, S., Ahmad, K.: Sentiment proxies: computing market volatility. In: Yin, H., Costa, J.A.F., Barreto, G. (eds.) IDEAL 2012. LNCS, vol. 7435, pp. 771–778. Springer, Heidelberg (2012)
10. De Long, J.B., Shleifer, A., Summers, L.H., Waldmann, R.J.: Noise trader risk in financial markets. J. Polit. Econ. 98, 703–738 (1990)
11. Campbell, J.Y., Grossman, S.J., Wang, J.: Trading volume and serial correlation in stock returns. Q. J. Econ. 108(4), 905–939 (1993)

Clusterization of Indices and Assets
in the Stock Market

Leszek J. Chmielewski, Maciej Janowicz, Luiza Ochnio,
and Arkadiusz Orłowski[✉]

Faculty of Applied Informatics and Mathematics (WZIM),
Warsaw University of Life Sciences (SGGW),
ul. Nowoursynowska 159, 02-775 Warsaw, Poland
{leszek_chmielewski,maciej_janowicz,luiza_ochnio,
arkadiusz_orlowski}@sggw.pl
http://www.wzim.sggw.pl

Abstract. K-means clustering algorithm has been used to classify major
world stock market indices as well as most important assets in the Warsaw stock exchange (GPW). In addition, to obtain information about
mutual connections between indices and stocks, the Granger-causality
test has been applied and the Pearson R correlation coefficients have
been calculated. It has been found that the three procedures applied
provide qualitatively different kind of information about the groups of
financial data. Not surprisingly, the major world stock market indices
appear to be very strictly interconnects from the point of view of both
Granger-causality and correlation. Such connections are less transparent
in the case of individual stocks. However, the "cluster leaders" can be
identified which leads to the possibility of more efficient trading.

Keywords: Technical analysis · Clustering · K-means · Granger causality · Pearson correlation

1 Introduction

The technical analysis of the stock market assets [1,2] belongs to the most controversial branches of analysis of financial data series. This is because that the
aim of technical analysis is, basically, no less than the approximate predictions of
trends and their corrections in the data which appear as realizations of a *random*
process.

On one hand, it has been declared a kind pseudoscience, which, because of
the incorrectness of its most important principles, cannot lead to any sustainable
increase of returns above the market level [3,4]. On the other hand, it has been
being applied rather blindly without any serious knowledge about the market
dynamics. More recent publications, e.g. [5–8] have lead to considerable revision
of the ultra-critical stand of the many experts regarding technical analysis,

As a part of a common knowledge about the stock market dynamics let us
notice that time series generated by the prices of stocks are *not* random walks,

K. Jackowski et al. (Eds.): IDEAL 2015, LNCS 9375, pp. 541–550, 2015.
DOI: 10.1007/978-3-319-24834-9_63

and at least the short-time correlations *are* present. Whether they can indeed be exploited with the purpose of maximization of returns is an open question. What we investigate here is meant to be a very small contribution to answer it.

One of the many possible strategies of "beating the market" which are close to the spirit of technical analysis consists of identification of groups of stocks with similar historical behavior of prices. Then, it may happen that within such groups certain "leaders" can be found. Such "leaders" should exhibit behavior which is, to some extent, emulated - with some time delay - by some other members of the group. It is exactly that time delay which might possibly lead to forecasting of the price movement of the whole cluster or at least one of its parts.

As a preliminary but useful exercise we have first performed such grouping, or clusterization, of 24 of the world stock market indices: ALL_ORD, AMEX_MAJ, BOVESPA, B-SHARES, BUENOS, BUX, CAC40, DAX, DJIA, DJTA, DJUA, EOE, FTSE100, HANGSENG, INTERNUS, MEXICIPS, NASDAQ, NIKKEI, RUSSEL, SASESLCT, SMI, SP500, TOPIX, TSE300. Among them there have been both the most important ones like S&P500, DJIA, NIKKEI, HANGSENG and DAX, and those related to smaller markets like Budapest's BUX and Amsterdam's EOE. Then, 30 stocks belonging to the "blue chips" of the Warsaw stock market (WIG30 group) have also been classified.

For the purpose of that classification we have employed: (i) a state-of-the-art clustering algorithm called K-means [9–13]; (ii) the Granger causality test [14] with lags equal to 1, 2, and 3 trading sessions; (iii) Pearson correlation coefficient [15] with lags; that is, we have computed the Pearson correlation coefficients between shifted closing prices of asset A ($P_{n-k}^{(A)}$ and unshifted closing prices of asset B ($P_n^{(B)}$, where k has been equal to 1, 2, or 3.

The main body of this work is organized as follows. In Sect. 2 we provide our preliminary results for the 24 stock market indices. Section 3 is devoted to analogous results for the stocks in the Warsaw stock market. Finally, Sect. 4 comprises some concluding remarks.

2   Classification of the Major World Stock Market Indices

From every index a we have obtained the time series $x_n^{(a)}$ in the following way. For each trading session we obtained the sequence of opening (O), maximum (X), minimum (N) and closing (C) values:

$$(O_k^{(a)}, X_k^{(a)}, N_k^{(a)}, C_k^{(a)}).$$

where k enumerates the trading sessions, $k = 1, 2, ..., N$. Then the series has been normalized by substracting the mean (over k) value of the closing values and dividing by the standard deviation of those values. This gave us the sequence:

$$\left(...(\bar{O}_k^{(a)}, \bar{X}_k^{(a)}, \bar{N}_k^{(a)}, \bar{C}_k^{(a)})...\right),$$

where the overbar denotes normalization in the above sense.

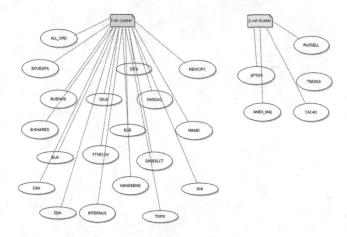

Fig. 1. Two clusters obtained from the K-means algorithm as applied to 24 world stock market indices.

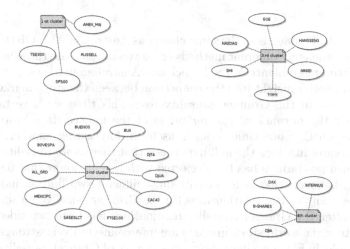

Fig. 2. The same as in Fig. 1 but with four clusters.

The series $x_n^{(a)}$ is obtained by flattening of the above sequence.

The K-means algorithm which has been applied to series generated in this way requires the number of clusters as an input parameter. We have chosen two, four, and six clusters.

The results are displayed in Figs. 1, 2 and 3.

We have been somewhat astonished by the fact that the clusters visible in Figs. 1, 2 and 3 do not appear to follow any pattern associated with geographical locations of the market to which an index corresponds. Also, our hopes to see, say, Euromarican indices contrasted with the East Asian ones have obviously failed. In fact, we have had some difficulties to provide any reasonable qualitative explanation of the clusterization results. For instance, it is not easy to understand why

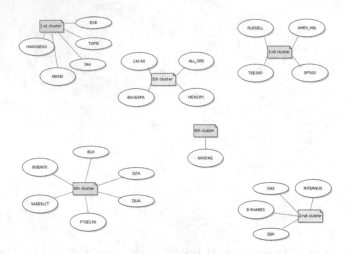

Fig. 3. The same as in Fig. 2 but with six clusters

the French CAC40 index is in the same cluster as Australian ALL ORDINARIES index given completely different methods employed to calculate those indices and rather doubtful resemblance of the French and Australian markets.

To obtain better insight into the connection between the stock market indices we have turned to the Granger causality test. This time we have taken into account only the *normalized closing values* of the indices. It is employed for determining whether one time series is useful in forecasting another. We have obtained all cases in which the null hypothesis that there is *no* causality relationships between two indices has been rejected with p-value lower that 0.01. It has turned out, however, that it is actually quite difficult two find an index which *does not* have any causal relationships (in the Granger sense) with some other. Also, very often the Granger-causality relationships has been two-sided. This is quite intuitive: the world stock markets are interconnected very strongly indeed. As an example, in Fig. 4 we have shown an example of Granger-causality relation with the lag equal to 2 between the Australian ALL ORDINARIES index and other indices. The arrows in Fig. 4 mean that the knowledge of the closing values of ALL ORDINARIES on the day $n - 2$ can be helpful to forecast the closing values of the indices displayed at the end of the arrow on the day n.

Our third classification procedure involved calculations of the Pearson correlation coefficient (PCC) between pairs of closing values of indices a and b with time delay. That is, we define the series $\bar{D}_n^{(a)} = \bar{C}_{n-m}^{(a)}$ and compute $PCC = PCC(m)$ as:

$$PCC(m) = \frac{cov(\bar{D}^{(a)})\bar{C}^{(b)}}{\sigma_D \sigma_C}$$

where *cov* denotes covariance and σ - standard deviation.

That is, covariances divided by the product of standard deviations have been computed for the series $\bar{C}_{n-m}^{(a)}$ and $\bar{C}_n^{(b)}$ where m has been equal to 1, 2, or 3.

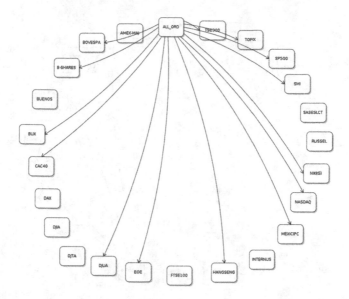

Fig. 4. Example of the Granger-causality relation: the arrows indicate that the Granger causality test rejected the null hypothesis (with p-value < 0.01) that there is no causality relation between the values of (normalized) ALL ORDINARIES index and other stock market indices.

Not surprisingly, we have found very strong correlation among all pairs of indices. In Fig. 5 we have displayed only those arrows which correspond to PCC larger than 0.98 with p-value smaller than 0.001. At last, we have found finally some connection between the geography and strong correlations. What is perhaps a bit amazing (and amusing) is the strong, one-sided correlation between the RUSSEL (i.e. Russian RTS) and BOVESPA indices.

3   Classification of the Most Important Stocks in Warsaw Stock Market

We have also employed three procedures described in Sect. 2 to those Polish stocks of which the WIG30 index has been composed in the beginning of June, 2015: ALIOR, ASSECOPOL, BOGDANKA, BORYSZEW, BZWBK, CCC, CYFRPLSAT, ENEA, ENERGA, EUROCASH, GRUPAAZOTY, GTC, HANDLOWY, INGBSK, JSW, KERNEL, KGHM, LOTOS, LPP, MBANK, ORANGEPL, PEKAO, PGE, PGNIG, PKNORLEN, PKOBP, PZU, SYNTHOS, TAURONPE and TVN.

All the prices have been normalized in the same way as in Sect. 2.

Firstly, the K-means algorithm has been employed to obtain the clustering for 2, 4, and 6 clusters. The results are shown in Figs. 6, 7 and 8.

In the case of the WIG30 stocks classification it is again difficult to find any regularities. For instance, it is not true that the financial intitutions appear

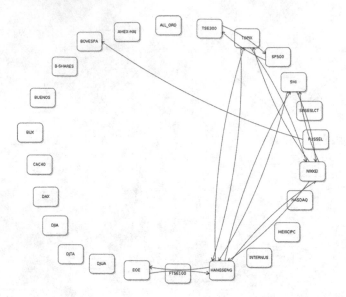

Fig. 5. Example of high correlations among the world indices: the arrows indicate that the Pearson correlation coefficient between the lagged ($m = 3$) and unlagged indices are larger than 0.98 with p-value smaller than 0.01. The beginning of each arrow is at the lagged value of an index.

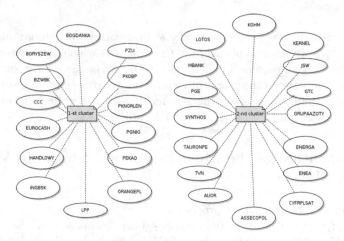

Fig. 6. Two clusters obtained from the K-means algorithm as applied to 30 stocks belonging to WIG30 index.

together in a single cluster or two clusters. Nor can it be said about the oil companies or those associated with production and distribution of energy. Our attempts to associate the results of classification with fundamental analysis have also failed. It appears that the clusterization algorithms simply provide knowledge of the type quite different from intuitive correlations.

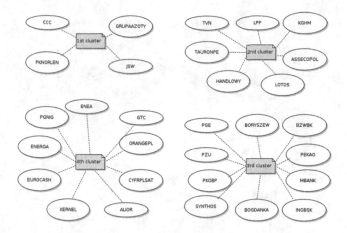

Fig. 7. The same as in Fig. 6 but with four clusters.

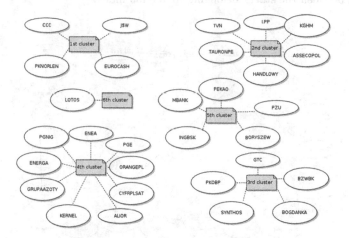

Fig. 8. The same as in Fig. 6 but with six clusters

The results of the Granger causality tests are displayed in Fig. 9. With the help of arrows we have shown the pairs of stocks which have passed the test with p-value smaller than 0.01.

Perhaps the most striking feature of the diagram shown in Fig. 9 is the very special status of the ENERGA stocks which seems to "influenced" (in the sense of Granger causality; there is of course no material influence) by eight other stocks. What we believe is also quite interesting is the fact the prices of INGBSK is in the Granger-causality relation with the prices of two other large banks, PKOBP and ALIOR. We believe that, with sufficient care, the diagram in Fig. 9 can be used to attempt forecasting the behavior of prices of stocks.

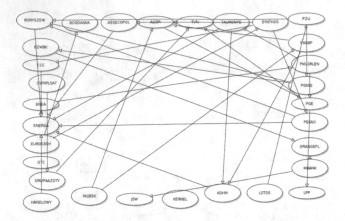

Fig. 9. Example of the Granger-causality relation: the arrows indicate that the Granger causality test rejected the null hypothesis (with p-value < 0.01) that there is no causality relation between the stocks belonging to WIG30 index.

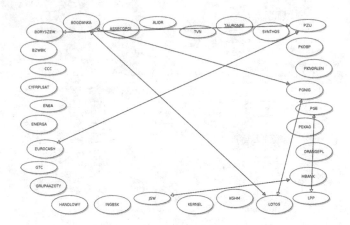

Fig. 10. Example of high correlations among the world indices: the arrows indicate that the Pearson correlation coeffcient between the retarded ($m = 3$) and unretarded indices are larger than 0.8 with p-value smaller than 0.01. The arrows start at the stock taken with retarded prices and end unretarded

The same may be true about the stocks for which strong correlation exists. In Fig. 10 we have displayed pairs for which PCC (computed from retarded ($m = 1$) and unretarded values of the closing prices) has been larger than 0.8 (with significance level 1 %).

As we can see, the strong correlations are often two-sided. That is, not only the sequence $\bar{C}_{n-1}^{(a)}$ is strongly correlated with $\bar{C}_n^{(b)}$ for, e.g., $a = EUROCASH$, and $b = PZU$, but the opposite is also true. Let us notice here that none of the time series analyzed in this work passes the augmented Dickey-Fuller test for the absence of the unit root, and, most likely, none of them is stationary.

4 Concluding Remarks

In this work we have performed classification of a group of important world stock market indices and major stocks in the Warsaw stock market using standard K-means clustering algorithm, Granger causality test, and Pearson correlation coefficients. We have found pairs of indices as well as pairs of stocks in the Warsaw stock market which are either very strongly Granger-causality related or strongly correlated in the Pearson's sense. Let us notice that the correlations obtained with the help of Spearman rank coefficient have not differed qualitatively from those obtained from the Pearson correlation coefficient. We believe that, with sufficient care, the diagrams similar to those obtained here may be used in practice, possibly even to enhance the predictive power of technical indicators for trading purposes.

It is to be noticed, however, that the above results are very preliminary and require careful reexamination. In particular, we have not yet provided any tests for the forecasting powers based on the above classifications. We hope to report results of such improved analysis in a forthcoming publication.

References

1. Murphy, J.: Technical Analysis of Financial Markets. New York Institute of Finance, New York (1999)
2. Kaufman, P.: Trading Systems and Methods. Wiley, New York (2013)
3. Malkiel, B.: A Random Walk Down the Wall Street. Norton, New York (1981)
4. Fama, E., Blume, M.: Filter rules and stock-market trading. J. Bus. **39**, 226–241 (1966)
5. Brock, W., Lakonishok, J., LeBaron, B.: Simple technical trading rules and the stochastic properties of stock returns. J. Finan. **47**(5), 1731–1764 (1992)
6. Lo, A., MacKinley, A.: Stock market prices do not follow random walks: evidence from a simple specification test. Rev. Finan. Stud. **1**, 41–66 (1988)
7. Lo, A., MacKinley, A.: A Non-Random Walk down Wall Street. Princeton University Press, Princeton (1999)
8. Lo, A., Mamaysky, H., Wang, J.: Foundations of technical analysis: computational algorithms, statistical inference, and empirical implementation. J. Finan. **55**(4), 1705–1765 (2000)
9. MacQueen, J.: Some methods for classification and analysis of multivariate observations. In: Cam, M.L., Neyman, J. (eds.) Proceedings of the 5th Berkeley Symposium on Mathematical Statistics and Probability, vol. 1, pp. 281–297. University of California Press, Berkeley (1967)
10. Steinhaus, H.: Sur la division des corps matériels en parties. Bull. Acad. Polon. Sci. **4**(12), 801–804 (1957)
11. Lloyd, S.: Least square quantization in PCM (1957) Bell Telephone Laboratories Paper

12. Forgy, E.: Cluster analysis of multivariate data: efficiency versus interpretability of classifications. Biometrics **21**(3), 768–769 (1965)
13. Hartigan, J.: Clustering Algorithms. Wiley, New York (1975)
14. Granger, C.W.J.: Investigating causal relations by econometric models and cross-spectral methods. Econometrica **37**(3), 424–438 (1969)
15. Pearson, K.: Notes on regression and inheritance in the case of two parents. Proc. R. Soc. Lond. **58**, 240–242 (1895)
16. Scikit-learn Community: Scikit-learn - machine learning in python (2015). http://scikit-learn.org
17. Pedregosa, F., Varoquaux, G., Gramfort, A., et al.: Scikit-learn: machine learning in Python. J. Mach. Learn. Res. **12**, 2825–2830 (2011)

Behaviour and Markets: The Interaction Between Sentiment Analysis and Ethical Values?

Jason A. Cook[✉] and Khurshid Ahmad

School of Computer Science and Statistics, ADAPT Centre, Trinity College,
University of Dublin, Dublin, Ireland
jcook@tcd.ie

Abstract. Financial scandals are prominently covered by the media, during which time the media becomes heavily involved in condemning the perpetrators. The language used to describe these scandals is often laden with negative sentiment, and much of this language carries ethical connotations. In this paper we explore the relationship between the use of emotive words and ethical words, focussing on financial crimes. Using the Enron Corporation as a case study, we show that evaluative terms describing things that are positive or negative are correlated with moral terms describing things that are virtuous or vicious. To explore the impact of these moral terms, we conduct a time series analysis and investigate the effect in changes of moral terms on Enron's share price.

Keywords: Sentiment analysis · Business ethics · Affect analysis · Enron

1 Introduction

The role of affect is important in business and finance, especially in the roller-coaster changes before and after a major economic crisis. Research in behavioural finance suggests that even "rational" investors can sometimes behave "irrationally" when motivated by feelings or emotions [9]. The objective of sentiment analysis is to quantify this emotion, using evaluative language as a proxy.

Sentiment analysis has its roots in *content analysis* - a procedure for "analysing the semantic content of messages" (OED). Early content analysis procedures were developed for analysing the content of political, military and ethical opinions relating to war and peace [7,8]. However, the subsequent development of computer-based content analysis led to researchers exploring the connection between irrational behaviour and market sentiment. Using the pioneering psychosocial dictionary and content analysis procedures in [11], it was found that (negative) sentiment is responsible for as many as 10–20 basis point changes in the movement of key financial indices, such as S&P 500 [13] and the Dow Jones Industrial Average [3], and around 5–10 basis point changes in the movement of firm-level equities [1].

K. Jackowski et al. (Eds.): IDEAL 2015, LNCS 9375, pp. 551–558, 2015.
DOI: 10.1007/978-3-319-24834-9_64

We have explored the basis of the psychosocial dictionary used in the studies of market movements described above. Specifically, we looked at major financial scandals, caused by questionable business practices. During the periods in and around major financial scandals, ethical language is often used to comment upon the moral aspects of a company. Words such as "violated", "greedy" and "deceive", for instance, have strong connotations with immoral behaviour. We explored the connection between moral language and firm-level sentiment, looking at how moral terms impact the standing of a company in the stock market. Intuitively, one would expect words of POSITIVE evaluation to be in the same category as VIRTUE words, and words of NEGATIVE evaluation to be in the same category as VICE words. We hypothesise that if evaluative language is psychologically grounded in ethics, then moral language may act as a proxy for investor sentiment. We measured the effect of moral language on a company's market value by using moral terms in an autoregressive model.

Using the infamous Enron Corporation as a case study, we built a corpus of 31,764 news and magazine documents (spanning the period 1997 through 2002), and measured the frequencies of moral and evaluative language over time using a variant of the General Inquirer (GI) dictionary [11]. We show that NEGATIVE and VICE terms exhibit a 46.30 % correlation, and that POSITIVE and VIRTUE terms have a 30.38 % correlation. The correlation between moral and evaluative terms leads us to believe that ethics can motivate what is regarded as positive and negative evaluation, and it is this relationship that we discuss in this paper.

2 Method

2.1 Introduction

It is believed that there is a causal link between words indicating a certain behaviour and market or company indices. Market indices form a discrete set of a time-ordered points - a time series - and certain branches of sentiment analysis assume that behavioural words can also be time-ordered. While market or company indices are associated with prices at an identifiable point in time, one cannot be certain that the sentiment found in a text prevails at the time it was published. At best, one can argue that stakeholders only notice the sentiment when the text is published. We have adopted this latter argument to create not only a sentiment time series, but also a time series of moral judgements of companies. We believe that the sentiment and moral judgements expressed are proxies, and this is the spirit in which we would like the reader to interpret them.

2.2 Bag-of-Words Model

Content analysts and behavioural finance scholars typically use the "bag-of-words" model to represent individual texts. A bag-of-words model is defined as *"an unordered set of words, ignoring their exact position"* [6]. In sentiment analysis studies words and multi-word expressions are placed in well-known but

arbitrarily defined categories, principally the categories of POSITIVE and NEGA-
TIVE evaluation. A text analysis system splits the text into its constituent tokens
and keeps cumulative frequencies of the individual words, normalized to the size
of the text. Every time a word is found, the system will update the frequency
count for both the word and its associated categories. In some dictionaries, such
as the GI dictionary, the categories may overlap, and a given word could belong
to one or more categories. In these cases the counts for each category are updated
simultaneously.

The overlap amongst words in the GI dictionary is quite striking: 93 % of
VICE terms are also tagged as NEGATIVE, and 84 % of VIRTUE terms are tagged
as POSITIVE. Conversely, there is no overlap between the VICE and POSITIVE categories, and only a minor overlap between the VIRTUE and NEGATIVE categories
(Table 1).

Table 1. Percentage of VICE and VIRTUE terms in the GI dictionary that are also
tagged as NEGATIVE or POSITIVE.

	VICE (N = 686)	VIRTUE (N = 719)
NEGATIVE (N = 2,291)	92.55 %	0.42 %
POSITIVE (N = 1,915)	0.00 %	84.14 %

To avoid double counting words appearing in both categories, we removed
the evaluative tag of POSITIVE or NEGATIVE from words that were tagged VICE
or VIRTUE. Thus the word "bribe", tagged as both NEGATIVE and VICE in the
GI dictionary, is counted only as a VICE word, but is not counted as a NEGATIVE
word. This ensures that any relationship found between the moral and evaluative
terms is not due to the overlap between the categories.

2.3 Price Returns

Rather than using a company's raw share price, corresponding to absolute frequency, we use the company's *returns*, which represents relative frequency. The
motivation behind this is that prices are highly correlated, and thus the price of
an asset at a given time period is dependent on the preceding time period(s).
Returns, however, are uncorrelated, making it easier to attribute the variation
due to investor sentiment. The notion of returns adopted in this study is that of
simple returns, defined as changes in the logarithms of prices:

$$r_t = \log\left(\frac{p_t}{p_{t-1}}\right) \tag{1}$$

2.4 Autoregression

We have noted studies in which researchers discuss the causality between returns
and sentiment. This causality could be bidirectional: companies with falling

returns may invoke a negative sentiment, but it is also possible that negative sentiment may precipitate a fall in returns. By extension, we propose that news about virtuous or vicious behaviour can act in a similar fashion, making returns rise or fall respectively.

The simplest form for expressing this bidirectional causality was developed by [10], who suggested that a variable could be dependent on its past history and on other variables. For instance, it has been shown in finance that the return on a share or equity depends to an extent on its value in the immediate past. Behavioural finance experts assume an additional dependence on sentiment, and we extend this assumption to the concept of morality.

The model used is a form of multiple regression, however the term *autoregression* is used because the dependent variable depends upon its historical values. Typically, there is a dependence of daily returns on the previous 5 days' values, and consequently our autoregressive equation comprises of 5 lags:

$$r_t = \alpha_0 + \sum_{i=1}^{5} \alpha_i r_{t-i} + \beta_1 s_t + \gamma_1 m_t + \varepsilon_t \tag{2}$$

where r_t denotes the return of a given company at time 't' and s and m are measures of sentiment and morality respectively. The values of the coefficients α, β and γ are to be estimated by the regression procedure, and ε_t refers to the error term, which is regarded as independent and identically distributed.

2.5   An Exemplar Case Study: Enron

Background. This research utilises the former Enron Corporation as a case study. Enron originated in Houston, Texas in 1985 as a gas pipeline company, before branching into other markets. The company played a pioneering role in the deregulation of energy markets worldwide, and was highly regarded for its philanthropic efforts and honouring of leading world figures.

Enron flourished throughout the 1990s, with share prices reaching a high of \$90 per share in August 2000. For six consecutive years it was labelled by Fortune Magazine as *"America's Most Innovative Company"*. However, in 2001 things turned sour: by November its share price had plummeted to less than \$1 per share and in December it was forced to file for bankruptcy protection - the largest in U.S. history at that time. A subsequent investigation by the U.S. Securities and Exchange Commission (SEC) found that key company leaders had been manipulating financial reports by inflating earnings and hiding losses from Enron's books [2]. A number of these executives were subsequently indicted and criminal penalties were imposed.

After its fall in 2001, the company was only remembered by morally negative terms - the so-called vice words. Its business practices, once key to deregulated markets, were regarded as reprehensible. However, the transition from a virtuous company to a vicious one did not happen overnight, and it is possible that there were traces of the illicit practices in the news before the share price plummeted. The question is whether that muted criticism involved the use of vice words, in contrast to the appreciation expressed previously in virtuous terms.

Data Used. The data used in this study comprises a collection of texts spanning the period 09/01/1997 through 31/12/2002, which encompasses the periods before, during and after news of the Enron scandal came to light. The corpus is made up of 31,774 documents (25 million tokens), taken from a variety of sources, including newspapers (both national and regional), newswire, press releases, magazine articles and industry and trade publications. The texts were extracted from LexisNexis, using the search terms *"Enron"*, *"InterNorth"*, and *"Northern Natural Gas"*, and restricting the search only to documents produced in the U.S. Share returns from the same time period comprise of 1,453 daily observations. We ignored news articles where there was no corresponding share price information on that day, such as weekends or public holidays, leading to an overall aggregated time series of 1,453 observations.

3 Results

3.1 Introduction

In this section, we explore the relationship between the moral terms of VICE and VIRTUE and the evaluative terms of POSITIVE and NEGATIVE, using words and categories from the GI dictionary. The frequency of these categories is computed using Rocksteady - a text analytic system for extracting sentiment from text.

3.2 Summary Statistics

Table 2 shows summary statistics for the Enron data set. On average, there are a higher number of POSITIVE and VIRTUE words in the corpus compared to NEGATIVE and VICE words, suggesting that positive news is more frequent than negative news. In addition, the average counts of POSITIVE and NEGATIVE words are higher than the average counts for VICE and VIRTUE words, although this is perhaps to be expected given that there are twice as many evaluative words in the GI dictionary. Also noteworthy is the standard deviations, which are higher for evaluative terms than moral terms, suggesting that there is more variability in the frequency of evaluative words than the frequency of moral ones.

In terms of returns, the distribution is clearly non-normal. The skewness of -6.71 suggests a distinct lack of symmetry in the distribution, and implies that there are more observations in the left-hand tail than would be expected in a normal distribution (by comparison, the hypothetical normal distribution has a skewness of 0). The returns also exhibit a high kurtosis of 202.95, indicating that the distribution is more peaked than would be expected in a normal distribution (which would have an expected kurtosis of 3). However, the literature argues that non-normality is a universal property (a so-called *stylized fact*) of *any* return distribution, irrespective of time, markets or instruments [12]. Thus, Enron's returns are exhibiting the typical properties expected of any return distribution.

Table 2. Summary statistics for returns, moral terms, and evaluative terms in the Enron corpus. The symbols $\bar{x}$, s, b and k, are the sample mean, standard deviation, skewness and kurtosis computed on the percentages of the various terms. The term f denotes the raw frequency.

	$\bar{x}$	s	b	k	f
Returns	−0.00	0.10	−6.71	202.95	N/A
Vice	0.43	0.29	1.37	5.76	159,630
Negative	0.93	0.50	0.94	4.34	334,622
Virtue	1.42	0.47	1.11	4.26	332,075
Positive	2.52	0.62	0.57	2.98	602,473

3.3 Correlations

Table 3 shows that there is a significant correlation between NEGATIVE and VICE terms ($r = 0.46, p < 0.01$), and between POSITIVE and VIRTUE terms ($r = 0.30, p < 0.01$). Conversely, there is a significant anti-correlation between POSITIVE and VICE terms ($r = −0.13, p < 0.01$), and between VIRTUE and NEGATIVE terms ($r = −0.08, p < 0.01$). This suggests that the use of evaluative terms is statistically associated with the use of moral terms, and adds further support to our hypothesis that evaluative words are psychologically grounded in ethics.

Table 3. Correlation matrix of returns, POSITIVE, NEGATIVE, VICE and VIRTUE categories.

	Negative	Positive	Vice	Virtue
Returns	−4.21	−0.38	−2.79	1.95
Negative		−5.95	46.30	−7.99
Positive			−13.29	30.38
Vice				−12.08

In terms of the correlations with returns, the NEGATIVE and VICE terms exhibit similar behaviours in that they are both anti-correlated with returns. Contrariwise, the VIRTUE terms exhibit a positive correlation with returns, but the strength of this association is not as high as that of the NEGATIVE or VICE terms. The correlation between POSITIVE terms and returns is practically zero.

3.4 Autoregressions

To investigate the impact of moral terms on a company's market value, we fitted four autoregressive models, using returns as the dependent variable and a mixture of lagged returns, NEGATIVE terms and VICE terms as the independent

Table 4. The four autoregressive models used.

Model	Equation	Independent variables
1	$r_t = \alpha_0 + \alpha L_5 r_t + \varepsilon_t$	5-day lagged returns
2	$r_t = \alpha_0 + \alpha L_5 r_t + \beta neg_t + \varepsilon_t$	Model 1 + NEGATIVE terms
3	$r_t = \alpha_0 + \alpha L_5 r_t + \beta vice_t + \varepsilon_t$	Model 1 + VICE terms
4	$r_t = \alpha_0 + \alpha L_5 r_t + \beta neg_t + \beta vice_t + \varepsilon_t$	Model 1 + NEGATIVE + VICE terms

Table 5. Results of the four autoregressive models. Coefficients are listed in rows, with statistical significance marked by * $(P < 0.05)$, ** $(P < 0.01)$ or *** $(P < 0.001)$.

Variable	1	2	3	4
Constant	-0.005^*	0.091	0.052	0.102
r_{t-1}	-0.083^{**}	-0.086^{**}	-0.085^{**}	-0.086^{**}
r_{t-2}	0.023	0.021	0.022	0.020
r_{t-3}	-0.075^{**}	-0.0788^{**}	-0.077^{**}	-0.078^{**}
r_{t-4}	-0.089^{***}	-0.092^{***}	-0.091^{***}	-0.092^{***}
r_{t-5}	0.013	0.001	0.011	0.009
βneg_t	N/A	-0.102^*	N/A	-0.087
$\beta vice_t$	N/A	N/A	-0.128	-0.058
F-stat.	6.15^{***}	5.61^{**}	5.3^{***}	4.9^{***}
R^2 (Adj.)	0.017	0.019	0.017	0.018

variables. The four models and accompanying regression results are shown in Tables 4 and 5 respectively.

Adding NEGATIVE terms to the regression equation (Model 2) improves the model beyond the baseline (Model 1), suggesting that NEGATIVE terms are able to capture some of the variability in r_t that is not accounted for by the historical returns. However, adding VICE terms (Model 3) to the model does not offer any significant improvement over the baseline model. Adding both NEGATIVE and VICE terms together (Model 4) also fails to yield any significant improvement, although this could be due to the high correlation between the two categories.

4   Conclusions

In this paper we investigated whether negative sentiment has a moral foundation. Using Enron as a case study and utilizing words and categories from the GI dictionary, we discovered that there were significant correlations between NEGATIVE and VICE terms and between POSITIVE and VIRTUE terms. To explore the impact of moral terms on a company's value, we regressed the number of NEGATIVE and VICE terms against Enron's daily returns over a 6-year period. While adding NEGATIVE terms to the regressive model improved the fit, adding VICE terms did not.

The results appear to yield conflicting interpretations. On the one hand, moral terms appear to have little impact in explaining the variability in returns. On the other, there is clearly a relationship between NEGATIVE and VICE terms, as evidenced by the overlap in the GI dictionary and the observed correlations between the categories. It could be the case that the GI dictionary is too general, in the sense that words tagged as VICE do not have vicious connotations in a financial context. However, the mixed results could also be due to the fact that there are a much larger number of NEGATIVE terms than VICE terms. Future work will therefore consist of expanding the list of VICE terms in the GI dictionary, to see whether an expanded set of moral terms is more useful in determining movements in asset prices. We will also expand the research to incorporate other case studies, such as WorldCom and Bernie Madoff Investment Securities.

Acknowledgements. This research is supported by Science Foundation Ireland through the CNGL Programme (Grant 07/CE/I1142) in the ADAPT Centre (www.adaptcentre.ie) at Trinity College, University of Dublin. In this study we used the sentiment analysis system Rocksteady, developed as part of the Faireacháin project for monitoring, evaluating and predicting behaviour of markets and communities (2009–2011). Support for Rocksteady's development was provided by Trinity College, University of Dublin and Enterprise Ireland (Grant #IP-2009-0595).

References

1. Ahmad, K., Kearney, C., Liu, S.: No news is good news: a time-varying story of how firm-specific textual sentiment drives firm-level performance. In: The European Financial Management Association (2013)
2. Baker, R., Christensen, T., Cottrell, D.: Advanced Financial Accounting, 9th edn. McGraw-Hill, London (2011)
3. Dougal, C., Engelberg, J., Garcia, D., Parsons, C.: Journalists and the stock market. Rev. Finan. Stud. **25**(3), 639–679 (2012)
4. Garcia, D.: Sentiment during recessions. J. Finan. **68**(3), 1267–1300 (2012)
5. Johnson, E., Fleischman, G., Valentine, S., Walker, K.: Managers' ethical evaluations of earnings management and its consequences. Contemp. Acc. Res. **29**(3), 910–927 (2011)
6. Jurafsky, D., Martin, J.: Speech and Language Processing, 2nd edn. Pearson Education Inc., New Jersey (2008)
7. Lasswell, H.: The measurement of public opinion. Am. Polit. Sci. Rev. **25**(2), 311–326 (1931)
8. Lasswell, H.: Language of Politics: Studies in Quantitative Semantics. George W. Stewart Publisher Inc., New York (1949)
9. Shiller, R.: Irrational Exuberance. Princeton University Press, Princeton (2000)
10. Sims, C.: Macroeconomics and reality. Econometrica **48**(1), 1–48 (1980)
11. Stone, P., Dunphy, D., Smith, M.S., Ogilvie, D.: The General Inquirer: A Computer Approach to Content Analysis (1966)
12. Taylor, S.: Asset Price Dynamics, Volatility, and Prediction. Princeton University Press, Princeton (2005)
13. Tetlock, P.: Giving content to investor sentiment: the role of media in the stock market. J. Finan. **62**(3), 1139–1168 (2007)

Author Index

Printed in the United States
By Bookmasters